ANTICANCER AGENTS BASED ON NATURAL PRODUCT MODELS

MEDICINAL CHEMISTRY

A Series of Monographs

A complete list of titles in this series appears at the end of this volume.

ANTICANCER AGENTS BASED ON NATURAL PRODUCT MODELS

Edited by
JOHN M. CASSADY
Department of Medicinal Chemistry and Pharmacognosy
School of Pharmacy and Pharmacal Sciences
Purdue University
West Lafayette, Indiana

JOHN D. DOUROS
Natural Products Branch
Developmental Therapeutics Program
National Cancer Institute
Silver Spring, Maryland

ACADEMIC PRESS   1980
A Subsidiary of Harcourt Brace Jovanovich, Publishers
New York   London   Toronto   Sydney   San Francisco

ACADEMIC PRESS, INC.
111 Fifth Avenue, New York, New York 10003

United Kingdom Edition published by
ACADEMIC PRESS, INC. (LONDON) LTD.
24/28 Oval Road, London NW1 7DX

Library of Congress Cataloging in Publication Data
Main entry under title:

Anticancer agents based on natural product models.

   (Medicinal chemistry series ;)
   Includes bibliographies and index.
   1. Cancer––Chemotherapy.   2. Antineoplastic agents
––Testing.   3. Antibiotics––Testing.   I. Cassady,
John M.   II. Douros, John, Date   III. Series:
Medicinal chemistry, a series of monographs
[DNLM: 1. Antineoplastic agents.   2. Models, Chemical.
W1 ME64 / QV269 A629]
RC271.C5A67   616.99'4061   79–6802
ISBN 0–12–163150–8

PRINTED IN THE UNITED STATES OF AMERICA

80 81 82 83   9 8 7 6 5 4 3 2 1

To the late Professor S. Morris Kupchan,
our colleague and friend,
and a pioneer in
the discovery of novel
naturally occurring
anticancer agents

and to T.J.

Contents

14. Miscellaneous Natural Products with Antitumor Activity

MATTHEW SUFFNESS and JOHN DOUROS

List of Contributors

Numbers in parentheses indicate the pages on which the authors' contributions begin.

Federico Arcamone (1), D. R. Ricerca Chimica, Farmitalia, Carlo Erba, Milano 20146, Italy

W. T. Bradner (43), Antitumor Biology Department, Bristol Laboratories, Syracuse, New York 13201

John M. Cassady (201), Department of Medicinal Chemistry and Pharmacognosy, School of Pharmacy and Pharmacal Sciences, Purdue University, West Lafayette, Indiana 47907

John D. Douros (465), Natural Products Branch, Developmental Therapeutics Program, National Cancer Institute, Silver Spring, Maryland 20910

T. W. Doyle (43), Medicinal Chemical Research, Bristol Laboratories, Syracuse, New York 13201

Koert Gerzon (271), Department of Pharmacology, Indiana University School of Medicine, Indianapolis, Indiana 46223

Ian Jardine (319), Department of Pharmacology, Mayo Foundation, Mayo Graduate School of Medicine, Rochester, Minnesota 55901

Toyokazu Kishi (353), Microbiological Research Laboratory, Central Research Division, Takeda Chemical Industries, Ltd., Yodogawa-ku, Osaka 532, Japan

Yasuo Komoda (353), Division of Medicinal Chemistry, Institute for Medical and Dental Engineering, Tokyo Medical and Dental University, Kanda, Chiyoda-ku, Tokyo 101, Japan

Kenneth L. Mikolajczak (391), Northern Regional Research Center, Agricultural Research, U.S. Department of Agriculture, Peoria, Illinois 61604

Masaji Ohno (73), Faculty of Pharmaceutical Sciences, University of Tokyo, Bunkyo-ku, Tokyo 113, Japan

Richard G. Powell (391), Northern Regional Research Center, Agricultural Research, U.S. Department of Agriculture, Peoria, Illinois 61604

William A. Remers (131), Department of Pharmaceutical Sciences, University of Arizona, Tucson, Arizona 85721

John P. Rosazza (437), Division of Medicinal Chemistry/Natural Products, College of Pharmacy, The University of Iowa, Iowa City, Iowa 52242

Cecil R. Smith, Jr. (391), Northern Regional Research Center, Agricultural Research, U.S. Department of Agriculture, Peoria, Illinois 61604

Matthew Suffness (201, 465), Plant and Animal Products Section, Natural Products Branch, National Cancer Institute, Silver Spring, Maryland 20910

Hamao Umezawa (147), Institute of Microbial Chemistry, Shinagawa-ku, Tokyo, Japan

Monroe E. Wall (417), Chemistry and Life Sciences Group, Research Triangle Institute, Research Triangle Park, North Carolina 27709

Mansukh C. Wani (417), Chemistry and Life Sciences Group, Research Triangle Institute, Research Triangle Park, North Carolina 27709

Paul F. Wiley (167), Cancer Research, The Upjohn Company, Kalamazoo, Michigan 49001

Preface

The use of crude natural products to treat cancer can be traced back to antiquity. Jonathon Hartwell, the former Section Head of the Natural Products Drug Development Program at the National Cancer Institute, for example, has found over 3000 species of plants that were reported to be used as some form of cancer treatment. The modern use of pure natural drugs in cancer chemotherapy has a shorter history tracing back about 30 years. As a result of extensive drug development programs, largely coordinated by the drug development program of the Natural Products Branch, National Cancer Institute, there are currently about 50 clinically active drugs for the treatment of cancer and another 100 in various phases of preclinical development. A number of these drugs originate from natural sources or were designed and synthesized based on an active natural product model.

The need for novel and more selective anticancer drugs continues, despite continuing efforts to understand how cancer develops and functions. If efforts to prevent the disease by identifying causal agents or by using chemical modifiers were immediately instituted and successful, it is likely that cancer would continue to occur at its present rate for several decades. Rational design of novel and effective anticancer drugs is limited by our still scant knowledge of cancer biology and biochemistry. Thus it appears likely that development of novel agents from a variety of sources based on screening in animal tumor models will remain as an important research area in the forseeable future.

Natural products, especially higher plants and microorganisms, have served as a rich source of novel drugs, including anticancer drugs. One of the strongest arguments for initiating a major effort to uncover novel drugs from higher and lower plants is the discovery of the antimitotic and anticancer alkaloids from *Catharanthus roseus,* the "Vinca" alkaloids, vincristine and vinblastine. These compounds are typical of natural drugs having novel and complex chemical and stereochemical features and potent biological effects. No drug is perfect, and so the medicinal chemist can use these prototype molecules, developed empirically, as a starting point in the synthetic or semi-synthetic development of related compounds which are potential drugs in their own right and which help establish the basis for understanding the molecular mechanism of action of the drug class. Finally these potent biologically active agents can serve as tools for the biochemist and pharmacologist in studying cell biology and biochemistry.

This book is divided into four major parts. The two major areas covered are antitumor agents from higher plants and from lower plants (microorganisms). In these subsections the chapters are subdivided based on the important drug prototypes. For example, chapters are included on the anthracyclines, trichothecanes, mitomycins, streptozocins, and bleomycins. Other lower plant-derived agents include the maytansinoids and nucleosides. Drug types from higher plants include the *Catharanthus* alkaloids, harringtonines, podophyllotoxins, camptothecins, and maytansinoids. In addition, a chapter is included on various terpenoid tumor inhibitors. We have also covered microbial transformation of anticancer drugs and finally there is a chapter on miscellaneous agents.

The emphasis we hope to achieve is on novel lead development, structure–activity relationships, and molecular pharmacology. All of these topics are designed to point out to the pharmacologist or medicinal chemist areas that are in active development or areas that should be further developed in our continuing search for new and better cancer chemotherapeutic agents.

John M. Cassady
John D. Douros

CHAPTER 1

The Development of New Antitumor Anthracyclines

FEDERICO ARCAMONE

I. INTRODUCTION

All important classes of antibiotics have been, and still are, the object of extensive investigations aimed at the optimization of their pharmacological properties. Among these investigations, the development of new analogs by chemical manipulation of the biosynthetic product or by total synthesis are major routes to the final objective.

Daunomycin and related antibiotics, among which doxorubicin is the most important because of its outstanding value in the chemotherapic treatment of human tumors, represent a structurally defined group within the family of the anthracyclines.

The structure and stereochemistry of daunomycin as determined by means of chemical and physicochemical studies (Arcamone *et al.*, 1964a,b, 1968a,b, 1970b) is indicated in (**1**).

Doxorubicin was originated from mutant strains of the daunomycin-producing microorganism, *Streptomyces peucetius* (Arcamone *et al.*, 1969d). Structure (**2**)

Anticancer Agents Based on Natural Product Models
Copyright © 1980 by Academic Press, Inc.
All rights of reproduction in any form reserved.
ISBN 0-12-163150-8

was attributed to doxorubicin on the basis of chemical degradations and physicochemical studies (Arcamone *et al.*, 1969a), and confirmed by semisynthesis (Arcamone *et al.*, 1974b). Recently, the reverse transformation of doxorubicin to daunomycin has also been carried out (Penco *et al.*, 1977).

Daunomycin-related compounds have been isolated from microbial cultures: the 4-*O*-demethyl analogue, carminomycin I (**3**) (Pettit *et al.*, 1975; Wani *et al.*, 1975), the 13-dihydro analogues of daunomycin (duborimycin) and of carminomycin (Lunel and Preud'homme, 1970; Arcamone *et al.*, 1970a; Cassinelli *et al.*, 1978), the disaccharide derivative daunosaminyldaunomycin (Arcamone *et al.*, 1970a), rhodinosyldaunomycin (Fedorova *et al.*, 1970), and the baumycins (Takahashi *et al.*, 1977).

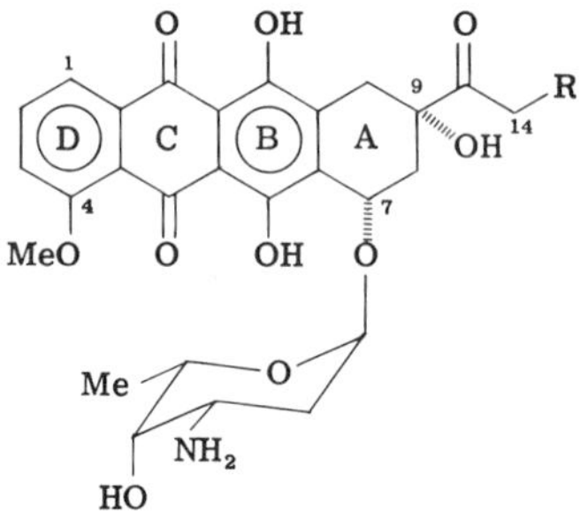

1 : R = H
2 : R = OH
3 : OH at C-4, R = H

II. NEW ANALOGUES MODIFIED AT C-9, C-13, AND C-14

Many chemical modifications of the antitumor anthracyclines are concerned with the C-9 sidechain of these antibiotics. There are different reasons for this. First, the presence of the ketone function allows the use of known carbonyl reactions for derivatization, and regioselectivity is assured by the different chemical behavior of the quinone carbonyls. In addition, substitutions on the C-14 methyl group easily afford doxorubicin derivatives, such as esters, or other analogues, such as the 14-aminodaunomycins. Second, according to the intercalation model (Di Marco and Arcamone, 1975), the sidechain should not be directly involved in the drug–receptor interaction. On the other hand, its importance for the stabilization of a given conformation of ring A and as a site of the aldehyde reductase action is clearly discernible (Felsted *et al.*, 1977).

A. Modification at C-13

1. Reduction of the C-13 Carbonyl Group

The 13-dihydro derivatives of daunomycin and of carminomycin have already been mentioned as biosynthetic representatives of the anthracycline glycosides, and reduction of the C-13 carbonyl group of daunomycin (**1**) and carminomycin (**3**) has also been carried out with a variety of microorganisms (Florent *et al.*, 1975; Florent and Lunel, 1976; Aszalos *et al.*, 1977). The glycosides bearing the C-13 secondary hydroxyl group have, however, also been obtained upon complex borohydride treatment of the corresponding carbonyl compounds. The reaction was described for daunomycin and doxorubicin, an 82% yield being reported for the reduction of the former with potassium borohydride (Jolles and Ponsinet, 1972). The regioselectivity of the reaction is assured by the air reoxidation of the reduced quinone. The biological activity of 13-dihydrodaunomycin in four different experimental tumors of mice was found to be similar to that of the parent compound, although 50% inhibition of HeLa cell proliferation *in vitro* occurred at a nearly 10 times higher concentration (Di Marco *et al.*, 1973). In a more recent study (Cassinelli *et al.*, 1978), the optimal doses (the most effective doses showing less than 10% toxic deaths among treated mice) of 13-dihydrodaunomycin and of 13-dihydrocarminomycin were not very different from those of the parent compounds (Table I). Although no information is available concerning the absolute stereochemistry at C-13 in these derivatives, it is reasonable to consider both the biosynthetic derivatives and the enzymically reduced ones as being constituted by a single stereoisomer. The same is not true for the chemically originated 13-dihydrodaunomycin, as deduced on the basis of the optical rotation shown by bisanhydro-13-dihydro-

TABLE I

Activity of 13-Dihydro Derivatives on L-1210 Leukemia in Mice at Optimal Nontoxic Doses[a]

Compound	Dose (mg/kg)	T/C (%)[b]
Daunomycin (**1**)	2.9	147
13-Dihydrodaunomycin	4.4	138
Carminomycin (**3**)	0.6	150
13-Dihydrocarminomycin	0.4	162

[a] Data from Cassinelli *et al.* (1978).

[b] Average survival time of treated animals expressed as percentage of survival time of untreated controls.

daunomycinone (**4**) when obtained from natural or from semisynthetic 13-dihydrodaunomycin (unpublished data from the author's laboratory). However, no differences could be found in the biological behavior of the biosynthetic and the chemically derived glycoside (A. Di Marco, personal communication).

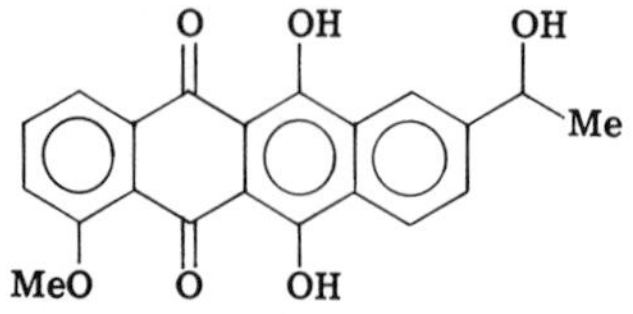

4

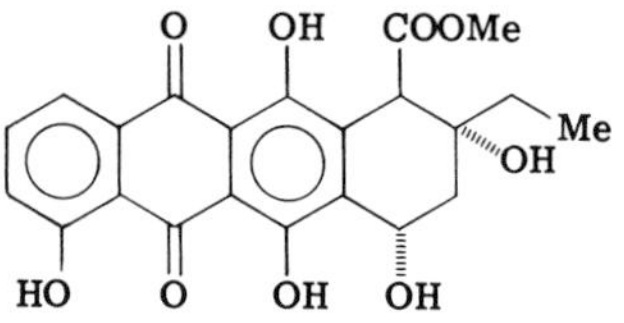

5

The synthesis of the 13-deoxy analogues of the antitumor anthracyclines was of obvious interest because of the presence of the ethyl C-9 sidechain in the majority of known naturally occurring anthracyclines. This aim was pursued by Smith *et al.* (1976). The authors first attempted, without success, the decarbomethoxylation of ε-rhodomycinone (**5**), a known abundant by-product of the daunomycin fermentation (Kern *et al.*, 1977). However, daunomycinone tosylhydrazone (**6**), obtained in 85% yield from daunomycinone, was selectively reduced to 13-deoxydaunomycinone (18% yield) with sodium cyanoborohydride in 1:1 sulfolane and dimethylformamide at 100°. The reaction was then performed with daunomycin and with doxorubicin tosylhydrazones with excess reducing agent to afford the 13-deoxy analogues (**7**) and (**8**) in reasonable yields. The new analogues retained substantial antitumor activity on P-388 leukemia in mice, indicating that the C-13 oxygen is not a strict requirement for the exhibition of the biological activity.

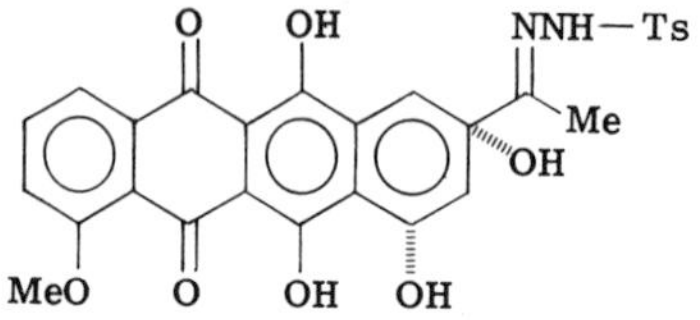

6

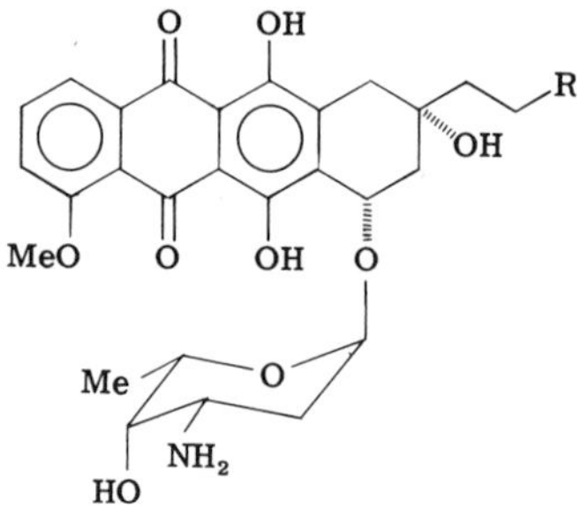

7 : R = H
8 : R = OH

2. Derivatives of the C-13 Ketone Function

The easy derivatization of daunomycin with carbonyl reagents has prompted the preparation of a number of derivatives described in the scientific literature (Yamamoto *et al.*, 1972) and in the patent (Jolles, 1969, 1973) literature. Daunomycin semicarbazone, thiosemicarbazone, oxime, and related derivatives were less effective, both *in vitro* and *in vivo*, when compared with the parent compound (Di Marco *et al.*, 1973; Yamamoto *et al.*, 1972). On the other hand, daunomycin benzoylhydrazone (rubidazone) has been found to be endowed with similar antitumor activity on experimental tumors, but with a lower toxicity than daunomycin and doxorubicin (Maral *et al.*, 1972). The compound has also been used for the treatment of human tumors (Jacquillat, 1976; Benjamin *et al.*, 1977).

B. C-14 Derivatives

Doxorubicin (2) is the only known biosynthetic anthracycline substituted at C-14, and the most important clinically. For this reason, and also because of the availability of 14-halogeno derivatives (9) and (10) of daunomycin, a range of 14-substituted analogues have been prepared.

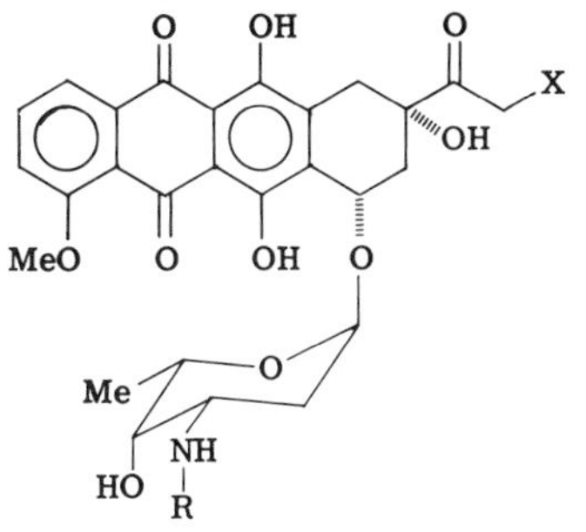

9 : R = H, X = Br
10 : R = COCF$_3$,
         X = J

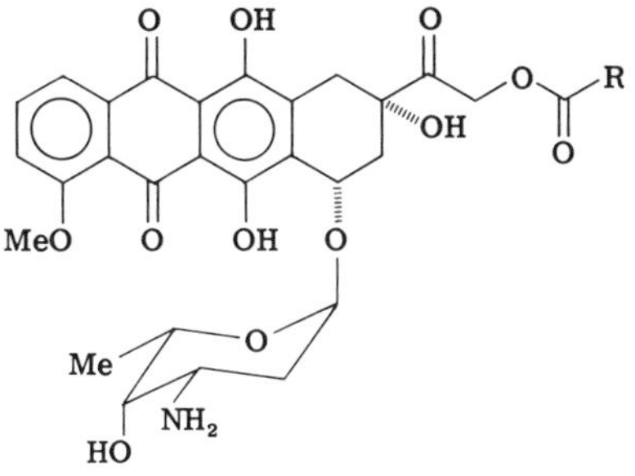

11 : R = Me
12 : R = Et
13 : R = (CH$_2$)$_6$Me
14 : R = Ph
15 : R = CH$_2$Ph

16 : R =

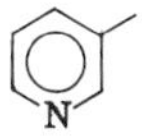

1. Doxorubicin Esters and Thioesters

Doxorubicin esters (11)–(16) were obtained by reaction of (9) with the sodium or potassium salt of the corresponding carboxylic acids (Arcamone *et al.*, 1974a). Other derivatives of this type have been synthesized, owing to the high level of antitumor activity retained by compounds (11)–(16). Such new compounds are double esters (17)–(20) (Patelli *et al.*, 1976) and thioesters (21) and (22) (Arcamone *et al.*, 1977b).

Compounds (11)–(16) were found to retain substantial activity in cell cultures and in tumor-bearing animals when compared with doxorubicin itself (Arcamone

17 : R = $(CH_2)_4Me$, X = O
18 : R = $(CH_2)_{10}Me$, X = O
19 : R = $(CH_2)_{16}Me$, X = O
20 : R = $(CH_2)_4Me$, X = S

et al., 1974a). In the same study it was also shown, on the basis of experiments with tritium-labeled adriamycin 14-octanoate, that esterification of the sidechain hydroxyl leads to modification of the distribution characteristics of the drug and that a conversion of the said ester to doxorubicin takes place in mouse serum samples and homogenates from liver and from other tissues. According to the above-mentioned authors and to a subsequent report of Lenaz *et al.* (1974), doxorubicin esters display a somewhat lower antiproliferative but, with the exception of (**9**), higher antimitotic activity on cultured HeLa cells, when compared with doxorubicin. In the animals tests, the octanoate (**13**) appeared generally the most effective, but no clear relationship existed between biological activity and partition coefficient in a 1-butanol aqueous buffer system in this series. The activity of doxorubicin esters on Gross leukemia is presented in Table II.

Doxorubicin 14-octanoate and doxorubicin 14-*O*-glycolate were each found able to form a complex with DNA (Arlandini *et al.*, 1977), with stability constants, respectively, 1.3×10^5 and 2.2×10^5 (doxorubicin 3.7×10^5). On the basis of this information and of the cell culture data, it could be concluded that the esters may possess biological activity per se, but enzymic formation of dox-

21 : R = Me
22 : R = Ph

TABLE II

Activity of Doxorubicin Esters on Gross Leukemia in Mice at Optimal Nontoxic Doses[a]

Compound	Structure	Dose[b] (mg/kg)	T/C (%)
Daunomycin	1	3.25	176
Doxorubicin	2	2	187
Doxorubicin, 14-acetate	9	2.5	168
Doxorubicin, 14-propionate	12	4	147
Doxorubicin, 14-octanoate	13	4.1	214
Doxoribicin, 14-benzoate	14	2.5	138
Doxorubicin, 14-phenylacetate	15	3.25	156
Doxorubicin, nicotinate	16	4	227

[a] Data from Arcamone *et al.* (1974a).
[b] Treatment intravenous on days 3 to 5.

orubicin in body fluids or intracellularly may play a role in the exhibition of antitumor activity. In their report concerning the comparison of the efficacy of doxorubincin and related compounds in different experimental tumors, Goldin and Johnson (1975) demonstrated that esters (**11**)–(**16**) were not more active than the parent drug in the L-1210 and P-388 test systems. However, in view of their retention of activity with differences in solubility and potential differences in pharmacological characteristics, such compounds were considered by the authors as worthy of further investigation and possibly of clinical application. Quite recently, Hill and Price (1977) have reported that doxorubicin 14-octanoate was superior to doxorubicin in reducing the viability of L-5178Y lymphoblasts resistant to methotrexate.

Compounds (**17**)–(**20**) exhibited antitumor activity comparable to that shown by doxorubicin, although at a somewhat higher dosage (Arcamone *et al.*, 1977a). On the other hand, thioesters (**21**) and (**22**) were clearly less effective. Data obtained under the auspices of the National Cancer Institute are presented in Table III.

Toxicology of doxorubicin octanoate in the rat was studied by Zbinden and Brändle (1975). The compound displayed a lower general toxicity than both doxorubicin and daunomycin, hematotoxicity being of the same order as that of the former. The compound also showed a minimal cumulative cardiotoxic dose (MCCD) in the rat of 18–20 mg/kg, substantially higher than that shown by doxorubicin (11–12 mg/kg). The MCCD value, which appears to be a useful screening parameter for the classification of the cardiotoxic properties of the new analogues (Zbinden *et al.*, 1978b), has also been determined for the nicotinate and the stearoylglycolate esters, the value being 32 mg/kg for the former and 40

TABLE III

Activity of Doxorubicin Esters and Thioesters on P-388 Leukemia in Mice[a]

Compound	Structure	Dose[b] (mg/kg)	T/C (%)
Doxorubicin[c]	**2**	1.08	215
Doxorubicin, 14-octanoate	**13**	2	247
Doxorubicin, 14-nicotinate	**16**	2	229
Doxorubicin, 14-stearoylglycolate	**19**	3.13	202
Doxorubicin, 14-hexanoylthioglycolate	**20**	3.13	201
Doxorubicin, 14-thioacetate	**21**	9.4	145
Doxorubicin, 14-thiobenzoate	**22**	3.13	141

[a] NCI data (screener A.D. Little).
[b] Dose affording the highest T/C value in the QD 1–9 treatment schedule.
[c] Average of 60 experiments.

mg/kg for the latter (G. Zbinden, personal communication). These results are clearly indicative of the potential clinical value of the 14-O-esters of doxorubicin.

2. Doxorubicin Ethers

A group of new doxorubicin derivatives is represented by compounds (**23**)–(**25**). These compounds have been prepared starting from 14-

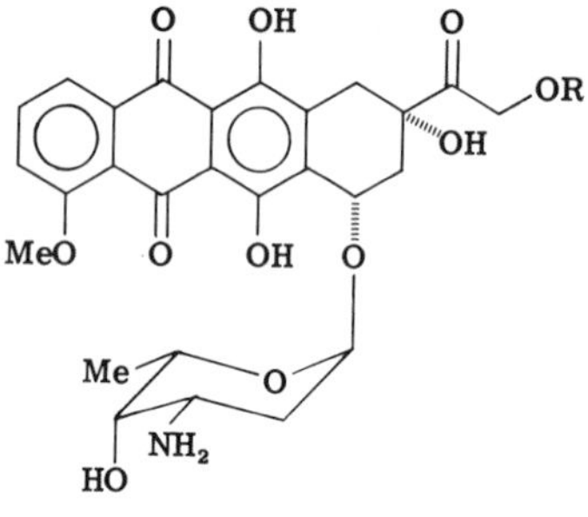

23 : R = Me
24 : R = Et
25 : R = iPr
26 : R = Ph

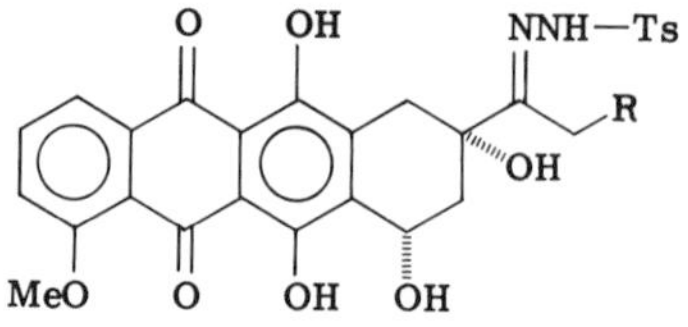

27 : R = Br
28 : R = OMe
29 : R = OEt
30 : R = OiPr
31 : R = OPh

bromodaunomycinone (Arcamone *et al.*, 1969c), which was converted to the tosylhydrazone (**27**), and the latter transformed into the corresponding 14-alkoxy derivatives (**28**)–(**30**) upon treatment with an alcohol in the presence of silver tri-fluoromethansulphonate (silver triflate). Refluxing of (**28**)–(**30**) in acetone in the presence of *p*-toluene-sulphonic acid afforded deblocking of the C-13 ketone group. Glycosidation of the resulting substituted daunomycinone with *N,O*-ditrifluoroacetyldaunosaminyl chloride in the presence of silver triflate, followed by removal of the trifluoroacetyl groups, gave the desired glycosides (**23**)–(**25**) (Masi *et al.*, 1977).

Compounds (**23**), (**24**), and (**26**), when tested in the P-388 system (Geran *et al.*, 1972), appeared to be less effective than doxorubicin and showed responses similar to those obtained with daunomycin, albeit at doses three times higher (Arcamone, 1978).

3. 14-Aminodaunomycins

Reaction of (**9**) with amines afforded the introduction of a substituted amino group at C-14. This type of substitution was of interest in connection with the anionic character of DNA and the possibility of increasing affinity for the

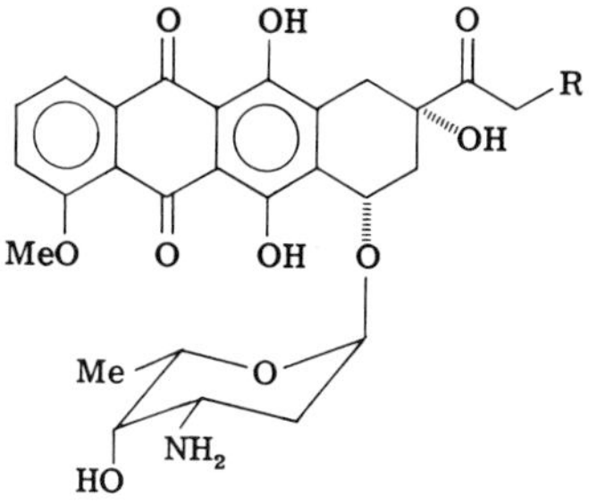

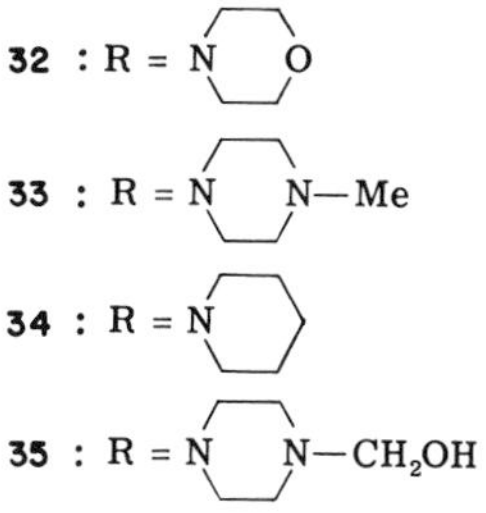

32 : R = N　O

33 : R = N　N—Me

34 : R = N

35 : R = N　N—CH$_2$OH

36 : R = N(CH$_2$CH$_2$OH)$_2$

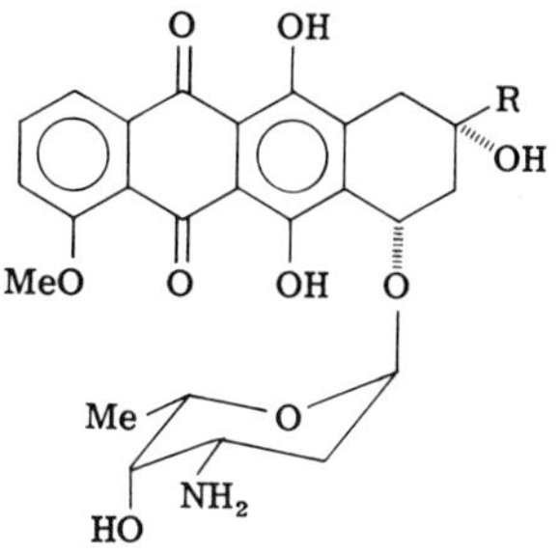

37 : R = COOH
38 : R = COOMe
39 : R = CH$_2$OH

receptor because of the presence of an additional basic function in the noninter-calating part of the drug molecule. Compounds (**32**)–(**36**) were prepared and tested for biological activity in cultured cells and in tumor-bearing mice (Ar-camone *et al.*, 1976g). The new compounds were 10 to 50 times less effective on the proliferation of mouse-embryo fibroblasts (MEF) when compared with (**1**) or (**2**), but (**32**) and (**33**) displayed antiviral activity on murine sarcoma virus (Moloney) in MEF at, respectively, sixfold and twofold higher doses than doxorubicin. *In vivo* activity on ascites sarcoma 180 was noticeable at optimal doses, but considerably lower than that of (**1**) or (**2**) on a weight basis. Compound (**32**) was found to form a DNA complex with stability constant 2.1×10^5 (Arlandini *et al.*, 1977).

C. Derivatives Arising by Sidechain Oxidative Degradation

Periodate oxidation of doxorubicin gave the amphoteric compound (37) which was also converted to the methyl ester (38). Both compounds were devoid of antitumor activity (Tong *et al.*, 1976). On the other hand, compound (39) originated upon periodate oxidation of 13-dihydrodoxorubicin *N*-trifluoroacetate (Penco *et al.*, 1977), followed by reduction of the resulting C-9 formyl compound to a C-9 hydroxymethyl derivative and de-N-trifluoroacetylation, retained substantial activity in the P-388 system (Penco *et al.*, 1978; Arcamone, 1978).

Oxidative degradation of 13-dihydrodaunomycin *N*-trifluoroacetate with sodium metaperiodate afforded the ketone (40). Reduction of (40) with sodium cyanoborohydride gave the two epimeric alcohols (41) and (43), with the *cis*

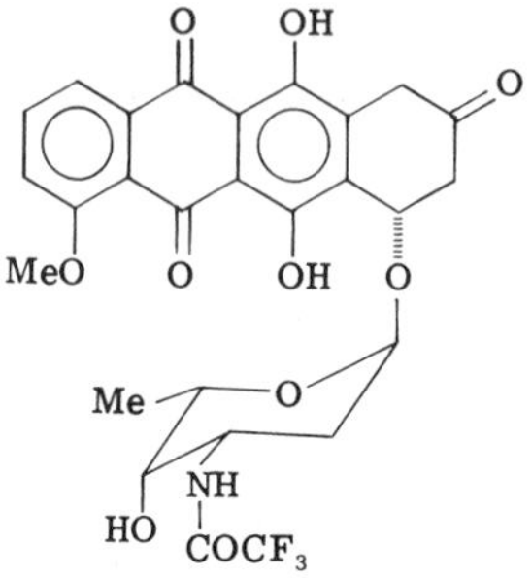

40

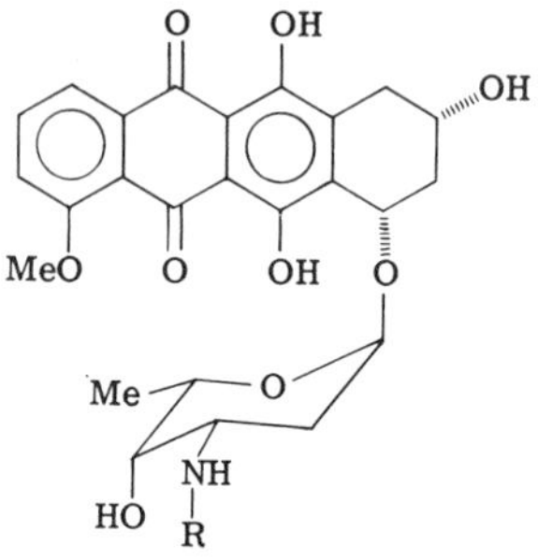

41 : R = COCF₃
42 : R = H

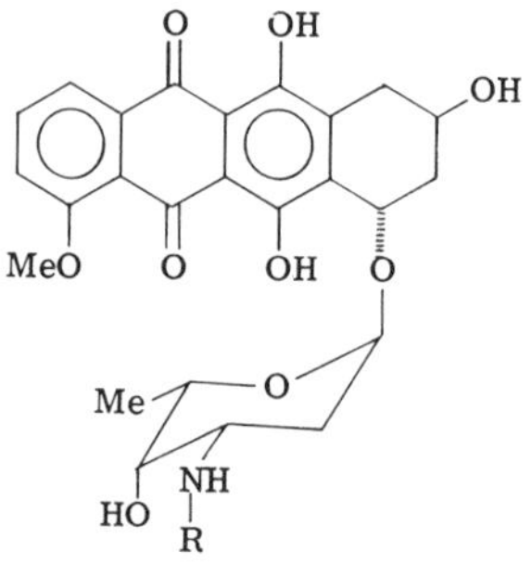

43 : R = COCF₃
44 : R = H

compound (41) being the main product of the reaction. The glycosides (42) and (44) were obtained upon alkaline treatment, and the stereochemistry of the compounds was deduced on the basis of the proton magnetic resonance (PMR) spectrum of the corresponding tetraacetylated aglycones (Penco *et al.*, 1977). In the same paper, antitumor activity and DNA-binding ability of (42) and (44) are reported. The *cis* compound exhibited an higher affinity for DNA as deduced from the stability constants of the drug–DNA complex (*cis*, 2.2×10^5; *trans*, 0.8×10^5) and from the greater enhancement of DNA viscosity shown by (42) in comparison to (44). The two epimers also showed different efficacy in the P-388 system, the *cis* compound giving (at optimal doses) a T/C percentage of 228 (dose 12.5 mg/kg) and the *trans* a T/C percentage of 171 (dose 6.25 mg/kg). The treatment schedule was Quoque Die (QD) 1–9. This study indicated that a higher selectivity of biological activity was associated with the 9(*S*)-*cis* in comparison with the 9(*R*)-*trans* compound, and that the effect of stereochemistry on pharmacological activity paralleled that on DNA-complexing ability.

III. ANALOGUES MODIFIED IN THE AMINOSUGAR RESIDUE

An important structural feature of the daunomycin–doxorubicin group of antibiotics is the presence of the previously unknown aminosugar daunosamine, 3-amino-2,3,6-trideoxy-L-lyxo hexose (Arcamone *et al.*, 1964b). In fact, the other anthracyclines have the *N*-dimethyl derivative rhodosamine as the aminosugar component, but also other non-aminated sugars can be present (Brockmann, 1963). The aminosugar moiety of the antitumor anthracyclines contains four of the six chiral centers of the molecule. To the said centers, the absolute stereochemistry $1'(R)$, $3'(S)$, $4'(S)$, and $5'(S)$, corresponding to the L-lyxo α-glycoside configuration was assigned. Daunosamine (45) is one of the eight 3-amino-2,3,6-trideoxyhexoses presently known as being present in antibiotic molecules. The others are acosamine (46) from actinoidin (Lomakina *et al.*, 1973), ristosamine (47) from ristomycin (Bognar *et al.*, 1974), the already mentioned rhodosamine (48), actinosamine (49) from actinoidin (Lomakina *et al.*, 1973), vancosamine (50), from vancomycin (see Williams and Kolman, 1976, and references cited therein), D-rhodosamine (51) from the megalomycins (Mallarms, 1969), and angolosamine (52) from angolamycin (Brufani and Keller Schierlein, 1966).

Analogues of the antitumor anthracyclines showing modifications in the sugar moiety are of importance because alterations in sugar stereochemistry and substitution are expected to modify those parameters of anthracycline efficacy which reside on active transport or facilitated diffusion and on enzyme reactions, namely tissue distribution and metabolism of the drug, on the basis of the known dependence of such processes on structure and stereochemistry of carbohydrate derivatives. Because of the intracellular location of the therapeutic target of the antitumor anthracyclines, cellular uptake and also intracellular distribution are

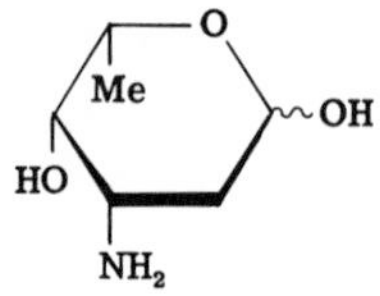

45

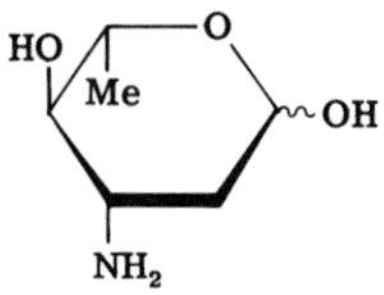

46

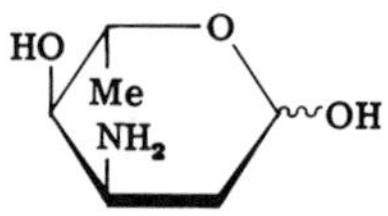

47

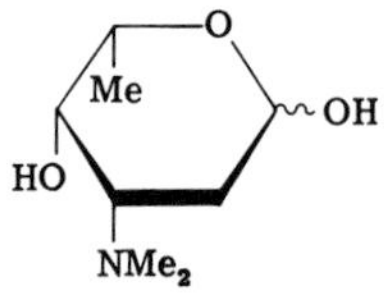

48

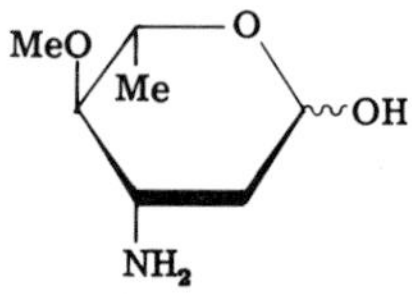

49

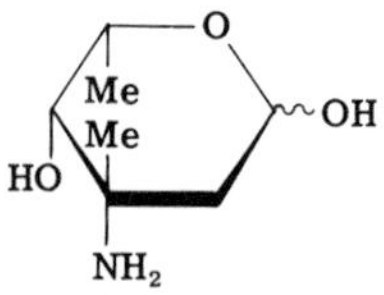

50

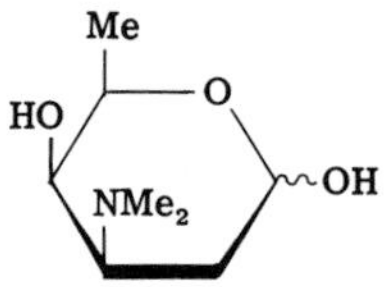

51

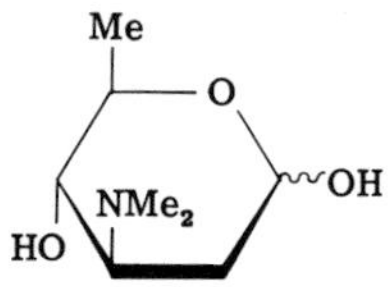

52

decisive factors in the pharmacological properties of these drugs (see Skovsgaard, 1977, and references cited therein).

A. Configurational Analogues

Configurational analogues of the antitumor anthracyclines, differing from the parent drugs in the absolute stereochemistry at one of the four chiral centers $1'$, $3'$, $4'$, or $5'$, have been synthesized, and each one's biological activity has been reported in the majority of cases. The new compounds have been obtained following a general scheme including the synthesis of the desired aminosugar in appropriately derivatized form in order to carry out the subsequent glycosylation of daunomycinone or of a protected adriamycinone derivative.

1. β-Daunomycin and β-Doxorubicin

Koenigs-Knorr condensation of daunomycinone with N,O-ditrifluoroacetyl-α-daunosaminyl chloride (**53**) afforded two products which could be separated by preparative chromatography after removal of the O-trifluoroacetyl group by methanolysis. The two reaction products were the α

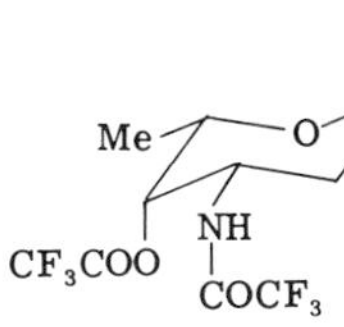

53

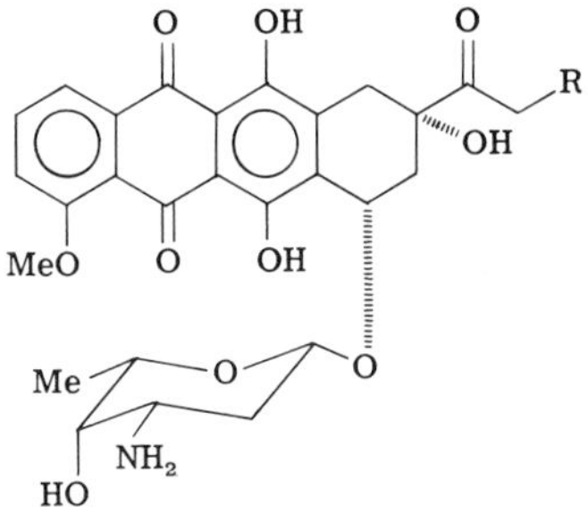

54 : R = H
55 : R = OH

and β glycosides arising from the coupling of daunomycinone with N-trifluoroacetyldaunosamine. The α glycoside was formed in greater amount than the β glycoside. Alkaline removal of the N-trifluoroacetyl group gave two anomeric glycosides, one being identical to biosynthetic daunomycin, the other

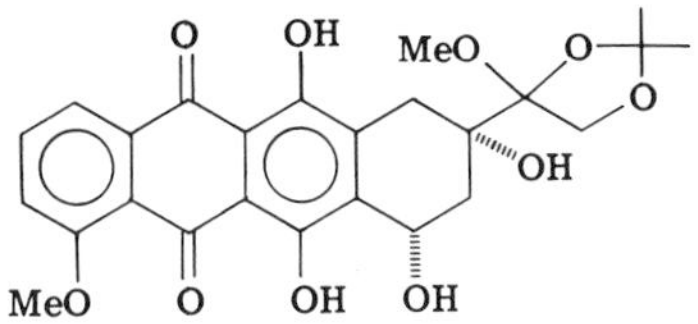

56

representing the corresponding β anomer (**54**). In a similar way, but starting with adriamycinone derivative (**56**), the corresponding α and β glycosides derived from the coupling of (**56**) with N-trifluoroacetyldaunosamine were obtained. After chromatographic separation the products were treated with alkali and then with acid to give doxorubicin and its β anomer (**55**).

As a diagnostic, PMR spectroscopy was used. The α-glycoside equatorial anomeric proton gives rise to a signal at about 5.5 δ, appearing as a broad singlet with a half-band width of about 6 Hz. The β-glycoside axial anomeric proton resonates at higher field (about 0.5 δ) with respect to the α anomers and appears as a pair of doublets, one with a low, and the other with a high (~ 10 Hz) coupling constant. On the other hand, the H-7 signal appears in the range 5.0 δ–5.2 δ in the α series and at about 5.5 δ in the β series (Arcamone et al., 1975b).

The β anomer of doxorubicin displayed a lower association constant ($K = 0.2 \times 10^5$) with heat-denatured calf-thymus DNA than did doxorubicin ($K = 1.5 \times 10^5$) (Arlandini et al., 1977), and was much less active in the inhibition of E. coli DNA polymerase I and RNA polymerase, of tritiated thymidine incorporation in mouse-embryo fibloblasts, and on cell proliferation than the parent antibiotic (Di Marco et al., 1976b). The lower affinity for DNA of the β anomers as compared with the compounds with the natural configuration can be explained by examination of the molecular model of an hypothetic complex with the DNA double helix. In the lower energy conformation, obtained by minimizing non-bonded interactions around the C(7)–O bond and the glycosidic linkage, the distance between the amino group and the second DNA phosphate away from the intercalation site is larger for β than for α anomers.

2. The L-Arabino Analogues

The first synthesis of an acosamine derivative in the anthracycline series was carried out starting from methyl-N-trifluoroacetyl-α-daunosaminide (**57**). Conversion of (**57**) to the 4-keto derivative (**58**) and stereoselective reduction of

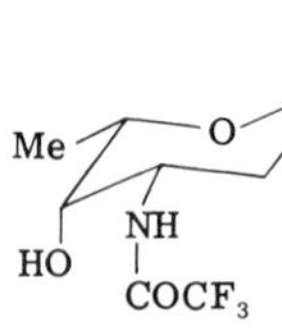

| 57 | 58 |

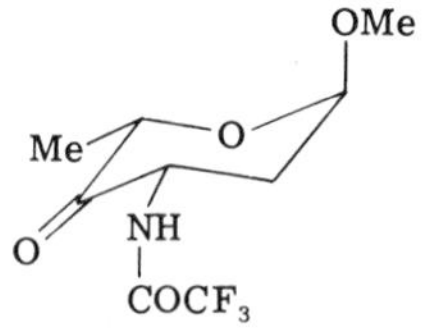

the latter to give the equatorial alcohol (**59**) were essential steps of the synthesis. Acid hydrolysis of (**59**) afforded N-trifluoroacetylacosamine, which was converted to (**60**) and condensed with daunomycinone or with (**56**) to afford, after

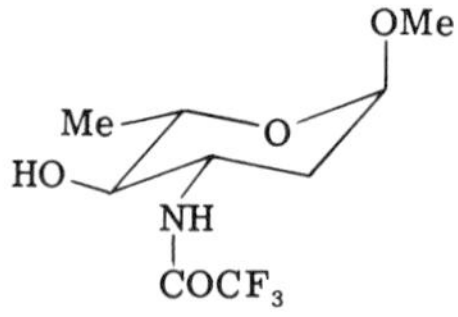

| 59 | 60 |

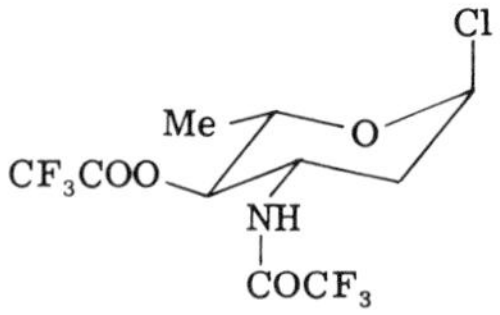

removal of the protecting groups, 4′-epidaunomycin (**61**) and its β anomer (**63**), or 4′-epidoxorubicin (**62**) and its β anomer (**64**), respectively. In the conditions used for this glycosidation, which was carried out in the presence of HgO,

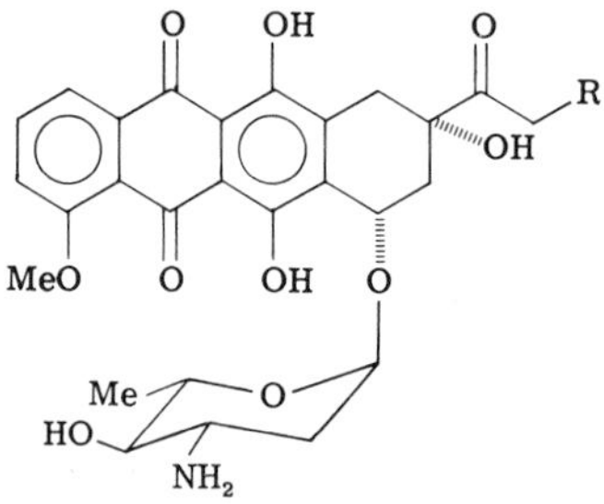

61 : R = H
62 : R = OH

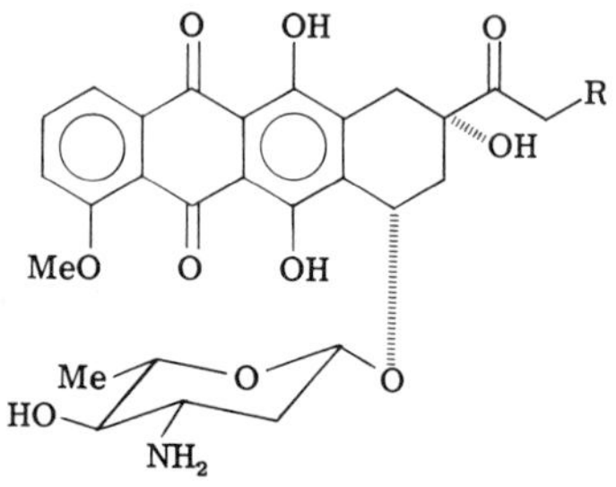

63 : R = H
64 : R = OH

HgBr$_2$, and molecular sieve in boiling dichloromethane, the ratio α to β anomers was approximately 7:3 (Arcamone *et al.*, 1975b).

Other syntheses of acosamine have been reported. A synthesis started from L-rhamnose which was converted to L-rhamnal (**65**) according to known proce-

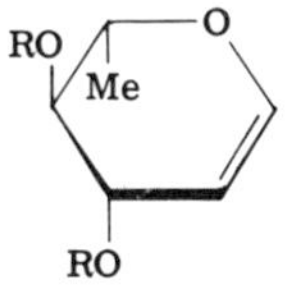

65 : R = H
66 : R = Ac

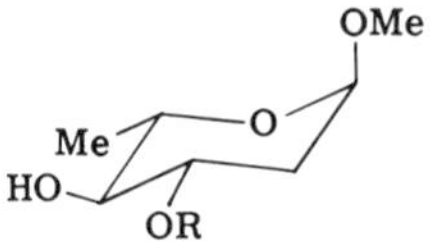

67 : R = H
68 : R = Ts

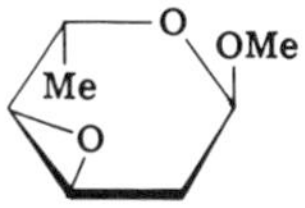

69

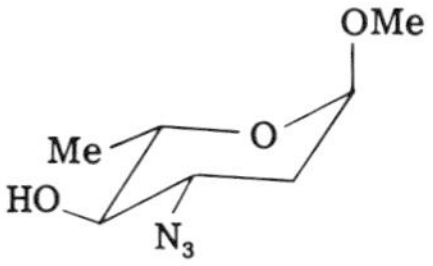

70

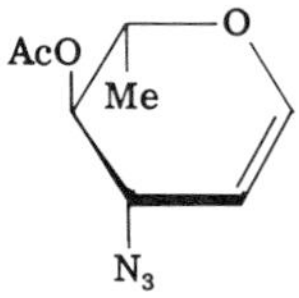

71

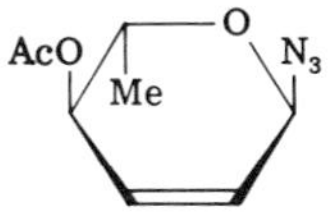

72

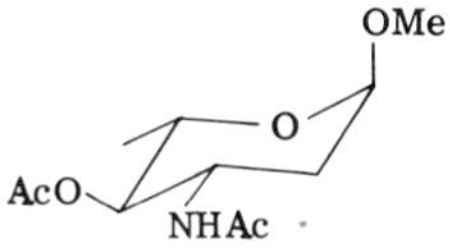

73

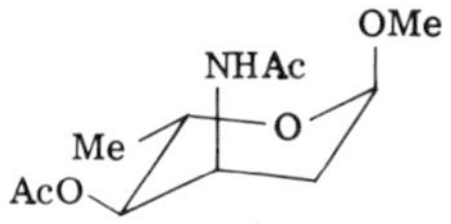

74

dures. Methoxymercuriation of (**66**) followed by a reduction step afforded the 2,6-dideoxy derivative (**67**). The 3-*O*-tosyl derivative (**68**) was converted to the epoxide (**69**), which was opened with azide to afford (**70**). Catalytic reduction of (**70**) afforded methyl-α-acosaminide (Gupta, 1974; Lee *et al.*, 1975b). L-Rhamnal was also the starting material in a new synthesis leading to both acosamine and ristosamine derivatives. Treatment of di-*O*-acetyl-L-rhamnal (**66**) with sodium azide and boron trifluoride etherate gave mainly (**71**) in equilibrium with (**72**). Iodomethylation of the equilibrium mixture followed by catalytic hydrogenation and acetylation afforded (**73**) as the major product, accompanied by the L-ribo stereoisomer (**74**) (Heyns *et al.*, 1976).

Both compounds (**61**) and (**62**) displayed antitumor activity similar to, and at the same dose levels as, that of the parent glycosides on L-1210 experimental leukemia (Table IV) and on other experimental tumors in mice (Arcamone *et al.*,

TABLE IV

Antitumor Activity of Configurational Analogues of Daunomycin and Doxorubicin on L-1210 Experimental Leukemia in Mice with Treatment i.p. on Day 1[a]

Compound	Structure	Configuration	Optimal dose (mg/kg)	T/C (%)[b]
Daunomycin	1	L-lyxo	4	150
Daunomycin, 4'-epi-	61	L-arabino	4	143
Daunomycin, 3',4'-diepi-	75	L-ribo	50	137
Daunomycin, 3'-epi-	89	L-xylo	33[c]	120
Doxorubicin	2	L-lyxo	5	166
Doxorubicin, 4'-epi-	56	L-arabino	5	150
Doxorubicin, 3',4'-diepi	76	L-ribo	60	161

[a] Data from A. Di Marco and A. M. Casazza, Istituto Nazionale dei Tumori, Milan, Italy.
[b] Average survival time expressed as percentage of survival time of controls.
[c] Maximal dose tested.

1975a). Modification of the absolute stereochemistry at C-4' did not reduce the ability to inhibit the *E. coli* DNA polymerase I and RNA polymerase catalyzed reactions, the incorporation of tritiated thymidine in mouse embryo fibroblasts (MEF) as well as MEF proliferation (Di Marco *et al.*, 1976b). DNA-complexing properties, as measured by the equilibrium dialysis experiments, by viscosimetric measurements (Arlandini *et al.*, 1977), and by the effect on thermal transition temperature of double-helical DNA (Di Marco *et al.*, 1976b), were also comparable with those displayed by the parent antibiotics.

In comparative studies (see Arcamone *et al.*, 1977a, and references cited therein), 4'-epidoxorubicin was shown to have the same antitumor activity as doxorubicin at the same doses on L-1210 and P-388 leukemia in mice, the former exhibiting, however, a reduced toxicity. In the P-388 system, the therapeutic ratio (LD_{10}/ED_{10}) was found to be 1.81 for (**62**) and 1.09 for (**2**). It has been suggested that the behavior of 4'-epidoxorubicin in respect to doxorubicin could be a consequence of the lower pK_a value [$pK_a = 8.08$ for (**62**) and 8.34 for (**2**)], resulting in a reduced DNA binding of the drug at the physiological pH in normal tissues, but in the expression of full intercalative properties in tumor cells, characterized by a lower cytoplasmic pH, and therefore in an enhanced selectivity of action. In addition, (**62**) was equally active, but less toxic, than (**2**) in mice bearing transplanted mammary carcinoma, and the lower toxicity of (**62**) allowed a higher dosage resulting in a more prolonged inhibition of tumor growth on MSV-induced sarcoma in mice. 4'-Epidoxorubicin was also shown to display an outstanding activity, and was superior to doxorubicin, on experimental metastatic diseases such as Lewis lung carcinoma and MS-2 sarcoma. These results, and the lower cardiotoxicity of (**62**) when compared with (**2**) in the chronic experiment in the rabbit (Casazza *et al.*, 1977) and in the rat (G. Zbinden, personal communication), the latter giving MCCD = 16 mg/kg for (**62**) and 12 mg/kg for (**2**), indicated 4'-epidoxorubicin as a good candidate for clinical trials.

The β-anomeric analogues (**63**) and (**64**) displayed a strongly reduced inhibition of DNA synthesis in cell cultures and of the *in vitro* activity of DNA-dependent bacterial polymerases, consistent with the lower cytotoxicity and lower antitumor activity in respect to the corresponding α anomers, as already mentioned for the L-lyxo series (Di Marco *et al.*, 1976b); Arcamone *et al.*, 1975a).

3. The L-ribo, L-xylo, and D-ribo isomers

The synthesis of the L-ribo configurational isomer (**75**) of daunomycin was performed by condensing 4-*O*-*p*-nitrobenzoyl-3-*N*-trifluoroacetylris-

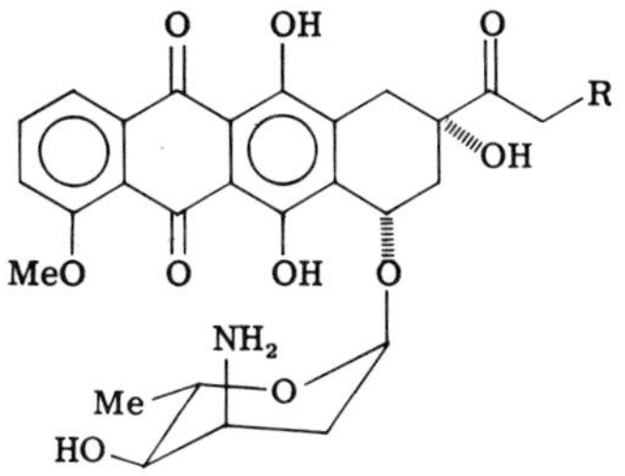

75 : R = H
76 : R = OH

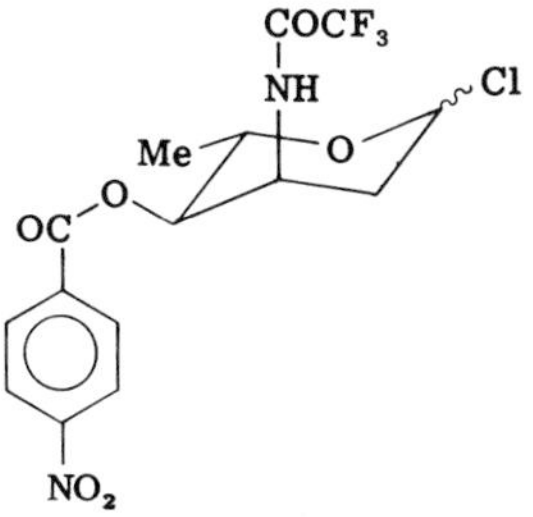

77

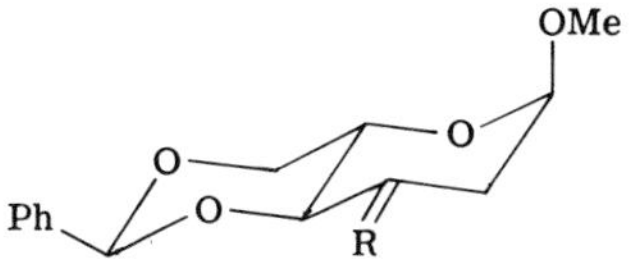

78 : R = O
79 : R = NOH

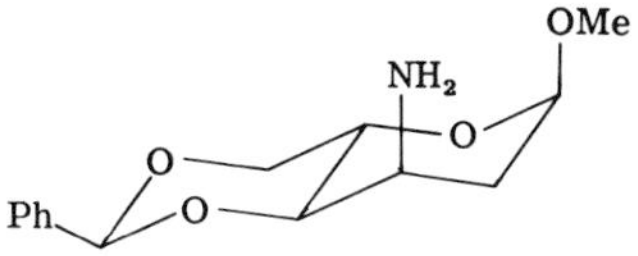

80

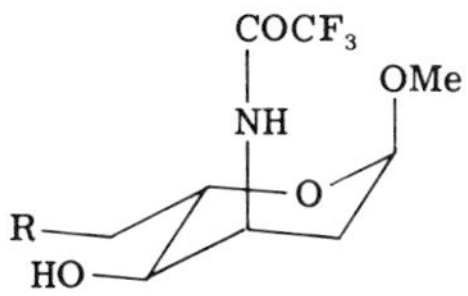

81 : R = OH
82 : R = Br
83 : R = H

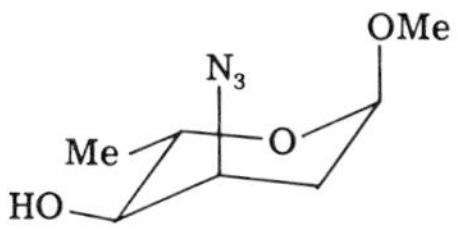

84

tosaminyl chloride (77) with daunomycinone in the presence of silver triflate, followed by removal of the protecting groups with alkali (Arcamone *et al.*, 1976a). Hydroxylation of (75) via the 14-bromoderivative gave 3',4'-diepidoxorubicin (76). The ristosamine moiety was obtained from methyl- 4,6-*O*-benzylidene-2-deoxy-α-L-erythro-hexopyranoside-3-ulose (78), whose oxime (79) was reduced to give the L-ribo amine (80) as the major product. Removal of the benzylidene group and N-trifluoroacetylation afforded (81), which was converted to the 6-bromoderivative (82) and hydrogenated catalytically to methyl-*N*-trifluoroacetylristosaminide (83). Compound (83) was used for the preparation of the glycosylating agent, (77), *via* hydrolysis of the methyl glycoside, *p*-nitrobenzoylation, and treatment with dry hydrogen chloride (Arcamone *et al.*, 1976a).

Other syntheses of ristosamine derivatives have been reported. Intermediate (68), synthesized from L-rhamnal, was treated with azide to give (84), which upon catalytic hydrogenation followed by acid hydrolysis gave (47) (Lee *et al.*, 1975a). A similar synthesis leading to *N*-benzylristosamine has also been reported (Sztaricskai *et al.*, 1975). A different reaction sequence leading to ristosamine as a minor product together with acosamine has already been mentioned (Heyns *et al.*, 1976).

The synthesis of the L-xylo analogue of daunosamine, showing an inversion of the chiral center at C-3, was performed starting from (83), which was converted

85 : R = COPh
86 : R = H

87

to (85) by O-mesylation followed by nucleophilic displacement with sodium benzoate in boiling dimethylformamide. Compound (85) afforded methyl-*N*-trifluoroacetyl-3-epidaunosaminide (86) by sodium methoxide catalyzed methanolysis. Hydrolysis of the methylglycoside followed by O-*p*-nitrobenzoylation and treatment with dry hydrogen chloride afforded the chlorosugar (87), which was condensed with daunomycinone to give (88) and, after removal of the protecting groups, 3'-epidaunomycin (89) (Bargiotti *et al.*, 1978).

It is of interest to note that the preferred conformation of the aminosugar residue in the above-mentioned L-xylo derivatives was found to be the 1C conformation, as indicated on the basis of 90 MHz PMR studies. For instance, compound (89) showed (CDCl$_3$) H-1' at 5.58 δ (W½ ~ 7 Hz), and H-4' at 5.03 δ (W½ ~ 8 Hz), the values of the coupling constant indicating the equatorial

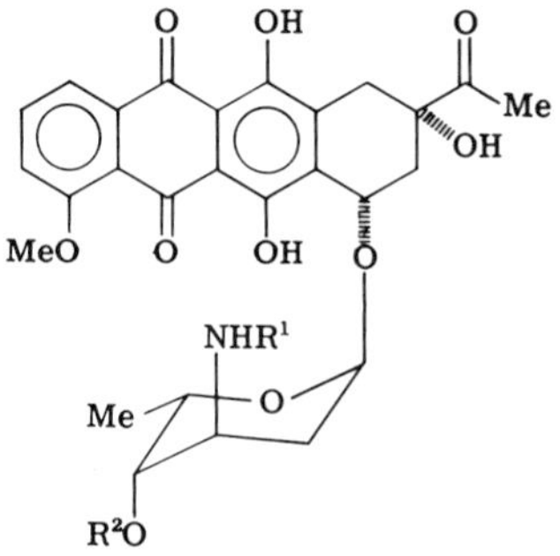

88 : $R^1 = COCF_3$,

$R^2 = CO$—⟨○⟩—NO_2

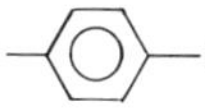

89 : $R^1 = R^2 = H$

orientation of the corresponding protons and consequently the α configuration of the glycoside bond and the 1C conformation of the pyranose ring. In agreement with the findings reported above, H-7 resonated at 5.22 δ (W½ ~ 9 Hz) (unpublished data from author's laboratory).

The C-5 epimer of daunosamine and the enantiomer of ristosamine, namely 3-amino-2,3,6-trideoxy-D-ribo-hexose, has been synthesized in different laboratories. Starting materials were D-mannose (Horton and Weckerle, 1976) and D-glucose (Pelyvás et al., 1977; Baer and Georges, 1977).

The two configurational analogues (75) and (89) show a lower affinity for DNA (2.6 × 10^5 and 1.3 × 10^5, respectively) when compared with (1) (unpublished results from author's laboratory).

This effect of the configurational modification of the sugar moiety could be related to the axial orientation of the amino group, whose importance in the stabilization of the DNA complex has already been stressed, and, for (89), with the trans relationship of the substituents at C-3' and C-4' altering their participation to the hydrogen bridges postulated for the said molecular interactions. Whatever the cause, 3',4'-diepidaunomycin and the corresponding doxorubicin analog are less effective in the inhibition of tumor growth than the parent compounds, 3'-epidaunomycin (89) appearing devoid of activity even at high dosages (Table IV).

B. Structural Modifications

The persistence of an outstanding level of biological activity in the 4'-epi derivatives of the antitumor anthracyclines, coupled with the antitumor efficacy of a known biosynthetic compound, 4'-daunosamininyldaunomycin (Arcamone et al., 1970a; Di Marco et al., 1973), have prompted the synthesis of C-4'-modified analogues of (1) and (2). Other structural modifications of the aminosugar moiety have been reported and include hydroxylated analogues, analogues in which the said moiety is substituted with a different aminosugar or with a nonaminated carbohydrate residue and N-acetylated derivatives. With the exception of the *N*-acyl derivatives, synthesis of the desired glycosylating agent was performed in all other cases, followed by coupling with the aglycone to afford the desired new 7-O-substituted anthracyclinones.

1. The 4′-Deoxy Analogues

Daunosamine was the starting material for the synthesis of the *N*-trifluoroacetyl derivative (**90**) of the new aminosugar, 3-amino-2,3,4,6-tetradeoxy-L-threo-hexopyranose. More specifically, methyl-*N*-trifluoroacetyl-α- daunosaminide (**57**) was converted to the 4- *O*- *p*- bromobenzenesulphonate, which was submitted to nucleophilic displacement with iodide to afford (**90**) after catalytic hydrogenation and hydrolysis of the glycoside. Compound (**90**) was transformed to (**91**) via the 1-*p*-nitrobenzoate, which was treated with dry hydrogen chloride to give (**91**). Condensation of (**91**) with daunomycinone followed by de-N-trifluoroacetylation gave 4′-deoxydaunomycin (**92**), whose sidechain was also functionalized to (**93**) (Arcamone *et al.*, 1976c). Compounds (**92**) and (**93**)

90 : R = OH
91 : R = Cl

92 : R = H
93 : R = OH

exhibited association constants with DNA of 3.1×10^5 and 4.4×10^5, respectively, indicating, especially for the latter, an affinity for the biopolymer of the same order of magnitude as that of the parent antibiotics. This trend was also confirmed by viscosity measurements (Arlandini *et al.*, 1977).

The substitution of the C-4′ hydroxyl with an hydrogen atom is not unfavorable for the exhibition of biological activity. The 4′-deoxy analogues (**92**) and (**93**) were found to retain the antitumor efficacy of the parent drugs on both the L-1210 leukemia and solid sarcoma 180 systems (Arcamone *et al.*, 1976c). In another report, the 4′-deoxy compounds were found to be as active as the parents in inhibiting the DNA template function in the enzymic reactions catalyzed *in vitro* by bacterial-DNA-dependent RNA and DNA polymerases; more effective, on a weight basis, on L-1210 leukemia in mice; and 16–18 times more active in the inhibition of colony-forming capacity of cultured HeLa cells. In the same study it was also proved that (**92**) and (**93**) were taken up to a greater extent than (**1**) and (**2**) in cultured L-1210 cells and that this property could be related with the higher lipophilicity of the new derivatives (Di Marco *et al.*, 1977b). Similarly to 4'-epidoxorubicin, 4'-deoxydoxorubicin is less cardiotoxic than doxorubicin in the rat (MCCD = 16 mg/kg) in the Zbinden test (G. Zbinden, personal communication).

2. The 4'-O-Methyl Derivatives

Owing to the favorable results obtained upon modification of the sugar moiety at C-4', as exemplified in the 4'-epi and 4'-deoxy analogues, 4'-O-methyl derivatives of the antitumor anthracyclines have been prepared (Cassinelli *et al.*, 1977). To this end, compound (**57**) was treated with diazomethane and boron trifluoride etherate to afford (**94**), which was converted to the sugar halide (**97**) via (**95**) and (**96**). Glycosylation of daunomycinone with (**97**) and N-deacylation of the condensation product gave 4'-O-methyl daunomycin (**98**), from which 4'-O-methyldoxorubicin (**99**) was also obtained. Following a similar synthetic route, but starting from the corresponding derivative in the acosamine series (**59**), 4'- epi- 4'- O- methyl analogues (**90**) and (**91**) were prepared (Cassinelli *et al.*, 1979).

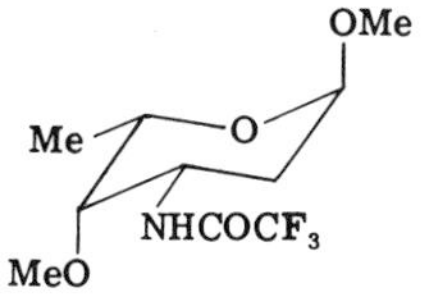

94

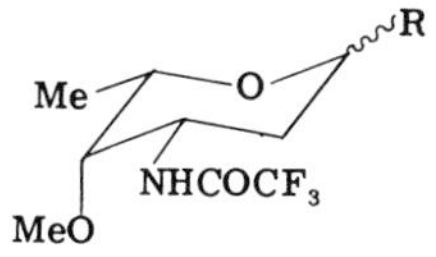

95 : R = OH

96 : R = OCO—⟨◯⟩—NO$_2$

97 : R = Cl

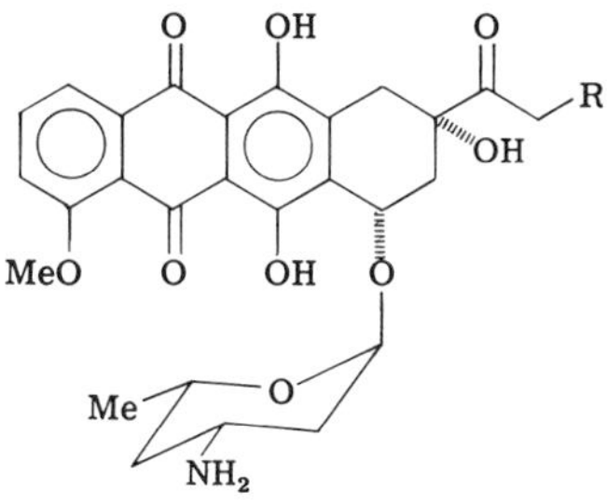

98 : R = H
99 : R = OH

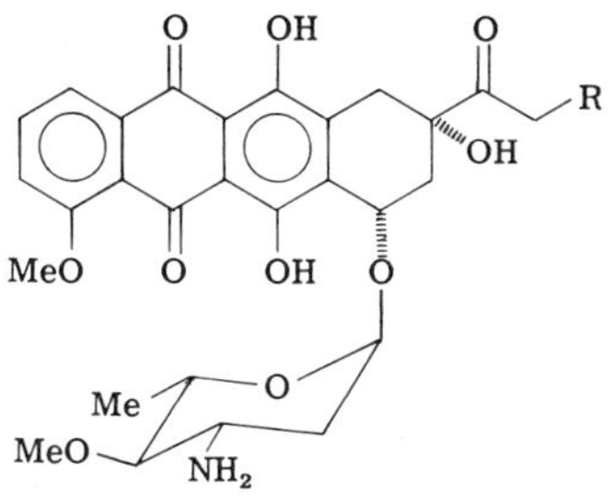

100 : R = H
101 : R = OH

The 4'-methyl compounds (98)–(101) showed remarkable antitumor activity in animal models. The activity of (99) on L-1210 leukemia is of particular interest Table V), since it is much higher than that shown by doxorubicin. Comparison of 4' modified doxorubicin in the P-388 system is shown in Table VI.

3. Hydroxylated Derivatives

Daunosamine, the aminosugar constituent of the antitumor anthracyclines, contains two deoxy groups at C-2 and C-6. In order to determine the relevance of this structural feature for the biological activity of the abovementioned antibiotics, the corresponding 6'-hydroxylated glycosides with L-arabino, L-lyxo, L-ribo, and L-xylo configuration, as well as 2 (R)-2-hydroxydaunomycin (L-talo configuration), were synthesized and tested.

For the synthesis of the sugar moieties possessing the L-arabino and L-lyxo configuration compound (102), obtained from L-glucose in 30% overall yield following procedures described in the literature for the D series, was converted to (108) via the 3-O-mesyl derivative (103) and the azide (104). Removal of the protecting groups afforded 6-hydroxyacosamine (107), which was converted to (108) and finally to (109) by mild alkaline treatment of an intermediate tri-O-p-nitrobenzoate. Reaction of (109) with daunomycinone in the presence of p-toluensulphonic acid as catalyst followed by deacylation of the product gave 4'-epi-6'-hydroxydaunomycin (110), which was also used for the preparation of the corresponding doxorubicin analogue (111) (Arcamone et al., 1976c). The amine (105) was also used for the synthesis of 6'-hydroxydaunomycin (112). Conversion of (105) to (106), followed by hydrolysis of the benzylidene group, afforded (113), which was tritylated to (114) in order to allow ruthenium tetroxide oxidation to (115) and subsequent stereoselective reduction to the desired L-lyxo derivative (116), which was easily transformed to 6-hydroxydaunosamine (117) on removal of the protecting groups. Trifluoroacetylation of (117) followed by treatment with isopropanol afforded (118), which was di-p-nitrobenzoylated

TABLE V

Activity of 4'-O-Methyl Analogues of Doxorubicin on Experimental L-1210 Leukemia in Mice[a]

Compound	Structure	Dose (mg/kg)	T/C (%)
Doxorubicin	**2**	6.6	175
4'-O-Methyldoxorubicin	**99**	4.4	300
4'-Epi-4'-O-methyldoxorubicin	**101**	10.0	187

[a] Single i.p. treatment on day 1. Optimal nontoxic doses are given. Data from Cassinelli et al. (1977).

TABLE VI

Activity of Antitumor Anthracyclines on P-388 Lymphocytic Leukemia in Mice[a]

Compound	Structure	Day 1 only		QD 1-9		QD 5, 9, 13	
		Dose[b]	T/C (%)[c]	Dose[b]	T/C (%)[c]	Dose[b]	T/C (%)[c]
Doxorubicin	**2**	9.00[d]	221[d]	1.08[e]	214[e]	9.33[d]	166[d]
Doxorubicin, 4'-epi-	**56**	7.80[f]	229[f]	0.99[f]	205[f]	12.50[f]	165[f]
Doxorubicin, 4'-deoxy-	**93**			1.17	253	3.14[f]	166[f]
Doxorubicin, 4'-O-methyl	**99**					3.13	166

[a] NCI data.
[b] Mg/kg body weight.
[c] Average survival time expressed as percentage of survival time of controls.
[d] Average of four to seven experiments.
[e] Average of 60 experiments.
[f] Mean value of two experiments.

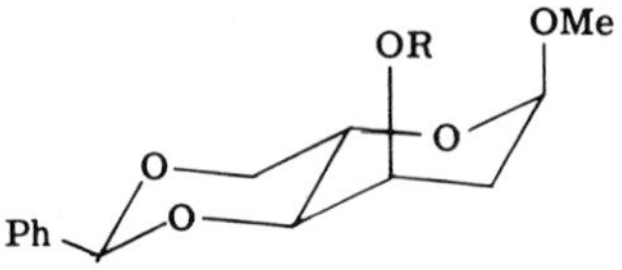

OR   OMe

Ph

102 : R = H
103 : R = Ms

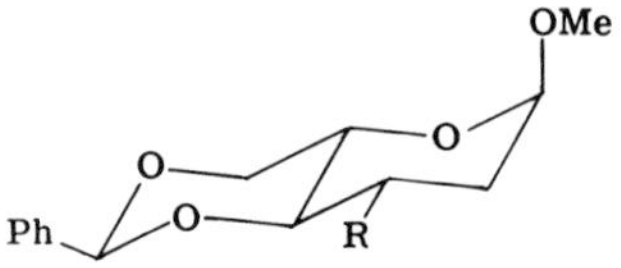

OMe

Ph   R

104 : R = N$_3$
105 : R = NH$_2$
106 : R = NHCOCF$_3$

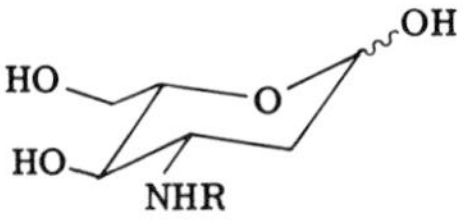

OH

HO

HO
NHR

107 : R = H
108 : R = COCF$_3$

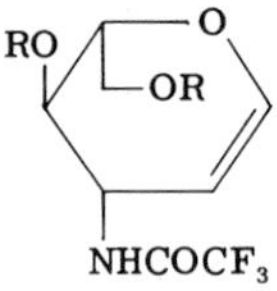

RO   O
OR

NHCOCF$_3$

109 : R = CO—⟨    ⟩—NO$_2$

O   OH   O
R
'''OH

MeO   O   OH   O

HO

HO
NH$_2$

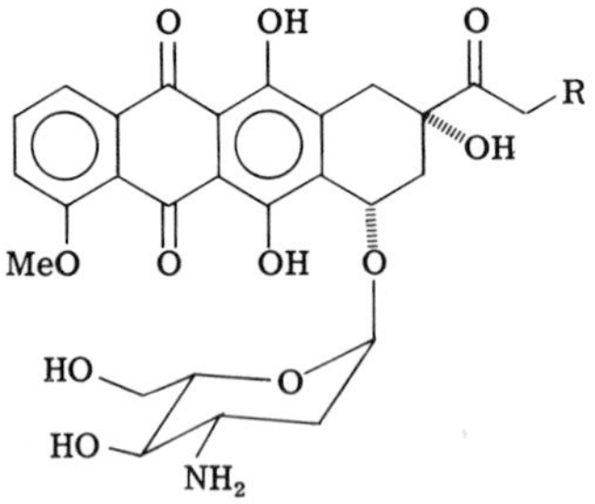

110 : R = H
111 : R = OH

O   OH   O
Me
'''OH

MeO   O   OH   O

HO

HO
NH$_2$

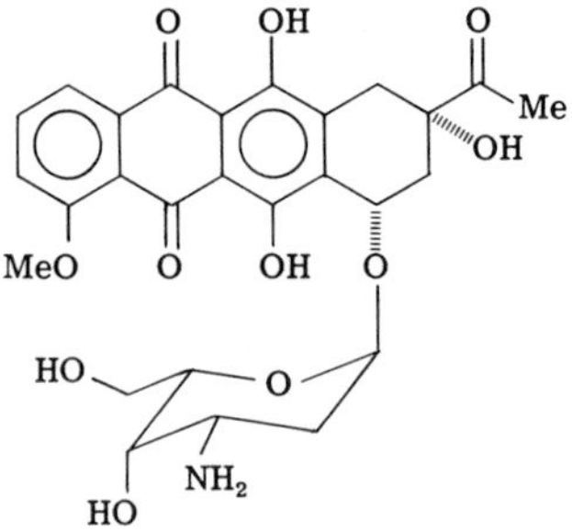

112

and finally converted to the halosugar (**119**). Condensation of (**119**) with daunomycinone by the silver triflate method and application of standard procedures gave the desired (**112**) (Bargiotti *et al.*, 1977; Arcamone *et al.*, 1978a).

Access to the L-ribo and L-xylo analogues was provided by the availability of (**81**), which was converted to the halosugar (**120**) in four steps, the latter compound being condensed with daunomycinone to give, after deacylation, the L-riboanalogue (**121**) (Arcamone *et al.*, 1976b). This compound was also converted to the corresponding doxorubicin analogue (**122**). On the other hand, conversion of (**81**) to the corresponding 4,6-di-O-mesyl derivative and nucleophilic displacement of the O-mesyl groups with sodium benzoate afforded (**123**), which was methanolyzed to (**124**). Treatment of (**124**), with p-nitrobenzoyl chloride in pyridine gave (**125**), and subsequent acid hydrolysis

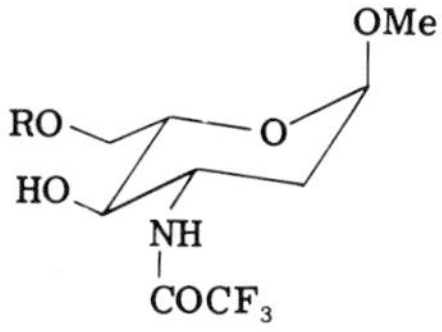

113 : R = H
114 : R = Tr

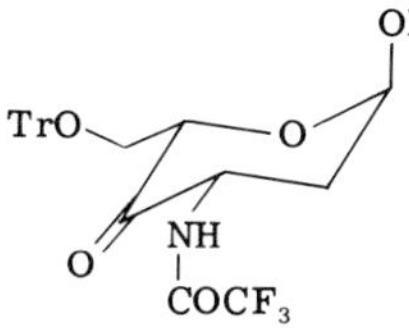

115

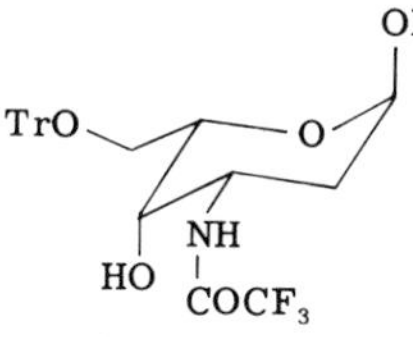

116

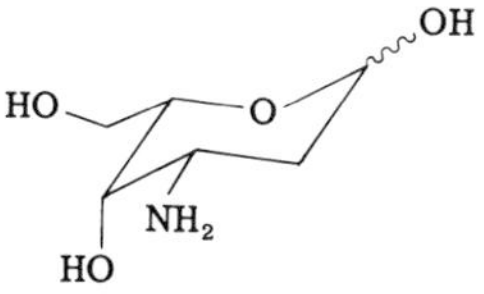

117

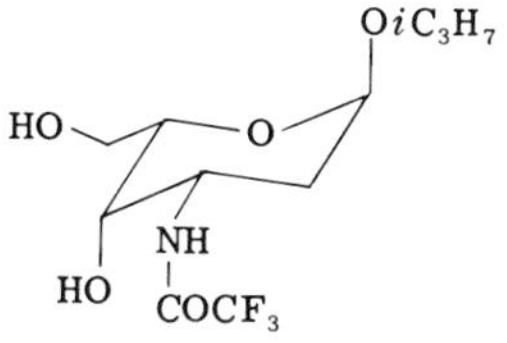

118

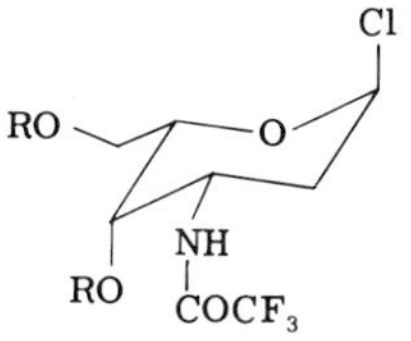

119 : R = CO—⟨◯⟩—NO₂

followed by *p*-nitrobenzoylation gave (**126**), which was converted with dry hydrogen chloride in chloroform to the corresponding protected sugar halide (**127**). Condensation of the latter with daunomycinone followed by hydrolytic removal of the protecting groups afforded (**128**), further transformed to (**129**), via the 14-bromo derivative (Bargiotti *et al.*, 1978).

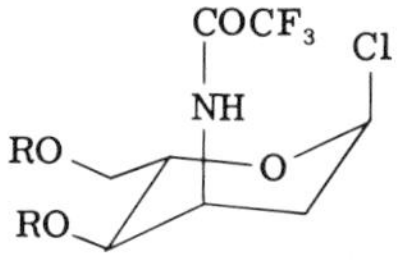

120 : R = CO—⟨◯⟩—NO₂

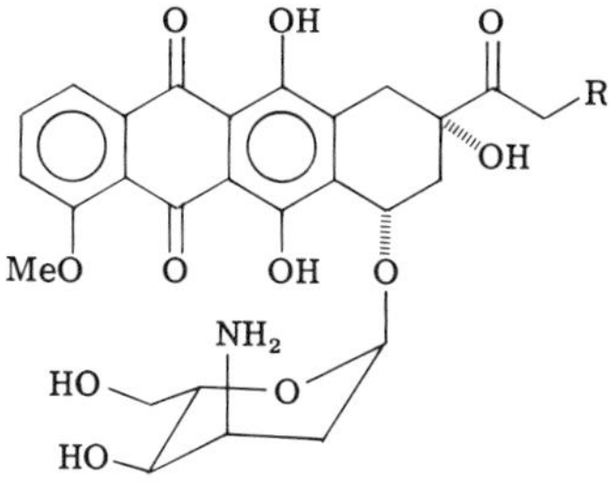

121 : R = H
122 : R = OH

The daunomycin analogue containing an axial hydroxyl group at C-2′ has been prepared. The corresponding known aminosugar (**130**) was obtained by a literature method starting from methyl α-L-rhamnoside (Richardson and McLauchlan, 1962), and was converted to bromosugar (**133**) via (**131**) and (**132**). Condensation of (**133**) with daunomycinone in methylene chloride in the presence of mercuric oxide and mercuric bromide, followed by alkaline treatment of the

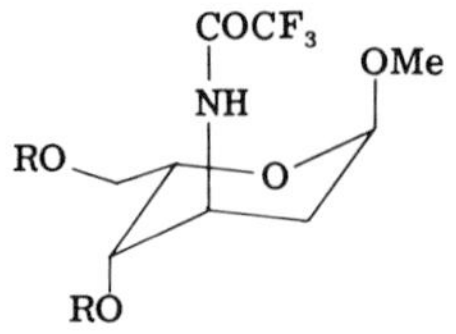

123 : R = COPh
124 : R = H

125 : R = CO—⟨○⟩—NO₂ → $125 : R = CO$—⟨○⟩—NO_2

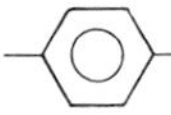

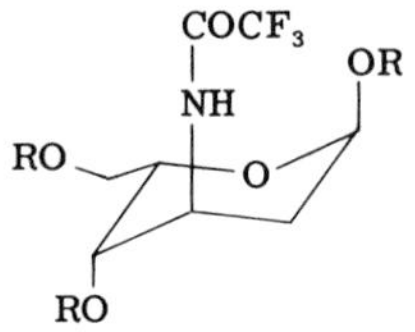

126 : R = CO—⟨○⟩—NO_2

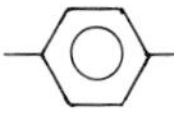

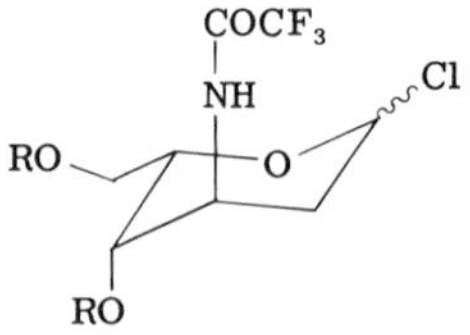

127 : R = CO—⟨○⟩—NO_2

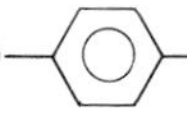

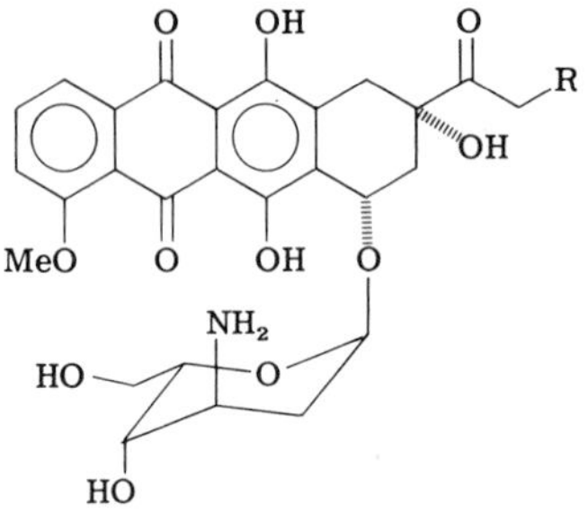

128 : R = H
129 : R = OH

reaction product, afforded 2'(R)-2'-hydroxydaunomycin (**134**) (Penco *et al.*, 1978).

The presence of the 6'-OH, as in the L-arabino and L-lyxo analogues (**110**), (**111**), and (**112**), did not abolish the DNA-complexing ability typical of the parent compounds, as the values of the stability constants of the corresponding complexes with native DNA were in the range of 45% to 90% of those of the nonhydroxylated compounds (Arlandini *et al.*, 1977; Di Marco *et al.*, 1977b). In the latter study, it was also shown that the same glycosides behaved similarly to the parent drugs as inhibitors of *in vitro* reactions catalyzed by bacterial DNA-

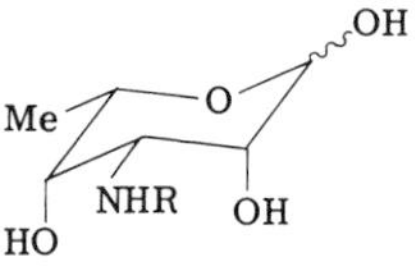

130 : R = H
131 : R = COCF₃

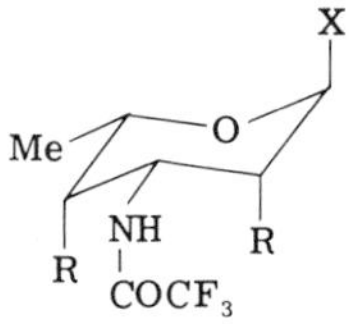

132 : R = X = OCO—⟨○⟩—NO_2

133 : R = OCO—⟨○⟩—NO_2, X = Br

134

dependent polymerases. On the other hand, compounds (**110**), (**111**), and (**112**) displayed a distinctly lower antitumor efficacy *in vivo* when compared with the corresponding 6′-deoxy glycosides (Arcamone *et al.*, 1976d; Arcamone, 1977; Di Marco *et al.*, 1977b). The difference in the biological activity was even more remarkable in cultured cells systems (Arcamone *et al.*, 1976d), indicating that the transport mechanism, adversely influenced by the hydroxylation at C-6′, is possibly related to the difference observed in comparison to the parents.

Less data are available on the biological activity of the L-ribo glycosides (**121**) and (**122**) or of the L-xylo analogues (**128**) and (**129**). These compounds appear to be endowed with a low level of bioactivity in the animal tests (Arcamone, 1977; Arcamone *et al.*, 1977a).

Compound (**134**) retained the ability to complex double-stranded DNA, albeit showing a lower stability constant than (**1**), and also inhibited DNA-dependent bacterial polymerases *in vitro* at 50% higher concentration when compared with (**1**). On the other hand, the activity on L-1210 leukemia was distinctly lower, if even present (Di Marco *et al.*, 1977b).

4. Analogues with Other Sugar Moieties

Daunomycinone glycosides with D-glucose and with D-glucosamine were prepared by condensing 2,3,4,6-tetracetyl-α-D-glucosyl bromide or 3,4,6-triacetyl-*N*-trifluoroacetyl-D-glucosaminyl bromide with daunomycinone in 1,2-dichloroethane and in the presence of silver carbonate, and subsequent alkaline removal of the acyl groups. Only one product was formed in the glycosidation reaction, and the β configuration was assigned to the glycosidic linkage of (**135**) and (**136**) on the basis of the known effect of participating groups at C-2 on the steric course of this type of reaction (Penco, 1968). Compound (**136**) displayed, however, a low tendency to form complexes with DNA when compared to (**1**) (Zunino *et al.*, 1972), and no activity in cultured cells or in tumor-bearing animals was detected (Di Marco *et al.*, 1973).

Two other new glycosidic derivatives of daunomycinone should be mentioned. The synthesis of 7-*O*-(2,6-dideoxy-α-L-lyxo-hexopyranosyl)daunomycinone, a

135 : R = OH
136 : R = NH₂

compound differing from (**1**) by the replacement of the amino function in the sugar moiety by a hydroxyl group, has been described (Fuchs *et al.*, 1977), but no biological data concerning this compound have been published. *N,N*-Dimethyldoxorubicin is an interesting derivative in which daunosamine is replaced by the strictly related rhodosamine. This compound was found to exhibit an enhanced cardiotoxicity in rats when compared with doxorubicin, apparently because of a higher accumulation in heart muscle tissue (Zbinden *et al.*, 1978a).

C. N-Acyl Derivatives

The aminoglycoside function present in the antitumor anthracyclines, as well as in most other anthracyclines and in other antibiotics belonging to different structural groups, is such an important structural feature of these compounds that it appears very unlikely that the blocking of the same, as in an acyl derivative, might result in compounds with remarkable biological activity. However, *N*-acyl derivatives, in which the acyl group is represented by the residue of a simple organic acid or of a derivative thereof, have been prepared in different laboratories. With the exception of those derivatives which were obtained as protected intermediates, as for instance the *N*-trifluoroacetates in the first reaction sequence leading from daunomycin to doxorubicin (Arcamone *et al.*, 1969b) or in glycosidations (see previous section), and those prepared for characterization purposes, such as *N*-acetyldaunomycin used for PMR studies (Arcamone *et al.*, 1970b), others have been synthesized probably with the hope of a subsequent *in vivo* N-deacetylation or of discovering agents with a different mechanism of action. In fact (Di Marco and Arcamone, 1975) the basic amino group is a necessary molecular requirement for a stable DNA-intercalation complex and probably for other molecular interactions as well.

In addition to the already mentioned *N*-acetyl and *N*-trifluoroacetyl derivatives of the antitumor anthracyclines, a number of simple *N*-acyl derivatives have been reported. Compounds (**137**)–(**145**) and (**147**) have been prepared and tested for antitumor activity (Yamamoto *et al.*, 1972). *N*-Guanidino-acetyldaunomycin

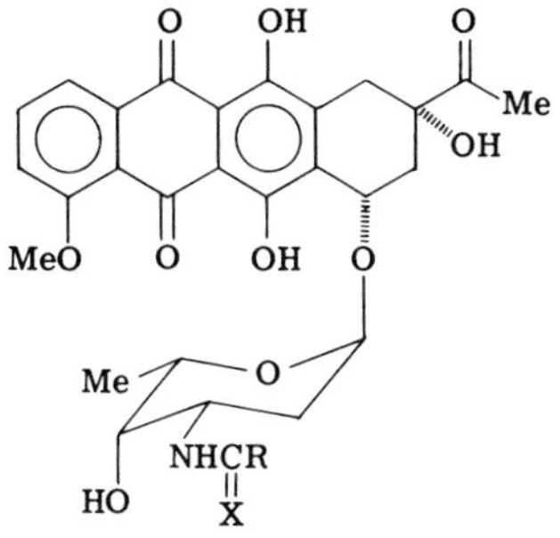

137 : R = Me, X = O
138 : R = Et, X = O
139 : R = Pr, X = O
140 : R = NHMe, X = O
141 : R = NH-nBu, X = O
142 : R = NHMe, X = S
143 : R = NH-nBu, X = S
144 : R = NHPh, X = S
145 : R = OMe, X = O
146 : R = CH$_2$NHC(NH)NH$_2$, X = O

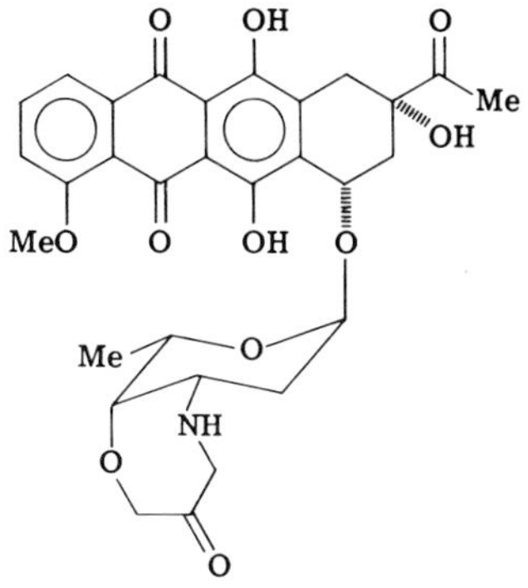

147

(146) was also available for biological studies (Di Marco *et al.*, 1973). Peptide derivatives (148), (149), and (150) were prepared by the mixed anhydride method (Wilson *et al.*, 1976), and their DNA-complexing ability was studied (Gabbay *et al.*, 1976). *N*-Aminoacyl derivatives (151)–(153) have also been synthetized (Bouchandon and Jolles, 1969).

The *N*-trifluoroacetyl derivatives of a number of doxorubicin 14-esters have been prepared by nucleophilic displacement of the halogen atom in the 14-iodo derivative of *N*-trifluoroacetyldaunomycin. Among these, *N*-trifluoroacetyldoxorubicin-14-valerate has received considerable attention (Israel *et al.*, 1975).

The *N*-acyl derivatives of the antitumor anthracyclines do show biological

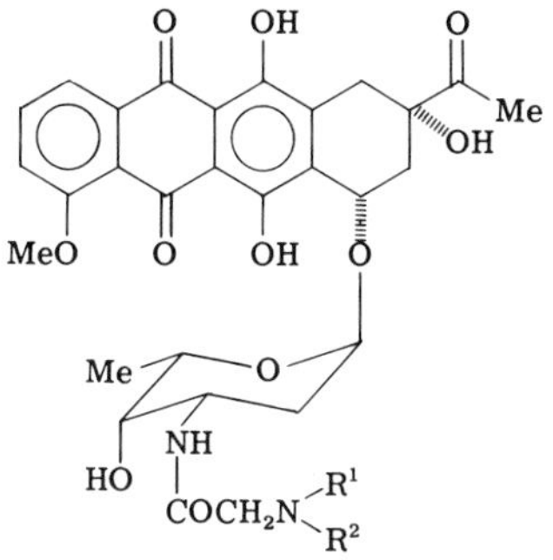

148 : R^1 = R^2 = Me
149 : R^1 = R^2 = Pr
150 : R^1 = H, R^2 = COCH$_2$NMe

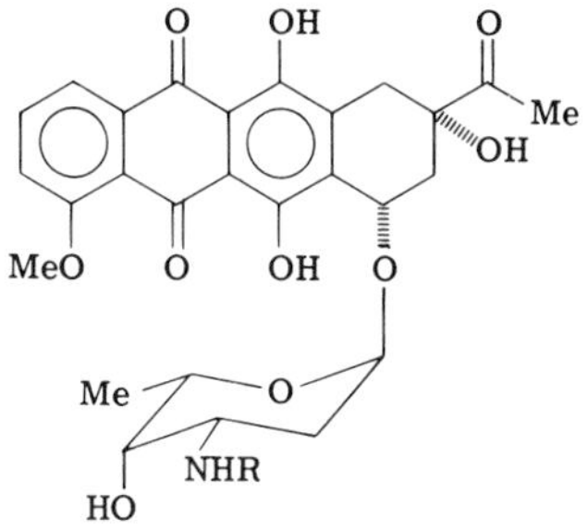

151 : R = L-Leu
152 : R = D-Leu
153 : R = L-Phe

activity in different systems. Optimal antitumor activity of (**137**) in mice bearing L-1210 leukemia under the 1,5,9-days schedule resulted in a T/C percentage of 126 at 40 mg/kg daily dose, whereas in the same conditions the parent compound (**1**) afforded a T/C percentage of 139–144 at 1 mg/kg dosage. Other regimens afforded similar indications of low, albeit noticeable, activity. Other derivatives at the amino group, such as (**147**) showed no activity (Yamamoto *et al.*, 1972). *N*-Acetyldaunomycin was found to bind to DNA, but the stability constant of the complex was two orders of magnitude smaller than that of (**1**) (Zunino *et al.*, 1972), and it displayed no effect on HeLa cell proliferation at concentrations 10 times greater than daunomycin ED_{50} (Di Marco *et al.*, 1971). *N*-Acetyl-daunomycin was not toxic when administered daily for 20 days at a total comulative dose equal to six times a toxic dose of daunomycin (Zbinden and Brändle, 1975).

N-Acetyldoxorubicin displayed considerable activity on P-388 leukemia in mice, but the dose giving the highest survival time of the animals was 50 times higher than that of the parent drug in the Q4D 5,9,13 schedule (Table VII). The *N*-trifluoroacetate also appeared to be endowed with an antitumor effect similar to that of the parent compound, but the dose required was lower than that of the *N*-acetyl analog. *N*-Trifluoroacetyldoxorubicin-14-valerate also shows outstanding activity in the P-388 test, the dose giving the highest T/C value being similar to that of *N*-trifluoroacetyldoxorubicin, in agreement with the results obtained with adriamycin esters (see above). The compound has been reported to show remarkable activity on experimental leukemias in mice (Israel *et al.*, 1975). Evaluation of antitumor activity *in vivo* of this as well as of other derivatives, requiring the use of emulsifiers or dispersing agents because of poor water solubility, appears particularly difficult, as the presence of such agents may substantially alter the pharmacological response (Di Marco *et al.*, 1976a; Casazza *et al.*, 1978; Aoshima *et al.*, 1976, 1977). This and other methodological reasons may explain different results obtained in different laboratories (Pratesi *et al.*, 1978).

Peptide derivatives of daunomycin exhibited a reduced activity on P-388 leukemia in mice when compared with the parent compound, in agreement with the lower inhibition of RNA polymerase and with the higher rate of dissociation of the DNA complex in respect to (**1**) (Wilson *et al.*, 1976; Gabbay *et al.*, 1976).

IV. ANALOGUES MODIFIED IN THE ANTHRAQUINONE CHROMOPHORE

Structural variations of the chromophoric moiety in the antitumor anthracyclines may modify the stability of the DNA complex, the selectivity towards different DNAs or other receptor macromolecules, and the redox potential

TABLE VII

Activity of N-Acyl Derivatives of Doxorubicin on P-388 Leukemia in Mice Treated i.p. on Days 5, 9, and 13 after Tumor Inoculation[a]

Derivative	Experiment no.	Dose (mg/kg)[b]	T/C (%)	Doxorubicin	
				Dose (mg/kg)	T/C (%)
N-Acetyl	5597	400	159	8	148
N-Trifluoroacetyl	5765	100	189	8	187
N-Trifluoroacetyl-14-valerate	6242	90	221	15	195

[a] Screener A.D. Little. Results obtained by courtesy of NCI.
[b] Dose giving the highest T/C (%) value.

of the drugs. Therefore, a total synthetic approach to daunomycinone derivatives, suitable precursors of new glycosides modified in the substitution on ring
D, has been developed. The glycosidated products have shown interesting
biological properties, as is reported below.

The availability of synthetic methods for the preparation of daunomycinone
analogues should not rule out the possibility that potential antitumor agents be
obtained starting from biosynthetically originated aglycones. ε-Rhodomycinone,
an inactive by-product of daunomycin fermentations (Di Marco and Arcamone,
1975; Kern *et al.*, 1977), has been used as starting material for the synthesis of
new daunomycin related glycosides, but other analogues will be obtained as a
consequence of the likely availability of new biosynthetic anthracyclinones.

A. The 4-Demethoxy Analogues

The synthetic scheme for the access to daunomycinone-related compounds originally developed by Wong *et al.* (1971, 1973) has been used for the
preparation of daunomycin and doxorubicin analogues showing different substitutions on ring D in the aromatic chromophore (Arcamone *et al.*, 1976e, 1978b;
Di Marco *et al.*, 1978). In its recently improved version (Arcamone *et al.*,
1978b), (−)-6-acetyl-1,4-dimethoxy-6-hydroxytetraline (**154**), obtained by resolution of the racemic compound into the two enantiomorphic forms upon fractional
crystallization of the corresponding Schiff bases with 1-(−)-phenylethylamine,
was heated with phthalic anhydride to give 7-deoxy-4-demethoxydaunomy-
cinone (**155**) in reasonable yields. Subsequent conversion of (**155**) to the
13-ethylenedioxyketal, followed by radical bromination and hydrolysis of the
7-bromo-13-ketal with water, afforded mainly 4-demethoxy-7-epidaunomycinone

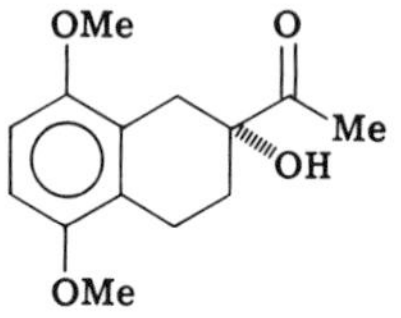

154

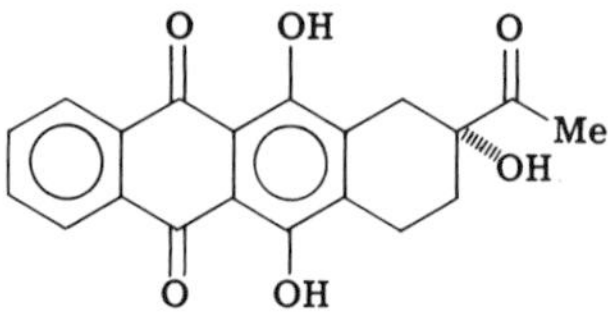

155

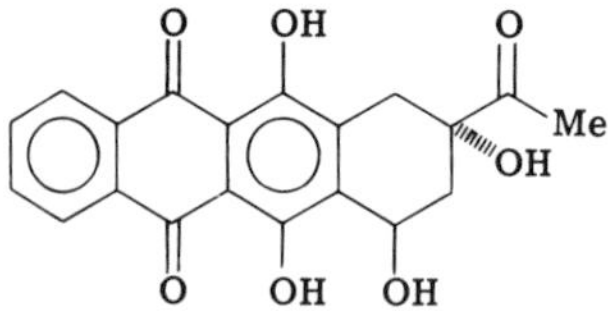

156

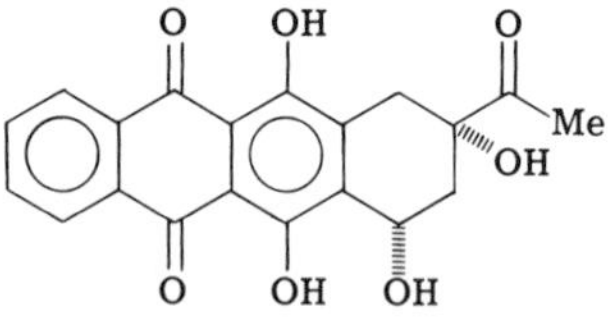

157

(156). Treatment of (156) with trifluoroacetic acid and then with aqueous ammonium hydroxide gave 4-demethoxydaunomycinone (157) in optically pure form. Condensation of (157) with *N,O*-ditrifluoroacetyl-α-daunosaminyl chloride followed by hydrolysis of the protecting groups afforded 4-demethoxydaunomycin (158). Using a similar procedure, but starting from appropriately substituted phthalic anhydride derivatives, analogues (159)–(163) have been prepared. Functionalization of (158) and of (159) afforded the corresponding doxorubicin analogues (164) and (165), while glycosidation of (157) with *N,O*-ditrifluoro-acetyl-α-acosaminyl chloride followed by detrifluoroacetylation gave 4-dem-

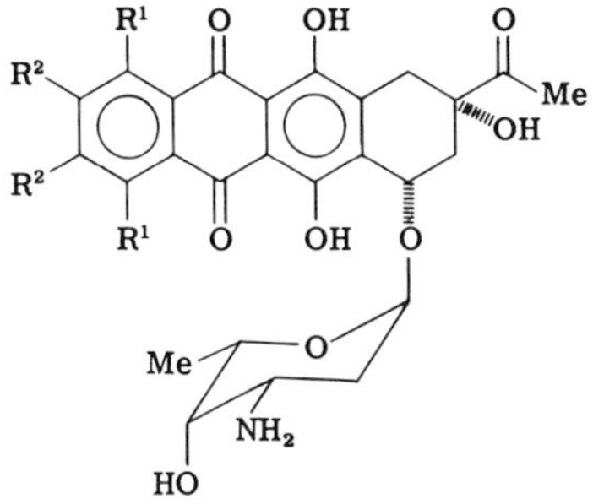

158 : $R^1 = R^2 = H$
159 : $R^1 = H$, $R^2 = Me$
160 : $R^1 = Me$, $R^2 = H$
161 : $R^1 = H$, $R^2 = Cl$
162 : $R^1 = Cl$, $R^2 = H$

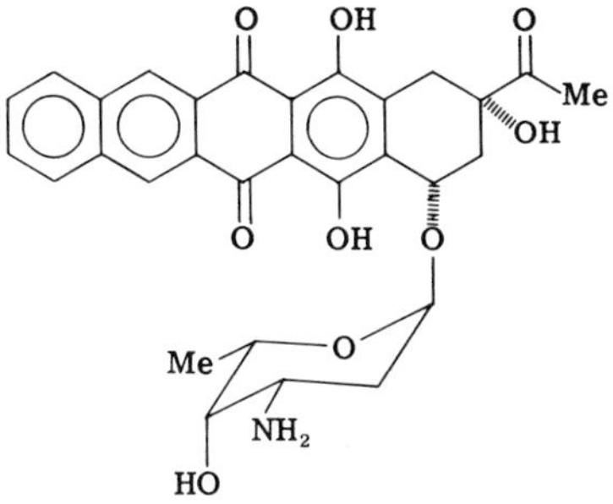

163

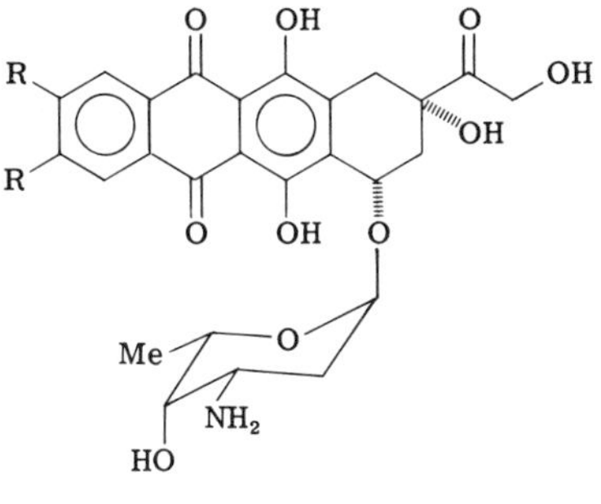

164 : R = H
165 : R = Me

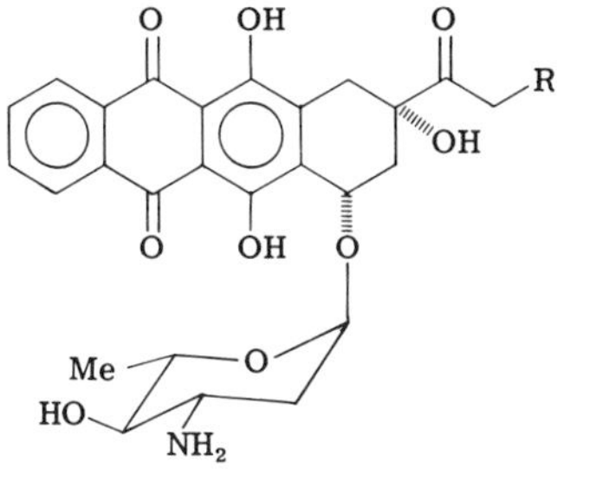

166 : R = H
167 : R = OH

ethoxy-4'-epidaunomycin (166), from which the 4'-epidoxorubicin analogue (167) was obtained (Arcamone *et al.*, 1976a).

The above-mentioned daunomycin analogues were obtained using silver triflate as the chloride acceptor in the condensation reactions of (158) with the sugar halide, a method giving rise essentially to the α anomers. When instead the

168   169

reaction had been performed using HgBr$_2$ and HgO as condensing agents, both
α- and β-anomeric glycosides were formed. A similar result was obtained starting from the 7(R),9(R) enantiomeric form of (**152**). However, preferential formation of one anomer was recorded when the coupling was performed in the
presence of silver triflate (Arcamone *et al.*, 1976f). Structures of the β
anomer of 4-demethoxydaunorubicin and of the α and β anomers of the 7(R),9(R)
diastereoisomeric form of the same daunomycin analogues are, respectively,
represented by (**168**), (**169**), and (**170**).

170

Biological activity of daunomycin analogues modified on ring D is summarized in Table VIII, showing results obtained in a cultured cell system and in
an animal test. 4-Demethoxydaunomycin exhibits a powerful cytotoxic effect
clearly related to the much greater extent of the uptake of this compound in cells
with respect to (**1**) (Supino *et al.*, 1977). On the other hand, affinity of (**158**) for
DNA was substantially of the same order of magnitude as that of (**1**), although
the former displayed a greater enhancement of the melting temperature of
double-stranded DNA than did daunomycin (Zunino *et al.*, 1976). No substantial
difference was found between the two compounds with regard to the inhibition of
nucleic-acid-polymerizing enzymes such as *E. coli* DNA-dependent DNA and
RNA polymerases *in vitro*, but in cultured cells (**158**) appeared more effective on
tritiated thymidine incorporation into DNA when compared with (**1**). In agreement with these findings, demethoxydaunomycin was as active as the parent

TABLE VIII

Daunomycin Analogues: Inhibition of Colony-Forming Ability of Cultured HeLa Cells after 24 hours Exposure to the Drug, and Anticumor Activity in L-1210 Experimental Leukemia in Mice[a]

Compound	Structure	Cytotoxicity (EC$_{50}$, ng/ml)	Antitumor activity Dose[b] (mg/kg)[b]	T/C (%)[c]
Daunomycin	1	10.00, 21.2[d]	2.9, 4[e]	144, 150[e]
4-Demethyxydaunomycin	158	0.15, 0.3[d]	1.0, 0.6[e]	150, 150[e]
4-Demethoxy-2,3-dimethyl-	159	5.80	1.25	131
4-Demethoxy-1,4-dimethyl-	160	10.05	6.6	147
4-Demethoxy-2,3-dichloro-	161	25.00	33.7	111
4-Demethoxy-1,4-dichloro-	162	7.15	20.0	116
4-Demethoxy-(2,3-a)benzo-	163	27.00	10.0	135
4-Demethoxy-β-anomer-	168	13.2[d]	8.0[e]	150[e]
4-Demethoxy-7,9-diepi-	169	>200[d]	8.0[e,f]	100[e]
4-Demethoxy-7,9-diepi-β-anomer	170	>200[d]	10.0[e,f]	100[e]

[a] Arcamone et al. (1978b).

[b] Optimal nontoxic dose. Treatment i.p. on day 1.

[c] Average survival time as percentage of survival time of controls.

[d] From Supino et al. (1977).

[e] From Arcamone et al. (1976g).

[f] Highest dose tested.

compound on L-1210 leukemia in mice (Table VIII), as well as on other experimental tumors, but the dose required was up to eight times less (Arcamone *et al.*, 1976d). Interestingly, (**158**) appeared active also when administered by the oral route (Di Marco *et al.*, 1977a).

All substituted 4-demethoxydaunomycins (**159**)–(**163**) exhibited a high level of cytotoxic activity, indicating that different substitutions on ring D of the chromophoric system are compatible with biological activity. The 2,3- and 1,4-dimethyl derivatives (**159**) and (**160**) were also effective *in vivo*, although less potent than (**158**) on a weight basis. The lack of efficacy of chloro derivatives (**161**) and (**162**) in the experimental tumor system could be ascribed to different metabolism *in vivo*, although no information at this regard be presently available (Arcamone *et al.*, 1978b; Di Marco *et al.*, 1978b).

The effects of stereochemical variations in this series are of interest. The β anomer (**168**) was clearly less active on a weight basis that the corresponding α anomer (**158**), but the antitumor efficacy of the former was comparable to that of the latter and of daunomycin itself. This result can be explained on the basis of the observation that the compound was as effective as (**1**) in enhancing the temperature of the helix-coil transition of calf-thymus DNA (Zunino *et al.*, 1976), and in the inhibition of tritiated thymidine incorporation in cultured cells (Supino *et al.*, 1977). Although β anomers display a lower affinity for DNA than the corresponding α anomers, molecular models show that, owing to the free rotation around the (C)-7-0 and the glycosidic bonds, β anomers can intercalate into the DNA double helix, the distance of the amino group and the second phosphate residue away from the intercalation site being, however, larger than with the α anomers (Arlandini *et al.*, 1977). On the other hand, no bioactivity is shown by the $7(R),9(R)$ analogues, in agreement with their low affinity for DNA (Zunino *et al.*, 1976; Arlandini *et al.*, 1977) and with their low activity on DNA synthesis in cultured cells, notwithstanding the high level of uptake of the same compounds by the cells (Supino *et al.*, 1977).

4-Demethoxydoxorubicin (**164**) and the corresponding 4'-epi derivative (**167**) were found to be more potent than the parent antibiotic in both cultured cell and *in vitro* tests (Table IX) (Di Marco *et al.*, 1978). A higher potency was also shown by the two analogues on tritiated thymidine incorporation in cultured cells, 4-demethoxydoxorubicin being taken up to a greater extent than doxorubicin in the same cells (Supino *et al.*, 1977). Compound (**167**) appeared more effective than doxorubicin in ascitic L-1210 leukemia and in systemic Gross leukemia (Di Marco *et al.*, 1978), confirming the outstanding properties of this class of new anthracycline analogues.

B. Semisynthetic ε-Rhodomycinone Glycosides

The aglycone, ε-rhodomycinone, contains the structural features of many typical anthracyclinones, which are the ethyl sidechain at C-9 and the

TABLE IX

Doxorubicin Analogues: Inhibition of Colony-Forming Ability of Cultured HeLa Cells after 24 hours Exposure to the Drug, and Antitumor Activity in L-1210 Experimental Leukemia in Mice[a]

Compound	Structure	Cytotoxicity (EC_{50}, ng/ml)	Antitumor activity	
			Dose (mg/kg)[b]	T/C (%)[c]
Doxorubicin	**2**	15.00, 43.0[d]	2.9	141
4-Demethoxydoxorubicin	**164**	0.10, 0.2[d]	0.5	166
4-Demethoxy-2,3-dimethyl-doxorubicin	**165**	7.00	4.4	173
4-Demethoxy-4′-epidoxorubicin	**167**	0.1[d]	0.75	189

[a] Di Marco et al. (1978a); Arcamone et al. (1978b).
[b] Optimal nontoxic dose. Treatment i.p. on day 1.
[c] Average survival time as percentage of survival time of controls.
[d] From Supino et al. (1977).

methylcarboxy group at C-10. The C-4 methoxyl, a characteristic feature of daunomycinone and related compounds, is replaced by a hydroxy group, but the substitution pattern in the aromatic chromophore is the same as in daunomycinone.

The coupling of ε-rhodomycinone with daunosamine, the aminosugar moiety of the biosynthetic antitumor anthracyclines, has been reported by Smith et al. (1976). The resulting aminoglycoside appeared, however, to be only moderately active, in contrast to the high biological potency of carminomycin, thus indicating the relevance of ring A substitution for the exhibition of antitumor activity.

The synthesis of a number of ε-rhodomycinone glycosides has been carried out by El Khadem et al. (1977) by using a Koenigs–Knorr type reaction in the presence of mercuric cyanide and mercuric bromide, followed by deblocking of the O-p-nitrobenzoyl protecting groups. The monosaccharide units so introduced were D-glucose, 2-deoxy-L-fucose, 2-deoxy-L-rhamnose, 2-deoxy-D-ribose, L-glucose, L-arabinose, D-ribose, D-xylose, and L-lyxose. As expected, and also because of the absence of an amino group on the sugar moiety, the new compounds did not exhibit appreciable activity in the P-388 leukemia test.

Acknowledgment

Thanks are due to Professor Bruno Camerino for the revision of the manuscript and to Mara Chierichetti for the typing work. Interest and cooperation in anthracycline research by Dr. Harry B. Wood and Dr. Ven L. Narayanan of the National Cancer Institute, Bethesda, Maryland, is also acknowledged.

REFERENCES

Aoshima, M., Tsukagoshi, S., Sakurai, Y., Ohishi, J., Ishida, T., and Kobayashi, H. (1976). Cancer Res. 35, 2726-2732.

Aoshima, M., Tsukagoshi, S., Sakurai, Y., Ishida, T., and Kobayashi, H. (1977). Cancer Res. 37, 2481-2486.

Arcamone, F. (1977). Lloydia 40, 45-66.

Arcamone, F. (1978). Int. Cancer Conf., 12th, Buenos Aires, In "Advances in Medical Oncology, Research and Education" (B. W. Fox, ed.), Vol. 5, pp. 21-32. Pergamon, Oxford.

Arcamone, F., Cassinelli, G., Orezzi, P., Franceschi, G., and Mondelli, R. (1964a). J. Am. Chem. Soc. 86, 5335-5336.

Arcamone, F., Franceschi, G., Orezzi, P., Cassinelli, G., Barbieri, W., and Mondelli, R. (1964b). J. Am. Chem. Soc. 86, 5334-5335.

Arcamone, F., Cassinelli, G., Franceschi, G., Orezzi, P., and Mondelli, R. (1968a). Tetrahedron Lett., 3353-3356.

Arcamone, F., Franceschi, G., Orezzi, P., Penco, S., and Mondelli, R. (1968b). Tetrahedron Lett., 3349-3352.

Arcamone, F., Franceschi, G., Penco, S., and Selva, A. (1969a). Tetrahedron Lett., 1007-1010.

Arcamone, F., Barbieri, W., Franceschi, G., and Penco, S. (1969b). Chim. Ind. (Milan) 51, 834-835.

Arcamone, F., Franceschi, G., and Penco, S. (1969c). Ger. Patent 1,917,874 [*C.A.* **73**, 45799 (1970)].

Arcamone, F., Cassinelli, G., Fantini, G., Grein, A., Orezzi, P., Pol, C., and Spalla, C. (1969d). *Biotechnol. Bioeng.* **11**, 1101–1110.

Arcamone, F., Cassinelli, G., Penco, S., and Tognoli, L. (1970a). Ger. Patent 1,923,885 [*C.A.* **72**, 140416 (1977)].

Arcamone, F., Cassinelli, G., Franceschi, G., Mondelli, R., Orezzi, P., and Penco, S. (1970b). *Gazz. Chim. Ital.* **100**, 949–989.

Arcamone, F., Franceschi, G., Minghetti, A., Penco, S., Redaelli, S., Di Marco, A., Casazza, A. M., Dasdia, T., Di Fronzo, G., Giuliani, F., Lenaz, L., Necco, A., and Soranzo, C. (1974a). *J. Med. Chem.* **17**, 335–337.

Arcamone, F., Franceschi, G., and Penco, S. (1974b). U.S. Patent 3,803,124.

Arcamone, F., Penco, S., Vigevani, A., Redaelli, S., Franchi, G., Di Marco, A., Casazza, A. M., Dasdia, T., Formelli, F., Necco, A., and Soranzo, C. (1975a). *J. Med. Chem.* **18**, 703–707.

Arcamone, F., Penco, S., and Vigevani, A. (1975b). *Cancer Chemother. Rep.* **6**, 123–129.

Arcamone, F., Bargiotti, A., Di Marco, A., and Penco, S. (1976a). Ger. Patent 2,618,822 [*C.A.* **86**, 140416 (1976)].

Arcamone, F., Bargiotti, A., Cassinelli, G., Penco, S., and Hanessian, S. (1976b). *Carbohydr. Res.* **46**, C3–C5.

Arcamone, F., Penco, S., Redaelli, S., and Hanessian, S. (1976c). *J. Med. Chem.* **19**, 1424–1425.

Arcamone, F., Bargiotti, A., Cassinelli, G., Redaelli, S., Hanessian, S., Di Marco, A., Casazza, A. M., Dasdia, T., Necco, A., Reggiani, P., and Supino, R. (1976d). *J. Med. Chem.* **19**, 733–734.

Arcamone, F., Bernardi, L., Giardino, P., Patelli, B., Di Marco, A., Casazza, A. M., Pratesi, G., and Reggiani, P. (1976e). *Cancer Treat. Rep.* **60**, 829–834.

Arcamone, F., Bernardi, L., Patelli, B., and Penco, S. (1976f). Belg. Patent 842,930 [*C.A.* **87**, 85201 (1977)].

Arcamone, F., Bernardi, L., Patelli, B., and Di Marco, A. (1976g). Ger. Patent 2,557,537 [*C.A.* **85**, 177886 (1976)].

Arcamone, F., Di Marco, A., and Casazza, A. M. (1977a). *Symp. Princess Takamatsu, 8th, Cancer Res. Found., Tokyo, In* "Advances in Cancer Chemotherapy" (H. Umezawa *et al.*, eds.), pp. 297–312. Japan Sci. Soc., Tokyo.

Arcamone, F., Bernardi, L., and Patelli, B. (1977b). Ger. Patent 2,713,745 [*C.A.* **88**, 23347 (1978)].

Arcamone, F., Di Marco, A., and Lazzari, E. (1978a). Br. patent 1,502,121.

Arcamone, F., Bernardi, L., Patelli, B., Giardino, P., Di Marco, A., Casazza, A. M., Soranzo, C., and Pratesi, G. (1978b). *Experientia* **34**, 1255–1256.

Arlandini, E., Vigevani, A., and Arcamone, F. (1977). *Farmaco, Ed. Sci.* **32**, 315–323.

Aszalos, A. A., Bachur, N. R., Hamilton, B. K., Langlykke, A. F., Roller, P. P., Sheikn, M. Y., Sutpin, M. S., Thomas, M. C., Wareheim, D. A., and Wright, L. H. (1977). *J. Antibiot.* **30**, 50–58.

Baer, H. H., and Georges, F. F. Z. (1977). *Carbohydr. Res.* **55**, 253–258.

Bargiotti, A., Cassinelli, G., Franchi, G., Gioia, B., Lazzari, E., Redaelli, S., Vigevani, A., Arcamone, F., and Hanessian, S. (1977). *Carbohydr. Res.* **58**, 353–361.

Bargiotti, A., Cassinelli, G., and Arcamone, F. (1978). Ger. Patent 2,752,115.

Benjamin, R. S., Keating, M. J., McCredie, K. B., Bodey, G. P., and Freireich, E. J. (1977). *Cancer Res.* **37**, 4623–4628.

Bognar, R., Sztaricskai, F., Munk, M. F., and Tomas, J. (1974). *J. Org. Chem.* **39**, 2971–2974.

Bouchandon, J., and Jolles, J. (1969). Ger. Patent 1,813,518 [*C.A.* **71**, 91866 (1969)].

Brockmann, H. (1963). *Fortschr. Chem. Org. Naturst.* **21**, 121–182.

Brufani, M., and Keller Schierlein, W. (1966). *Helv. Chim. Acta* **49**, 1962–1970.

Casazza, A. M., Di Marco, A., Bertazzoli, C., Formelli, F., Giuliani, F., and Pratesi, F. (1977). *Int. Congr. Chemother., 10th, Zurich* Abstr. No. 502.

Casazza, A. M., Pratesi, G., Giuliani, F., Formelli, F., and Di Marco, A. (1978). *Tumori* **64**, 115–129.

Cassinelli, G., Ruggeri, D., and Arcamone, F. (1979). *Med. Chem.* **22**, 121–123.

Cassinelli, G., Arcamone, F., and Di Marco, A. (1977). Ger. Patent 2,757,102 [C.A. **89**, 17891 (1978)].

Cassinelli, G., Grein, A., Masi, P., Suarato, A., Bernardi, L., Arcamone, F., Di Marco, A., Casazza, A. M., Pratesi, G., and Soranzo, C. (1978). *J. Antibiot.* **31**, 178–184.

Di Marco, A., and Arcamone, F. (1975). *Arzneim.-Forsch.* **25**, 368–375.

Di Marco, A., Zunino, F., Silverstrini, R., Gambarucci, C., and Gambetta, R. A. (1971). *Biochem. Pharmacol.* **20**, 1323–1328.

Di Marco, A., Casazza, A. M., Dasdia, T., Giuliani, F., Lenaz, L., Necco, A., and Soranzo, C. (1973). *Cancer Chemother. Rep.* **57**, 269–274.

Di Marco, A., Casazza, A. M., and Pratesi, G. (1976a). *IRCS Med. Sci.* **4**, 452.

Di Marco, A., Casazza, A. M., Gambetta, R., Supino, R., and Zunino, F. (1976b). *Cancer Res.* **36**, 1962–1968.

Di Marco, A., Casazza, A. M., and Pratesi, G. (1977a). *Cancer Treat. Rep.* **61**, 893–894.

Di Marco, A., Casazza, A. M., Dasdia, T., Necco, A., Pratesi, G., Rivolta, P., Velcich, A., Zaccara, A., and Zunino, F. (1977b). *Chem.-Biol. Interact.* **19**, 291–302.

Di Marco, A., Casazza, A. M., Giuliant, F., Pratesi, G., Arcamone, F., Bernardi, L., Franchi, G., Giardino, P., Ptaelli, B., and Penco, S. (1978a). *Cancer Treat. Rep.* **62**, 375–380.

Di Marco, A., Casazza, A. M., Soranzo, C., and Pratesi, G. (1978b). *Cancer Chemother. Pharmacol.* **1**, 249–254.

El Khadem, H. S., Swartz, D. L., and Cermak, R. R. (1977). *J. Med. Chem.* **20**, 957–960.

Fedorova, G. B., Brazhnikova, M. G., Mezentsev, A. S., and Ksheinsky, I. (1970). *Antibiotiki (Moscow)* **15**, 403–406.

Felsted, R. L., Richter, D. R., and Bachur, N. R. (1977). *Biochem. Pharmacol.* **26**, 1117–1124.

Florent, J., and Lunel, J. (1976). Ger. Patent 2,610,557 [C.A. **86**, 153947 (1977)].

Florent, J., Lunel, J., and Renaut, J. (1975). Ger. Patent 2,456,139 [C.A. **83**, 112355 (1975)].

Fuchs, E. F., Horton, D., and Weckerle, W. (1977). *Carbohydr. Res.* **57**, C36–C39.

Gabbay, E. J., Grier, D., Fingerle, R. E., Reimer, R., Levy, R., Pearce, S. W., and Wilson, W. D. (1976). *Biochemistry* **15**, 2062–2070.

Geran, R. L., Greenberg, N. H., MacDonald, M. M., Schumacher, A. M., and Abbot, B. J. (1972). *Cancer Chemother. Rep., Part 3* **3**, No. 2, 9.

Goldin, A., and Johnson, R. K. (1975). *In* "Adriamycin Review" (M. Staquet *et al.*, eds.), Part I, pp. 37–54. Eur. Press Medikon, Ghent.

Gupta, S. K. (1974). *Carbohydr. Res.* **37**, 381–383.

Heyns, K., Lim, M., and Park, J. I. (1976). *Tetrahedron Lett.*, 1477.

Hill, B. T., and Price, L., *A. J. Natl. Cancer Inst.* **59**, 1311–1314.

Horton, D., and Weckerle, W. (1976). *Carbohydr. Res.* **46**, 227–235.

Ishizu, K., Dearman, H. H., Huang, M. T., and White, J. R. (1968). *Biochim. Biophys. Acta* **165**, 283–285.

Israel, M., Modest, E. J., and Frei, E. (1975). *Cancer Res.* **35**, 1365–1368.

Jacquillat, C. (1976). *Cancer (Philadelphia)* **37**, 653–659.

Jolles, G. (1969). Ger. Patent 1,803,892 [C.A. **71**, 70907 (1969)].

Jolles, G. (1973). Ger. Patent 2,327,211 [C.A. **82**, 171381(1975)].

Jolles, G., and Ponsinet, G. (1972). Ger. Patent 2,202,690 [C.A. **77**, 164320 (1972)].

Kern, D. L., Bunge, R. H., French, J. C., and Dion, H. W. (1977). *J. Antibiot.* **30**, 432–434.

Lee, W. W., Wu, H. Y., Marsh, J. J., Jr., Mosher, C. W., Acton, E. M., Goodman, L., and Henry, D. W. (1975a). *J. Med. Chem.* **18**, 767–768.

Lee, W. W., Wu, H. Y., Christensen, J. E., Goodman, L., and Henry, D. W. (1975b). *J. Med. Chem.* **18**, 768-769.

Lenaz, L., Necco, A., Dasdia, T., and Di Marco, A. (1974). *Cancer Chemother. Rep.* **38**, 769-776.

Lomakina, N. N., Spiridonova, I. A., Sheinker, Y. N., and Vlassova, T. F. (1973). *Khim. Prir. Soedin.* **9**, 101-107 [*C.A.* **78**, 14817 (1973)].

Lunel, J., and Preud'homme, J. (1970). Ger. Patent 1,911,240 [*C.A.* **72**, 20603 (1970)].

Mallarms, A. K. (1969). *J. Am. Chem. Soc.* **91**, 7505-7506.

Maral, R., Ponsinet, G., and Jolles, G. (1972). *C. R. Acad. Sci., Ser.* D **275**, 301-304.

Masi, P., Suarato, A., Giardino, P., Bernardi, L., and Arcamone, F. (1977). Ger. Patent 2,735,455.

Patelli, B., Bernardi, L., Arcamone, F., and Di Marco, A. (1976). Ger. Patent 2,627,146 [*C.A.* **86**, 190419 (1977)].

Pelyvás, I., Sztaricskai, F., Bognár, R., and Buitás, G. (1977). *Carbohydr. Res.* **53**, C17.

Penco, S. (1968). *Chim. Ind. (Milan)* **50**, 908.

Penco, S., Angelucci, F., Vigevani, A., Arlandini, E., and Arcamone, F. (1977). *J. Antibiot.* **30**, 764-766.

Penco, S., Angelucci, F., and Arcamone, F. (1978). Ger. Patent 2, 757,057 [C.A. **89**, 197892 (1978)].

Pettit, G. R., Einck, J. J., Herald, C. L., Ode, R. H., Von Dreerle, R. H., Brown, P., Brazhnikova, M. C., and Gause, G. F. (1975). *J. Am. Chem. Soc.* **97**, 7387-7388.

Pratesi, G., Casazza, A. M., and Di Marco, A. (1978). *Cancer Treat. Rep.* **62**, 105-110.

Richardson, A. C. and McLaughlan, K. A. (1962). *J. Chem. Soc.* 2499-2506.

Skovsgaard, T. (1977). *Biochem. Pharmacol.* **26**, 215-222.

Smith, T. H., Fujiwara, A. N., and Henry, D. W. (1976). *Natl. Meet. Am. Chem. Soc., 172nd, San Francisco, Calif.* Abstr. MEDI-88.

Supino, R., Necco, A., Dasdia, T., Casazza, A. M., and Di Marco, A. (1977). *Cancer Res.* **37**, 4523-4528.

Sztaricskai, F., Pelyvás, L., Bognár, H., and Bujtás, G. (1975). *Tetrahedron Lett.*, 1111-1114.

Takahashi, Y., Nagasawa, H., Takeuchi, T., Umezawa, H., Komiyama, T., Oki, T., and Inui, T. (1977). *J. Antibiot.* **30**, 622-624.

Tong, C., Lee, W. W., Black, D. R., and Henry, D. W. (1976). *J. Med. Chem.* **19**, 395.

Wani, M. C., Taylor, H. L., Wall, M. E., McPhail, A. T., and Onan, K. D. (1975). *J. Am. Chem. Soc.* **97**, 5955-5956.

Williams, D. H., and Kolman, J. R. (1976). *Tetrahedron Lett.*, 4829-4830.

Wilson, D. W., Grier, D., Reimer, R., Bauman, J. D., Preston, J. F., and Gabbay, E. J. (1976). *J. Med. Chem.* **49**, 381-384.

Wong, C. M., Popieu, D., Schwenk, R., and Te Raa, J. (1971). *Can. J. Chem.* **49**, 2712-2718.

Wong, C. M., Schwenk, R., Popieu, D., and Ho, T. (1973). *Can. J. Chem.* **51**, 466-467.

Yamamoto, K., Acton, E. M., and Henry, D. W. (1972). *J. Med. Chem.* **15**, 872-875.

Zbinden, G., and Brändle, E. (1975). *Cancer Chemother. Rep.* **59**, 707-715.

Zbinden, G., Pfister, M., and Holderegger, C. (1978a). *Toxicol. Lett.* **1**, 267-274.

Zbinden, G., Bochmann, E., and Holderegger, C. (1978b). *Antibiot. Chemother.* **23**, 255-270.

Zunino, F., Gambetta, R., Di Marco, A., and Zaccara, A. (1972). *Biochim. Biophys. Acta* **277**, 489-498.

Zunino, F., Gambetta, R., Di Marco, A., Luoni, G., and Zaccara, A. (1976). *Biochem. Biophys. Res. Commun.* **69**, 744-750.

CHAPTER 2
Trichothecanes

T. W. DOYLE AND W. T. BRADNER

I. INTRODUCTION

A. Historical

The original stimulus for the isolation of the trichothecane sesquiterpenes (Fig. 1) from fungi was provided by observations of their antifungal and phytotoxic effects. Thus Freeman and Morrison (1948) reported the isolation of trichothecin (7) (Fig. 2) from *Trichothecium roseum* and commented on the antifungal activity. A later paper by Freeman (1955) expanded on these initial observations. Somewhat later, Gláz *et al.* (1959, 1960) reported the isolation of a new antifungal compound structurally related to but different from trichothecin. Yet another member of this class was reported by Brian *et al.* (1961), who isolated anguidine (diacetoxyscirpenol) (12n) (Fig. 3) from *Fusarium scirpi* and commented on its phytotoxic effects. Soon after, a series of papers appeared disclosing the isolation of a number of structurally related macrocyclic esters of the trichothecane verrucarol (5) exhibiting cytotoxic effects (Harri *et al.*, 1962; Bohner *et al.*, 1965). The isolation of trichodermin (4) by Godtfredsen and

Anticancer Agents Based on Natural Product Models

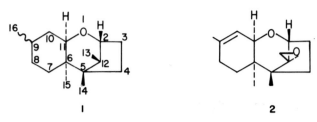

Figure 1.   Basic skeleton and numbering system for the trichothecanes.

Vangedal (1965) and subsequent structural determination by Abrahamsson and Nilsson (1964, 1966) using x-ray diffraction analysis led, together with the earlier chemical work, to a realization that all of the compounds discovered to that date were structurally related.

Subsequent to these initial discoveries, the implication of various tricho-thecane-producing fungi in a number of animal and human toxicoses due to the ingestion of moldy cereal grains led to the isolation and characterization of over 40 members of the class. Trichothecanes were shown to be the causative agent or likely to be the causative agent(s) in fescue foot disease (Yates *et al.*, 1968); moldy corn toxicoses (Gilgan *et al.*, 1966; Bamburg *et al.*, 1969; Hsu *et al.*, 1972); stachybotrytoxicosis (Ueno *et al.*, 1971a; Eppley and Bailey, 1973); food-refusal phenomena (Vesonder *et al.*, 1973, 1976); and human alimentary toxic aleukia (Mayer, 1953; Forgacs and Carll, 1962).

The first indications that members of this class might possess some utility as antitumor agents came from the reports of Harri *et al.* (1962) on the activity of the verrucarins (**10a–g**) and roridins (**10h–j**) in mouse tumor cell (P-815) tissue-culture experiments. Soon thereafter (Loeffler *et al.*, 1967), diacetoxyscirpenol (anguidine) was reported to possess cytotoxic activity, and claims for its utility as an antitumor agent were made. Recently, anguidine completed phase I clinical trials in the United States (Helman and Slavik, 1976) and is about to begin phase II clinical trials.

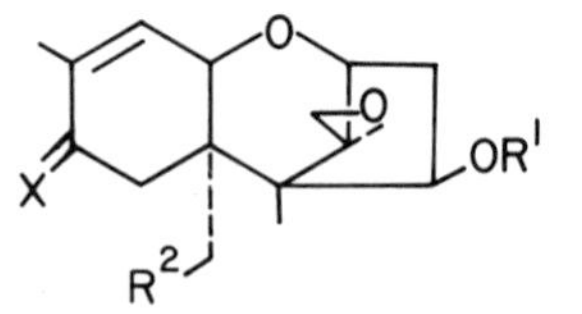

3	$X = H_2$, $R^1 = R^2 = H$	Trichodermol (Roridin C)
4	$X = H_2$, $R^1 = Ac$, $R^2 = H$	Trichodermin
5	$X = H_2$, $R^1 = H$, $R^2 = OH$	Verrucarol
6	$X = O$, $R^1 = R^2 = H$	Trichothecolone
7	$X = O$, $R^1 = COCH = CHCH_3$, $R^2 = H$	Trichothecin

Figure 2.   Structure of trichothecin.

Since the initial reports of Harri *et al.* (1962) and Loeffler *et al.* (1967) the majority of the trichothecanes have been reported to possess cytotoxic activity in tissue culture. The purpose of this chapter is to review the data bearing on the utility of these compounds and to provide an overview of the state of the art with respect to structure–activity relationships of the trichothecanes as antitumor agents.

B. Sources

With the sole exception of the baccharins (**10o–s**), the trichothecanes to be reviewed in this chapter are products of the culture of various species of imperfect fungi. The organisms, common names of the toxins produced, and parent alcohols of the toxin complexes are listed in Table I. As is evident from the table, any individual toxin may be produced by a number of species (e.g., anguidine is produced by no less than eight species of *Fusarium*), and many species produce more than one toxin (usually closely related esters).

The isolation of trichothecanes from a higher plant (*Baccharis megapotamica*) (Kupchan *et al.*, 1976, 1977) represents a departure from the usual sources and may possibly be due to the collection of a plant parasitized by fungi. Indeed the majority of trichothecane-producing fungi were first isolated as plant parasites responsible for various outbreaks of cattle poisonings from eating contaminated feed.

C. Structure

Godtfredsen *et al.* (1967) have proposed the trivial name trichothecane for this family of compounds after trichothecin, the member first isolated by Freeman and Morrison (1948). The basic skeleton and numbering system for the trichothecanes are as illustrated in structure (**1**) of Fig. 1. In addition to the features evident in structure (**1**), the majority of the structures with which we shall be concerned also have a 9,10 double bond and a 12,13-epoxy function as shown in structure (**2**).

The efforts of a number of research groups to elucidate the structures of the trichothecanes led to a thorough exploration of the chemistry of the system (Freeman *et al.*, 1959; Fishman *et al.*, 1960; Brian *et al.*, 1961; Tamm and Gutzwiller, 1962; Harri *et al.*, 1962; Gutzwiller and Tamm, 1963). Ultimately, the structure for trichodermol (**3**) was elucidated by Godtfredsen and Vangedal (1964), using chemical and x-ray analysis (Abrahamsson and Nilsson, 1964, 1966). Several other trichothecanes were shown to have the same ring system as trichodermol (Gutzwiller *et al.*, 1964), thus leading to the resolution of the structures of trichothecin and verrucarol as well as trichodermin (Fig. 2). The history of the discovery and structure determination of the majority of naturally

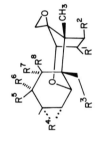

	R^1	R^2	R^3	R^4	R^5	R^6	R^7	R^8
12a Trichodermol	H	H	H	—	H	H	H	H
b Trichodermin	H	OH	H	—	H	H	H	H
c	H	OAc	H	—	H	H	H	H
d Verrucarol diacetate	H	OAc	OAc	—	H	H	H	H
e Calonectrin	OAc	H	OAc	—	H	H	H	H
f 15-Desacetylcalonectrin	OAc	H	OH	—	OH	H	H	H
g	H	OH	H	—	H	H	H	H
h Scirpen-3,4,15-triol	OH	OH	OH	—	H	H	H	H
i	OAc	OH	OH	—	H	H	H	H
j	OH	OAc	OH	—	H	H	H	H
k	OH	OH	OAc	—	H	H	H	H
l	OAc	OAc	OH	—	H	H	H	H
m	OAc	OH	OAc	—	H	H	H	H
n Anguidine	OH	OAc	OAc	—	H	H	H	H
o	OAc	OAc	OAc	—	H	H	H	H

Compound							
p 9,10-Dihydroanguidine	OH	OAc	OAc	H₂	H	H	H
q 9,10-Epoxyanguidine	OH	OAc	OAc	O	H	H	H
r 8β-Hydroxyanguidine	OH	OAc	OAc	—	H	OH	H
s HT-2 Toxin	OH	OH	OAc	—	O-i-Val	H	H
t T-2 Toxin	OH	OAc	OAc	—	O-i-Val	H	H
u T-2 Toxin	OH	OAc	OAc	—	O-i-Val	H	H
v Neosolaniol	OH	OAc	OAc	—	OH	H	H
w	OH	OAc	OH	—	OAc	H	H
x	OH	OH	OAc	—	OAc	H	OH
y	OH	OAc	OAc	—	H	H	OH
z	OH	OAc	OAc	—	OAc	H	OH
aa	OAc	OAc	OAc	—	OAc		OH
13a Nivalenol	OH	OH	OH	—	O		OH
b Fusarenon-X	OH	OAc	OH	—	O		OH
c Nivalenol diacetate	OH	OAc	OAc	—	O		OAc
d Nivalenol tetraacetate	OAc	OAc	OAc	—	O		OH
e 4-Deoxy nivalenol	OH	H	OH	—	O		OH
f	OAc	H	OH	—	O		OH
g Dihydro nivalenol	OH	OH	OH	H₂	O		OH
h Trichothecin	H	OCrot	H	—	H		H
14 Crotocin	H	OCrot	H	—	H	O	H

Figure 3. Structures of the trichothecane polyols.

TABLE I

Naturally Occurring 12,13-Epoxytrichothec-9-enes

Producing organisms	Toxins	Parent alcohols	Reference[a]
Baccharis megapotamica	Baccharins	9β,10β-epoxy-4β,15-dihydroxy	1, 2
		4β,8β,15-trihydroxy	
		3α,4β,15-trihydroxy	
Calonectria nivalis	Roridins D and E	4β,15-dihydroxy	
Cephalosporium crotocinegenum	Calonectrin	3α,15-dihydroxy	3
	Crotocin	7β,8β-epoxy-4β-hydroxy	4, 5, 6
Cylindrocarpon sp. PF-60	Isororidins	7β,8β-epoxy-4β,15-dihydroxy	7, 8
	Epoxy isororidins		
	Roridins	4β,15-dihydroxy	
Fusarium diversisporum, equiseti, oxysporum sambucinum, scirpi, solani, tricinctum, sulphureum	Anguidine	3α,4β,15-trihydroxy	9, 10
			11, 12
			13, 14
			15, 50
Fusarium concolor, roseum	Desacetylanguidine	3α,4β,15-trihydroxy	16
			17, 18
Fusarium culmorum, nivale, graminearum	Nivalenol	3α,4β,7β,15-tetrahydroxy-8-oxo	19, 20
	Fusarenon		21, 22, 23
	Vomitoxin		24, 25
Fusarium poae,	T-2	3α,4β,8α,15-tetrahydroxy	26, 11, 12

Organism	Compound	Structure	References
solani, sporotrichiodes tricinctum, oxysporum	HT-2	3α,7α,15-trihydroxy-8-oxo	13, 27
	Neosolaniol	3α,4β,7α,15-tetrahydroxy	28, 29
	Acetyl T-2	3α,4β,15-trihydroxy	30, 47, 48
Fusarium culmorum, roseum	4-Deoxynivalenol		19, 31
Fusarium lateritium 5036	—		14, 49
	Anguidine		
Fusarium equiseti	—	3α,4β,7α,8α,15-pentahydroxy	14, 45, 46
Myrothecium roridin	Roridins	4β,15-dihydroxy	32
Myrothecium verrucaria	Verrucarins	4β,15-dihydroxy	32–37
Stachybotry atra	Satratoxins	4β,15-dihydroxy	38, 39
Trichoderma viride, lignorum	Trichodermin	4β-hydroxy	40, 41
Trichothecium roseum	Trichothecin	Parent compound	42
		4β-hydroxy-8-oxo	42, 43
		4β,8α-dihydroxy	42
Verticimonosporium diffractum	Vertisporin	4β,15-dihydroxy	44

[a] References: (1) Kupchan et al. (1976); (2) Kupchan et al. (1977); (3) Gardner et al. (1972); (4) Gláz et al. (1959); (5) Gláz et al. (1966); (6) Gyimesi and Melera (1967); (7) Matsumoto et al. (1977a); (8) Matsumoto et al. (1977b); (9) Brian et al. (1961); (10) Sigg (1965); (11) Bamburg et al. (1968); (12) Bamburg and Strong (1969); (13) Ishii et al. (1971); (14) Ishii (1975); (15) Ghosal et al. (1976); (16) Pathre et al. (1976); (17) Loeffler et al. (1967); (18) Bamburg and Strong (1970); (19) Blight and Grove (1974); (20) Uneo et al. (1971a); (21) Uneo et al. (1971b); (22) Uneo et al. (1968); (23) Tatsuno (1968); (24) Vesonder et al. (1973); (25) Vesonder et al. (1976); (26) Gilgan et al. (1966); (27) Yates et al. (1968); (28) Uneo et al. (1972a); (29) Uneo et al. (1972b); (30) Yoshizawa and Morooka (1973); (32) Harri et al. (1962); (33) Fetz et al. (1965); (34) Bohner et al. (1965); (35) Guarino et al. (1968); (36) Nespiak et al. (1961); (37) Okuchi et al. (1968); (38) Eppley and Bailey (1973); (39) Eppley et al. (1977); (40) Gotfredsen and Vangedal (1964); (41) Gotfredsen and Vangedal (1965); (42) Machida and Nozoe (1972); (43) Freeman and Morrison (1948); (44) Minato et al. (1975); (45) Grove (1970); (46) Lansden et al. (1978); (47) Ilus et al. (1977); (48) Ishii et al. (1978); (49) Uneo et al. (1977); (50) Steyn et al. (1978).

occurring trichothecanes has been reviewed and will not be repeated here (Bamburg and Strong, 1971; Bamburg, 1972, 1976; Ciegler, 1975, 1978; Tamm, 1974). Table I (Section I,B) lists the naturally occurring trichothecanes with references to the original articles in which their discovery, isolation, and characterization were first described.

In Fig. 4, two conformations of the trichothecane skeleton are illustrated. Examination of molecular models indicates that conformer A is undoubtedly the more stable of the two. It should be noted, however, that some reactions of the trichothecanes proceed as though the molecule adopts conformation B; e.g., Sigg *et al.* (1965) have reported the conversion of 3α, 4β, 15-triacetoxy-12,13-epoxytrichothec-9-ene (**8**) to (**9**) on refluxing in water (Fig. 5).

To date, the isolations of 44 naturally occurring trichothecanes have been reported. The structures of these compounds, as well as others derived via biotransformation or semisynthesis, are shown in the accompanying Figs. 3, 6, and 7. Ueno *et al.* (1973a) have suggested a classification of the trichothecanes into four types based on structure: the macrocyclic trichothecanes (**10a–s**) and (**11a–f**); the polyhydroxylated trichothecanes (**12a–aa**); the keto trichothecanes (**13a–h**); and the $7\beta,8\beta$-epoxytrichothecane crotocin (**14**).

As can be seen from an examination of the figures, a great deal of structural diversity exists for this class of antitumor agent. The main features of trichothecane biosynthesis have been reviewed (Tamm, 1974; Bamburg, 1976) and will not be discussed here. In the remaining sections the biological activity of this class of antitumor agents will be reviewed.

II. BIOLOGY

A. Mechanism of Action

A good deal of work on the mechanism of action of the trichothecanes has appeared. Ueno *et al.* (1968) reported the effects of nivalenol (**13a**) on

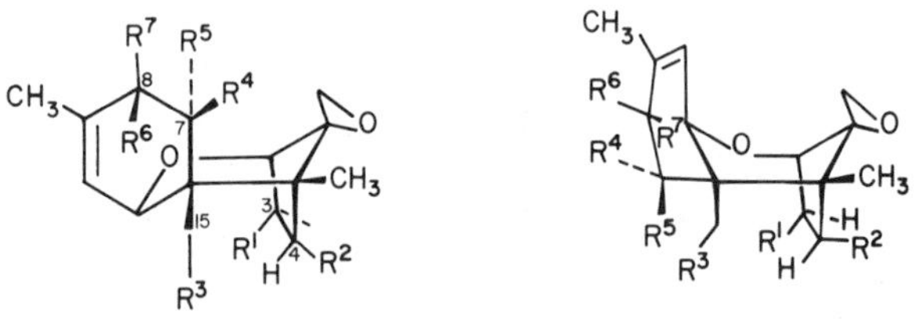

A   B

Figure 4.   Conformations of the 12,13-epoxytrichothec-9-ene system.

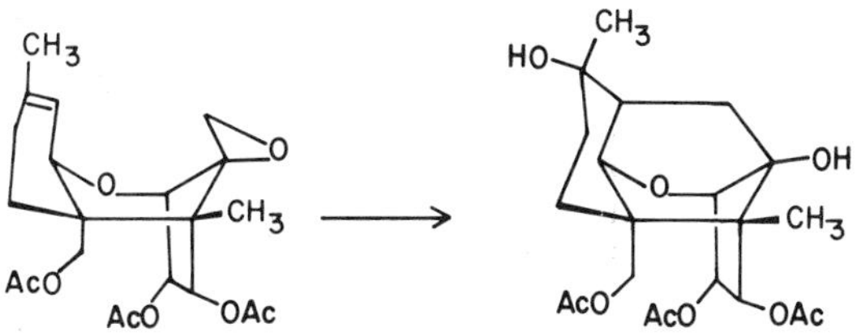

Figure 5. Double-bond participation in solvolytic epoxide opening.

macromolecule synthesis in both cultured cells and cell-free systems. Nivalenol was shown to inhibit protein synthesis in the rabbit reticulocyte and rat liver ribosomal systems. This compound *in vitro* was shown to inhibit both poly(U) directed polyphenylalanine synthesis and poly(A)-directed polylysine synthesis. The authors also reported that nivalenol had no effect on protein synthesis in bacterial systems either in whole cells or cell extracts. In a later paper, Ueno and Fukushima (1968) extended these observations to Ehrlich ascites tumor cells, in which protein synthesis inhibition was demonstrated in addition to DNA synthesis inhibition following a lag phase.

Additional studies on the effect of another trichothecane on protein synthesis appeared somewhat later. Ohtsubo *et al.* (1972) reported the effect of fusarenon (**13b**) on polyribosome profiles using cultured mouse fibroblasts. Complete runoff of the polysomes was observed at low concentrations of (**13b**) ($ID_{50} = 0.2$ mg/ml). Similar effects on polyribosome profiles of rabbit reticulocytes using T-2 toxin (**12t**), anguidine (**12n**), and neosolaniol (**12v**) were reported by Ueno *et al.* (1973a). This group showed the effect to be reversible upon washing the cells free of the toxins. On the basis of their effect on protein synthesis and their structures, the compounds were divided into two broad classes by Ueno *et al.* (1973a): those having multiple hydroxyl functions (Class A), such as T-2 toxin, HT-2 toxin, neosolaniol and anguidine; and those bearing an 8-keto function (Class B), such as nivalenol, fusarenon-X, and trichothecin. The effects of the Class A agents on protein synthesis, cytotoxicity, and polysomal breakup was greater than the effects of Class B agents in whole cell preparations, while the order was reversed in cell-free preparations, thus indicating transport to be an important consideration.

The behavior of the compounds on polyribosome profiles implied that their effect was primarily on the initiation stages of protein synthesis. A series of papers appeared which seemed to contradict the earlier findings. Carrasco *et al.* (1973), using human tonsil ribosomes and yeast cytoplasmic ribosomes (from *S. cerevisiae*), showed that trichodermin (**12c**) inhibited the peptide-bond-formation step of protein synthesis and did not affect translocation. It was suggested that

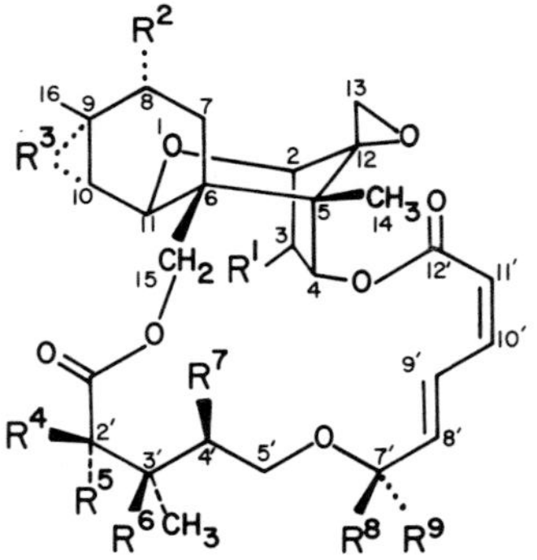

10

		R^1	R^2	R^3	R^4	R^5	R^6	R^7	R^8	R^9
a	Verrucarin A	H	H	—	H	OH	H	H	O	
b	2'-Dehydroverrucarin A	H	H	—	H	O	H	H	O	
c	Verrucarin A 9,10-epoxide	H	H	O	H	OH	H	H	O	
d	Verrucarin A 9,10-epoxide-2'acetate	H	H	O	H	OAc	H	H	O	
e	Verrucarin B	H	H	—	H	O			O	
f	Verrucarin B 9,10-epoxide	H	H	O	H	O			O	
g	Verrucarin J	H	H	—	H				O	
h	Roridin A	H	H	—	H	OH	H	H	H	$CHOHCH_3$
i	Roridin D	H	H	—	H	O		H	H	$CHOHCH_3$
j	Roridin E	H	H	—	H	O		H	H	$CHOHCH_3$
k	Isororidin E	H	H	—	H	O		H	H	$CH(CH_3)OH$
l	Roridin A, 8β-Hydroxy	H	OH	—	H	OH	H	H	H	$CHOHCH_3$
m	Isororidin A, 8β-Hydroxy	H	OH	—	H	OH	H	H	H	$CH(CH_3)OH$
n	Roridin A, 2'-Deoxy-4',8β-Dihydroxy	H	H	—	H	H	H	H	H	$CHOHCH_3$
o	Baccharinol	H	OH	—	H	O		OH	H	$CHOHCH_3$
p	Isobaccharinol	H	OH	—	H	O		OH	H	$CH(CH_3)OH$
q	Baccharin	H	H	O	H	O		OH	H	$CHOHCH_3$
r	Isobaccharin	H	H	O	H	O		OH	H	$CH(CH_3)OH$
s	Baccharin 5	OH	H	—	H	O		OH	H	$CHOHCH_3$

Figure 6. General structure of the macrocyclic trichothecenes

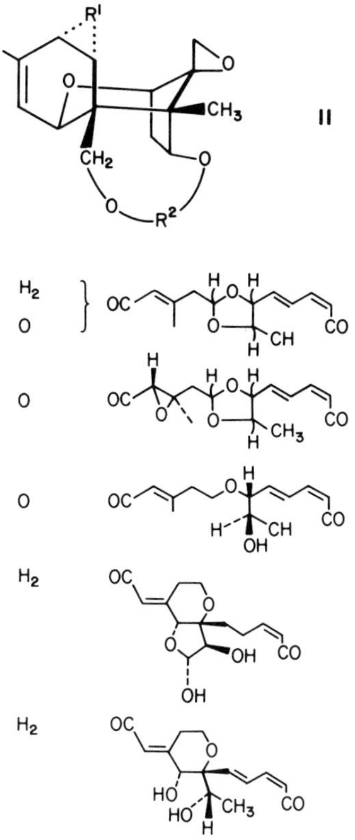

Figure 7. Structures of several macrocyclic trichothecanes.

trichodermin, trichodermol, verrucarin A, fusarenon-X, and trichothecin all have a similar mode of action. Tate and Caskey (1973) studied the effect of trichodermin and T-2 toxin on the release of formyl[³H]methionine from the [³H]formylmet-tRNA ribosome with reticulocyte release factors, poly(U)-directed polyphenylalanine synthesis, and the puromycin reaction. The results of these experiments suggested a mechanism of action for these compounds whereby the elongation or termination steps were inhibited. Wei *et al.* (1974b) suggested that trichodermin is a specific inhibitor of the termination process. Hansen and Vaughan (1973) showed, using intact HeLa cells, that protein synthesis is reversible using trichodermin, that trichodermin stabilized polyribosomes, and that it does not prevent reformation of polyribosomes, all of which supported the suggestions that trichodermin acts via inhibition of the termination step. Barbacid and Vasquez (1974) carried out competitive binding studies using trichodermin and concluded that the effect on protein synthesis was a general one of the peptidyl transferase center and not just specific to the termination step.

Some of the contradictions were cleared up by Cundliffe *et al.* (1974), who showed that trichodermin antagonized the effect of nivalenol, T-2 toxin, and verrucarin A on polyribosome profiles. On this basis, the trichothecanes could be divided into three classes: elongation inhibitors (E-types), such as trichodermin and trichothecin; initiation inhibitors (I-types), such as nivalenol, T-2 toxin, and verrucarin A; and termination inhibitors (T-types). The earlier work of Carrasco *et al.* (1973), implying verrucarin A to be an E-type, could be explained as due to the fact that they used monosomes rather than polyribosomes in their study. Further work permitted the assignment of a number of agents as either E-types or I-types (Schindler, 1974). This author suggested that I-types can only bind to ribosomes free of long nascent peptide chains (>2–3 amino acids), and that verrucarin A (an I-type) and trichodermin (an E-type) share a common binding site on the ribosome. In another paper, Schindler *et al.* (1974) showed that T-2 toxin, verrucarin A, and nivalenol exhibited cross resistance with trichodermin to a mutant trichodermin resistant strain of *S. cerevisiae.*

Support for Schindler's hypothesis comes from the work of Smith *et al.* (1975), who showed that while T-2 toxin was capable of inhibiting the f-met-puromycin reaction, there was no effect of T-2 toxin on the release of nascent peptides by puromycin.

Some of the structural requirements for I-type of E-type activity were defined by Wei and McLaughlin (1974). For E-type activity, substitution of the 4β-position by an acylated alcohol is necessary. If the α face of the molecule is substituted at C_{15} by an acylated alcohol, the compounds exhibit I-type activity. Substitution of both the 3α and the 15 positions by hydroxyl results in I-type activity, whereas the 4β,15-diol verrucarol is an E-type compound.

In a further study, the binding of trichodermin to intact 80 S ribosome and the 60 S and 40 S subunits was examined (Wei *et al.*, 1974a). Free 80 S reticulocyte ribosomes and 80 S ribosomes in polysomes were found to bind trichodermin with similar affinities, while the 60 S subunits had only a low affinity and the 40 S subunits none at all. The strong blocking of trichodermin binding by tricothecanes which inhibit *in vivo* protein synthesis suggested the possibility of competitive binding at a single ribosomal site.

In a key paper, Carter *et al.* (1976) reported I-type behavior (polysomal runoff) when trichodermin was used at low concentrations. Thus at 25 μg/ml in the cell-free reticulocyte system, trichodermin totally inhibits protein synthesis and exhibits a total E-type response, while at 0.25 μg/ml there is 75% inhibition of protein synthesis and a 30–35% partial runoff of polyribosomes. This mitigates against the possibility that inhibition of protein synthesis by trichodermin is exclusive to the termination steps (T-type behavior). The authors suggested that initially trichodermin binds to recently initiated ribosomes *due to easier access to the active site,* and only at higher concentrations were the other sites filled.

In a further paper, the same group demonstrated that while trichodermin was capable of inhibiting protein synthesis whenever it was added, T-2 toxin was incapable of inhibiting protein synthesis in ribosomes carrying nascent peptide chains of more than four amino acid residues (Cannon *et al.*, 1976a). Additional clarification was provided by Cannon *et al.* (1976b). They measured the dissociation constants of trichodermin to polyribosomes and monoribosomes and showed binding to be three times more efficient to the latter than to the former. The ability of a series of trichothecanes to bind competitively to monosomes and polysomes were measured. While all members of the series competed equally well with trichodermin in binding to monosomes, the competition with trichodermin for binding to polysomes was dependent on the structure of the trichothecane (e.g., trichodermol > trichothecin > fusarenon-x>T-2 toxin > verrucarin A). The ability of a ribosome to bind trichodermin is independent of whether the "P" or "A" site is occupied by tRNA. Both the E-class and I-class trichothecanes inhibit the fragment reaction between puromycin and cacca-Leu-Ac as catalyzed by the peptidyl transferase center, probably by inhibition of cacca-Leu-Ac binding to the donor site of the peptidyl transferase.

Liao *et al.* (1976) demonstrated that certain I-type trichothecanes when given at high concentrations begin to exhibit E-type behavior (freezing of polysome profiles). They also showed that the agents do not act by preventing the binding of tRNA to the ribosomes. That the agents do not block formation of the 80 S ribosomes was demonstrated by Fresno *et al.* (1976), who showed an enhancement in the formation of the initiation complex on the ribosome but not on the 40 S subunit.

Differentiation of I-types into two classes has been proposed by Cundliffe and Davies (1977). The macrocyclic trichothecanes verrucarin A and roridin A (I_1-types) have been shown to behave somewhat differently than I_2-types (T-2 toxin, etc.), in that the former appear to be able to prevent formation of 80 S initiation complexes in addition to inhibiting the functions of intact ribosomes. These authors have also demonstrated true T-type behavior for trichodermol and trichodermone.

Finally, Carter and Cannon (1978) have demonstrated for a number of trichothecanes that the expression of E-type or I-type activity is dependent upon the concentration at which the compound is given, thus permitting division of these drugs into further subsets: pure I-types, I-types showing partial E-type activity at high concentrations, pure E-types, and E-types showing partial I-type activity at low concentrations.

The foregoing suggests that all these compounds act by the same or very similar mechanisms at the peptidyl transferase center in the 60 S subunit of intact 80 S ribosome·mRNA complexes. The variable in each case is the size of the nascent peptide chain present on the tRNA in the "P" or "A" site. In general, one can postulate that the larger this chain is, the greater the inhibition of binding of the trichothecane to the active site (either due to steric or other effects). Thus, the "smaller" trichothecanes (having either no substituent at C_{15}, or at worst a hydroxyl group) can bind to peptidyl transferase centers in most of the ribosomal units of the polyribosome. Since peptide-bond formation is prevented, the polyribosomal profile is frozen (E-type activity). When low concentrations are used, preferential binding to the least sterically hindered peptidyl transferase centers occurs, thus permitting runoff of the uncomplexed, more sterically hindered, ribosomal units (I-type activity). The smaller I-type inhibitors are prevented at normal concentrations from interacting with peptidyl transferase centers in which a nascent peptide of more than about four amino acids is present. At high concentrations, however, these can interact thus giving an E-type response. Larger I-type inhibitors are unable to do this at any concentration.

The reason that the parent system and trichodermol exhibit T-type activity at lower concentrations may be specific inhibition of the termination step. It is possible that only these sterically quite small trichothecanes are able to fit the peptidyl transferase (or other site) in ribosomes near the end of translation. The fact that the smallest E-type known, trichodermin, binds to polyribosomes with less than one molecule trichodermin per ribosomal unit suggests that all trichothecanes equal to or larger than trichodermin in size may be excluded from ribosomes at the end of translation.

The structural features which determine whether or not a molecule will exhibit I-type, E-type, or T-type activity may be summarized as follows (refer to Fig. 2). In Table II are listed those trichothecanes which have been classified according to their behavior in the rabbit reticulocyte protein synthesis bioassay.

1. T-Type Activity

Compounds which exhibit this activity are either unsubstituted ($R^1=R^2 =R^3=R^4=R^5=R^6=R^7=H$) or carry only a small substituent at C_4 ($R^2=OH$, $R^1=R^2=R^4=R^5=R^6=R^7=H$). While both trichodermone and 4-epitrichodermol are T-type inhibitors, the potency is greatly diminished so the 4β configuration is probably necessary. Whether or not trichothecolone (8-keto-4β-ol) or crotocol exhibit T-type behavior is not known, but on the basis of present data on T- and E-type compounds these will probably prove to be T-types.

2. E-Type Activity

Compounds substituted by an ester at C_4 are E-types (trichodermin). That a C_4 substitute is not a necessary prerequisite to E-type activity is shown by the 3,15-didesacetyl and 15-desacetyl calonectrin. Acylation at C_{15} leads to E-I_2-type activity as shown by 3-desacetyl calonectrin and 15-acetoxyscirpenediol.

TABLE II

Classification of Trichothecanes by Mechanism of Action

Structure	Initiation inhibitors	Subgroup[a]
10a,e,g	Verrucarin A, B, J,	I_1
10h	Roridin A	I_1
12t	T-2 toxin	I_2
12s	HT-2 toxin	—
12v	Neosolaniol	—
12n	Anguidine	I_2–E
12e	Calonectrin	I_2–E
13a	Nivalenol	—
13b	Fusarenon-X	I_2–E
	Elongation inhibitors	
12h	Scirpenetriol	E–I_2
12k	15-Acetoxyscirpendiol	E–I_2
	3-Desacetylcalonectrin	E–I_2
12f	15-Desacetylcalonectrin	E
	3,15-didesacetylcalonectrin	E
12c	Trichodermin	E
13h	Trichothecin	E
	Trichothecolone	—
14	Crotocin	—
	Cortocol	—
	Verrucarol	—
	Termination inhibitors	
12b	Trichodermol	—
	Trichodermone	—
	4-Epitrichodermol	—

[a] I_1, macrocyclic initiation inhibitors; I_2, pure initiation inhibitor; I_2–E, primarily I_2-type, causes partial polysome stabilization at high drug concentration; E–I_2, primarily E-type, causes partial runoff at low drug concentrations; E, pure E-type.

3. I-Type Activity

If in addition to an acylated alcohol at C_{15} either the C_3 or C_4 alcohol groups are acylated, the compounds become I-type inhibitors. Anguidine and calonectrin show I_2–E-type activity. The addition of more substitution at C_8 (the α-isovaleroxy group), as in T-2 toxin, gives pure I activity. The distinction between the macrocyclic I-types and the I_2-types is not absolutely clear, although these probably belong in a class by themselves.

One possible explanation of the mechanism by which the trichothecanes interfere with the peptidyl transferase center may be that they react with thiol residues in the active site. Ueno (1977a) has presented evidence that trichothecanes are capable of reacting with thiol residues in a number of thiol containing enzymes.

While the studies of protein-synthesis inhibition by the trichothecanes have served to define the mechanism of action of this interesting drug class, they are of only limited value in quantitatively predicting gross toxicity or cytostatic activity in general, probably due to transport differences (Ueno, 1977a). Ueno and co-workers have made extensive use of the rabbit reticulocyte assay to guide the fractionation of new agents from natural sources. In view of the marked lack of antibiotic activity of these compounds, this has proven quite useful. A discussion of assay methods follows (Section II,D).

B. Metabolism

To date, only a few papers have appeared concerning the metabolic fate of trichothecanes in mammals. Ueno *et al.* (1971b) have shown that fusarenon-x (**13b**) is rapidly distributed to tissues and also rapidly eliminated via the kidneys in the mouse. Examination of the urine from dosed animals indicated that the majority of the label resided not in intact fusarenon-x, but rather in a more polar metabolite. This was later shown to consist of nivalenol (**13a**), in which the 4-acetyl function was removed (Ueno, 1977a). Similarly it has been shown that the major metabolite of T-2 toxin (**12t**) was HT-2 toxin (**12s**), in which the 4-acetyl function has been removed (Ueno, 1977a). Ellison and Kotsonis (1974) have shown that T-2 toxin is hydrolyzed by liver esterases in a number of species, including man. In all cases the 4-β ester was hydrolyzed.

C. Toxicity

As a group, the trichothecanes exhibit a broad range of activities, including radiomimetic, dermatological, phytotoxic, insecticidal, and hematological effects. Bamburg and Strong (1971) have reviewed much of the early literature. In Table III are listed the LD_{50}s of a number of trichothecanes in rats and mice. For comparative purposes, the ID_{50}s of the reticulocyte protein-synthesis inhibition assay as well as the cytotoxicity assay on HEp2 cells are included. As is readily seen, while the correlation is not exact there is a rough parallel between LD_{50} and both cytotoxicity and protein-synthesis inhibition.

The toxic manifestations of trichothecane poisoning have been studied for a number of these compounds, most notably T-2 toxin, anguidine (4β,15-diacetoxy-3α-hydroxy-12,13-epoxytrichothec-9-ene), nivalenol, and fusarenon-x

TABLE III

Selected Biological Activities of Trichothecanes

Compound	Toxicity LD$_{50}$ (mg/kg) in rat* or mouse		ID$_{50}$[a] protein synthesis (μg/ml)	Cytotoxicity[b] HEp2 cells (ng/ml)	Tetrahymena[a] pyriformes ID$_{100}$ (μg/ml)
Verrucarin A	0.5	(ip)	0.01	1	
Verrucarin B	7.0	(iv)		5	
Verrucarin J	0.5	(ip)			
Roridin A	0.0	(iv)	0.01		
Roridin H					
Trichodermol				25	
Trichodermin				250	
Verrucarol	500	(sc)	1.0	75	
Scirpentriol				7000	
15-Acetoxyscirpendiol	0.81	(ip)*		75	
4,15-Diacetoxyscirpenol	0.75	(ip)*		2.5	
3,4,15-Triacetoxyscirpene	1.1	(ip)*	0.03	5.0	0.05
9,10-Dihydro-4,15-diacetoxyscirpenol	18	(ip)*		25	
T-2 Toxin	3.0	(ip)	0.03	100	
HT-2 Toxin	9.0	(ip)	0.03	1	0.05
Neosolaniol	14.5	(ip)	0.25		0.5
12y	5	(ip)	0.4		0.5
12z	1.02	(ip)			
12aa				10	
Trichothecolone	100	(iv)	20.0	5000	
Trichothecin	300	(iv)	0.15	75	
Nivalenol	4.0	(ip)	3.0	225	25
Fusarenon-X	3.3	(ip)	0.25		5
4,15-Diacetoxynivalenol	9.0	(ip)	0.10	30	1
Tetraacetoxynivalenol				250	
4-Deoxynivalenol	70	(ip)			
3-Acetoxy-4-deoxynivalenol	46	(ip)			5
Crotocin	810	(ip)	<1.0	250	150
Crotocol	200	(ip)			

[a] Ueno (1977a).
[b] Grove and Mortimer (1969).

(Rusch and Stahelin, 1965; Stahelin *et al.*, 1968; Uneo *et al.*, 1971a; Sato *et al.*, 1975; Uneo, 1977a,b; Marasas *et al.*, 1969).

The toxicity of anguidine has been studied in greater detail because of interest in it as an antitumor agent. This was facilitated by the independent discovery of a high-yielding fungus in a government-supported antitumor antibiotic program (NIH contract #PH43-64-1159) at Bristol Laboratories in 1968 (W. T. Bradner, J. A. Bush, and D. E. Nettleton, unpublished research). The material isolated was originally thought to be novel because it had no recognizable antibiotic activity. However, chemical analyses established that it was anguidine, which had significant activity on L-1210 and P-388 leukemia in mice. A sample was submitted to the National Cancer Institute, U.S.A., in 1970, and the tumor-inhibiting effects were confirmed in many other laboratories and in other systems. In August 1971, the National Cancer Institute determined that the antitumor activity was of sufficient interest to warrant consideration for clinical trial and requested scaled-up production and dose formulation. Preclinical toxicological studies were performed by the International Research and Development Corporation, Mattawan, Michigan, in dogs and monkeys. The side effects observed were:

1. *Gastrointestinal* toxicity was manifested by vomiting and diarrhea, usually within one hour of treatment.

2. *Hepatic* toxicity was suggested by elevated SGOT, SGPT, and serum alkaline phosphatase levels, although no histopathology was seen in the liver.

3. *Cardiovascular* toxicity was manifested as erythema and dehydration.

4. *Hematopotetic* toxicity was evidenced by anemia with the presence of nucleated red blood cells in the peripheral circulation.

5. *Lymphoid system* toxicity was observed as lymphopenia with lymphoid necrosis in the spleen, lymph nodes, and intestinal tract.

The highest nontoxic dose of drug given daily for five days was 0.32 mg/m^2 for beagle dogs and 1.5 mg/m^2 for rhesus monkeys. The lethal doses were 5.0 mg/m^2 and 6.0 mg/m^2 for dogs and monkeys, respectively.

Phase I studies in advanced cancer patients have been conducted at two clinical centers under the auspices of the NCI. Murphy *et al.* (1976) observed nausea and vomiting as well as chills and fever in many patients treated with 2 to 5 mg/m^2 intravenously daily for 5 days. Myelosuppression was universal among those receiving 5 mg/m^2. This group recommended an upper limit of 4–5 mg/m$^2 \times 5$ every two to three weeks, with close observation of renal function and hematologic parameters. Goodwin and co-workers (1978) found that hypotension, CNS, and gastrointestinal toxicity could be reduced by using 4–8-hour i.v. infusion rather than rapid infusion (5–10 min). No antitumor effects were reported in these studies.

Lindenfelser *et al.* (1974) studied skin tumor induction in mice and found that anguidine or T-2 toxin failed to induce papillomas, whereas aflatoxin B was tumorigenic. Anguidine did act as a weak promoter in mouse skin treated with 7,12-dimethylbenzanthracene (DMBA), but not in mice treated with aflatoxin B_1. Anguidine was not mutagenic in the Ames test using five strains of *Salmonella*, both with and without activation (Bradner, 1978). Similar results were reported by Wehner *et al.* (1978).

Similarly, Ueno *et al.* (1977) failed to demonstrate mutagenicity for fusarenon-x and T-2 toxin, although crotocin was a weak mutagen in this test. The authors suggested that the diepoxide in crotocin might be responsible for the activity noted.

D. Bioassay Methods

In the course of the isolation of the naturally occurring trichothecanes, a variety of bioassay techniques have been used. While these compounds generally lack antibacterial activity, they have been shown to possess antifungal and antiprotozoal effects. Ueno and Yamakawa (1970) have used the activity against *Tetrahymena pyriformis* as a method for following isolation of these agents and ranking their potency. They report a good correlation with the rabbit reticulocyte protein-synthesis inhibition bioassay as well as with a number of other tests. Other tests which have been used are the rat skin irritation, insect feeding inhibition, and phytotoxicity tests (Bamburg and Strong, 1971). A number of these tests are rather difficult to employ and are probably of limited relevance to ranking antitumor agents quantitively.

Bioassays based on inhibition of cell growth in tissue culture have been used extensively using both normal cells (baby hamster kidney cells, BHK) and cells derived from tumor tissue (KB, P815, HeLa, HEp2, and Ehrlich ascites cells). By far the greatest number of compounds to be tested in any one cell line were reported by Grove and Mortimer (1969), using BHK and HEp2 cells. The results reported by these workers are summarized in Table III (Section II,C). In addition, results for a limited series of compounds against HeLa cells have been reported by Ueno (1977a). These generally parallel those seen for HEp2 cell line. By far the most potent agents in cell tissue culture have been the macrocyclic antitumor agents, a number of which are inhibitory in the nanogram per milliliter range (Harri *et al.*, 1962). Typical values for the baccharin family of antitumor agents are 10^{-3}–10^{-4} mcg/ml (Kupchan *et al.*, 1977). Similar results are seen for vertisporin (Minato *et al.*, 1975).

The most extensive animal tumor testing has been performed under the auspices of the National Cancer Institute and more recently in our laboratories (Claridge *et al.*, 1978a). The P-388 lymphatic leukemia in mice which is the current NCI primary screen for natural products has been used most extensively. Several transplanted mouse tumors which make up the NCI tumor panel for critical evaluation of antitumor effects are: L-1210 leukemia, B16 melanoma, colon 26, colon 38 and Lewis lung carcinoma.

III. Structure–Activity Relationships

Despite the fact that the potential utility of the trichothecanes as antitumor agents was reported as early as 1962 (Harri et al., 1962), very few reports have appeared detailing the activities of this class of compounds. The majority of the data recorded in Tables IV and V have been made available by the U.S. National Cancer Institute (J. Douros, personal communication). Where other sources have been used, these are acknowledged in the footnotes to the tables. Data for the P-388 and L-1210 leukemias, as well as for one solid tumor (B16 melanoma), are included. While a number of agents have been tested in other tumors, the data are sparse and will be given in the text where relevant.

In the evaluation of data of this kind, several factors are important: activity, therapeutic ratio, and potency. The activity (T/C in survival-based models) is a measure of a drug's selectivity for tumor cells versus host cells; the higher the T/C, the better, all other things being equal. The therapeutic ratio is a measure of the drug's activity compared to its safety. For the most part, such data are not available. However, in our experience, those trichothecanes possessing high activity (T/C > 180) usually have quite acceptable therapeutic ratios. The potency is simply a measure of the amount of drug required to produce a desired biological effect. In this context, it is the dose to produce the optimum T/C.

Earlier work on the activity of the trichothecanes as dermatological agents has resulted in a number of general observations which may be valid in the context of antitumor activity (Bamburg and Strong, 1971). Reduction of the 9,10 double bond has been shown to result in great loss in potency, and removal of the 12,13-epoxy function gives complete loss of activity. In addition, Grove and Mortimer (1969) have demonstrated that the trichothecanes in which the 3,4 carbon–carbon bond is cleaved or in which rearrangement of the skeleton has occurred are inactive (in cell tissue culture).

Because there appears to be mechanistic differences (Section II,A) between the macrocyclic trichothecanes and the other trichothecanes, each class will be treated separately in the following subsections.

A. Macrocyclic Trichothecanes

The data for the macrocyclic trichothecanes are recorded in Table IV. An examination of the data permits one to draw some conclusions concerning the structure–activity relationships, both with respect to changes in the trichothecane nucleus and the macrocyclic portion. Since data for all the compounds in the

TABLE IV

Antitumor Effects of Macrocyclic Trichothecanes

Compound	NSC no.	Name	P-388 Dose[b]	P-388 T/C (%)	L-1210 Dose[b]	L-1210 T/C (%)	B16 Dose[b]	B16 T/C (%)
10a	200736	Verrucarin A	2.0	127	e	—	e	
10c	283445	Verrucarin A-9,10-epoxide	8.0	210	—	—	—	
10d		Verrucarin A-9,10-epoxide-2′-acetate[g]	5.0	172	—	—	—	
10f		Verrucarin B-9,10-epoxide	10.0	157	—	—	—	
10g	272704	Verrucarin J[f]	0.8	150	e	—	0.39	146
10h	200737	Roridin A	0.08	128	e	—	0.15	143
10i		Roridin D	e	—	e	—	—	
10l	269755	Roridin A-8β-hydroxy	3.75	205c	—	—	—	
10m	269759	Isororidin A-8β-hydroxy	1.25	157	—	—	—	
11a	274540	Roridin H	12.5	131	e	—	—	
10	269757	Baccharin	7.5	246d	40.0	157	6.25	139
10o	269756	Baccharinol	2.5	185	0.62	119	5.4	138h
10p	269758	Isobaccharinol[g]	2.5	166	—		2.5	170
10s		Baccharin 5	e	—	—		—	
10n	269753	Roridin A-2′-deoxy-4′-hydroxy-8β-hydroxy	0.75	175d	na		2.5	180

Tumor model[a]

[a] All data are for ip dosing using nine daily injections (QD 1 → 9) unless otherwise noted.
[b] mg/kg/injection.
[c] Average of two experiments.
[d] Average of three experiments.
[e] Inactive.
[f] J. French, Parke-Davis Co. (personal communication).
[g] B. Jarvis, University of Maryland (personal communication).
[h] QD 1, 5, 9.

TABLE V

Antitumor Effects of Nonmacrocyclic Trichothecanes

Compound	NSC no.	Name	Tumor model[a] P-388 Dose	P-388 T/C	L-1210 Dose	L-1210 T/C	B16 Dose	B16 T/C
12h	269142	Scirpene-3α,4β,15-triol	8.0	163	0.8	143	—	—
12i	298222	3α-Acetoxyscirpen-4β,15-diol	12.8	133	6.4	122	—	—
12j	281805	4β-Acetoxyscirpen-3α,15-diol	6.4	133	10.0	138	5.0	116
12k	267030	15-Acetoxyscirpen-3α,4β-diol	0.8	233	0.4	157	—	—
12l	283150	3α,4β-Diacetoxyscirpen-15-ol	1.6	128	3.2	121	—	—
12m	301462	3α-15-Diacetoxyscirpen-4β-ol	3.2	133	1.6	133	—	—
12n	141537	4β,15-Diacetoxyscirpen-3α-ol (Anguidine)	2.5	210[b]	0.25	166	1.0	141
12o	267031	3α,4β,15-Triacetoxyscirpene	0.8	144	1.6	167	—	—
12p	267032	9,10-Dihydro-4β,15-Diacetoxyscirpen-3α-ol	[e]	[e]	—	—	—	—
12t	138780	T-2 Toxin	1.5	228[c]	1.0	162	2.0	156
12s	278571	HT-2 Toxin	1.25	168[d]	4.0	147	1.0	156
12v	197212	Neosolaniol	2.5	184[d]		—	2.5	167
13b	197211	Fusarenon-X	0.8	165[d]		—	1.6	146
13a	269143	Nivalenol		—	1.5[f]	159	—	—
13c	267034	Nivalenol diacetate	[e]	—	—	—	—	—
13d	267035	Nivalenol tetraacetate	[e]	—	—	—	—	—
13g	—	Dihydronivalenol	[e]	—	—	—	—	—
13e	267036	4-Deoxynivalenol 3-acetate		—	5.0[f]	221	—	—
12c	073846	Trichodermin	[e]	—	[e]	—	—	—
14a	283835		80	134	—	—	—	—
14b	283834		[e]	—	—	—	—	—

[a] All data are for ip dosing using nine daily injections (QD 1 → 9) unless otherwise indicated. The dose is in mg/kg/injection.
[b] Average of 12 experiments.
[c] Average of five experiments.
[d] Average of two experiments.
[e] Not active.
[f] Shirasu et al. (1969).

table are available in the P-388 leukemia model, the remarks which follow are with regard to this tumor alone. In general, however, these compounds which show good P-388 activity are usually also active in the L-1210 and B16 systems.

It is evident that conversion of the 9,10 double bond to an epoxide results in greatly increased activity, e.g., compare verrucarin A (NSC 200736) with its epoxide (NSC 283445). Similarly, a great increase of activity was also seen on introduction of an 8β-hydroxy function, e.g., roridin A (NSC 200737) versus 8β-hydroxy roridin A (NSC 269755). In both cases, however, the increase in activity is accompanied by a loss in potency. A comparison of the 8β-hydroxy modification with the 9,10-epoxy modification is also available, e.g., baccharin (NSC 269757) versus baccarinol (NSC 269756). In this case the 9,10-epoxido derivative appears to be the more active although baccharinol may possess greater B16 melanoma activity. While a direct comparison of baccharin with a compound which possesses a 9,10 double bond as well as the macrocycle present in baccharin is not available, it is notable that baccharin 5 is inactive (B. Jarvis, University of Maryland, personal communication).

As is evident from Fig. 8, the variations in the macrocyclic portion of the trichothecane macrocycles occur for the most part at positions 2′, 3′, 4′ and 7′. The presence of a double bond between 2′ and 3′, as in verrucarin J (R^3=H, R^4=R^5=O), confers greater potency and activity that when the 2′ position is hydroxylated, e.g., verrucarin A (R^1=OH, R^2=R^3=H, R^4=R^5=O). Acetylation of the 2′-hydroxyl group results in a slight potency increase accompanied by a loss in activity, e.g., verrucarin A 9,10-epoxide versus verrucarin A 9,10-epoxide-2′-acetate. When the 2′-oxygen is tied back into a 2′,3′-epoxide function, an even less active compound results, e.g., verrucarin A 9,10-epoxide versus verrucarin B 9,10-epoxide (R^1=R^2=O, R^3=H, R^4=R^5=O). Substitution of the 2′, 3′-epoxido function as in baccharinol [R^1=R^2=O, R^3=OH, R^4=H, R^5=CH(OH)CH$_3$] by hydrogen, as in roridin A-2′-deoxy-4′-8-dihydroxy, results in a less potent, although equiactive compound. Unfortunately, these variations do not occur in a single series with a common nucleus and thus may be of only limited value in predicting the effects of macrocycle substitution on activity and potency.

Comparison of verrucarin A (R^4=R^5=O) with roridin A (R^4=H, R^5=CHOHCH$_3$) indicates that the latter compound is considerably more potent, although neither has more than borderline activity. The configuration at $C_{7'}$ in roridin A is R, and from Table IV a comparison of both configurations about the

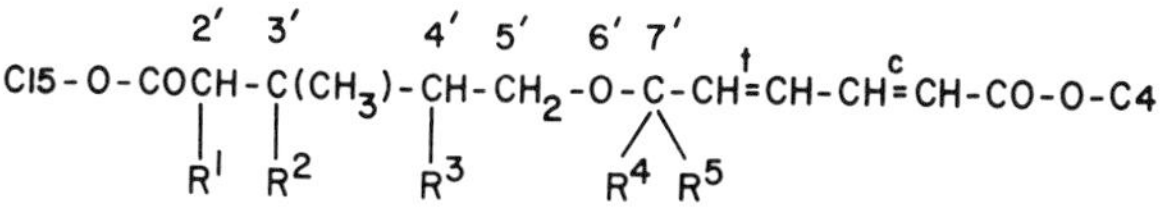

$$\overset{2'\quad 3'}{} \quad \overset{4'\ 5'}{} \quad \overset{6'\ 7'}{}$$
$$\text{C15-O-COCH-C(CH}_3\text{)-CH-CH}_2\text{-O-C-CH}\overset{t}{=}\text{CH-CH}\overset{c}{=}\text{CH-CO-O-C4}$$
$$\underset{R^1}{|}\quad\underset{R^2}{|}\qquad\underset{R^3}{|}\qquad\overset{/\backslash}{\underset{R^4\ R^5}{}}$$

Figure 8. Variations in the C_4–C_{15} bridge of the macrocyclic trichothecanes.

chiral center of the hydroxyethyl sidechain is available (e.g., 8β-hydroxy roridin A versus 8β-hydroxy isororidin A and baccharinol versus isobaccharinol.

The 7'R,8'R configuration appears to confer slightly greater activity than the 7'R,8'S configuration, although in both examples the potency remains the same. When the $C_{8'}$ hydroxyl group is linked through an ether linkage to $C_{5'}$ a loss in potency is observed, e.g., roridin A versus roridin H.

B. Trichothecane Polyols

The data for the polyoxygenated trichothecanes not having a macrocyclic bridge are recorded in Table V. In this instance, data on the effects of different oxygen substitution patterns and the effect of acetylation of the hydroxyl functions are available. The compound which has received the most attention as an antitumor agent is anguidine (4β,15-diacetoxyscirpen-3α-ol). Recently, all possible acetates of scirpenetriol have been described by Claridge and Schmitz (1979), as well as their antitumor activity (Claridge *et al.,* 1978, 1979). Structure–activity relationships on the effect of trichothecanes in the rabbit reticulocyte protein synthesis assay (Section II,A) had suggested that substitution of the 3α-hydroxyl results in loss of potency. A similar result is observed for the antitumor potency and activity in the present series. All derivatives of scirpenetriol containing the 3-acetyl function showed much reduced potency and activity. It was of interest that the most potent compound in the series was 15-acetyl scirpenediol—this compound being two-fold to four-fold more potent than anguidine itself. In view of the fact that 4-acetyl trichothecanes have been shown to undergo rapid cleavage in the presence of liver esterases (Section II,B), the possibility that the 15-acetoxy-scirpen-4β-3α-diol is responsible for the antitumor efficacy of anguidine must be considered.

The lack of activity seen for the 9,10-dihydroanguidine bears out the observation that an intact double bond at this position may be necessary for optimal activity.

Incorporation of an 8-α-valeryloxy function into anguidine resulted in a compound, T-2 toxin, which is slightly more active and potent. In this instance, however, removal of the 4β-acetoxy function resulted in a less active analog, HT-2 toxin. Hydrolysis of the 8α-valeryloxy function also gave a less active compound, neosolaniol.

The 8-keto-7-α-hydroxy compound fusarenon-x was active against P-388 leukemia at a dose comparable to the 15-acetoxy-scirpen-3α-4β-diol, but with much less activity. Removal of the 15-acetyl function to give nivalenol resulted in complete loss of activity, while addition of a 4-acetyl function to fusarenon-x gave an inactive compound as well. The tetraacetate of nivalenol was also inactive. It is notable that nivalenol has been reported to be active against L-1210 leukemia and that reduction of the 9,10 double bond, while it does result in a lowering in the potency, gives a more active compound (Shirasu *et al.*, 1969).

The simple trichothecane, trichodermin (Fig. 9), was inactive on L-1210 leukemia. Marginal activity for the totally synthetic analog (**15a**) was observed while the isomer (**15b**) was inactive.

C. Conclusions

The differences noted between the macrocyclic trichothecanes and the nonmacrocyclic trichothecanes, on the basis of the reticulocyte protein synthesis bioassay, appear to be supported by the results of *in vivo* tumor models. In the macrocyclic case, 9,10-epoxidation or 8β-hydroxylation appears to confer high activity albeit at the expense of potency. Variations in the sidechain also appear to affect the potency and activity, with a slight advantage seen for the 7'-hydroxyethyl sidegroup over the 7'-carbonyl. Unfortunately, the *in vivo* tumor data are available for only 14 of the 27 known macrocyclic trichothecanes.

Earlier, Ueno *et al.* (1973a) had noted differences between the 8-keto-7α-hydroxylated series and the other polyhydroxylated tricothecanes, and suggested that these might well differ in their transport properties. The lack of activity of nivalenol, its 4α,15-diacetate, and the poorer activity of fusarenon-x (relative to 15-acetoxyscirpen-3α,4β-diol), would seem to indicate that this substitution pattern confers less activity. This is probably not due to poorer absorption, since fusarenon-x has been shown to be rapidly absorbed whether given orally, subcutaneously, intravenous, or interperitoneally (Uneo *et al.*, 1971b). The observation that these compounds are much more active in cell-free systems in the reticulocyte protein synthesis bioassay (Ueno *et al.*, 1973a) would implicate cell penetration as the problem.

It is evident that the trichothecanes have been inadequately studied in mouse tumor models to permit any sweeping conclusions. Although literally hundreds of derivatives have been prepared in the course of structure elucidation studies, very few of these have been tested. A systematic study of activity variations with structural change is needed, especially in the macrocyclic series.

A number of synthetic studies on the trichothecanes have been carried out.

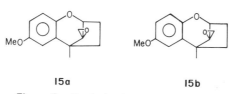

15a   15b

Figure 9.   Synthetic trichothecane analogues.

Colvin *et al.* (1971, 1973) have reported the total synthesis of trichodermin and, more recently, have described an approach to the synthesis of verrucarol (Colvin *et al.*, 1978). The parent compound, 12,13-epoxytrichothec-9-ene (**1**), has been synthesized by Fujimoto *et al.* (1974). A biomimetic synthesis of (**1**) has also been reported by Masuoka and Kamikawa (1976). In addition to these successful syntheses, a number of other authors have reported work in progress on the synthesis of trichothecanes (Goldsmith *et al.*, 1973; Welch and Wong, 1972a,b; Snider and Amin, 1978; Trost and Rigby, 1978). While these studies may lead to the syntheses of a limited number of trichothecane analogues, it is unlikely that an extensive structure–activity relationship can be developed on the basis of totally synthetic analogues with the complexity of the natural product models. In what might prove to be a more successful synthetic approach, a number of workers have begun to explore the synthesis of analogs incorporating some features of the natural product. Anderson *et al.* (1977) have reported the synthesis of the epoxide (**15a**) which retains marginal P-388 activity (Table V). A very simple series of analogues has been reported by Fullerton *et al.* (1976), in which only the spiro epoxide is retained (Fig. 10). A number of these compounds showed marginal P-388 activity at very high doses. The most active analogue was (**16**), followed by (**17b**) and the monomethyl derivatives (**17c**) and (**17d**). None of the analogues having the ''wrong'' epoxide configuration as in (**18**) were active. The authors noted that the more sterically hindered epoxides (**17b–d**) were active, while (**17a**) was inactive. These results show that the synthesis of analogues possessing only part of the basic trichothecane skeleton may be a useful approach.

A systematic study using one or more of the natural products as a starting point and varying the substitution at all accessible positions should be carried out and would probably be most useful in the establishment of structure–activity relationships. This could lead to synthesis of an optimal drug for use against neoplastic disease in man.

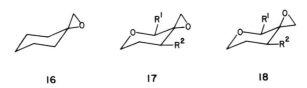

Figure 10.   Synthetic trichothecane analogues.

Acknowledgments

The authors are grateful to J. French and B. Jarvis, who kindly supplied data on research in progress. We also thank J. Douros for supplying much of the *in vivo* tumor data.

REFERENCES

Abrahamsson, S., and Nilsson, B. (1964). *Proc. Chem. Soc., London* 188.

Abrahamsson, S., and Nilsson, B. (1966). *Acta Chem. Scand.* **20**, 1044.

Anderson, W. K., LaVoie, E. J., and Lee, G. E. (1977). *J. Org. Chem.* **42**, 1045.

Bamburg, J. R. (1972). *Clin. Toxicol.* **5**, 495.

Bamburg, J. R. (1976). *In* "Mycotoxins and Other Fungal Related Food Problems" (J. V. Rodricks, ed.), Advances in Chemistry Series, Vol. 149, pp. 144–162. Am. Chem. Soc., Washington, D.C.

Bamburg, J. R., and Strong, F. M. (1969). *Phytochemistry* **8**, 2405.

Bamburg, J. R., and Strong, F. M. (1971). *In* "Microbial Toxins" (S. Kadis, A. Ciegler, and S. J. Ajl, eds.), Vol. VII, pp. 207–292. Academic Press, New York.

Bamburg, J. R., Riggs, N. V., and Strong, F. M. (1968). *Tetrahedron* **24**, 3329.

Bamburg, J. R., Strong, F. M., and Smalley, E. B. (1969). *J. Agric. Food Chem.* **17**, 443.

Barbacid, M., and Vazquez, D. (1974). *Eur. J. Biochem.* **44**, 437.

Barbacid, M., Fresno, M., and Vazquez, D. (1975). *J. Antibiot.* **28**, 453.

Blight, M. M., and Grove, J. F. (1974). *J.C.S. Perkin I* 1691.

Bohner, B., and Tamm, C. (1966). *Helv. Chim. Acta* **49**, 2527.

Bohner, B., Fetz, E., Harri, E., Sigg, H. P., Stoll, C., and Tamm, C. (1965). *Helv. Chim. Acta* **48**, 1079.

Bradner, W. T. (1978). *Antibiot. Chemother. (Basel)* **23**, 4.

Brian, P. W., Dawkins, A. W., Grove, J. E., Hemming, H. G., Lowe, D., and Norris, G. L. F. (1961). *J. Exp. Bot.* **12**, 1.

Cannon, M., Smith, K. E., and Carter, C. J. (1976a). *Biochem. J.* **156**, 289.

Cannon, M., Jiminez, A., and Vazquez, D. (1976b). *Biochem. J.* **160**, 137.

Carrasco, L., Barbacid, M., and Vazquez, D. (1973). *Biochim. Biophys. Acta* **312**, 368.

Carter, C. J., and Cannon, M. (1977). *Biochem. J.* **166**, 399.

Carter, C. J., and Cannon, M. (1978). *Eur. J. Biochem.* **84**, 103.

Carter, C. J., Cannon, M., and Smith, K. E. (1976). *Biochem. J.* **154**, 171.

Ciegler, A. (1975). *Lloydia* **38**, 21.

Ciegler, A. (1978). *J. Food Prot.* **41**, 399.

Claridge, C. A., and Schmitz, H. (1978). *Appl. Environ. Microbial.* **36**, 63.

Clardige, C. A., and Schmitz, H. (1979). *Appl. Environ. Microbiol.* **37**, 693–696.

Claridge, C. A., Bradner, W. T., And Schmitz, H. (1978). *J. Antibiot.* **31**, 485.

Claridge, C. A., Bradner, W. T., and Schmitz, H. (1979). *Cancer Chemother. Pharmacol.* **2**, 181–182.

Colvin, E. W., Raphael, R. A., and Roberts, J. S. (1971). *J.C.S. Chem. Commun.* 858.

Colvin, E. W., Malchenko, S., Raphael, R. A., and Roberts, J. S. (1973). *J.C.S. Perkin I* 1989.

Colvin, E. W., Malchenko, S., Raphael, R. A., and Roberts, J. S. (1978). *J.C.S. Perkin I* 658.

Cundliffe, E., and Davies, J. E. (1977). *Antimicrob. Agents Chemother.* **11**, 491.

Cundliffe, E., Cannon, M., and Davies, J. (1974). *Proc. Natl. Acad. Sci. U.S.A.* **71**, 30.

Ellison, R. A., and Kotsonis, F. N. (1974). *Appl. Microbiol.* **27**, 423.

Eppley, R. M., and Bailey, W. J. (1973). *Science* **181**, 758.

Eppley, R. M., Mazzola, E. P., Highet, R. J., and Bailey, W. J. (1977). *J. Org. Chem.* **42**, 240.

Fetz, E., Bohner, B., and Tamm, C. (1965). *Helv. Chim. Acta* **48**, 1669.

Fishman, J., Jones, E. R. H., Lowe, G., and Whiting, M. C. (1960). *J. Chem. Soc.* 3948.

Forgacs, J., and Carll, W. T. (1962). *Adv. Vet. Sci.* **7**, 273.

Freeman, G. G. (1955). *J. Gen. Microbiol.* **12**, 213.

Freeman, G. G., and Morrison, R. I. (1948). *Nature (London)* **162**, 30.

Freeman, G. G., Gill, J. E., and Waring, W. S. (1959). *J. Chem. Soc.* 1105.

Fresno, M., Carrasco, L., and Vazquez, D. (1976). *Eur. J. Biochem.* **68**, 355.

Fujimoto, Y., Yokura, S., Nakamura, T., Morikawa, T., and Tatsuno, T. (1974). *Tetrahedron Lett.*, 2523.

Fullerton, D. S., Chen, C. M., and Hall, I. H. (1976). *J. Med. Chem.* **19**, 1391.

Gardner, D., Glen, A. T., and Turner, W. B. (1972). *J.C.S. Perkin I*, 2576.

Ghosal, S., Chakrabarti, D. K., and Chaudhary, K. C. B. (1976). *J. Pharm. Sci.* **65**, 160.

Gilgan, M. W., Smalley, E. B., and Strong, F. M. (1966). *Arch. Biochem. Biophys.* **114**, 1.

Gláz, E. T., Scheiber, E., Gyimesi, J., Horvath, I., Steczek, K., Szentirmai, A., and Bohus, G. (1959). *Nature (London)* **184**, 908.

Gláz, E. T., Scheiber, E., and Járfás, K. (1960). *Acta Physiol. Acad. Sci. Hung.* **18**, 225.

Gláz, E. T., Csany, E., and C: imesi, J. (1966). *Nature (London)* **212**, 617.

Godtfredsen, W. O., and Vangedal, S. (1964). *Proc. Chem. Soc., London* 188.

Godtfredsen, W. O., and Vangedal, S. (1965). *Acta Chem. Scand.* **19**, 1088.

Godtfredsen, W. O., Grove, J. F., and Tamm, C. (1967). *Helv. Chim. Acta* **50**, 1666.

Goldsmith, D. J., Lewis, A. J., and Still, W. C., Jr. (1973). *Tetrahedron Lett.*, 4807.

Goodwin, W., Haas, C. D., Fabian, C., Heller-Bettinger, J., and Hoogstraten, B. (1978). *Cancer (Philadelphia)* **42**, 23.

Grove, J. F. (1970). *J. Chem. Soc. C* 378.

Grove, J. F., and Mortimer, P. H. (1969). *Biochem. Pharmacol.* **18**, 1473.

Guarino, A. M., Mendillo, A. B., and DeFeo, J. J. (1968). *Biotechnol. Bioeng.* **10**, 457.

Gutzwiller, J., and Tamm, C. (1963). *Helv. Chim. Acta* **46**, 1786.

Gutzwiller, J., Mauli, R., Sigg, H. P., and Tamm, C. (1964). *Helv. Chim. Acta* **47**, 2234.

Gyimesi, J., and Melera, A. (1967). *Tetrahedron Lett.*, 1665.

Hansen, B. S., and Vaughan, M. H., Jr. (1973). *Fed. Proc., Fed. Am. Soc. Exp. Biol.* **32**, 494.

Harri, E., Loeffler, W., Sigg, H. P., Stahelin, H., Stoll, C., Tamm, C., and Weisinger, D. (1962). *Helv. Chim. Acta* **45**, 839.

Helman, L. J., and Slavik, M. (1976). "Anguidine Clinical Brochure." Invest. Drug Branch, Natl. Cancer Inst., Bethesda, Maryland.

Hsu, I. C., Smalley, E. B., Strong, F. M., and Ribelin, W. E. (1972). *Appl. Microbiol.* **24**, 682.

Ilus, T., Ward, P. J., Nummi, M., Adlercruetz, H., and Gripenberg, J. (1977). *Phytochemistry* **16**, 1839.

Ishii, K. (1975). *Phytochemistry* **14**, 2469.

Ishii, K., Sakai, K., Ueno, Y., Tsunoda, H., and Enomoto, M. (1971). *Appl. Microbiol.* **22**, 718.

Ishii, K., Pathre, S. V., and Mirocha, C. J. (1978). *J. Agric. Food Chem.* **26**, 649.

Kotsonis, F. N., Ellison, R. A., and Smalley, E. B. (1975). *Appl. Microbiol.* **30**, 493.

Kupchan, S. M., Jarvis, B. B., Dailey, R. G., Jr., Bright, W., Bryam, R. F., and Shizuri, Y. (1976). *J. Am. Chem. Soc.* **98**, 7092.

Kupchan, S. M., Streelman, D. R., Jarvis, B. B., Dailey, R. G., Jr., and Sneden, A. T. (1977). *J. Org. Chem.* **42**, 4221.

Lansden, J. A., Cole, R. J., Dorner, J. W., Cox, R. H., Cutler, H. G., and Clark, J. D. (1978). *J. Agric. Food Chem.* **26**, 246.

Liao, L., Grollman, A. P., and Horowitz, S. B. (1976). *Biochim. Biophys. Acta* **454**, 273.

Lindenfelser, L. A., Lillcho, E. B., and Burmeister, H. R. (1974). *J. Natl. Cancer Inst.* **52**, 113.

Loeffler, W., Mauli, R, Ruesch, M. E., and Stahelin, H. (1967). Ger. Patent 1,233,089; Br. Patent 1,063,255.

Machida, Y., and Nozoe, S. (1972). *Tetrahedron* **28**, 5105, 5113.

Marasas, W. F. O., Bamburg, J. R., Smalley, E. B., Strong, F. M., Ragland, W. L., and Degurse, P. E. (1969). *Toxicol. Appl. Pharmacol.* **15**, 471.

Masuoka, N., and Kamikawa, T. (1976). *Tetrahedron Lett.*, 1691.

Matsumoto, M., Minato, H., Tori, K., and Ueyama, M. (1977a). *Tetrahedron Lett.*, 4043.

Matsumoto, M., Minato, H., Uotani, N., Matsumoto, K., and Kondo, E. (1977b). *J. Antibiot.* **30,** 681.

Mayer, C. F. (1953). *Mil. Surg.* **113,** 173, 295.

Minato, H., Katayama, T., and Tori, K. (1975). *Tetrahedron Lett.,* 2579.

Murphy, K., Burgess, M. A., Valdivieso, M., and Bodey, G. P. (1978). *Proc. Am. Assoc. Cancer Res.* **19,** 411.

Nespiak, A., Koćor, M., and Siewinski, A. (1961). *Nature (London)* **192,** 138.

Ohtsubo, K., Kaden, P., and Mittermayer, C. (1972). *Biochim. Biophys. Acta* **287,** 520.

Okuchi, M., Itoh, M., Kaneko, Y., and Doi, S. (1968). *Agric. Biol. Chem.* **32,** 394.

Pathre, S. V., Mirocha, C. J., Christensen, C. M., and Behrens, J. (1976). *J. Agric. Food Chem.* **24,** 97.

Rusch, M. E., and Stahelin, H. (1965) *Arzneim.-Forsch.* **15,** 893.

Sato, N., Uneo, Y., and Enomoto, M. (1975). *Japan J. Pharmacol.* **25,** 263–270.

Schindler, D. (1974). *Nature (London)* **249,** 38.

Schindler, D., Grant, P., and Davies, J. (1974). *Nature (London)* **248,** 535.

Shirasu, Y., Moritz, Y., Fujimoto, Y., Tatsuno, T., and Ueno, Y. (1969). *Chem. Pharm. Bull.* **17,** 406.

Sigg, H. P., Mauli, R., Flury, E., and Hauser, D. (1965). *Helv. Chim. Acta* **48,** 962.

Smith, K. E., Cannon, M., and Cundliffe, E. (1975). *FEBS Lett.* **50,** 8.

Snider, B. B., and Amin, S. G. (1978). *Synth. Commun.* **8,** 117.

Stahelin, H., Kalberer-Rusch, M. E., Signer, E., and Lazary, S. (1968). *Arzneim.-Forsch.* **18,** 989.

Steyn, P. S., Vieggaar, R., Rabie, C. J., Kreik, N. P. J., and Harrington, J. S. (1978). *Phytochemistry* **17,** 949.

Tamm, C. (1974). *Fortschr. Chem. Org. Naturst.* **31,** 63.

Tamm, C., and Gutzwiller, J. (1962). *Helv. Chim. Acta* **45,** 1726.

Tate, W. P., and Caskey, C. T. (1973). *J. Biol. Chem.* **248,** 7970.

Tatsuno, T. (1968). *Cancer Res.* **28,** 2393.

Trost, B. M., and Rigby, J. H. (1978). *J. Org. Chem.* **43,** 2940.

Ueno, Y. (1977a). *Pure Appl. Chem.* **49,** 1737.

Ueno, Y. (1977b). *Ann. Nutr. Aliment.* **31,** 885.

Ueno, Y., and Fukushima, K. (1968). *Experientia* **24,** 1032.

Ueno, Y., and Yamakawa, H. (1970). *Jpn. J. Exp. Med.* **40,** 385.

Ueno, Y., Hosoya, M., Morita, Y., Ueno, I., and Tatsuno, T. (1968). *J. Biochem. (Tokyo)* **64,** 479.

Ueno, Y., Ueno, I., Amakai, K., Ishikawa, Y., Tsunoda, H., Okubo, K., Saito, M., and Enomoto, M. (1971a). *Jpn. J. Exp. Med.* **41,** 507.

Ueno, Y., Ueno, I., Iitoi, Y., Tsunoda, H., Enomoto, M., and Ohtsubo, K. (1971b). *Jpn. J. Exp. Med.* **41,** 521.

Ueno, Y., Ishii, K., Sakai, K., Kanaeda, S., Tsunoda, H., Tanaka, T., and Enomoto, M. (1972a). *Jpn. J. Exp. Med.* **42,** 187.

Ueno, Y., Sato, N., Ishii, K., Sakai, K., and Enomoto, M. (1972b). *Jpn. J. Exp. Med.* **42,** 461.

Ueno, Y., Nakajima, M., Sakai, K., Ishii, K., Sato, N., and Shimada, N. (1973a). *J. Biochem. (Tokyo)* **74,** 285.

Ueno, Y., Sato, N., Ishii, K., Sakai, K., Tsunoda, H., and Enomoto, M. (1973b). *Appl. Microbiol.* **25,** 699.

Ueno, Y., Sato, N., Ishii, K., Tsunoda, H., and Enomoto, M. (1973c). *Appl. Microbiol.* **25,** 699.

Ueno, Y., Ishii, K., Sawano, M., Ohtsubo, K., Matsuda, Y., Tanaka, T., Kurata, H., and Ichinoe, M. (1977). *Jpn. J. Exp. Med.* **47,** 177.

Vesonder, R. F., Ciegler, A., and Jensen, A. H. (1973). *Appl. Microbiol.* **26,** 1008.

Vesonder, R. F., Ciegler, A., Jensen, A. H., Rohwedder, W. K., and Weisleder, D. (1976). *Appl. Microbiol.* **31,** 280.

Wehner, F. C., Marasas, W. F. O., and Thiel, P. G. (1978). *Appl. Environ. Microbiol.* **35,** 659.

Wei, C., and McLaughlin, C. S. (1973). *Fed. Proc., Fed. Am. Soc. Exp. Biol.* **32,** 494. (Abstr.)

Wei, C., and McLaughlin, C. S. (1974). *Biochem. Biophys. Res. Commun.* **57,** 838.

Wei, C., Campbell, I. M., McLaughlin, C. S., and Vaughan, M. H. (1974a). *Mol. Cell. Biochem.* **3,** 215.

Wei, C., Hansen, B. S., Vaughan, M. H., Jr., and McLaughlin, C. S. (1974b). *Proc. Natl. Acad. Sci. U.S.A.* **71,** 713.

Wei, R., Strong, F. M., Smalley, E. B., and Schnoes, H. K. (1971). *Biochem. Biophys. Res. Commun.* **45,** 396.

Welch, S. C., and Wong, R. Y. (1972a). *Tetrahedron Lett.,* 1853.

Welch, S. C., and Wong, R. Y. (1972b). *Synth. Commun.* **2,** 291.

Yates, S. G., Tookey, H. L., Ellis, J. J., and Burkhardt, H. J. (1968). *Phytochemistry* **7,** 139.

Yoshizawa, T., and Morooka, N. (1973). *Agric. Biol. Chem.* **37,** 2933.

Yoshizawa, T., and Morooka, N. (1975). *Appl. Microbiol.* **29,** 54.

CHAPTER 3

Nucleosides

MASAJI OHNO

I. INTRODUCTION

The nucleosides are undoubtedly closely connected with the most basic principles of all living systems as essential components of nucleic acids, and have been widely studied. Since the excellent review of nucleosides of microbial origin written by Suhadolnik (1970) was published, several comprehensive reviews or books have been written by experts in the field (Ikehara, 1969; Bloch, 1973; Daves and Cheng, 1976; Hanessian and Pernet, 1976; Robins, 1977; Mizuno, 1977; Saneyoshi, 1977). The design of biologically active nucleosides and the strategy and synthetic principles are also discussed in the recent literature. There are still a number of mechanistic, structural, and synthetic problems in this area which fascinate organic chemists and stimulate their imagination. However, relatively few academic chemists do research in this area compared with those found in more conventional areas such as alkaloids, terpenes, prostaglandins and β-lactam compounds.

One reason for the situation may be that the physical properties of nucleosides make them difficult to handle in comparison to the majority of organic compounds (Shapiro, 1977). Another reason may be that such a study is located in an interdisciplinary field of life science and requires close cooperation among biochemists, pharmacologists, biologists, and clinicians.

Now, it can be said that the chemistry of nucleosides is one of the most important areas to investigate, since it has been discovered that many potential anticancer agents are found among the nucleosides. Recently, extensive efforts have been directed toward the chemical synthesis of nucleosides based on an-

Anticancer Agents Based on Natural Product Models

timetabolite theory. The biological effects associated with metabolic processes and specific enzyme control mechanisms are diverse in naturally occurring nucleosides and synthetic analogues. Nucleosides exhibit several biological effects including antibiotic, anticancer and antiviral activity, cytokinin and immunosuppressive activities, cardiovascular and other effects (Bloch, 1973). Furthermore, it should be kept in mind that nucleoside analogues can assume other functional roles not as yet recognized, and further therapeutic application will be expected in the future.

This chapter covers very recent advances in the chemistry of nucleosides concerned mainly with anticancer and partly with antiviral activities from a synthetic point of view.

II. STRUCTURE-ACTIVITY RELATIONSHIPS

The wide-ranging biological activity of a number of natural and synthetic nucleosides makes it desirable to compare the biological effects of the compounds in relation to their structures. Such structure–activity relationships are summarized in Tables I, II, and III. The synthetic approaches to the nucleoside molecules are widely diversified, as shown in previous studies, and the starting point of the design is based on the choice of the prototype nucleoside possessing the special biological effect. The increasing interest in the chemistry of C-nucleosides is especially noteworthy and will be reviewed. Next, a variety of synthetic modification of N-nucleosides is reviewed. N-Nucleoside modifications can be grouped according to the following conventional classification: (1) modification of the base moiety and substituents; and (2) modification of the sugar moiety and substituents.

A. C-Nucleosides

A general procedure for the synthesis of pyrimidine C-5 nucleosides related to pseudouridine (6) was developed by Chu *et al.* (1976).

The key intermediate (3) was obtained from 2,3-O-isopropylidene-5-O-trityl-D-ribofuranose (1) through (2) and converted to the clinically important 5-(β-D-ribofuranosyl)isocytosine (4, pseudoisocytidine) and the α isomer, 5- (β- D-ribofuranosyl)- 2- thiouracil (5), and pseudouridine (6), the first C- nucleoside found in nature (Reichman *et al.*, 1976).

Pseudoisocytidine (4) shows significant antileukemic activities, especially against mouse leukemia resistant to ara-C, and is currently undergoing phase I clinical trials. Furthermore, compounds (4) and (6) were converted into 4,2'-anhydro-5- (β- D- arabinofuranosyl)isocytosine (7) and 4,2'- anhydro- 5- (β- D-arabinofuranosyl)uracil (8), respectively.

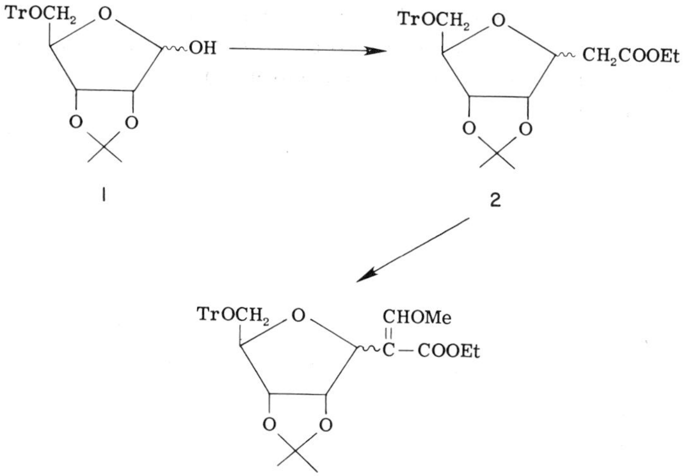
TrOCH2
O
OH
O
O
TrOCH2
O
CH2COOEt
O
O
TrOCH2
O
CHOMe
C
COOEt
O
O

1 2 3

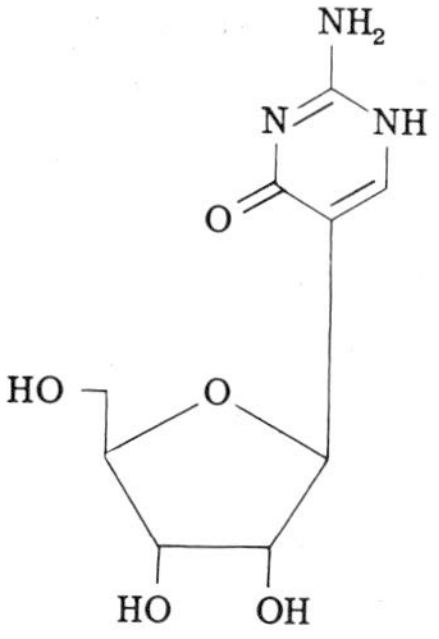
NH2
N
NH
O
HO
O
HO
OH

4

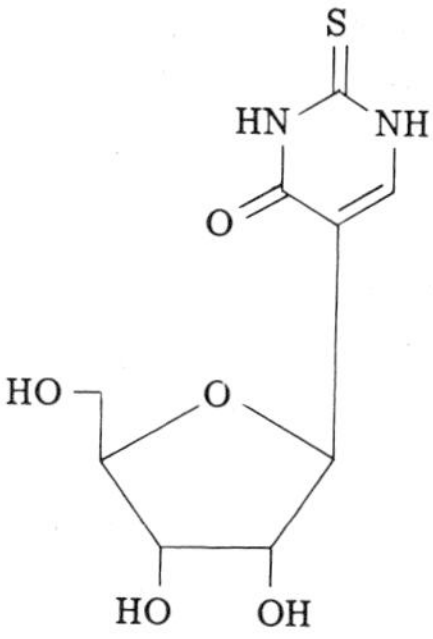
S
HN
NH
O
HO
O
HO
OH

5

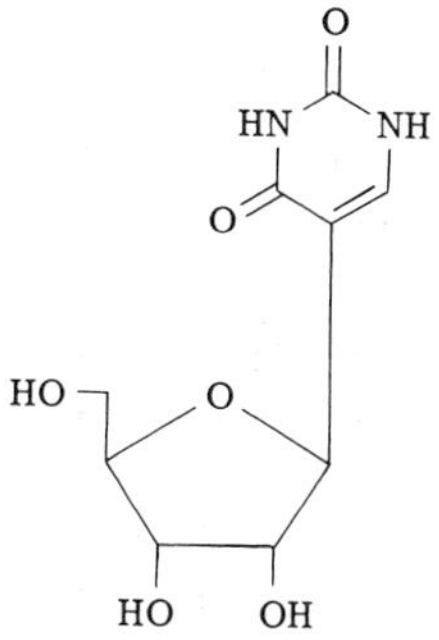
O
HN
NH
O
HO
O
HO
OH

6

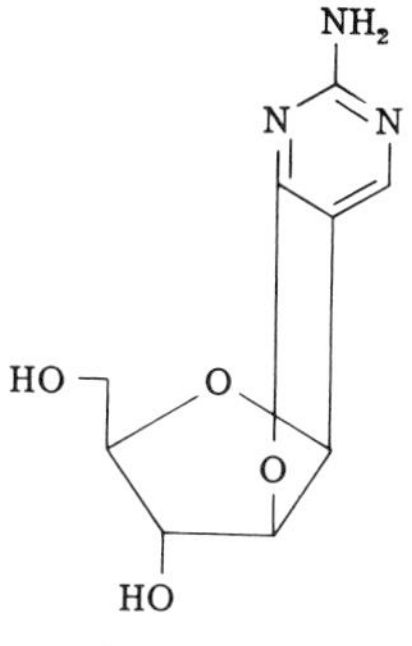
NH2
N
N
HO
O
O
HO

7

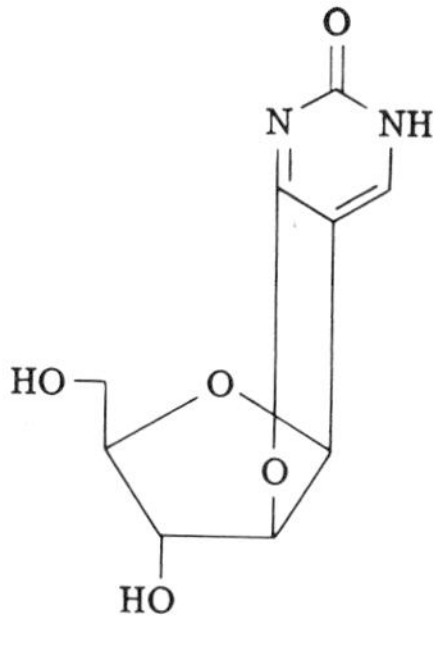
O
N
NH
HO
O
O
HO

8

2-(2,3-O-Isopropylidene-5-O-trityl-D-ribofuranosyl)-acetonitrile (**9**) was also used as a versatile intermediate to 2,4-diamino-5-(β-D-ribofuranosyl)pyrimidine (**10**) and 5-(β-D-ribofuranosyl)-2-thiocytosine (**11**, 2-thiopseudocytidine) (Chu *et al.*, 1977a).

    10   R = NH$_2$
    11   R = SH

However, Chu *et al.* (1978) reported that pseudo-ara-iso-C did not show any significant inhibitory activity against L-1210 cells *in vitro*. Other new nucleosides, (**8**), (**10**), and (**11**), also did not inhibit the growth of leukemic cells in culture.

Based on finding the antileukemic activity of pseudoisocytidine (**4**), 2'-deoxypseudouridine (**12**), 2'-deoxy-1-methylpseudouridine (**13**), and 2'-deoxypseudoisocytidine (**14**) were synthesized from pseudouridine (**6**) (Chu *et al.*, 1977b).

Preliminary studies of the 2'-deoxy-C-nucleosides against P-815 cells *in vitro* showed activities for **13** (an isostere of thymidine) and for **14** (an isostere of 2'-deoxycytidine), and no activity for **12**.

A unique synthesis of C-nucleosides has been elegantly developed by Noyori *et al.* (1978), using the polybromoketone-iron-carbonyl reaction as the key

    12   13   14

15   16

17   18   19   20

methodology. The starting material is a symmetrical bicyclic ketone (15), which is prepared easily from $\alpha,\alpha,\alpha',\alpha'$-tetrabromoacetone and furan. The ketone (15) was smoothly transformed to the lactone (16) with trifluoroperacetic acid, having a C-β-glycosidic structure. This product was hydrolyzed, resolved and relactonized. The optically active lactone (16) was then subjected to further transformation. Thus, the above method has been shown to be a general synthesis for various C-nucleosides by controlling the stereochemistry at the C-1' position, producing pseudouridine (6), 2-thiopseudouridine (5), pseudocytidine (17), pseudoisocytidine (4), showdomycin (18), and 6-azapseudouridines (19) and (20) (Sato et al., 1978).

A facile chemical synthesis of 1-methylpseudouridine (21) was accomplished by direct methylation of pseudouridine (6) (Earl and Townsend, 1977). It is a natural product, but characterization was limited to only chromatographic and ultraviolet spectral data.

A new synthesis of pyrazomycin (22) and pyrazomycin B (23) was reported (Bernardo and Weigele, 1976). The starting material is 2,3-O-isopropylidene-D-ribofuranose (α and β in a ratio of 1:9). It was condensed with diethyl acetonedicarboxylate, subjected to diazotization with p-toluenesulfonyl azide, and converted to pyrazomycin (22). Both pyrazomycin and its α epimer (23) showed antitumor and broad-spectrum antiviral activity.

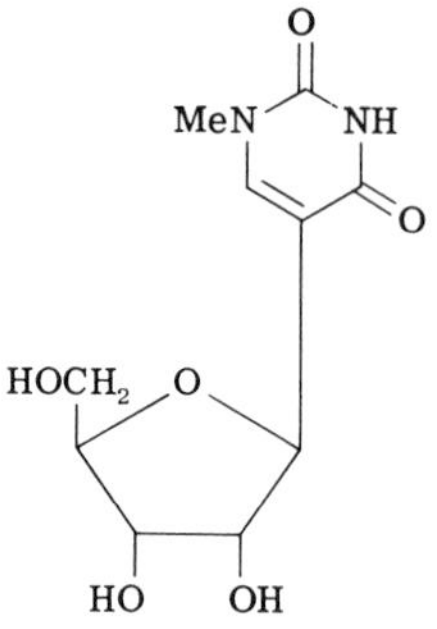

21

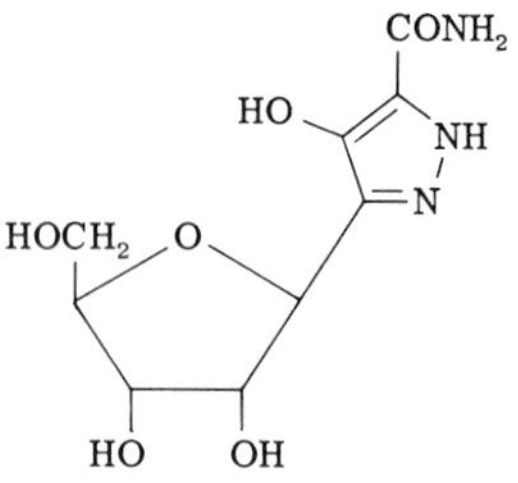

22

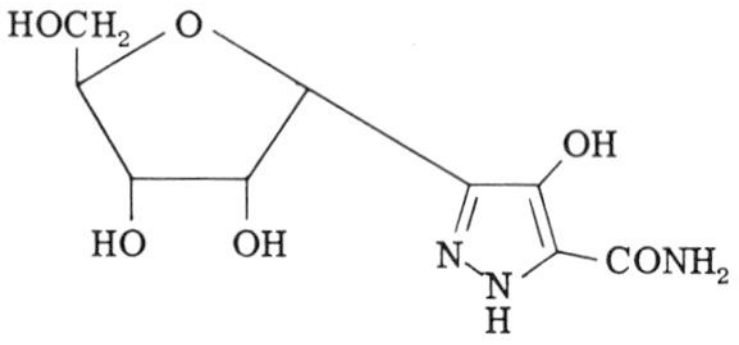

23

8-(β-D-Ribofuranosyl)pyrazolo[1,5-a]-1,3,5-triazine (**24**) isosteres of adenosine and inosine were synthesized. They proved to have moderate inhibitory
activity against mouse leukemia L-1210, L-5178Y, and P-815, and to be more
active than formycin (Tam *et al.*, 1976).

The synthesis of 4-amino-3-(β-D-ribofuranosyl)pyrazole-5-carboxamide (**25**),
a C- nucleoside analogue of AICA and a key intermediate for the synthesis of
bicyclic C-nucleosides, was accomplished by degradation of purine nucleosides
(Lewis *et al.*, 1976).

It has been shown that a β-ketoester (**26**) is a versatile intermediate which
might also serve as a precursor to various homo-C-nucleosides (**27**) (Secrist,
1978).

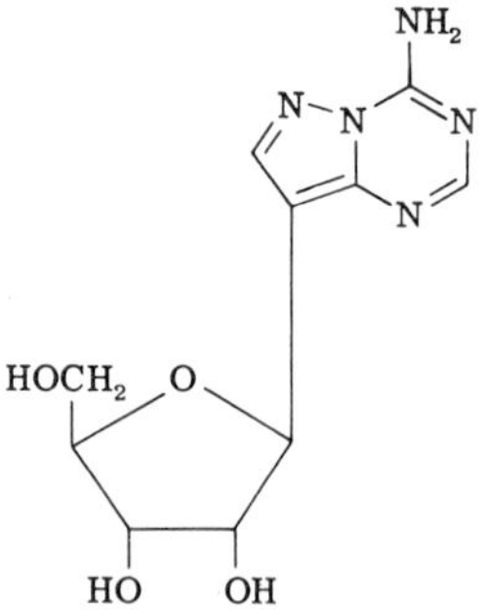

24

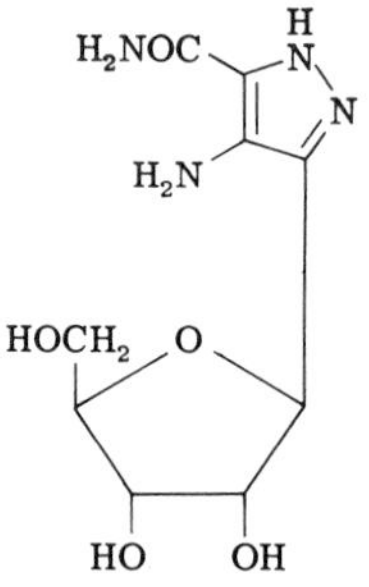

25

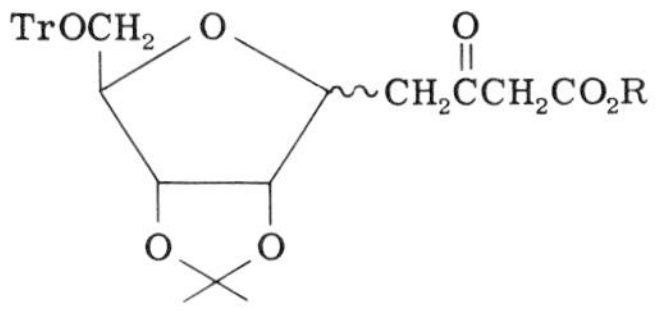

26

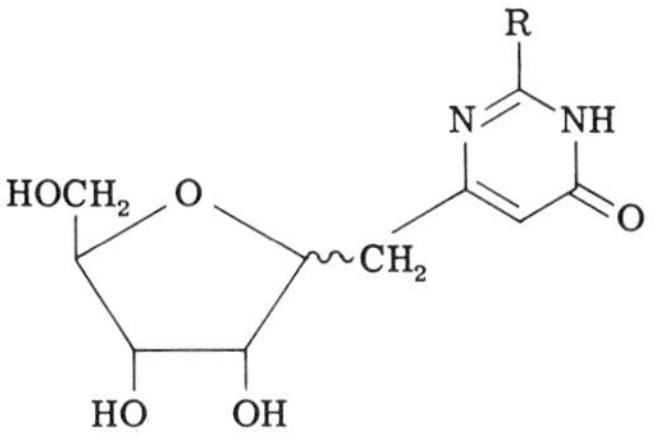

27

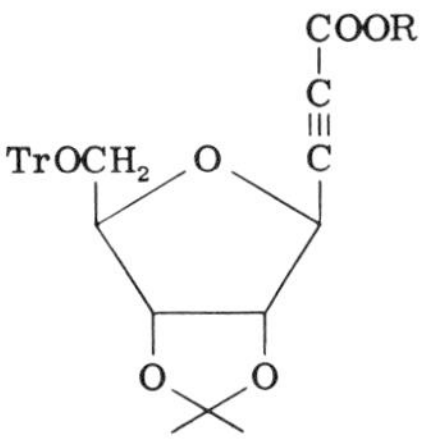

28

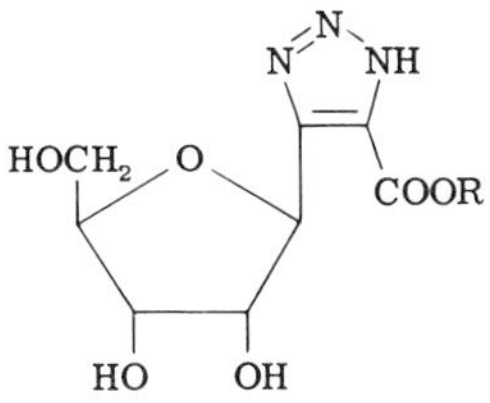

29

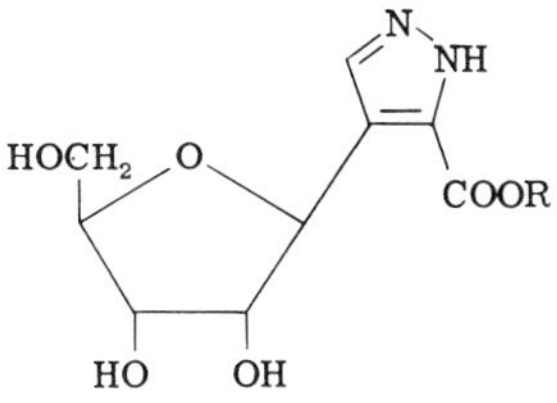

30

1,3-Dipolar cycloaddition of β-D-ribofuranosyl propiolate (**28**) with trimethylazide and diazomethane affords 1,2,3-triazole ester (**29**) and pyrazole ester (**30**), respectively (De las Heras *et al.*, 1976a).

Fox *et al.* further developed various C-nucleosides using ribofuranosyl acetylenic esters and ethyl 2-(D-ribofuranosyl)-2-formylacetate (De las Heras *et al.*, 1976b).

A C-nucleoside antibiotic oxazinomycin (**31**) has been synthesized starting with the 2′,3′-*O*-isopropylidene-5′-*O*-trityl-D-ribofuranosyl-acetonitriles (Bernardo and Weigele, 1977).

The structure and activity relationship of the C-nucleosides mentioned above is summarized in Table I, although most of them require further biological study.

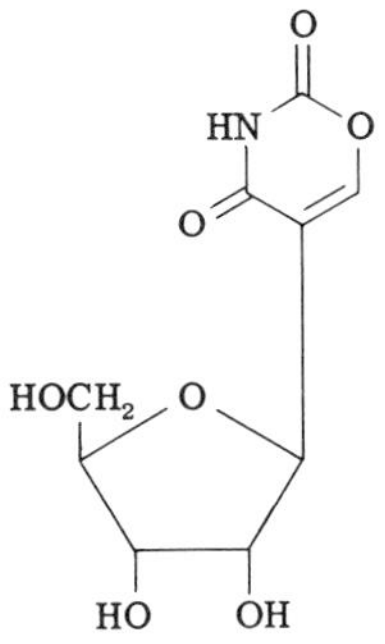

31

TABLE I

Structure–Activity Relationship of C-Nucleosides

Pyrimidines (pyrimidinone ring bearing R^1 at C-2, NR^2, R^3 at C-5, and =O)

R^1	R^2	R^3	Biological activity	Compound no.	References
NH_2	H	rib	Pseudoisocytidine) active against leukemic cells	4	Chu et al. (1978); Noyori et al. (1978); Hirota et al. (1978)
NH_2	H	ara	Inactive against leukemic cells		Reichman et al. (1976); Chu et al. (1978)
NH_2	H	2'-drib	$ID_{50} = 1.25$ µg/ml against P-815 cells	14	Chu et al. (1977b)
OH	CH_3	2'-drib	$ID_{50} = 4.9$ µg/ml against P-815 cells	13	Chu et al. (1977b)
OH	H	2'-drib		12	Bridges et al. (1977); Robins and Muhs (1978)
OH	H	rib	(Pseudouridine)	6	Chu et al. (1977); Noyori et al. (1978)
OH	H	ara	Inactive against leukemic cells		Reichman et al. (1976); Chu et al. (1978)
OH	H	2'-Cl-2'-drib	Inactive against leukemic cells		Reichman et al. (1976); Chu et al. (1978)
OH	H	3'-drib			Robins and Muhs (1978)
OH	H	rib		5	Chu et al. (1976); Noyori et al. (1978); Hirota et al. (1978)
SH	H	rib			
OH	CH_3	α,β-rib		21	Earl and Townsend (1977)

Pyrimidines (ring bearing R^1, two N, H_2N, and R^3)

R^1	R^2	R^3	Biological activity	Compound no.	References
NH_2		α,β-rib		10	Chu et al. (1977a)
NH_2		α,β-ara	Inactive against leukemic cells	17	Chu et al. (1978)
OH		ara			Chu et al. (1978)
OH		rib	(Pseudocytidine)		Chu et al. (1978); Noyori et al. (1978)
SH		α,β-rib		11	Chu et al. (1977a)

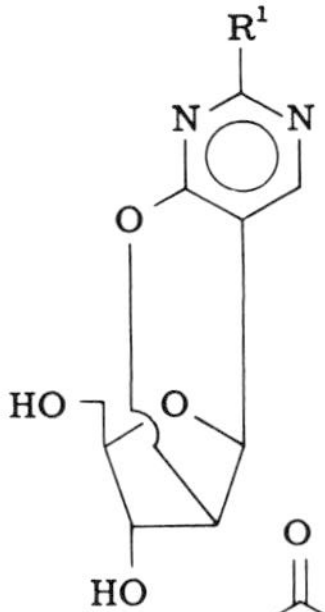

Structure			Compound no.	Biological activity	References
R^1	NH_2		7	Inactive against leukemic cells	Chu et al. (1978)
	OH		8	Inactive against leukemic cells	Chu et al. (1978)
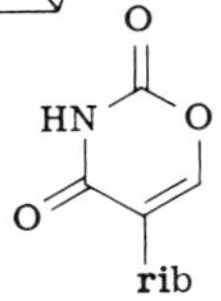 (rib structure)			31	(Oxazinomycin) inhibits transplantable tumors, antimicrobial against G(+), G(−)	Bernardo and Weigele (1977)

X	R	Compound no.	Biological activity	References
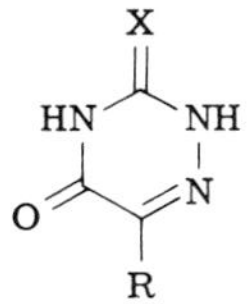 O	rib	19		Sato et al. (1978)
O	ara			Just and Ouellet (1976)
S	rib	20		Sato et al. (1978)
S	ara			Just and Ouellet (1976)

Purine	X	R	Compound no.	Biological activity	References
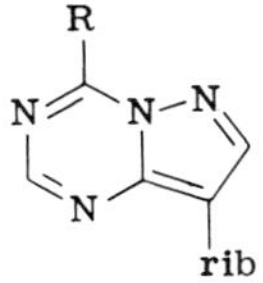 (Purine structure, rib)	NH_2		24	(ID$_{50}$ μg/ml): P-815, 0.3; L-1210, 0.1	Tam et al. (1976)
	OH				Tam et al. (1976)

5-Membered heterocycles

(*Continued*)

TABLE I (*Continued*)

Five-membered heterocycles	X	R	Compound no.	Biological activity	References
	O	OH	18	(Showdomycin) Active against Ehrlich ascites tumor, HeLa cells	Noyori et al. (1978)
	O	H			Just and Lim (1977)
	NCO_2CH_3	OH		Inactive against DNA, RNA virus	Just and Domini (1977)
	C	NH_2			De las Heras et al. (1976a); Just and Chalard-Faure (1976)
	C	OMe, OEt	30		De las Heras et al. (1976a); Just and Chalard-Faure (1976)
	N	OMe, NH_2	29		De las Heras et al. (1976a); Just and Chalard-Faure (1976)

Five-membered heterocycles (cont.)	X	R^1	R^2	Compound no.	Biological activity	References

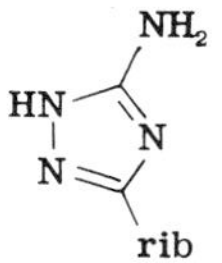

		X	R1	R2	Compound		References
OH	β-rib				**23**		Just and Chalard-Faure (1976)
OH	α-rib				**23**		Just and Chalard-Faure (1976)
OH	2'-drib						Bernardo and Weigele (1976)
NH₂	rib				**25**	Inactive against HSV, VV	Bernardo and Weigele (1976)
						Pyrazomycin A)	Just and Kim (1976)
						(Pyrazomycin B)	Lewis et al. (1976)
						Inactive against L-1210 at 10^{-4} M	

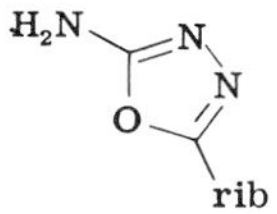

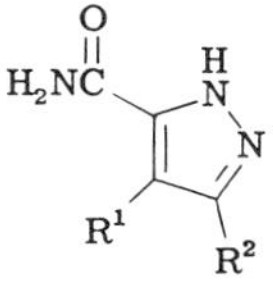

X	R1	R2	Compound	% Inhib.[b]	Virus rating			References
					HSV-1	PIV/3	RV/13	
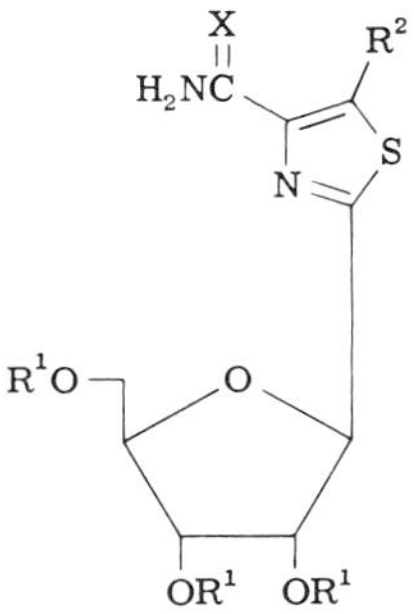 O	Ac	H		42	0.6	0.2	0.3	Srivastava et al. (1977b)
O	H	H	**83**	69	0.6	0.6	0.5	Srivastava et al. (1977b)
S	H	H		12	0.1	0.0	0.0	Srivastava et al. (1977b)
O	H	CONH₂	**82**					Fuertes et al. (1976)

(Continued)

83

TABLE I (*Continued*)

Five-membered heterocycles	R¹	R²	Compound no.	Biological activity	References
	H	OH	81		Huynh-Dinh et al. (1977); Poonian and Nowoswiat (1977)
	H	H	81	No antiviral activity	Huynh-Dinh et al. (1977); Poonian and Nowoswiat (1977)
	OH	H	81		Huynh-Dinh et al. (1977); Poonian and Nowoswiat (1977)
	H		27		Secrist (1978)
	CH₃				
	SH				
	NH₂				
	Ph				

[a] Abbreviations: Rib, D-ribose; ara, D-arabinose; 2′-drib, 2′-deoxy-D-ribose; 3′-drib, 3′-deoxy-D-ribose; α,β-rib, both anomers of D-riboside were obtained and separately evaluated; HSV-1, herpes simplex virus-1; VV, Vaccinia virus; HSF, human skin fibroblasts; PRK, primary rabbit kidney; HV/1, type 1 herpes virus; RV/13, type 13 rhino virus; PIV/3, type 3 parainfluenza virus; MDC, minimum degeneration concentration; H. Ep. No. 2, human epidermoid carcinoma No. 2 cells; ED₅₀, the concentration required to inhibit the growth of treated cells to 50% of that of untreated controls; ID₅₀, the concentration required to inhibit virus-induced cytopathogenicity by 50%; ILS, increase in life span.

[b] % Inhibition of guanidine nucleotide synthesis.

B. N-Nucleosides

The modification of the pyrimidine and purine heterocycles and their substituents is also under intense investigation. The following examples can provide an indication of the type and extent of activity associated with substituents on the base moiety, and certain patterns can be seen and will be helpful in guiding further synthesis, as exemplified in Table II.

1. Modification of the Base Moiety and Substituents

a. Pyrimidines   5-Formyl -2' -deoxyuridine (**32**), a potent inhibitor of thymidylate synthetase, was synthesized from 5-fluorouracil and a ribofuranosyl chloride in three steps (Kampf *et al.* 1976). 5-Acetyl-2'-deoxyuridine (**33**) was also prepared in a similar manner from 5-acetyluracil. Other derivatives (**34**) and (**35**) are also reported, although their biological effects are not yet investigated.

32   33   34 R = OMe
                                                35 R = N_3

3 -(β -D -Ribofuranosyl) -2,3 -dihydro -6H -1,3 -oxazine -2,6 -dione (**37**) was prepared starting from 2,3 -dihydro -6H -1,3 -oxazine -2,6 -dione (**36**, uracil anhydride). The uracil anhydride is a pyrimidine analogue with an oxygen isosteric replacement. Compounds (**36**) and (**37**) exert a moderate growth inhibition of mouse leukemia L -5178Y, and of HeLa and Novikoff hepatoma cells in culture, and produce weak inhibition of viral replication in HeLa cells (Chwang *et al.*, 1976).

A substitution reaction at the C -5 position of uridine and 2' -deoxyuridine was effected by reaction of olefins with organopalladium intermediates generated in situ from mercurinucleosides. Thus, 2' -deoxy -5 -(4,5 -dihydroxypentyl)uridine (**38**), methyl trans -3 -(5 -uridylyl)propenoate (**39**), and 5 -allyluridine (**40**) were prepared (Bergstrom and Ruth, 1976; Ruth and Bergstrom, 1978). 1,2 -Dihyo -1(2 -deoxy -β -D -erythropentofuranosyl) -2 -oxo -5 -methylpyrazine -4 -oxide (**41**) was synthesized by condensation of its corresponding base and sugar (Bobek and

TABLE II

Structure–Activity Relationship of N-Nucleosides

[Structure: pyrimidine N-nucleoside with substituents R¹ (position 5), R² and R³ on the deoxyribose ring]

Pyrimidines	R¹	R²	R³	Compound no.	Biological activity					References
					Thymidylate synthetase inhibitor					
					% Inhib. (μM)	Sarcoma 180	Vero cells	HSV-1	% Control HSV-1	
	CHO	H	OH	32						Kampf et al. (1976)
	COMe	H	OH	33						Kampf et al. (1976)
	CH$_2$OMe	H	OH	34						Kampf et al. (1976)
	NH$_2$N$_3$	H	OH	35						Kampf et al. (1976)
	F	OH	OH	50	400	100	100	84.3		Lin and Prusoff (1978a)
					0.1	100	100			Lin and Prusoff (1978a)
					0.01	100	None			Lin and Prusoff (1978a)
	I	OH	OH	48	50	100	100	99.5		Lin and Prusoff (1978a); Lin et al. (1976a)
	I	OH	NH$_2$		400	100	None		0.02	Lin and Prusoff (1978a); Lin et al. (1976a)
	F	OH	N$_3$		200	None	None	96.1	1.	Lin and Prusoff (1978a)
	Cl	OH	NH$_2$		400	100	100	5		Lin and Prusoff (1978a)
					400	None	None	74.3	29	Lin et al. (1976a)

R	R'	R''	% Inhib. (μM)	Sarcoma 180	L-1210	Vero cells	HSV-1	% Control HSV-1	References
Br	OH	NH_2	200	None	None		86.2	1	Lin and Prusoff (1978a); Lin et al. (1976a)
F	OH	NH_2	400	100	100	100	85.6	101	Lin et al. (1976a)
H	OH	NH_2	1	100	None	None	None		Lin et al. (1976a); Lin and Prusoff (1978b)
CH_3	OH	NH_2	400	None	None	None	98	3	Lin and Prusoff (1978b)
CH_3	NH_2	OH	[ED₅₀ μm]	5	1				Lin and Prusoff (1978b)
CH_3	NH_2	NH_2		Inactive	Inactive	Inactive	Inactive	Inactive	Lin and Prusoff (1978b)
CH_3	$NHC(=O)N(NO)CH_3$	OH	**102**		1.5				Lin et al. (1978)
CH_3	$NHC(=O)N(NO)CH_2CH_2Cl$	OH	**103**		1				Lin et al. (1978)
CH_3	OH	$NHC(=O)N(NO)CH_3$			95				Lin et al. (1978)
CH_3	OH	$NHC(=O)N(NO)CH_2CH_2Cl$			6.6				Lin et al. (1978)
CH_3	OH	$N(NO)C(=O)NHCH_2CH_2Cl$			4.2				Lin et al. (1978)
CH_3	OH	NH_2	Active against sarcoma 180 cells and HSV						Lin et al. (1976b)

Compound no.	R^1	R^2	ID₅₀ (μM) VV		HSV		References
			PRK cells	HSF	PRK cells	HSF	
54	$OCH_2CH=CH_2$	H	7×10^{-5}	3.5×10^{-4}	1.4×10^{-5}	1.4×10^{-4}	Torrence et al. (1978)
	$OCH_2C\equiv CH$	H	3.5×10^{-5}	1.4×10^{-4}	2.6×10^{-4}	3.5×10^{-6}	Torrence et al. (1978)

(Continued)

TABLE II (*Continued*)

Pyrimidines (cont.) R¹	R²	Compound no.	ID_{50} (μM) VV PRK cells	HSF	HSV PRK cells	HSF	References
OCH_2CONH_2	H		1.3×10^{-4}	1.3×10^{-4}	1.4×10^{-4}	2.6×10^{-4}	Torrence et al. (1978)
OCH_2COOH	H		$>6 \times 10^{-4}$	Inactive	Inactive	Inactive	Torrence et al. (1978)
OH	H		1.4×10^{-5}	4×10^{-5}	1.6×10^{-5}	3.5×10^{-6}	Torrence et al. (1978)
I	H	48	1.4×10^{-6}	1.4×10^{-6}	1.4×10^{-6}	3.5×10^{-7}	Torrence et al. (1978)
I	H	48	1.1×10^{-6}	1.1×10^{-6}	5.7×10^{-7}	5.7×10^{-7}	Torrence and Bhooshan (1977)
CN	H		1.6×10^{-5}	2.7×10^{-5}	1.6×10^{-4}	$>7 \times 10^{-4}$	Torrence and Bhooshan (1977)
CN	OH		$>3.7 \times 10^{-4}$	$>7 \times 10^{-4}$	3.7×10^{-4}	$>7 \times 10^{-4}$	Torrence and Bhooshan (1977)
CF_3	H		3.3×10^{-7}	1.3×10^{-6}	6.6×10^{-7}	1.3×10^{-6}	Torrence and Bhooshan (1977)
OCH_2Ph	H		$>3 \times 10^{-4}$	$>6 \times 10^{-4}$	$>1.2 \times 10^{-4}$	4.5×10^{-4}	Torrence et al. (1978)
OCH_2-(p-NO_2-phenyl)	H		$>2.6 \times 10^{-4}$	$>5.2 \times 10^{-4}$	$>2.6 \times 10^{-4}$	$>5.2 \times 10^{-4}$	Torrence et al. (1978)

Pyrimidines R¹	R²	Compound no.	Biological activity	References
$OCH_2CH=CH_2$	OH		Inactive against both VV and HSV	Torrence et al. (1978)
$OCH_2C\equiv CH$	OH			
OCH_2CONH_2	OH			
$CH_2CH=CH_2$	H		Active against HSV	Ruth and Bergstrom (1978)
$CH_2C\equiv CH$	H		Active against HSV transformed HeLa cells	Ruth and Bergstrom (1978)
Et	H		Active against HSV	Bergstrom and Ruth (1976)
Et	alOH			Bergstrom and Ruth (1976)
$CH=CHCO_2Me$	OH	39		Bergstrom and Ruth (1976)
$CH_2CH=CH_2$	OH	40		Bergstrom and Ruth (1976); Ruth and Bergstrom (1978)
$(CH_2)CHCH_2OH$ (OH)	H	38		Bergstrom and Ruth (1976)

Pyrimidines	R^1	R^2	R^3 or X	Compound no.	Biological activity	References
(cytosine 2′-deoxyribonucleoside structure)	$C{\equiv}CH$	H				Perman et al. (1976)
	$CH{=}CBr_2$	H			$ID_{50} = 2 \times 10^{-8}$ against L-1210 cells	Perman et al. (1976)
	I	H	OH	49		Lin and Prusoff (1978a)
	$C{\equiv}CH$	OH	OH			Jones et al. (1977)
	$CH_2CH{=}CH_2$	H	OH		Active against HSV-1,2 and inhibit nucleoside phosphorylase in HeLa cells	Ruth and Bergstrom (1978)
	$CH_2CH_2CH_3$	H	OH		Inhibit mitochondrial and cytoplasmic deoxythymidine kinases	Ruth and Bergstrom (1978)
	$CH{=}CH{-}CH_3$	H	OH			Ruth and Bergstrom (1978)
	$CH_2CH{=}CH_2$	OH	OH			Ruth and Bergstrom (1978)
	$CH{=}CHCH_3$	OH	OH			Ruth and Bergstrom (1978)
	$CH_2CH_2CH_3$	OH	OH			Ruth and Bergstrom (1978)
(2′-deoxyuridine, 5-CH_2R^1 structure)	$NHCH_2CH_2NMe_2$	H, PO_3^{2-}		42	} Thymidylate synthetase inhibitor	Edelman et al. (1977)
	NMe (—N—NMe piperazine)	H, PO_3^{2-}		43		Edelman et al. (1977)
	—N⟨NMe⟩ (N-methylpiperazinyl)	H, PO_3^{2-}		44		Edelman et al. (1977)
	—N⟨ ⟩ (pyrrolidinyl)	H, PO_3^{2-}		45		Edelman et al. (1977)

(Continued)

TABLE II (*Continued*)

Pyrimidines	R¹	R²	R³ or X	Compound no.	Biological activity	References
	OH	Br,I		**107**		Hollenberg et al. (1977)
	OH	N_3,SCN		**107**		Hollenberg et al. (1977)
	NH_2	Br,I		**108**		Hollenberg et al. (1977)
	NH_2	N_3,SCN		**108**		Hollenberg et al. (1977)
	$NHC(CH_2)_{m-2}CH_3$ (=O)	OH			Active against L-1210, resistant to deamination	Akiyama et al. (1978)
	Alkyl C_8–C_{12}	H	Cl	**59**	Active against DNA virus	Hanamura et al. (1976)
	Alkyl C_{16}–C_{22}	H	Cl	**59**	Active against L-1210	Hanamura et al. (1976)
	Alkyl	$= R^1$	Cl		Active against L-1210 and HeLa cells	Hanamura et al. (1976); Kondo and Inoue (1977)
	Alkyl	$= R^1$	BF_4		Antileukemic agents	Kondo and Inoue (1977)

Biological activity

Compound no.	R^1	LD_{50} (M) L-5178Y	HeLa	Novikoff hepatoma	Novikoff hepatoma lacking thymidine kinase	E. coli B23	References
36	H	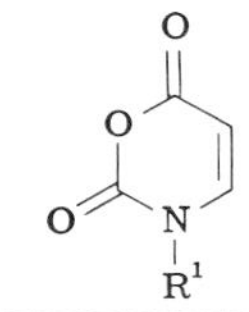 2.5×10^{-5}	5×10^{-5}	2×10^{-4}	6×10^{-5}	5×10^{-5}	Chwang et al. (1976)
37	rib	7×10^{-5}	2×10^{-4}	2×10^{-4}	1×10^{-4}	5×10^{-5}	Chwang et al. (1976)
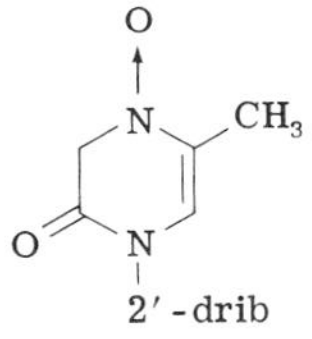 **41**		$ED_{50} = 9 \times 10^{-7}$ M against L-1210, 55% ILS at 400 mg/kg/day $\times$ 6					Bobek and Bloch (1977)

Pyrimidines (cont.)

Compound no.	R^1	R^2	R^3	Biological activity	References
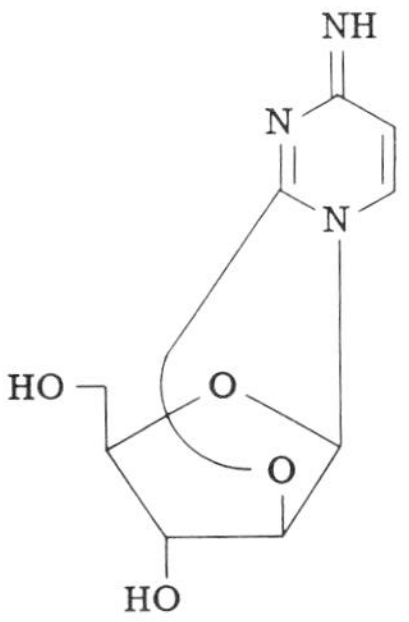 **55**				Antileukemic, immunosuppressive, anti-DNA viral agent	Kondo and Inoue (1977)

(Continued)

TABLE II (*Continued*)

Pyrimidines	R¹	R²	R³	Compound no.	Biological activity	References
	NH₂			**56**		Cook *et al.* (1977)
	OH			**57**		Cook *et al.* (1977)
	NH₂	H	rib	**60**	Active against L-1210 cells, stable to hydrolysis	Beisler *et al.* (1977)
	OH	CH₃	2'-drib		No antiviral and antimicrobial activity	Skulnick (1978)
					No antitumor activity	Long *et al.* (1976)
					No antiviral and antimicrobial activity	Skulnick (1978)

Purines	R¹	R²	R³	X	Compound no.	Biological activity	References

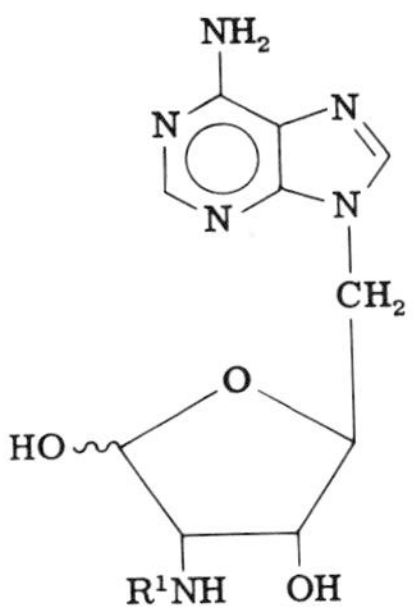

First structure (top):

Adenine–ribose nucleoside: purine bearing NH_2, N-glycoside CH_2 linkage to ribofuranose with OH, and R^1NH / HO substituents.

R¹ = $COCHCH_2Ph$ / NH_2

Second structure (lower):

2-Aminopurine nucleoside: R^1 / H_2N on purine ring bearing X; ribofuranose with R^2, HO, R^3O.

				No.	T/C (mg/kg/dose)	Reference
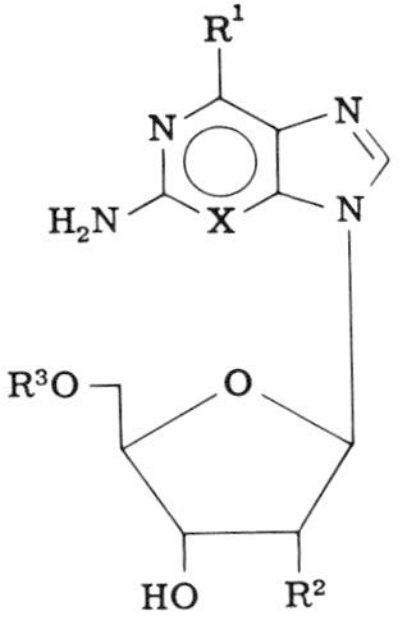 OH	H	C		**64**	Active against adenocarcinoma 755, DNA, RNA virus	Cook et al. (1976)
OH	PO_3^{2-}	C		**65**	Active against L-1210, adenocarcinoma 755, DNA, RNA virus	Cook et al. (1976)
SMe	H		N	**66**	L-1210 (128, 400)	Martinez et al. (1977)
$SCH_2CH{=}CH_2$	H		N	**66**	L-1210 (245, 400), P-388 (181, 200)	Martinez et al. (1977)
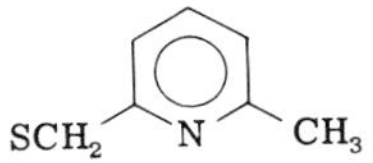 SCH_2-(2,6-dimethylpyridinyl)	H		N	**66**	L-1210 (153, 600), P-388 (>125, 400)	Martinez et al. (1977)
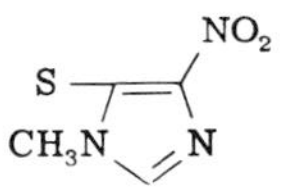 S-(1-methyl-NO_2-imidazolyl)	H		N	**66**		Martinez et al. (1977)
OH	NH_2	H	N	**99**	L-1210 (173, 12.5), P-388 (161, 6.25)	Ikehara et al. (1976)

(Continued)

93

TABLE II (*Continued*)

Purines	X	R¹	R²	R³	Compound no.	Biological activity	References
	SMe	(SCH₂—⟨C₆H₄⟩—NO₂)					Montero et al. (1977)
	SCH₂CH=CH₂	O_2N imidazolyl					Montero et al. (1977)
		S-imidazolyl-Me					Montero et al. (1977)
	NHMe						Montero et al. (1977)
	NMe₂						Montero et al. (1977)
	NHNH₂						Montero et al. (1977)
	NHOH						Montero et al. (1977)
	H						Ueda et al. (1978)
	H			H	H	Active against L-1210, NF-sarcoma, sarcoma 180 at 3–10 mg/kg	
				OH	PO_3^{2-}		

Purines	X	R_1	R_2	R_3	Compound no.	Biological activity	References
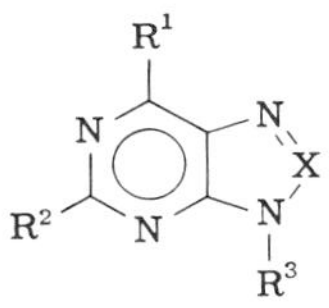	CH	NH_2	MeO	ara			Montgomery et al. (1977)
	CH	NH_2	MeS	ara			Montgomery et al. (1977)
	CH	NH_2	$NHNH_2$	ara			Montgomery et al. (1977)
	N	NH_2	H	ara			Montgomery et al. (1977)
	N	NH_2	MeO	ara	**69**	No antiviral activity	Montgomery et al. (1977)
	N	NH_2	MeS	ara			Montgomery et al. (1977)
	N	NH_2	$MeSO_2$	ara			Montgomery et al. (1977)
	N	NH_2	$NHNH_2$	ara			Montgomery et al. (1977)
	N	NH_2	N_3	ara			Montgomery et al. (1977)
	N	NH_2	H	rib		$ED_{50} = 0.8$ μM against H.EP. No. 2	Elliott and Montgomery (1977)
	N	SH	H	rib		$ED_{50} = 1$ μM against H.EP. No. 2	Elliott and Montgomery (1977)
	N	SMe	H	rib		$ED_{50} = 0.02$ μM against H.EP. No. 2	Elliott and Montgomery (1977)
	N	SEt	H	rib		$ED_{50} = 0.1$ μM against H.EP. No. 2	Elliott and Montgomery (1977)

Structure (left diagram):

R^1 — (purine ring with X, N, N, R^3, R^2 substituents)

Purines (cont.)	X	R^1	R^2	R^3	Biological activity	References
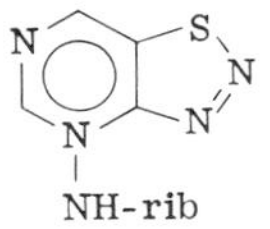	N	OMe	H	rib	$ED_{50} = 0.3$	Elliott and Montgomery (1977)
	N	OEt	H	rib	$ED_{50} = 0.1$	Elliott and Montgomery (1977)
	N	NHBu	H	rib	$ED_{50} = 0.3$	Elliott and Montgomery (1977)
	N	$NCH_2CH{=}CH_2$	H	rib	$ED_{50} > 3$	Elliott and Montgomery (1977)
					$ED_{50} = 0.1$	Elliott and Montgomery (1977)

$\}$ μM against H.EP. No. 2

Structure (bottom diagrams):

(thiazolo-pyrimidine) NH-rib

NH_2 — (imidazo-pyrimidine ring) R^1S ... rib

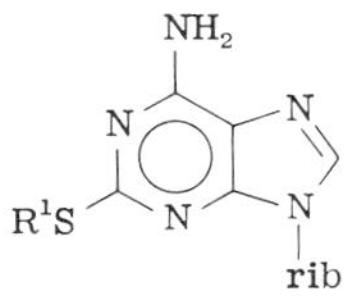

(Continued)

TABLE II (*Continued*)

Purines (cont.)	% Inhibition of rabbit platelet aggregation	References
	81	Kikugawa *et al.* (1977b)
	83	Kikugawa *et al.* (1977b)
	74	Kikugawa *et al.* (1977b)
PhCH₂—N⟨ ⟩N—CH₂CH₂—	96	Kikugawa *et al.* (1977b)
Cl—⟨ ⟩—CH₂—N⟨ ⟩N—CH₂CH₂—	96	Kikugawa *et al.* (1977b)
PhCH=CHCH₂—N⟨ ⟩N—CH₂CH₂—	89	Kikugawa *et al.* (1977b)
	96	Kikugawa *et al.* (1977b)

Purines (cont.)	R	Compound no.	Biological activity	References
H_2N purine—rib, R	H	**70**		Cook and Robins (1978)
	SH			Cook and Robins (1978)
	NH_2			Cook and Robins (1978)
H_2N purine—rib, R	H	**71**		Cook and Robins (1978)
	SH			Cook and Robins (1978)
	NH_2			Cook and Robins (1978)
NH_2 $CONH_2$ purine—rib, R	SMe	**73**	Active against leukemic cells	Anderson and Broom (1977)
	NY_2			Anderson and Broom (1977)
	H			Anderson and Broom (1977)
NHR purine—rib; $-(CH_2)_n-NH-$ imidazole—rib	$n = 1,2$		Inactive against L-1210	Zemlicka and Owens (1977)
	$CH_2CH_2NH_2$			Zemlicka and Owens (1977)
	$(CH_2)_4NH_2$		Inactive against L-1210	Zemlicka and Owens (1977)
	Et		Active against L-1210, sarcoma 180, carcinoma TA-3	Zemlicka and Owens (1977)
	Bu			Zemlicka and Owens (1977)
				Zemlicka and Owens (1977)
				Zemlicka and Owens (1977)

(*Continued*)

TABLE II (*Continued*)

Purines (cont.)

R	Biological activity	References
$NHNH_2$		Chattopadhyaya and Reese (1977)
NHMe		Chattopadhyaya and Reese (1977)
$NH(CH_2)_3Me$		Chattopadhyaya and Reese (1977)
$NHCH_2Ph$		Chattopadhyaya and Reese (1977)
(pyrrolidin-1-yl)		
N_3		Chattopadhyaya and Reese (1977)
Br		Chattopadhyaya and Reese (1977)
$NHNH_2$		Chattopadhyaya and Reese (1978)
$NMeNH_2$		Chattopadhyaya and Reese (1978)
NHMe		Chattopadhyaya and Reese (1978)
(piperidin-1-yl)		Chattopadhyaya and Reese (1978)
		Chattopadhyaya and Reese (1978)

Purines (cont.)

Compound no.	R^1	R^2	R^3	Biological activity (Adenosine) ID_{50} ($\times 10^{-4}\ m$) against leukemia	References
	NH_2	OH	OH	1.4	Rosowsky et al. (1976)
	NH_2	OH	H	2.7	Rosowsky et al. (1976)
	NH_2	H	OH	2.5	Rosowsky et al. (1976)
	NH_2	OH	CH_3	5.0	Rosowsky et al. (1976)
	NH_2	OH	Et	3.8	Rosowsky et al. (1976)
	NH_2	OH	Bu	0.63	Rosowsky et al. (1976)
	NH_2	OH	C_6H_{13}	0.42	Rosowsky et al. (1976)
	NH_2	NH_2	OH		Ikehara et al. (1976); Ranganathan (1977); Hobbs and Eckstein (1977)
98	NH_2	F	OH		Ranganathan (1977)

Purines (cont.)

	R¹	R²	Biological activity	References
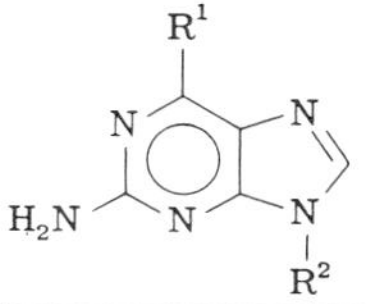 (structure)	Cl	H	79% ILS against leukemia	Elliott and Montgomery (1978)
	SH	rib	<15% ILS against leukemia	Elliott and Montgomery (1978)
	SMe	rib		Elliott and Montgomery (1978)
	Cl	rib	<15% ILS against leukemia	Elliott and Montgomery (1978)

	R		Biological activity	References
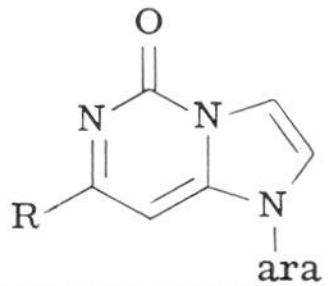 (structure, ara)	NH_2		No antitumor, antiviral, antimicrobial activity	Bartholomew et al. (1976)
	H			Bartholomew et al. (1976)

Purines (cont.)

	Compound no.		Biological activity	References
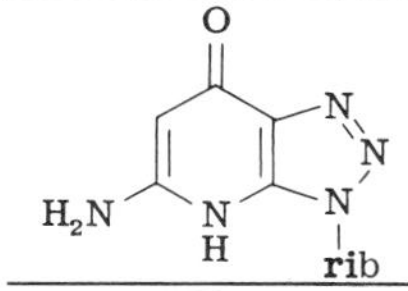 (structure, rib)				Cline et al. (1976)

	R		Biological activity	References
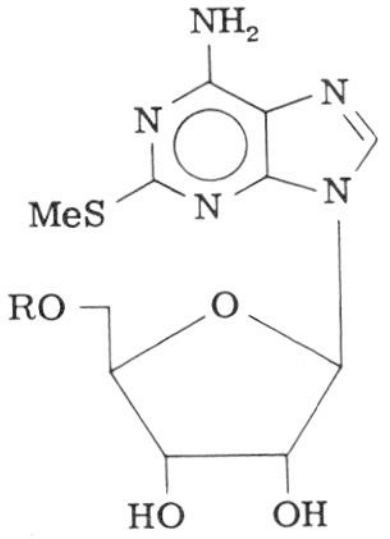 (structure)	SO_2NH_2		Platelet aggregation inhibitor	Gough et al. (1978)
	PSO_2^{-2}			Gough et al. (1978)

(Continued)

99

TABLE II (*Continued*)

Purines (cont.)	R	Compound no	Biological activity	References
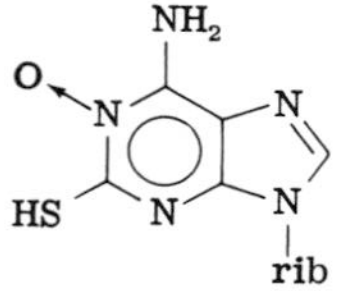			Platelet aggregation inhibitor, resistant to deamination	Kikugawa *et al*. (1977a)
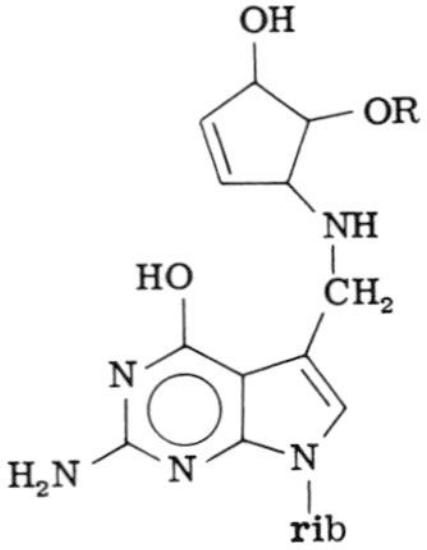	H		(Nucleoside Q)	Cheng *et al*. (1976); Ohgi *et al*. (1976, 1977); Townsend *et al*. (1976)
	glycosyl		(Nucleoside Q*)	Cheng *et al*. (1976); Ohgi *et al*. (1976, 1977); Townsend *et al*. (1976)
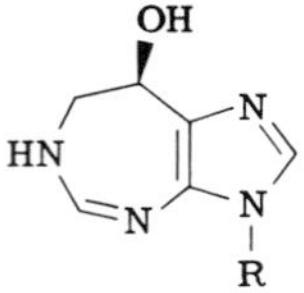	rib	**74**	(Coformycin) } adenosine deaminase inhibitor	Nakamura *et al*. (1974); Ohno *et al*. (1974); Sawa *et al*. (1967)
	2′-drib	**76**	(2′-Deoxycoformycin) } adenosine deaminase inhibitor	Woo *et al*. (1974)
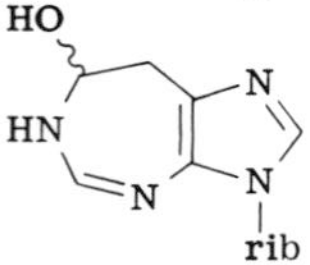		**75**	(Isocoformycin) } adenosine deaminase inhibitor	Shimazaki *et al*. (1979)

Five-membered heterocycles

H₂NC—, X=, N, N—R³, R¹ (structure)

Biological activity — % Inhibition of biosynthesis

X	R¹	R²	Compound no.	Adenine nucleotide	Guanine nucleotide	References
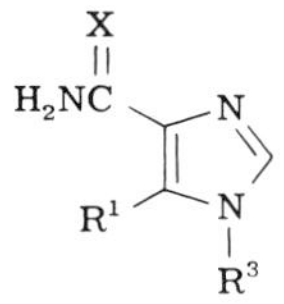 O	H	rib	79	16	63	Srivastava et al. (1976)
O	F	rib		0	40	Srivastava et al. (1976)
O	Cl	rib		23	53	Srivastava et al. (1976)
O	Br	rib		15	60	Srivastava et al. (1976)
O	I	rib		0	21	Srivastava et al. (1976)
S	SH	rib	78	0	34	Srivastava et al. (1976)

Biological activity — MIC (µmole/ml)

X	R¹	R²	Compound no.	Staph. aureus	C. albicans	T. mentagraphytes	References
O	Cl	rib		Inactive	0.3	Inactive	Srivastava et al. (1976)
O	Br	rib		Inactive	0.15	Inactive	Srivastava et al. (1976)
S	SH	rib	78	0.05	0.05	0.02	Srivastava et al. (1976)
S	Cl	rib		Inactive	Inactive	0.3	Srivastava et al. (1976)
S	Br	rib		0.3	Inactive	Inactive	Srivastava et al. (1976)

Biological activity — Virus rating

X	R¹	R²	Compound no.	HV/1	VV	RV/13	PIV/13	References
O	H	rib	79	0.9		0.4	0.3	Srivastava et al. (1976)
O	F	rib		0.7–1.1	0.3–0.6	0.3–0.6	0.2–0.4	Srivastava et al. (1976)
O	Cl	rib		0.6–0.8	0.5–0.6	0.4–0.9	0.2	Srivastava et al. (1976)

(Continued)

TABLE II (Continued)

Five-membered heterocycles	X	R¹	R²	Biological activity				References
	O	Br	rib	0.7–0.8	0.5	0.4–0.5	0.1	Srivastava et al. (1976)
	O	I	rib	0.6–0.7	0.3–0.4	0.0	0.1	Srivastava et al. (1976)
	S	NH₂	3′-drib					Miyoshi et al. (1976)
	O	NH₂	3′-drib					Miyoshi et al. (1976)
	O	NH₂	3′-Cl-3′-drib	No antitumor activity				Miyoshi et al. (1976)
	O	NH₂	2′,3′-anhydro-rib					Miyoshi et al. (1976)

Five-membered heterocycles (cont.)	R¹	R²	R³	Biological activity	Compound no.	References
	CONH₂	OMe	H	No antiviral activity		Dudycz et al. (1977)
	CONH₂	H	OMe			Dudycz et al. (1977)
	CONH₂	OH	NH₂	Inactive against P-388 mouse lymphoid leukemia, ID₅₀ >100 µg/ml		Narang and Vince (1977)
	CONHNH₂	OH	NH₂			Narang and Vince (1977)
	CONOH	OH	NH₂			Narang and Vince (1977)
	$\overset{N\cdot NH_2}{C}\!\!-NH_2$	OH	NH₂	Inactive against P-388 mouse lymphoid leukemia, ID₅₀ >100 µg/ml		Narang and Vince (1977)
	$\overset{NOH}{C}\!\!-NH_2$	OH	NH₂			Narang and Vince (1977)

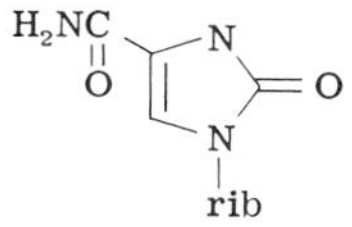

Srivastava et al. (1977a)

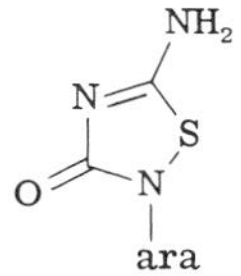

Revankar and Robins (1976)

80

Five-membered heterocycles (cont.)	R¹	R²	R³	R⁴	Compound no.	Biological activity	References
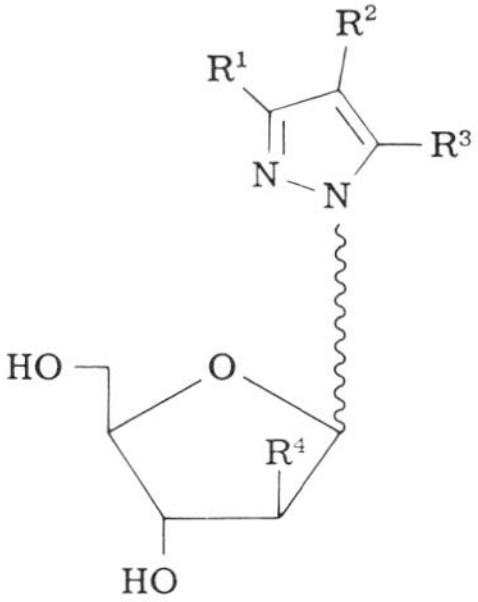	CONH₂	H	H	OH			Makabe et al. (1976)
	CONH₂	H	CONH₂	OH			Makabe et al. (1976)
	H	H	CONH₂	OH			Makabe et al. (1976)
	CONH₂	NH₂	CONH₂	OH			Makabe et al. (1976)
	CONH₂	H	H	H			Makabe et al. (1976)
	H	H	CONH₂	H			Makabe et al. (1976)
	CONH₂	H	CONH₂	H			Makabe et al. (1976)
	CONH₂	NO₂	CONH₂	H			Makabe et al. (1976)
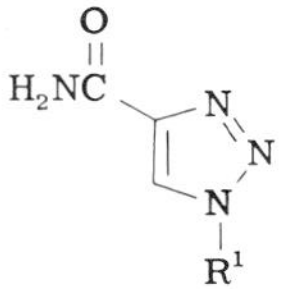	ara						Makabe et al. (1977)
	2'-drib						Makabe et al. (1977)

(Continued)

103

TABLE II (*Continued*)

Five-membered heterocycles (cont.)	Compound no.	Biological activity	References
A	A B }	Inactive against influenza A2/Asian/J305 virus infection in mice	Poonian *et al.* (1976) Poonian *et al.* (1976)
B	85 C D E }		Vasella (1977) Sasaki *et al.* (1978b)
C	D R=OMe E R=NH$_2$		

Miscellaneous	R	Biological activity	References
	H	(Wyosine)	Nakatsuka *et al.* (1978)
	CH$_2$CH$_2$C---H CO$_2$CH$_3$ NHCO$_2$CH$_3$	(Wybutosine)	Itaya and Ogawa (1978)
	CH$_2$CHC---H HOO CO$_2$CH$_3$ NHCO$_2$CH$_3$	(Wybutoxosine)	Itaya and Ogawa (1978)

			Inhibition of hypoxanthine phosphoribosyltransferase	
90	rib		68%	Pickering et al. (1977)
	H		82%	Pickering et al. (1977)
91	rib		0%	Pickering et al. (1977)
	H		5%	Pickering et al. (1977)
92	rib		72%	Pickering et al. (1977)
	H		1%	Pickering et al. (1977)

[a] Abbreviations: Rib, D-ribose; ara, T-arabinose; 2'-drib, 2'-deoxy-D-ribose; 3'-drib, 3'-deoxy-D-ribose; α,β-rib, both anomers of D-riboside were obtained and separately evaluated; HSV-1, herpes simplex virus-1; VV, Vaccinia virus; HSF, human skin fibroblasts; PRK, primary rabbitt kidney; HV/1, type 1 herpes virus; RV/13, type 13 rhino virus; PIV/3, type 3 parainfluenza virus; MDC, minimum degeneration concentration; H. Ep. No. 2, human epidermoid carcinoma No. 2 cells; ED₅₀, the concentration required to inhibit the growth of treated cells to 50% of that of untreated controls; ID₅₀, the concentration required to inhibit virus-induced cytopathogenicity by 50%; ILS, increase in life span.

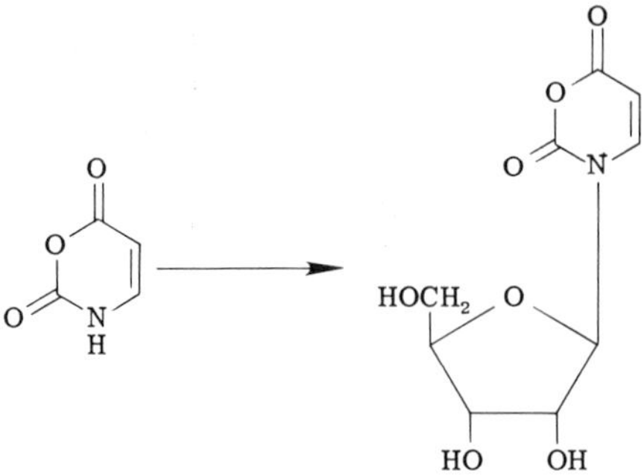

36   37 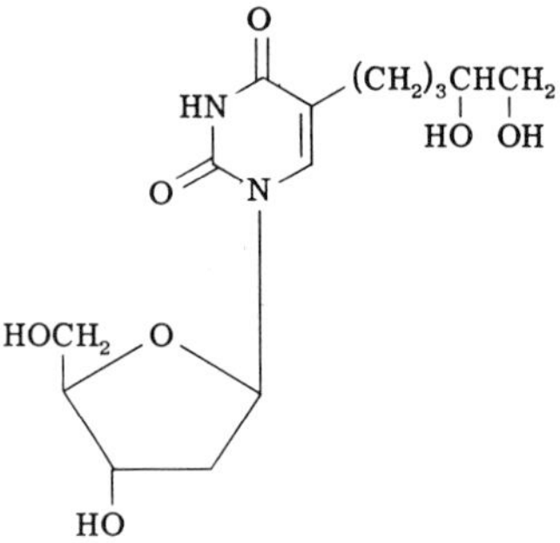   38

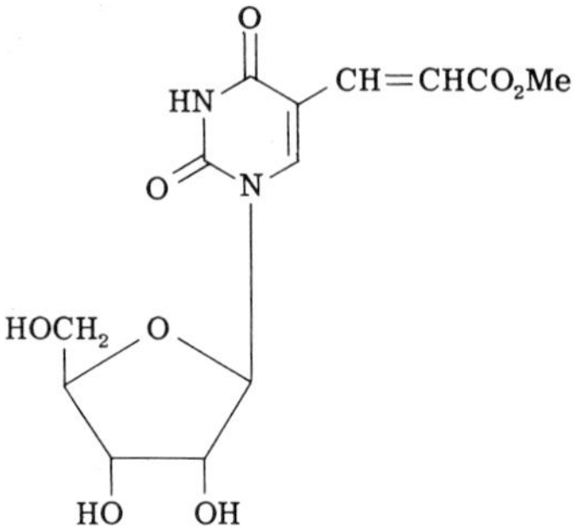

39 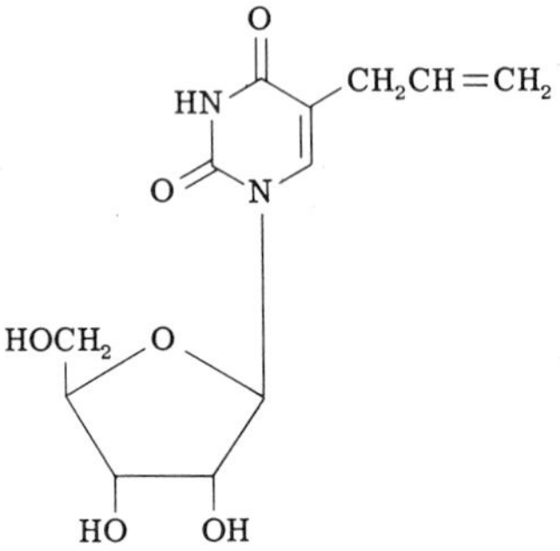   40

Bloch, 1977). The compound (41) inhibited the growth of leukemia L-1210 cells *in vitro*.

Based on the postulated formation of 5′-thymidylyltetrahydrofolic acid as an intermediate in the conversion of deoxyuridine 5′-phosphate to thymidine 5′-phosphate, a series of substituted 5-aminomethyl-2′-deoxyuridines (42, 43, 44, 45) was synthesized (Edelman *et al.*, 1977). The 5′-phosphate was found most active for substrate competitive inhibitors of thymidylate synthetase.

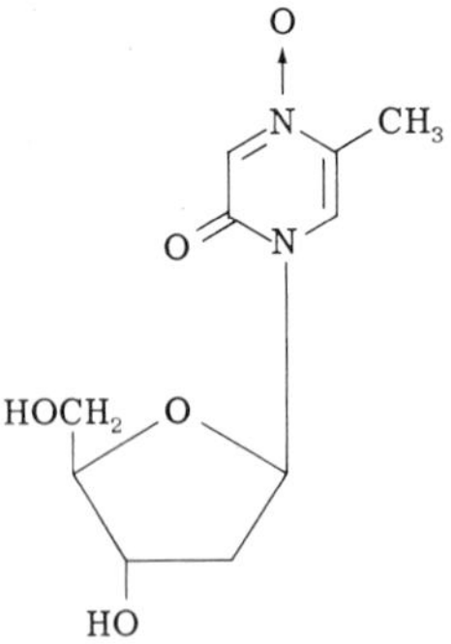

41

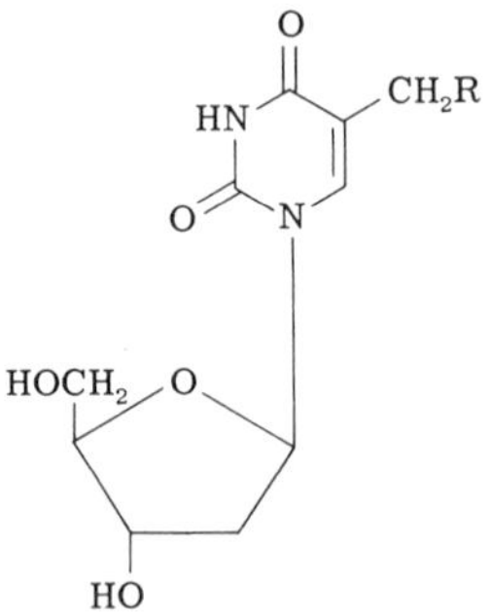

42 R = $-NHCH_2CH_2N\diagdown^{Me}_{Me}$

43 R = $-N\diagdown^{Me}_{Me}$

44 R = $-N\diagup\diagdown Me$

45 R = $-N\diagup\diagdown$

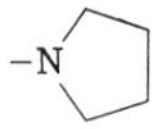

An improved synthesis of 5-cyanouridine and 5-cyano-2'-deoxyuridine was reported (Torrence and Bhooshan, 1977). A novel synthetic procedure has been developed for 5-chloro-, 5-bromo-, and 5-iodo-5'-amino-2',5'-dideoxyuridine (**46, 47, 48**), as well as for two new analogues (**49**) and (**50**) (Lin and Prusoff,

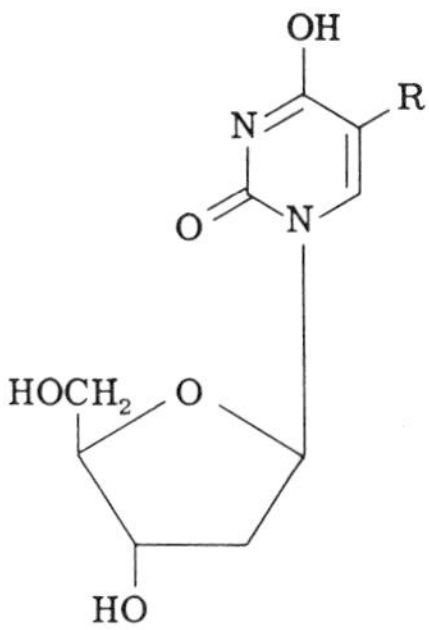

46 R = Cl
47 R = Br
48 R = I

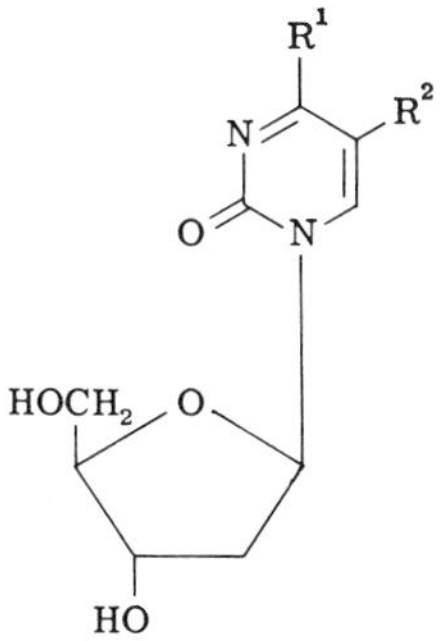

49 R^1 = NH_2, R^2 = I
50 R^1 = OH, R^2 = F

1978a). Their biological activities were studied on the replication of Sarcoma 180 cells, Vero cells, and Herpes Simplex virus Type 1 *in vitro*.

An interesting observation of a hydrolysis product of the antitumor nucleoside, 5-azacytidine, was noted with studies (Beisler, 1978) using high-pressure liquid chromatography. Thus, it was shown that N-(formylamino)-N'-β-D-ribofuranosylurea (**51**) in water solution readily equilibrates to 5-azacytidine (**52**) and more slowly deformylates to give 1-β-D-ribofuranosyl-3-guanylurea (**53**) irreversibly. It should be pointed out that (**51**) showed considerable antitumor activity against murine L-1210 leukemia.

As for 5-*O*-alkylated derivatives of 5-hydroxy-2'-deoxyuridine, 5-propynyloxy-2'-deoxyuridine (**54**) seems to be important for antiherpes activity (Torrence *et al.*, 1978), just as seen for ethynyl-2'-deoxyuridine (Perman *et al.*, 1976).

A facile one-step synthesis of anhydro-ara-C (**55**) was developed by Kondo and Inoue (1977).

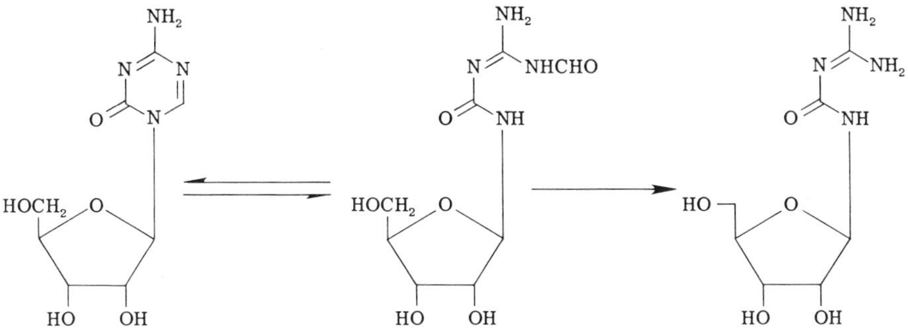

52   51   53

3-Deazacytidine (**56**) and 3-deazauridine (**57**) were prepared in high overall yields from 1 -methoxy -1 -buten -3 -yne (Cook *et al.*, 1977).

A new synthesis of 5 -ethynylcytosine and 5 -ethynylcytidine was also reported (Jones *et al.*, 1977).

The reactions of cytidine (**58**) with various 2 -*O* -acyloxyisobutyryl chlorides afforded the corresponding 2,2' -anhydro -1 -(3' -*O* -acyl -β -D -arabinofuranosyl) cytosine hydrochloride (**59**). Activity against DNA viruses (vaccinia and Herpes) in tissue culture was maximal in compounds with 8–12 carbon atoms in the acyl groups, and activity against L-1210 leukemia in mice varied markedly according to the length of the acyl groups and was high in the case of long-chain esters (C_{16}–C_{22}) (Hamamura *et al.*, 1976).

Borohydride reduction of 5-azacytidine (**52**) gave 5,6-dihydro-5-azacytidine (**60**), showing that **60** is stable at room temperature in aqueous solution and has approximately 80% of the antitumor efficacy shown by **52** (Beisler *et al.*, 1977).

b. Purines As an analogue of puromycin (**61**), which is an aminoacyl nucleoside having antitumor activity as well as antibiotic activity, a reversed aminoacyl nucleoside (**62**) was prepared starting from D-xylose (Nair and Emanuel, 1977).

61 R = COCHCH₂—⟨⟩—OCH₃
         |
        NH₂

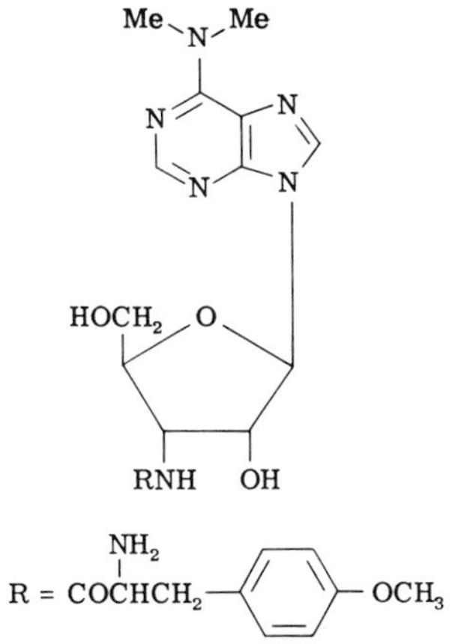

62 R = COCHCH₂—⟨⟩
         |
        NH₂

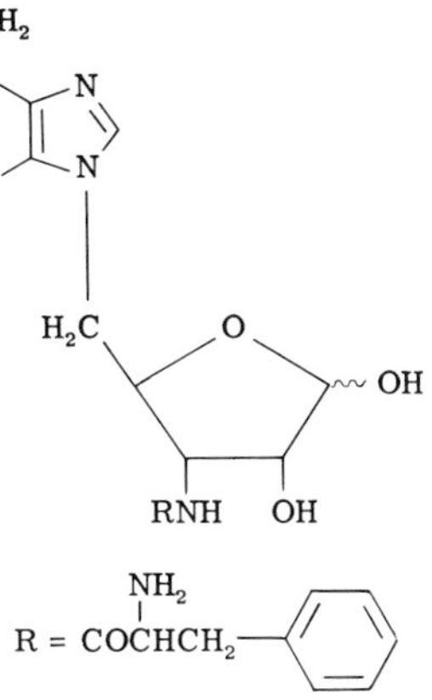

3-Deazaguanine (**63**), 3-deazaguanosine (**64**), and 3-deazaguanylic acid (**65**) have been synthesized for the first time from imidazole precursors (Cook *et al.*, 1976). They have demonstrated a potent, broad-spectrum *in vivo* activity against various DNA and RNA viruses, L-1210 leukemia, and adenocarcinoma 755 in mice.

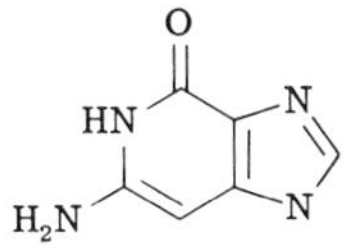

63 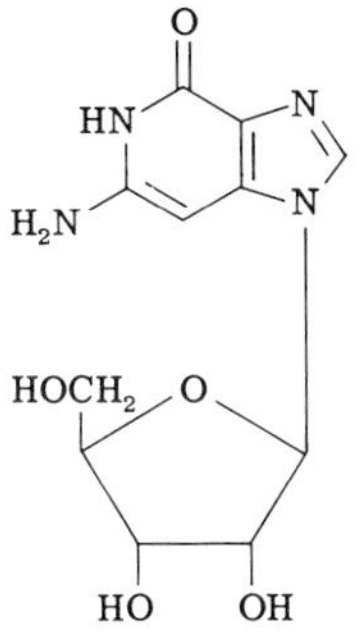   64 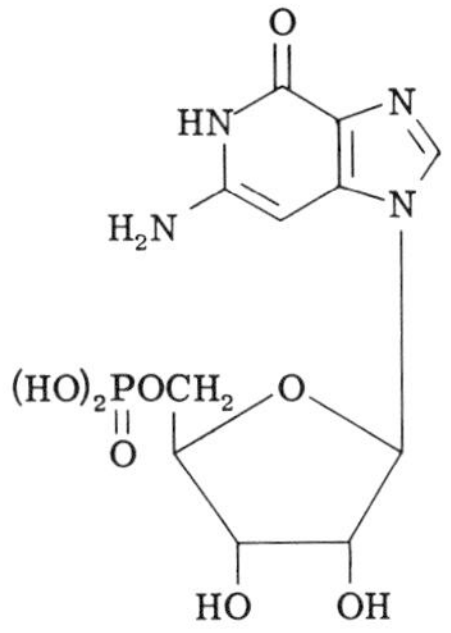   65

Since 9-(β-D-ribofuranosyl)purine-6-thione and 6-hydrazino-9-(β-D-ribofuran-osyl)purine possess anticancer activity, synthetic attempts in this area are continuing.

A series of S-substituted derivatives of the α and β anomers of 2'-deoxy-6-thioguanosine has been prepared by S-alkylation of the parent nucleosides and/or

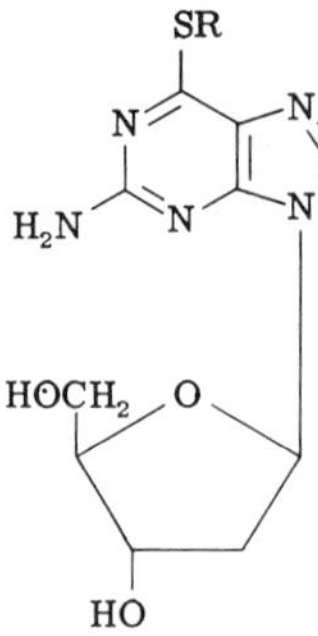

66

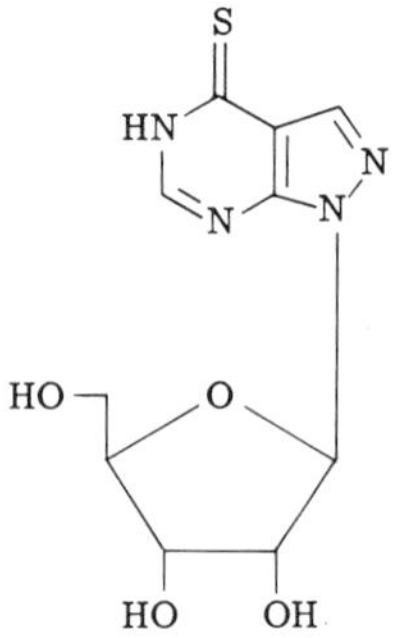

67

by mercaptide displacement reactions of 6-chloro intermediates. All β anomers (**66**) were active against L-1210 murine leukemia. Most S-alkyl-α-anomers were inactive (Martinez *et al.*, 1977).

1-(β-D-Ribofuranosyl)pyrazolo[3,4-d]pyrimidine-4-thione (**67**) was converted into the corresponding 4-alkylthio derivatives (Montero *et al.*, 1977), and then 4-alkylamino, 4-hydrazino, and 4-hydroxylamino derivatives were obtained.

Based on the α anomer of 9-β-D-arabinofuranosyladenine (**68**) and 9-β-D-arabinofuranosyl-8-azaadenine (**69**), some 2-substituted derivatives of (**68**) and (**69**) were prepared, but they showed no antiviral activity (Montgomery *et al.*, 1977).

Nucleosides of 3-deaza-6-thioguanine (**70**) and 3-deaza-2,6-diaminopurine (**71**) were prepared from the corresponding cyanomethylimidazole β-D-ribofuranosides (Cook and Robins, 1978), and converted to various nucleosides.

The synthesis of pyrido[2,3-d]pyrimidine ribonucleoside (**73**), structurally

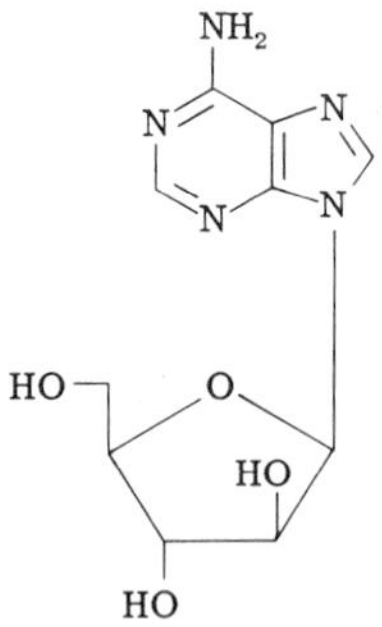

68

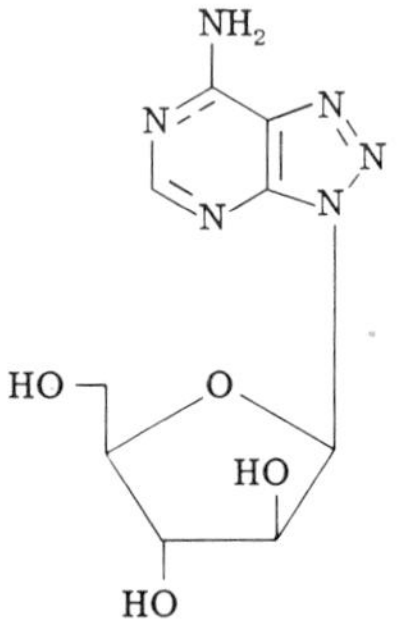

69

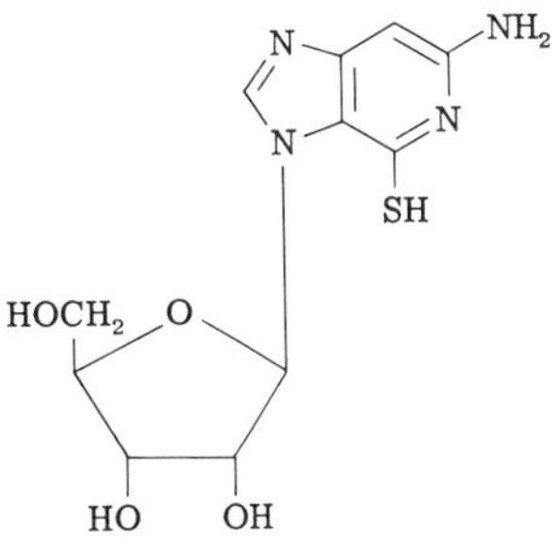

70 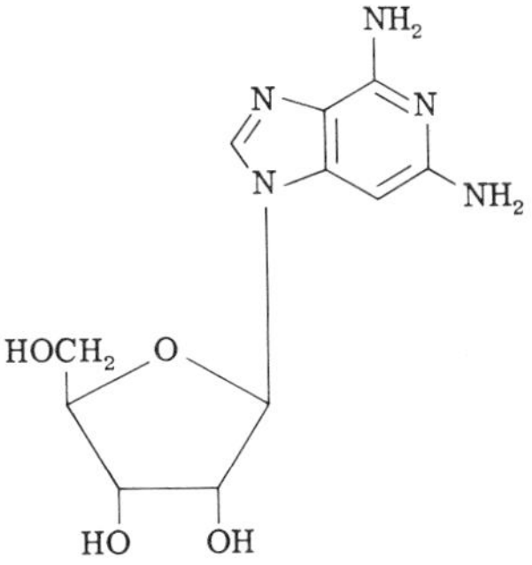   71

related to the antibiotic sangivamycin (**72**), was reported by Anderson and Broom (1977). Sangivamycin has potent antileukemic activity and is currently undergoing clinical trials in the U.S.A.

c. **Nucleosides with Expanded Ring Systems**   The identification of the antibiotic coformycin (**74**) as 3-(β-D-ribofuranosyl)6,7,8-trihydroimidazo[4,5-d][1,3]diazepin-8(R)-ol showed that expansion of purine moiety to a seven-membered ring can also give rise to biologically active nucleosides (Nakamura *et al.*, 1974; Ohno *et al.*, 1974). Coformycin showed a strong synergistic activity with formycin by inhibiting the growth of bacteria, and was confirmed to be the strongest inhibitor of adenosine deaminase (Sawa *et al.*, 1967). Isocoformycin (**75**) was synthesized and it also showed an inhibitory activity to adenosine deaminase (Shimazaki *et al.*, 1979). Another adenosine and ara-A deaminase inhibitor, (R)-3-(2'-deoxy-β-D-erythropentofuranosyl)-3,6,7,8-tetrahydroimidazo[4,5- d][1,3]- diazepin- 8- ol (**76**), was also obtained from a strain of *Streptomyces antibioticus* (Woo *et al.*, 1974), and its antitumor activity was investigated (Caron *et al.*, 1977).

d. **Five-Membered Nucleosides**   The broad-spectrum antiviral activity of 1- β- D- ribofuranosyl- 1,2,4- triazole- 3- carboxamide (**77**, ribavirin or

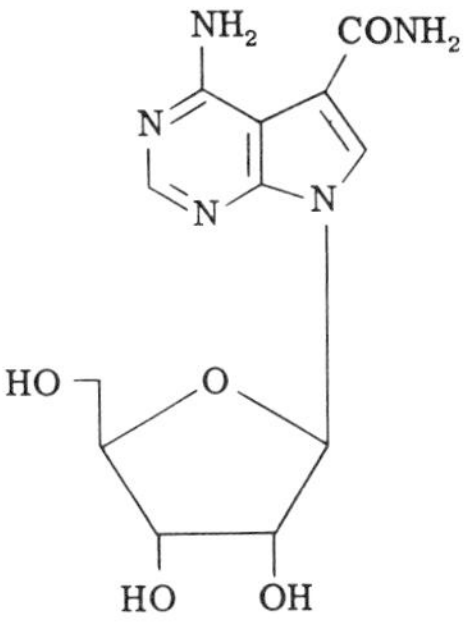

72 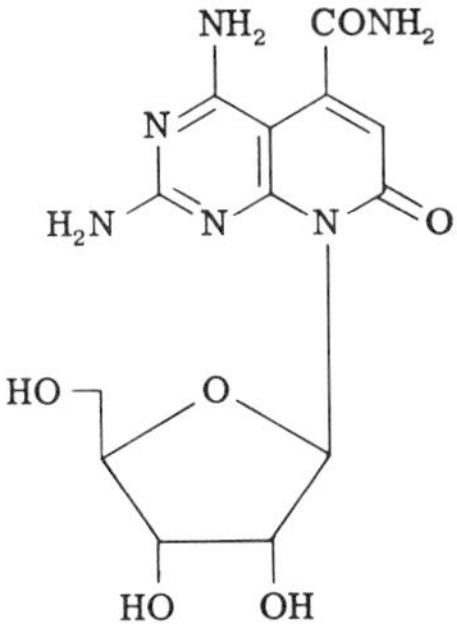   73

74   75   76

virazole) has stimulated a great deal of effort into the synthesis of five-membered heterocyclic nucleosides (Narang and Vince, 1977). Thus, the 3'-amino analogues of virazole were prepared, showing some activity against P-388 mouse lymphoid leukemia and no significant activity in antiviral testing. Synthesis of the analogues (**78**, (**79**), (**80**), and (**81**) (R=H or OH) were reported (Srivastava *et al.*, 1976, 1977a; Huynh-Dinh *et al.*, 1977). The thiazole C-nucleosides (**82**) and (**83**) were prepared, and interestingly, only the latter possessed significant antiviral activity (Srivastava *et al.*, 1977b).

As analogues of bredinin (**84**) and virazole (**77**), pyrrolidinoneribosides (**85**) were prepared from isoxazolidine nucleosides (Vasella, 1977).

77   78

79   80   8l

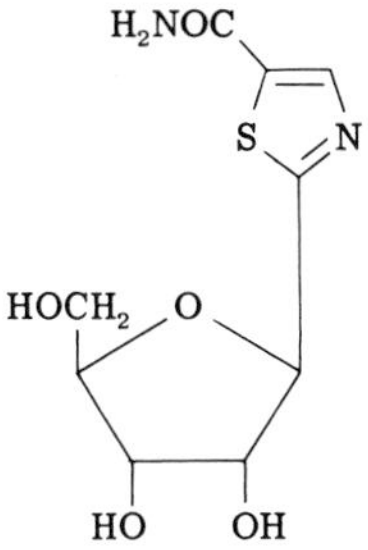

82 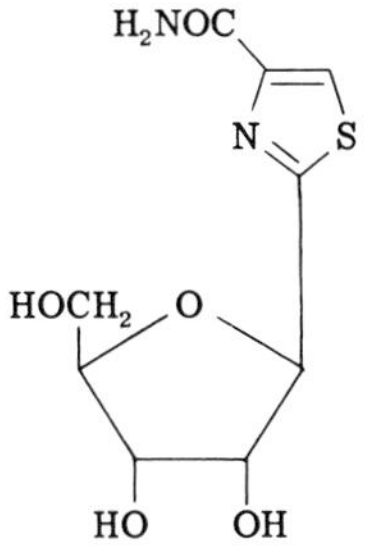   83 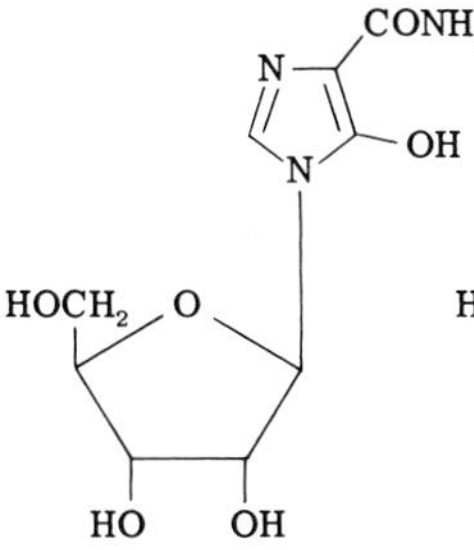   84 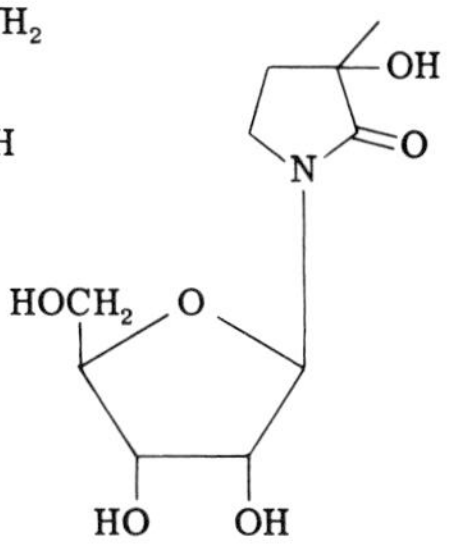   85

e. Miscellaneous   A new route to the regiospecific synthesis of 4-substituted triazole reversed nucleosides was exploited by Sasaki *et al.* (1978b). Thus, $N',5'$-anhydro-N^ω-($2',3'$-O-isopropylidene-β-D-ribofuranosyl)-4- allophanoyl- 1,2,3- triazole (**87**), 6,5′- imino- 1- (5′- deoxy- $2',3'$- O- isopropylidene- β- D- ribofuranosyl)uracil (**88**), and 9,5′- cyclo- 3- ($2',3'$- O- isopropylidene-β-D-ribofuranosyl)-8-azaxanthine (**89**) have been synthesized, starting from 1-(5′-azido-5′-deoxy-$2',3'$-O-isopropylidene-β-D-ribofur-anosyl)uracil (**86**).

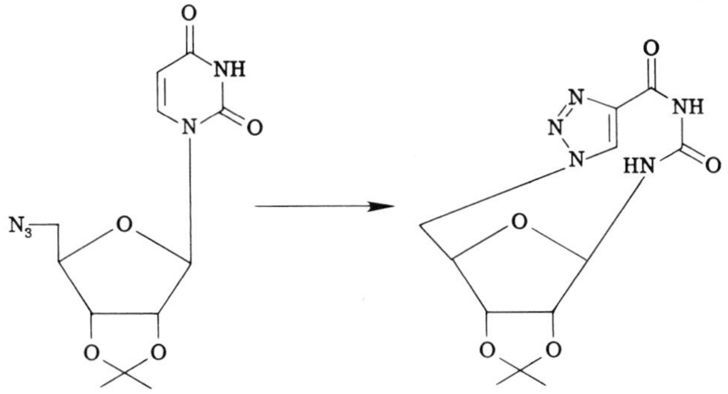

86   87

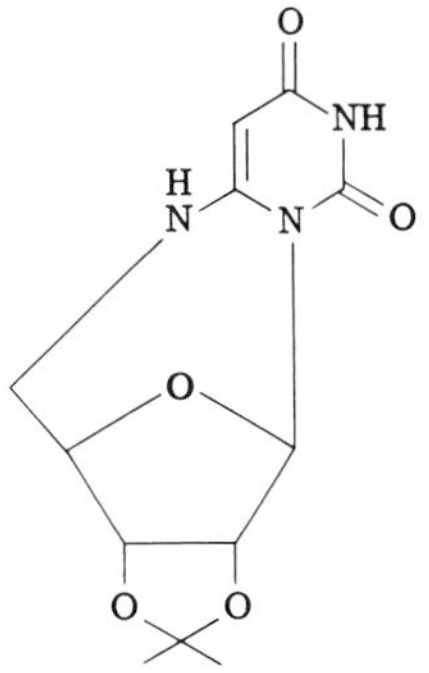

88 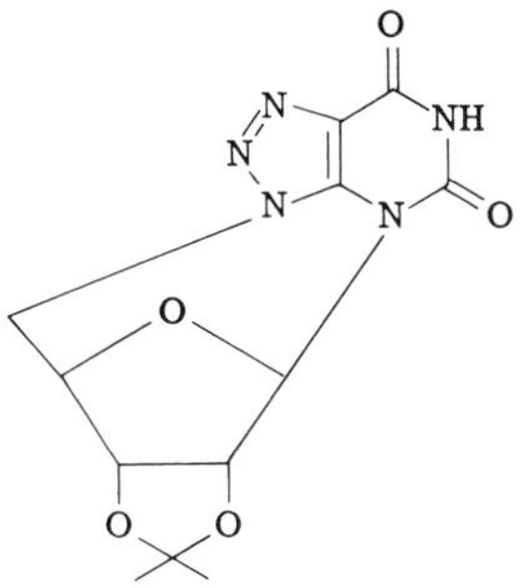   89

90   91   92

The nucleosides containing quinone heterocycles have been reported (Pickering *et al.*, 1977), and (**90**), (**91**), and (**92**) have been shown to be effective inhibitors of hypoxanthine phosphoribosyltransferase.

2. Modification of the Sugar Moiety and Substituents

The modification of the carbohydrate moiety is also a very active field with the investigators in search of potentially active nucleoside analogs, and many interesting papers have been published (Table III). However, the structure–activity relationships are still uncertain and require further subtle modification in the future.

The nucleoside antibiotic angustmycin A(**93**), which shows modest antitumor activity, was synthesized starting with D-fructose by multistep procedures (Prisbe *et al.*, 1976).

3'- Deoxy- 3'- C- dibromomethylidene- adenosine (**94**) was prepared from 3- deoxy- 3- C- dibromomethylidene- 1,2,5,6- di- *O*- isopropylidene- α- D- hexo-

93   94

furanose, and showed slow deamination by adenosine deaminase (Tronchet and Schwarzenbach, 1977).

By analogy to the 4',5' unsaturation found in decoyinine (angustmycin A)

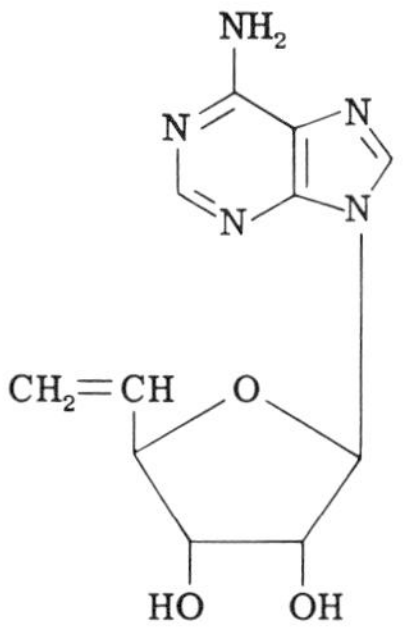

95 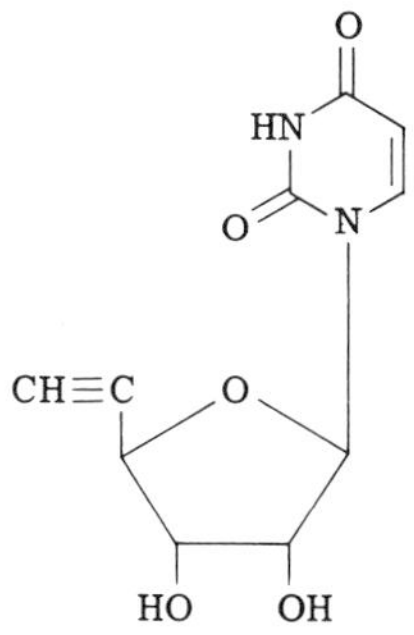   **96** 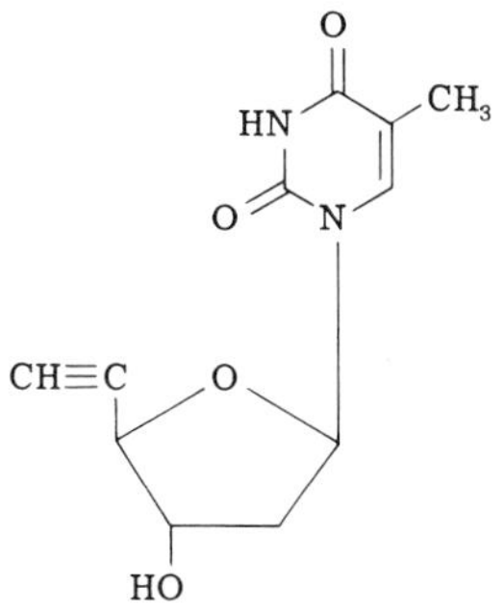   **97**

nucleosides, 9-(5,6-dideoxy-β-D-ribo-hex-5-enofuranosyl)adenine (**95**) and its enantiomers were prepared (Lerner, 1978).

Nucleosides bearing the acetylenic function at various positions in the carbohydrate moiety of pyrimidines and purines are considered to be of interest as potential antimetabolites. Thus, 1-(5,6-dideoxy-β-D-ribo-hex-5-ynofuranosyl)-uracil (**96**) and the corresponding 5-methyluracil (**97**) were synthesized (Sharma and Bobek, 1978).

The structure of 2'-amino-2'-deoxyguanosine (**98**), the first natural nucleoside having the 2'-amino ribose structure, was synthetically confirmed, and 2'-amino-2'-deoxyadenosine (**99**) was also obtained via cyclonucleosides (Ikehara et al., 1976).

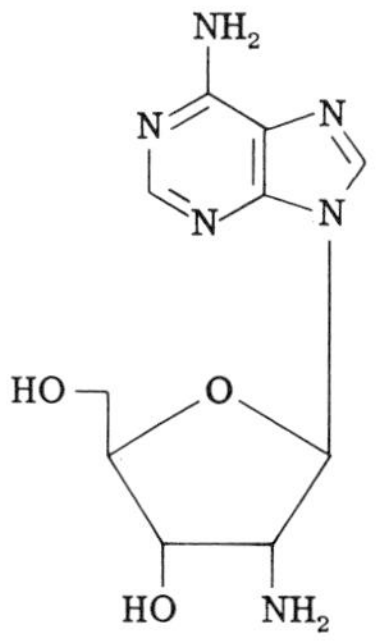

98 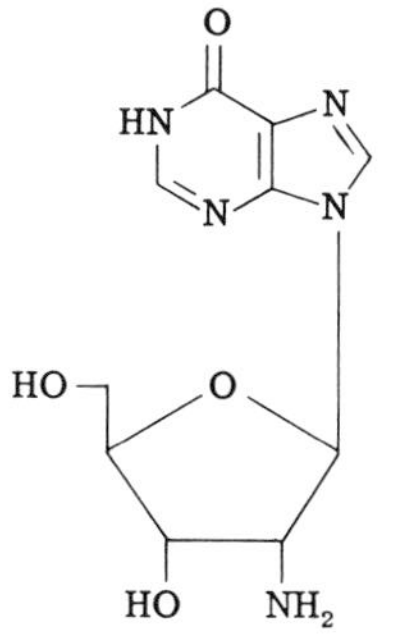   **99**

Eight carbocylic puromycin analogues, in which the furanosyl ring of puromycin is replaced with a cyclopentyl system, were synthesized. Two of them (**100** and **101**) shown here were found to be remarkably active against HeLa cells in a tissue culture (Suami et al., 1978).

A new class of chloroethyl- and methylnitrosourea analogs (**102**) and (**103**) of thymidine has been synthesized and two of them shown here inhibited L-1210 cell growth in culture (Lin et al., 1978).

TABLE III

Structure–Activity Relationship of Nucleosides of Modified Sugars

Nucleoside	R^1	R^2	R^3	R^4	R^5	R^6	Compound no.	Biological activity	References
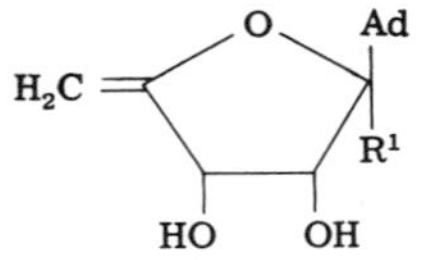 (structure)	CH_2OH H						93	(Angustmycin A)	Prisbe et al. (1976)
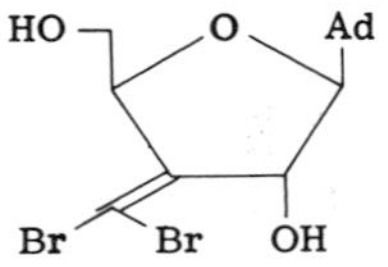 (structure)	CH=CH						94	Inactive against E. coli at 200 μg/ml	Tronchet and Schwarzenbach (1977)
(structure)	H	C≡CH	OH				95		Lerner (1978)
	CH_3	C≡CH	H				96		Sharma and Bobek (1978)
	F	CH_2I	OH				97		Sharma and Bobek (1978) Long et al. (1976)

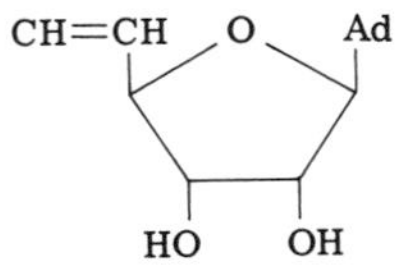

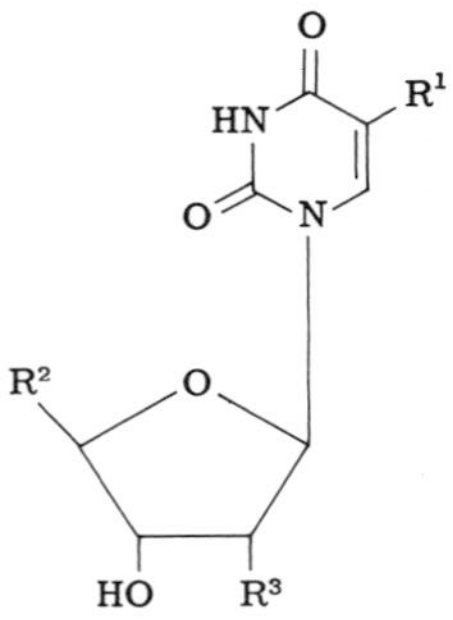

MCD μg/ml against HeLa cells

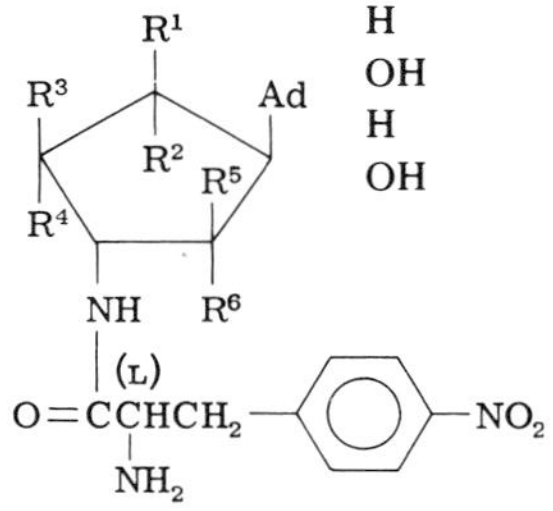

R¹	R²	R³	R⁴	R⁵	R⁶			
Ad	H	OH	H	H	OH	15.6	**100**	Suami et al. (1978)
OH	H	H	OH	H	OH	31.2	**101**	Suami et al. (1978)
H	OH	OH	OH	OH	H	1000		Suami et al. (1978)
OH	H	H	OH	OH	H	250		Suami et al. (1978)

Nucleoside	Compound no.	Biological activity	References
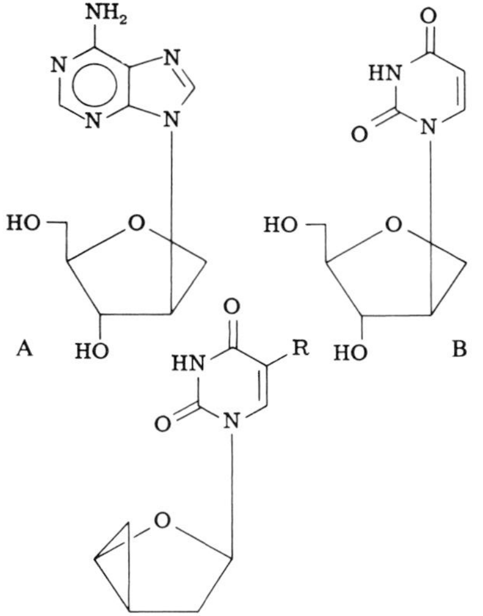 A	**109**	No antitumor activity	Montgomery and Thomas (1978)
B	**110**	Inactive against L-1210	Montgomery and Thomas (1978)
C	**111**		Adachi et al. (1977)
D	**111**		Adachi et al. (1977)

(C) R = CH₃
(D) R = H

(*Continued*)

TABLE III (*Continued*)

Nucleoside		Compound no.	Biological activity	References
	R¹			
	Cl	112		McCormick and McElhinney (1978)
	SH	112		McCormick and McElhinney (1978)
	NHCH₂Ph	112		McCormick and McElhinney (1978)
	OCH₂Ph	112		McCormick and McElhinney (1978)
	OMe	112		McCormick and McElhinney (1978)
	OH	112		McCormick and McElhinney (1978)
		113		McCormick and McElhinney (1978)
		α,114		Lerner (1976)
		β,115		Lerner (1976)

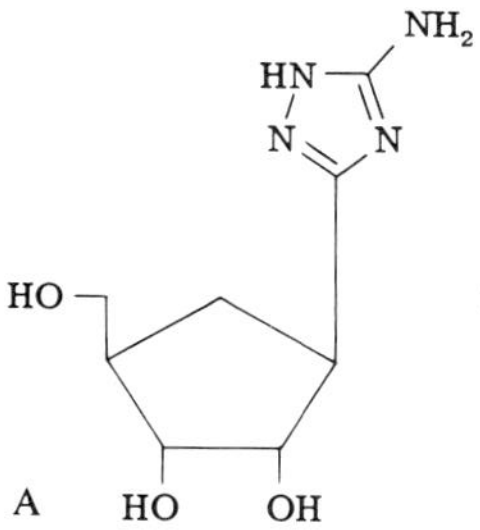

Just and Chalard-Faure (1976)
Just and Chalard-Faure (1976)

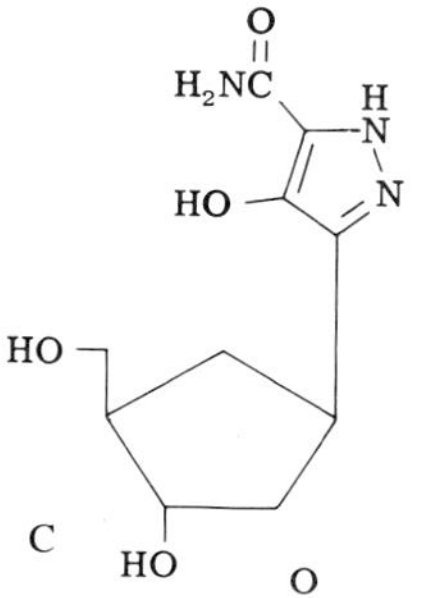

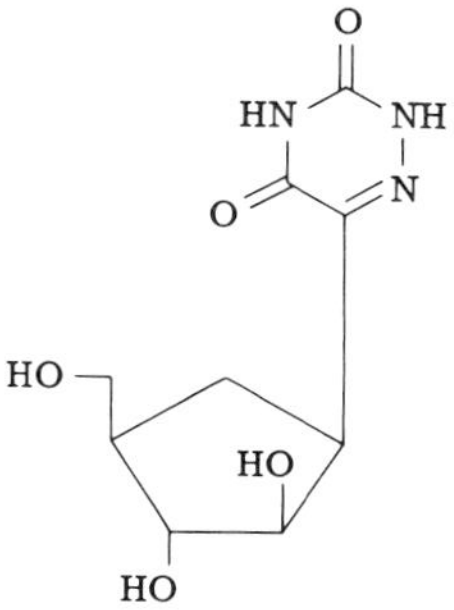

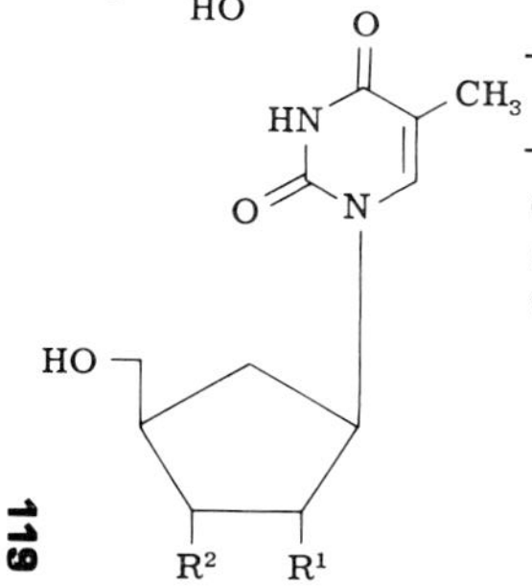

Just and Chalard-Faure (1976)
Just and Chalard-Faure (1976)

Shealy and O'Dell (1976)
Shealy and O'Dell (1976)
Shealy and O'Dell (1976)

116
118

119
117

120
120
120

A
B

C
D

E
F
G

A   B

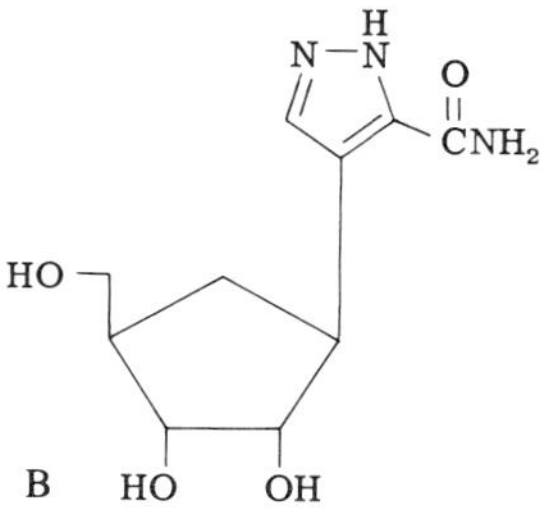

C   D

	R^1	R^2
(E)	H	OH
(F)	OH	H
(G)	OH	OH

(Continued)

119

TABLE III (Continued)

Nucleoside	R¹	R²	R³	R⁴	R⁵	Compound no.	Biological activity	References
	CH₃	H	OH	OH	H			Tadano et al. (1978)
	CH₃	H	OH	H	OH			Tadano et al. (1978)
	CH₃	OH	H	H	OH		Inactive against HeLa cells	Tadano et al. (1978)
	H	H	OH	OH	H			Tadano et al. (1978)
	H	H	OH	H	OH			Tadano et al. (1978)
	H	OH	H	H	OH			Tadano et al. (1978)

	R¹	R²	R³				Biological activity	References
	NH₂	OH	NH₂				Active against HSV-1, VV	Daluge and Vince (1978)
	NMe₂	NH₂	OH					Daluge and Vince (1976)

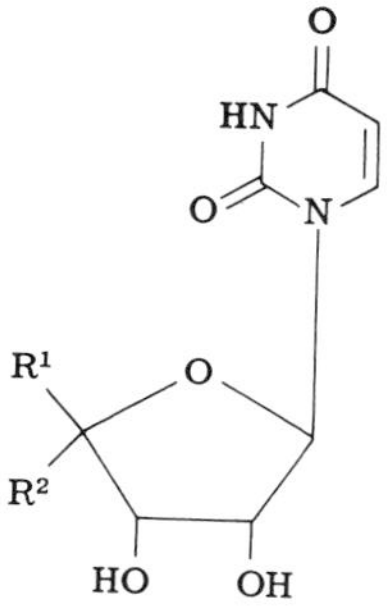

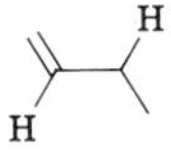

CH$_2$OH   H
CH$_2$OH   H
CH$_3$ CH$_2$OH

H
H$_3$C

H
H

H$_3$C

CH$_3$

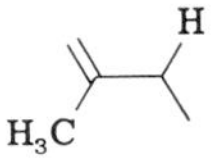

CH$_2$OH
CH$_2$OH
CH$_2$OH

H
H
H$_3$C
H
CH$_3$
H

O
HN
O
N
R^1
R^2
O
HO
OH

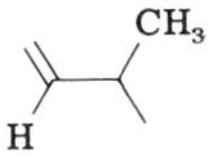

Secrist and Winter (1978)

Secrist and Winter (1978)

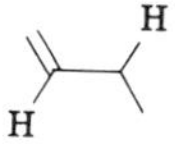

Secrist and Winter (1978)

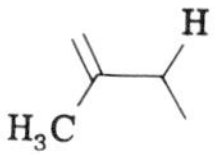

Secrist and Winter (1978)

Secrist and Winter (1978)

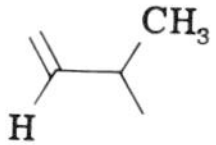

Secrist and Winter (1978)

H
Na
Me
Et

CH$_2$CH$_2$OH

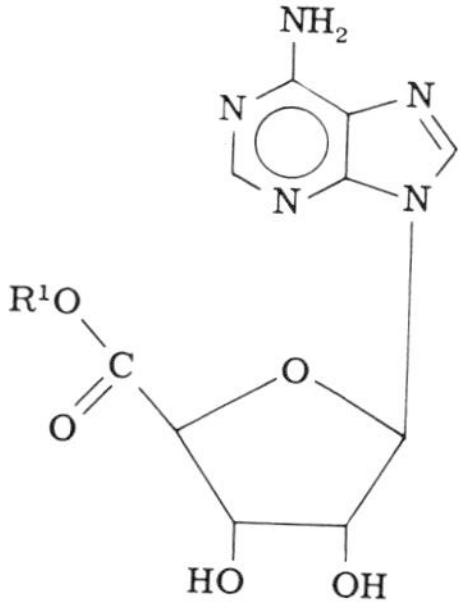

} Inactive as antiangial

} Active as antiangial

NH$_2$
N
N
N
N
R^{1}O
O
O
HO
OH

Prasad et al. (1976)
Prasad et al. (1976)
Prasad et al. (1976)
Prasad et al. (1976)

Prasad et al. (1976)

(Continued)

121

TABLE III (*Continued*)

Nucleoside	Compound no.	Biological activity	References
	A B	No antimicrobial activity	Just and Chalard-Faure (1976) Yamazaki *et al.* (1977)
	C (Partial component of oxamicetin)		Lichtenthaler and Kulikowski (1976)

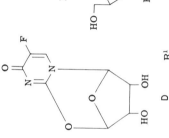

Long *et al.* (1976)

Cook (1977)

Sasaki *et al.* (1978a)
Sasaki *et al.* (1978a)
Sasaki *et al.* (1978a)

Robins and Muhs (1976)
Robins and Muhs (1976)

Inactive against sarcoma 180

D

E

E	R^1	R^2
	NH_2	NH_2
	NH_2	H
	OH	H

D

	R
	SCH_2Ph
	H

123

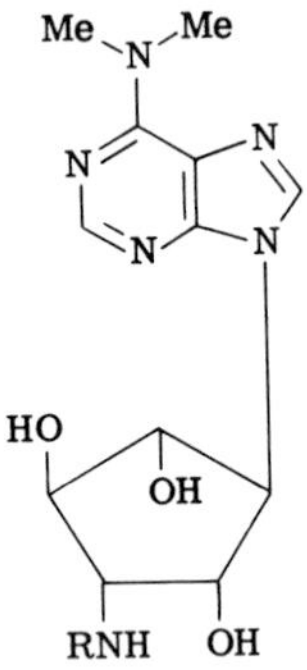

100 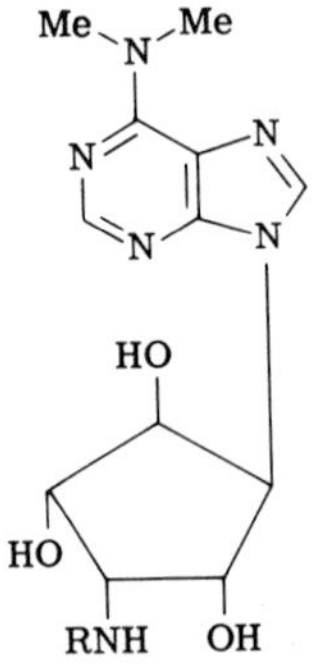   101

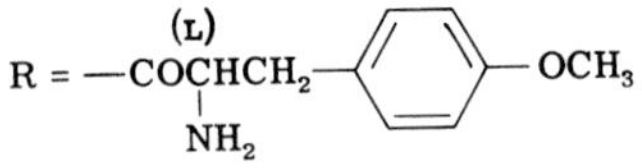

$$R = \underset{\underset{NH_2}{|}}{-COCHCH_2} - \underset{(L)}{} \!\!\!\!\!\!\! \bigcirc \!\!\! - OCH_3$$

2'-α-Substituted-2'-deoxyadenosine analogs (**104**) were synthesized from the corresponding trifluoromethanesulfonate with various nucleophiles (Ranganathan, 1977).

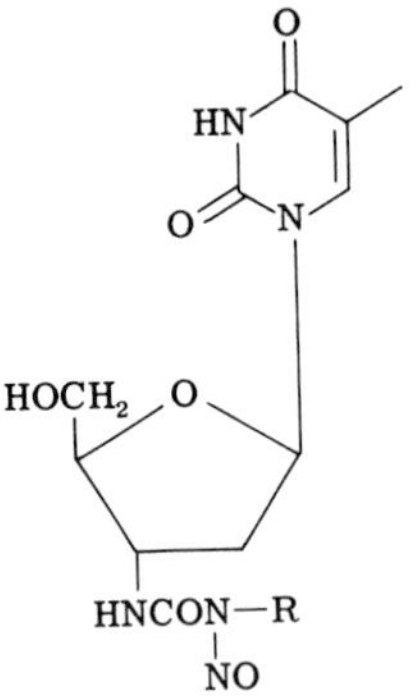

102 R = CH$_3$
103 R = CH$_2$CH$_2$Cl

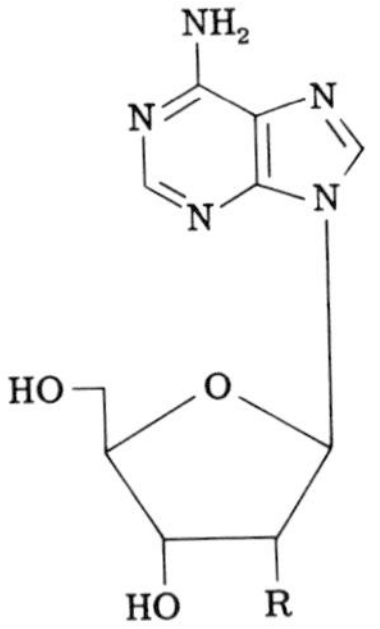

R = N$_3$, F, Cl, Br, I, SAc, OAc

104

Opening of oxirane rings (**105** and **106**) of 9-(2,3-anhydro-β-D-ribofuranosyl)adenine with various reagents afforded 2'- and 3'-azido, -amino, and -halo substituted deoxyadenosines (Mengel and Wiedner, 1976).

3'-Deoxy-3'-substituted arabinofuranosylpyrimidine nucleosides (**107**) and

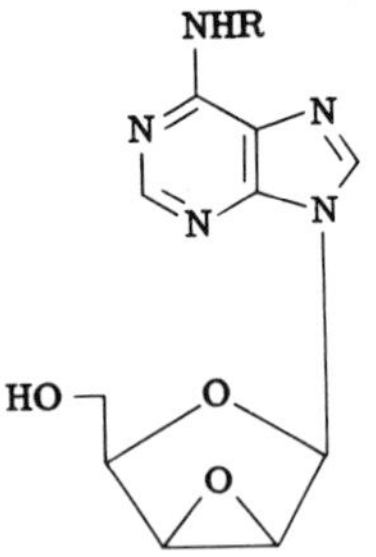

105 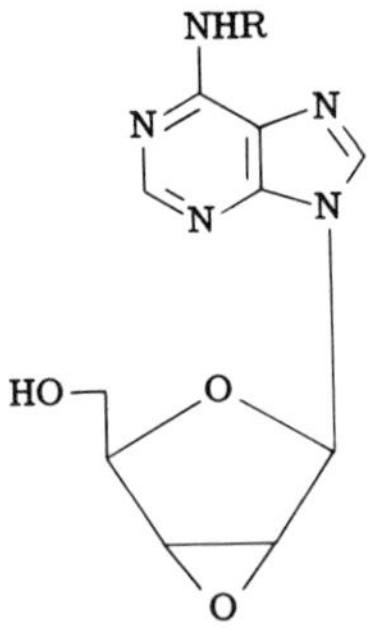   106

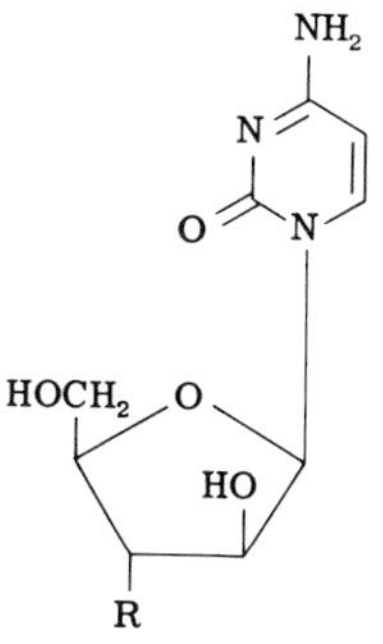

107 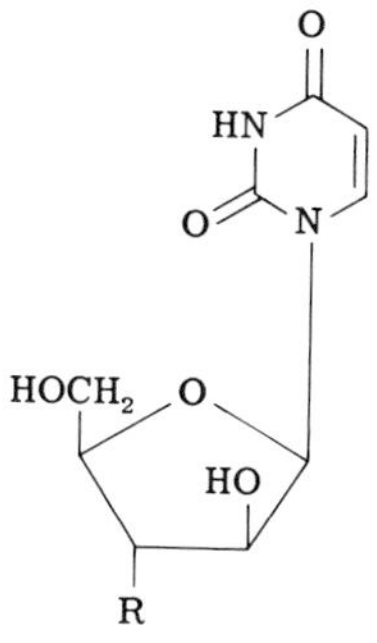   108

(**108**) showed no significant inhibitory properties against L-5178Y cells in culture (Hollenberg *et al.*, 1977).

Purine and pyrimidine derivatives of 1,4- anhydro- 2- deoxy- D- arabinitol (isonucleoside) (**109**) and (**110**) were reported by Montgomery and Thomas (1978).

Nucleosides (**111**) containing a fused cyclopropane ring were prepared (Adachi *et al.*, 1977).

4-Thiotetrafuranose nucleosides (**112**) and (**113**) were synthesized by McCormick and McElhinney (1978).

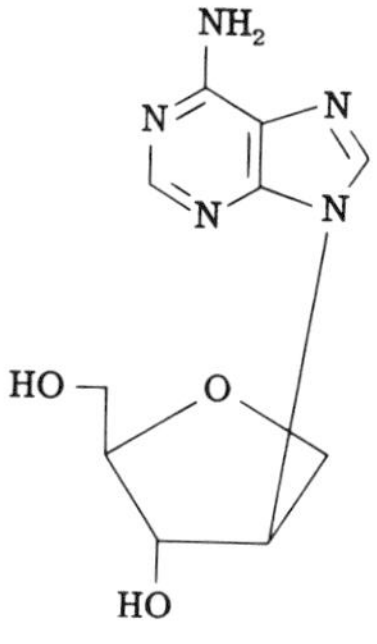

109 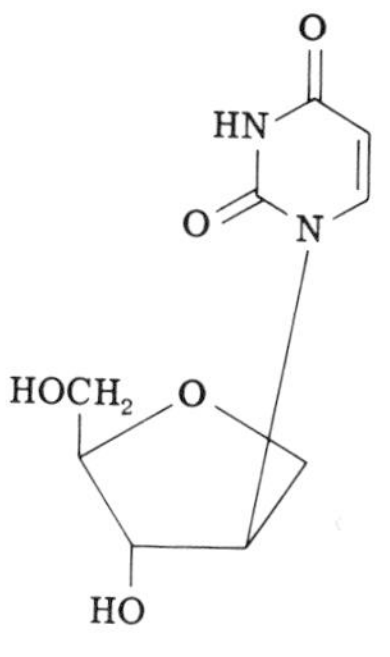   110 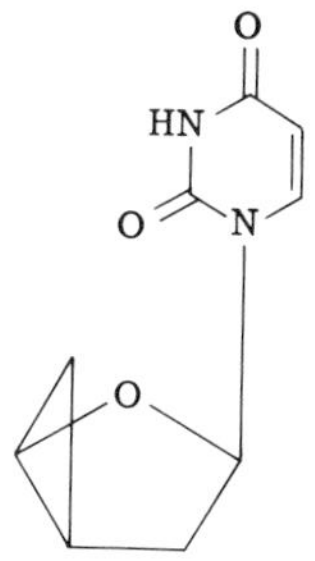   111

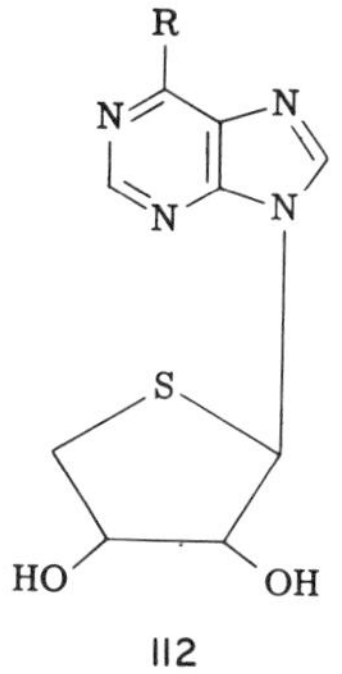

112 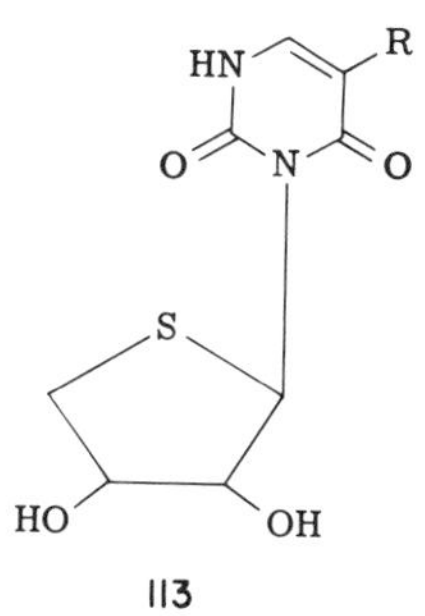   113

Adenine nucleosides (**114**) and (**115**) derived from 6-deoxyhexofuranoses were prepared as potential precursors for the synthesis of other compounds of biological interest by Lerner (1976).

114   **115**

Various carbocylic analogs of C-nucleosides (**116**)–(**119**) have been prepared starting from the Diels–Alder adducts of *trans-β-* bromoacrylic acid with cyclopentadiene by multistep reaction (Just and Chalard-Faure, 1976; Just and Donnini, 1977; Just and Kim, 1976; Just and Ouellet, 1976; Just *et al.*, 1976a,b).

Synthesis of the carbocyclic analogues of uracil nucleosides (**120**) showed no

116   **117**   **118**

R^1 = H or OH
R^2 = OH or H

119   **120**

activity against KB cells *in vitro* or L-1210 leukemia *in vivo* (Shealy and O'Dell, 1976).

A stereoselective synthesis of the "carbocyclic aminonucleoside" (**121**) and (**122**), analogues of puromycin aminonucleoside, 3'-amino-3'-deoxyadenoside and 3'-amino-3'-deoxyarabinosyladenine, was reported. The former showed highly significant antiviral activity (Daluge and Vince, 1976, 1978).

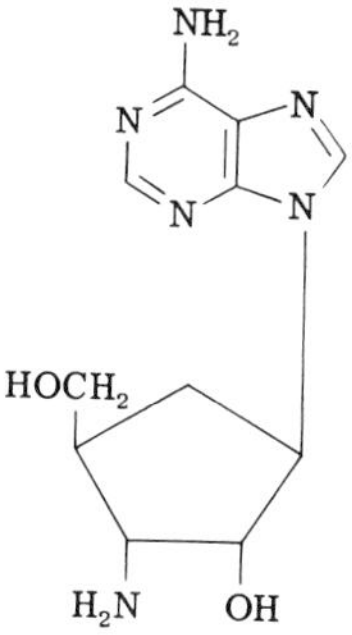

121

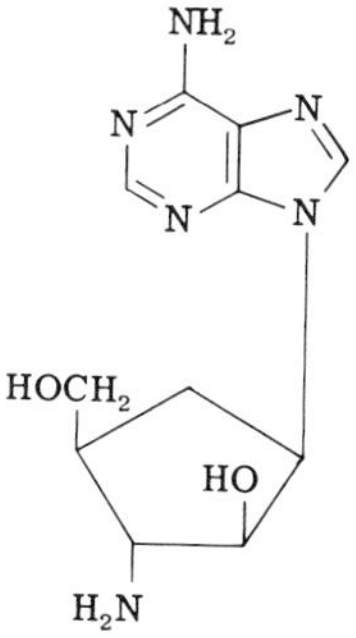

122

Acknowledgment

The author wishes to thank Masami Shimazaki for his invaluable assistance in helping him to prepare this chapter. He would especially like to acknowledge Hamao Umezawa for his invaluable comments and advice in this chapter.

REFERENCES

Adachi, T., Iwasaki, T., Miyoshi, M., and Inoue, I. (1977). *J.C.S. Chem. Commun.* 248.
Akiyama, M., Ohishi, J., Shirai, T., Akashi, K., Yoshida, K., Nishikido, J., Hayashi, H., Usubuchi, Y., Nichimura, D., Itoh, H., Shibuya, C., and Ishida, T. (1978). *Chem. Pharm. Bull.* **26**, 981.
Anderson, B. L., and Broom, A. D. (1977). *J. Org. Chem.* **42**, 997.
Bartholomew, D. G., Huffman, J. H., Matthews, T. R., Robins, R. K., and Revankar, G. R. (1976). *J. Med. Chem.* **19**, 814.
Beisler, J. A. (1978). *J. Med. Chem.* **21**, 204.
Beisler, J. A., Abbasi, M. M., Kelly, J. A., and Driscoll, J. S. (1977). *J. Med. Chem.* **20**, 806.
Bergstrom, D. E., and Ruth, J. L. (1976). *J. Am. Chem. Soc.* **98**, 1587.
Bernardo, S. D., and Weigele, M. (1976). *J. Org. Chem.* **41**, 287.
Bernardo, S. D., and Weigele, M. (1977). *J. Org. Chem.* **42**, 109.
Bloch, A. (1973). *In* "Drug Design" (E. J. Ariëns, ed.), Vol. 4, pp. 286–378. Academic Press, New York.
Bobek, M., and Bloch, A. (1977). *J. Med. Chem.* **20**, 458.
Bridges, S. D., Brown, D. M., and Ogden, R. C. (1977). *J.C.S. Chem. Commun.* 460.
Caron, N., Lee, S. H., and Kimball, A. P. (1977). *Cancer Res.* **37**, 3274.
Chattopadhyaya, J. B., and Reese, C. B. (1977). *J.C.S. Chem. Commun.* 414.

Chattopadhyaya, J. B., and Reese, C. B. (1978). *J.C.S. Chem. Commun.* 86.

Cheng, C. S., Hinshaw, B. C., Panzica, R. P., and Townsend, L. B. (1976). *J. Am. Chem. Soc.* **98,** 7870.

Chu, C. K., Wempen, I., Watanabe, K. A., and Fox, J. J. (1976). *J. Org. Chem.* **41,** 2793.

Chu, C. K., Reichman, U., Watanabe, K. A., and Fox, J. J. (1977a). *J. Org. Chem.* **42,** 711.

Chu, C. K., Reichman, U., Watanabe, K. A., and Fox, J. J. (1977b). *J. Heterocycl. Chem.* **14,** 1119.

Chu, C. K., Reichman, U., Watanabe, K. A., and Fox, J. J. (1978). *J. Med. Chem.* **21,** 96.

Chwang, T. L., Wood, W. F., Parkhurst, J. R., Nesnow, S., Daenberg, P. V., and Heidelberger, C. (1976). *J. Med. Chem.* **19,** 643.

Cline, B. L., Panzica, R. P., and Townsend, L. B. (1976). *J. Heterocycl. Chem.* **13,** 1365.

Cook, A. F. (1977). *J. Med. Chem.* **20,** 344.

Cook, P. D., and Robins, R. K. (1978). *J. Org. Chem.* **43,** 289.

Cook, P. D., Rousseau, R. J., Mian, A. M., Dea, P., Meyer, R. B., and Robins, R. K. (1976). *J. Am. Chem. Soc.* **98,** 1492.

Cook, P. D., Day, R. T., and Robins, R. K. (1977). *J. Heterocycl. Chem.* **14,** 1295.

Daluge, S., and Vince, R. (1976). *Tetrahedron Lett.,* 3005.

Daluge, S., and Vince, R. (1978). *J. Org. Chem.* **43,** 2311.

Daves, G. D., and Cheng, C. C. (1976). *Prog. Med. Chem.* **13,** 303.

De las Heras, F. G., Tam, S. Y. K., Klein, R. S., and Fox, J. J. (1976a). *J. Org. Chem.* **41,** 84.

De las Heras, F. G., Chu, C. K., Tam, S. Y. K., Klein, R. S., Watanabe, K. A., and Fox, J. J. (1976b). *J. Heterocycl. Chem.* **13,** 175.

Dudycz, L., Shugar, D., De Clercq, E., and Descamps, J. (1977). *J. Med. Chem.* **20,** 1354.

Earl, R. A., and Townsend, L. B. (1977). *J. Heterocycl. Chem.* **14,** 699.

Edelman, M. S., Barfknecht, R. L., Huet-Rose, R., Gosuslawaski, S., and Mertes, M. P. (1977). *J. Med. Chem.* **20,** 669.

Elliott, R. D., and Montgomery, J. A. (1977). *J. Med. Chem.* **20,** 116.

Elliott, R. D., and Montgomery, J. A. (1978). *J. Med. Chem.* **21,** 112.

Fuertes, M., Lopez, T. G., Munoz, G. G., and Stud, M. (1976). *J. Org. Chem.* **41,** 4074.

Gough, G. R., Nobbs, D. M., Middleton, J. C., Caredes, F. P., and Maguire, M. H. (1978). *J. Med. Chem.* **21,** 520.

Hamamura, E. K., Prystasz, M., Verheyden, J. P. H., Moffatt, J. G., Yamaguchi, K., Uchida, N., Sato, K., Nomura, A., Shiratori, O., Takase, S., and Katagiri, K. (1976). *J. Med. Chem.* **19,** 654, 663.

Hanessian, S., and Pernet, A. G. (1976). *Adv. Carbohydr. Chem. Biochem.* **33,** 111.

Hirota, K., Watanabe, K. A., and Fox, J. J. (1978). *J. Org. Chem.* **43,** 1193.

Hobbs, J. B., and Eckstein, F. (1977). *J. Org. Chem.* **42,** 714.

Hollenberg, D. H., Watanabe, K. A., and Fox, J. J. (1977). *J. Med. Chem.* **20,** 113.

Huynh-Dinh, T., Igolen, J., Bisagni, E., Marquet, J. P., and Civier, A. (1977). *J.C.S. Perkin I,* 761.

Ikehara, M. (1969). *Acc. Chem. Res.* **2,** 47.

Ikehara, M., Maruyama, T., and Miki, H. (1976). *Tetrahedron Lett.,* 4485.

Itaya, T., and Ogawa, K. (1978). *Tetrahedron Lett.,* 2907.

Jones, A. S., Serafinowski, P., and Walker, R. T. (1977). *Tetrahedron Lett.,* 2459.

Just, G., and Chalard-Faure, B. (1976). *Can. J. Chem.* **54,** 861.

Just, G., and Donnini, G. P. (1977). *Can. J. Chem.* **55,** 2998.

Just, G., and Kim, S. (1976). *Can. J. Chem.* **54,** 2935.

Just, G., and Lim, M. (1977). *Can. J. Chem.* **55,** 2993.

Just, G., and Ouellet, R. (1976). *Can. J. Chem.* **54,** 2925.

Just, G., Reader, G., and Chalard-Faure, B. (1976a). *Can. J. Chem.* **54,** 849.

Just, G., Ramjeesingh, M., and Liak, T. J. (1976b). *Can. J. Chem.* **54,** 2940.

Kampf, A., Pillar, C. J., Woodford, W. J., and Mertes, M. P. (1976). *J. Med. Chem.* **19**, 909.

Kikugawa, K., Suehiro, H., Yanase, R., and Aoki, A. (1977a). *Chem. Pharm. Bull.* **25**, 1959.

Kikugawa, K., Suehiro, H., and Aoki, A. (1977b). *Chem. Pharm. Bull.* **25**, 2624.

Kondo, K., and Inoue, I. (1977). *J. Org. Chem.* **42**, 2809.

Lerner, L. M. (1976). *J. Org. Chem.* **41**, 306.

Lerner, L. M. (1978). *J. Org. Chem.* **43**, 2469.

Lewis, A. F., Long, R. A., Roti, L. W., and Townsend, L. B. (1976). *J. Heterocycl. Chem.* **13**, 1359.

Lichtenthaler, F. W., and Kulikowski, T. (1976). *J. Org. Chem.* **41**, 600.

Lin, T. S., and Prusoff, W. H. (1978a). *J. Med. Chem.* **21**, 106.

Lin, T. S., and Prusoff, W. H. (1978b). *J. Med. Chem.* **21**, 109.

Lin, T. S., Neenan, J. P., Cheng, Y. C., Prusoff, W. H., and Ward, D. C. (1976a). *J. Med. Chem.* **19**, 495.

Lin, T. S., Chai, C., and Prusoff, W. H. (1976b). *J. Med. Chem.* **19**, 915.

Lin, T. S., Fischer, P. H., Shiau, G. T., and Prusoff, W. H. (1978). *J. Med. Chem.* **21**, 130.

Long, R. A., Matthews, T. R., and Robins, R. K. (1976). *J. Med. Chem.* **19**, 1072.

McCormick, J. E., and McElhinney, R. S. (1978). *J.C.S. Perkin I*, 500.

Makabe, O., Yajima, J., and Umezawa, S. (1976). *Bull. Chem. Soc. Jpn.* **49**, 3552.

Makabe, O., Suzuki, H., and Umezawa, S. (1977). *Bull. Chem. Soc. Jpn.* **50**, 2689.

Martinez, A. P., Lee, W. W., and Henry, D. W. (1977). *J. Med. Chem.* **20**, 341.

Mengel, R., and Wiedner, H. (1976). *Chem. Ber.* **109**, 433, 1359.

Miyoshi, T., Suzuki, S., and Yamazaki, A. (1976). *Chem. Pharm. Bull.* **24**, 2089.

Mizuno, Y. (1977). "Synthesis of Nucleosides and Nucleotides," p. 1. Maruzen, Tokyo. (In Jpn.)

Montero, J. L. G., Bhat, G. A, Panzica, R. P., and Townsend, L. B. (1977). *J. Heterocycl. Chem.* **14**, 483.

Montgomery, J. A., and Thomas, H. J. (1978). *J. Org. Chem.* **43**, 541.

Montgomery, J. A., Shortnacy, A. T., Arnett, G., and Shannon, W. M. (1977). *J. Med. Chem.* **20**, 401.

Nair, V., and Emanuel, D. J. (1977). *J. Am. Chem. Soc.* **99**, 1571.

Nakamura, H., Koyama, G., Iktaka, Y., Ohno, M., Yagisawa, N., Kondo, S., Maeda, K., and Umezawa, H. (1974). *J. Am. Chem. Soc.* **96**, 4327.

Nakatsuka, S., Ohgi, T., and Goto, T. (1978). *Tetrahedron Lett.*, 2579.

Narang, A. S., and Vince, R. (1977). *J. Med. Chem.* **20**, 1684.

Noyori, R., Sata, T., and Hayakawa, Y. (1978). *J. Am. Chem. Soc.* **100**, 2561.

Ohgi, T., Goto, T., Kasai, H., and Nishimura, S. (1976). *Tetrahedron Lett.*, 367.

Ohgi, T., Kondo, T., and Goto, T. (1977). *Tetrahedron Lett.*, 4051.

Ohno, M., Yagisawa, N., Shibahara, S., Kondo, S., Maeda, K., and Umezawa, H. (1974). *J. Am. Chem. Soc.* **96**, 4326.

Perman, J., Sharma, R. A., and Bobek, M. (1976). *Tetrahedron Lett.*, 2427.

Pickering, M. V., Dea, P., Streeter, D. G., and Witkowski, J. T. (1977). *J. Med. Chem.* **20**, 818.

Poonian, M. S., and Nowoswiat, E. F. (1977). *J. Org. Chem.* **42**, 1109.

Poonian, M. S., Nowoswiat, E. F., Blount, J. F., and Kramer, M. J. (1976). *J. Med. Chem.* **19**, 1017.

Prasad, R. N., Fung, A., Tietje, K., Stein, H. H., and Brondyk, H. D. (1976). *J. Med. Chem.* **19**, 1180.

Prisbe, E. J., Smejkal, J., Verheyden, J. P. H., and Moffatt, J. G. (1976). *J. Org. Chem.* **41**, 1836.

Ranganathan, R. (1977). *Tetrahedron Lett.* 1291.

Reichman, U., Chu, C. K., Wempen, I., Watanabe, K. A., and Fox, J. J. (1976). *J. Heterocycl. Chem.* **13**, 933.

Revankar, G. R., and Robins, R. K. (1976). *J. Heterocycl. Chem.* **13**, 169.

Robins, M. J. (1977). *In* "Bioorganic Chemistry" (E. E. van Tamelen, ed.), Vol. 3, p. 221. Academic Press, New York.

Robins, M. J., and Muhs, W. H. (1976). *J.C.S. Chem. Commun.* 269.

Robins, M. J., and Muhs, W. H. (1978). *J.C.S. Chem. Commun.* 677.

Rosowsky, A., Lararus, H., and Yamashita, A. (1976). *J. Med. Chem.* **19,** 1265.

Ruth, J. L., and Bergstrom, D. E. (1978). *J. Org. Chem.* **43,** 2870.

Saneyoshi, M. (1977). "Approaches to Cancer Chemotherapy." Kodansha, Tokyo. (In Jpn.)

Sasaki, T., Minamoto, K., and Itoh, H. (1978a). *J. Org. Chem.* **43,** 2320.

Sasaki, T., Minamoto, K., Suzuki, T., and Sugiura, T. (1978b). *J. Am. Chem. Soc.* **100,** 2248.

Sato, T., Ito, R., Hayakawa, Y., and Noyori, R. (1978). *Tetrahedron Lett.,* 1829.

Sawa, T., Fukagawa, Y., Homma, H., Takeuchi, T., and Umezawa, H. (1967). *J. Antibiot., Ser. A* **20,** 227.

Secrist, J. A., and Winter, W. J. (1978). *J. Am. Chem. Soc.* **100,** 2554.

Secrist, J. A. (1978). *J. Org. Chem.* **43,** 2925.

Shapiro, R. (1977). *In* "Bioorganic Chemistry" (E. E. van Tamelen, ed.), Vol. 3, p. 245. Academic Press, New York.

Sharma, R. A., and Bobek, M. (1978). *J. Org. Chem.* **43,** 367.

Shealy, Y. F., and O'Dell, C. A. (1976). *J. Heterocycl. Chem.* **13,** 1015.

Shimazaki, M., Kondo, S., Maeda, K., Ohno, M., and Umezawa, H. (1979). *J. Antibiot.* **32,** 537.

Skulnick, H. (1978). *J. Org. Chem.* **43,** 3188.

Srivastava, P. C., Streeter, D. G., Matthews, T. R., Allen, L. B., Sidwell, R. W., and Robins, R. K. (1976). *J. Med. Chem.* **19,** 1020.

Srivastava, P. C., Rousseau, R. J., and Robins, R. K. (1977a). *J.C.S. Chem. Commun.* 151.

Srivastava, P. C., Pickering, M. V., Allen, L. B., Streeter, D. G., Campbell, M. T., Witkowski, J. T., Sidwell, R. W., and Robins, R. K. (1977b). *J. Med. Chem.* **20,** 256.

Suami, T., Tadano, K., Ayabe, M., and Emori, Y. (1978). *Bull. Chem. Soc. Jpn.* **51,** 855.

Suhadolnik, R. J. (1970). "Nucleoside Antibiotics." Wiley (Interscience), New York.

Tadano, K., Horiuchi, S., and Suami, T. (1978). *Bull. Chem. Soc. Jpn.* **51,** 897.

Tam, S. Y. K., Hwang, T. S., De las Heras, F. G., Klein, R. S., and Fox, J. J. (1976). *J. Heterocycl. Chem.* **13,** 1305.

Torrence, P. F., and Bhooshan, B. (1977). *J. Med. Chem.* **20,** 974.

Torrence, P. F., Spencer, J. W., and Bobst, A. M. (1978). *J. Med. Chem.* **21,** 228.

Townsend, L. B., Tolman, R. L., Robins, R. K., and Milne, G. H. (1976). *J. Heterocycl. Chem.* **13,** 1363.

Tronchet, J. M. J., and Schwarzenbach, D. (1977). *Helv. Chim. Acta* **60,** 1989.

Ueda, T., Miura, K., and Kasai, T. (1978). *Chem. Pharm. Bull.* **26,** 2122.

Vasella, A. (1977). *Helv. Chim. Acta* **60,** 426.

Woo, P. W. K., Dion, H. W., Lange, S. M., Dahl, L. F., and Durham, L. J. (1974). *J. Heterocycl. Chem.* **11,** 641.

Yamazaki, T., Sugiyama, H., Matsuda, K., and Seto, S. (1977). *Bull. Chem. Soc. Jpn.* **50,** 3423.

Zemlicka, J., and Owens, J. (1977). *J. Org. Chem.* **42,** 517.

CHAPTER 4

Mitomycins

WILLIAM A. REMERS

I. DISCOVERY AND DEVELOPMENT

The first mitomycins were discovered in 1956 by Hata and collaborators (1956) at the Kitasato Institute. From an extract of the culture fluids of *Streptomyces caespitosus* containing many active fractions, they were able to isolate two crystalline compounds. These compounds, designated mitomycins A and B, showed highly potent antibacterial activity and moderate antitumor activity, but they were quite toxic in mice. Subsequent research by Wakaki and co-workers (1958) at the Kyowa Fermentation Industry showed that in some fermentations with *S. caespitosus* mitomycins A and B were not produced; however, the high biological activity remained. The complex mixture of products from these fermentations yielded a new compound which was named mitomycin C. This chance discovery of mitomycin C proved to be extremely valuable because it has much better antitumor activity than the other two mitomycins.

In 1960, a group at the Upjohn Company (De Boer *et al.*, 1961) reported the isolation of porfiromycin from *Streptomyces ardus*. Some mitomycin C also was obtained from cultures of this organism. The structural similarity between porfiromycin and mitomycin C (Fig. 1) was apparent from their behavior on chromatography and their ultraviolet absorption spectra. Porfiromycin is less potent than mitomycin C as an antitumor agent, but its toxicity is correspondingly lower. The net result is that mitomycin C has a slightly more favorable therapeutic index, at least in murine leukemias (Table I).

The final example of mitomycin discovery was furnished by Bohonus and

131

Anticancer Agents Based on Natural Product Models
Copyright © 1980 by Academic Press, Inc.
All rights of reproduction in any form reserved.
ISBN 0-12-163150-8

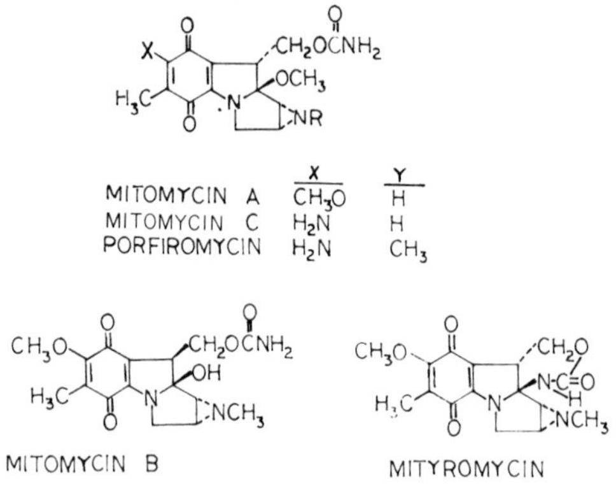

Figure 1. Structures of the natural mitomycins.

colleagues at the Lederle Laboratories (Lefemine *et al.*, 1962). They isolated mitomycins A, B, C, porfiromycin, and a less active compound named mityromycin from *Streptomyces verticillatus*. Lederle mounted a major effort under Webb (Webb *et al.*, 1962) to elucidate the mitomycin structures. Their successful effort, together with a confirmatory X-ray crystallographic analysis by Tulinsky (1962), was communicated in 1962. An independent structure determination for mitomycin C was reported subsequently by Stevens and co-workers (1964) at Wayne State University.

Mitomycins and porfiromycin have potent, broad-spectrum activities against bacteria. However, their toxicity has precluded clinical use in bacterial infections. Mitomycin C was rapidly developed as an anticancer drug in Japan (Frank and Osterberg, 1960). It was estimated that half of the patients receiving chemotherapy in Japan during the 1960s were given mitomycin C singly or in combination. In contrast, this agent was not well received in the United States. Clinicians were disappointed in the early trials by the unexpectedly high toxicity, especially leukopenia, and a more limited efficacy than had been reported by the Japanese investigators (Jones, 1959). Acceptance of mitomycin C came slowly and only after dosage schedules had been improved. Bristol Laboratories finally received approval in 1974 to introduce mitomycin C (Mutamycin®) into standard clinical practice. It has grown moderately, but steadily, in importance since then. At the present time, mitomycin C is used primarily for palliative treatment of adenocarcinomas of the stomach, pancreas, and colon. Secondary indications are carcinoma of the breast, malignant melanoma, and various squamous-cell carcinomas (Bristol Laboratories, 1974; Carter and Kershner, 1975).

Porfiromycin has not become an established drug, but there is continued interest in its potential. In phase I clinical studies, it produced responses against carcinomas of the cervix, ovary, stomach, colon, head, and neck (Izbicki *et al.*, 1972). It was highly toxic to bone marrow, with leukopenia and thrombocytopenia frequently observed (Izbicki *et al.*, 1972).

II. STRUCTURES AND CHEMICAL PROPERTIES

The mitomycins have unique chemical structures in which three different carcinostatic functions, aziridine, carbamate, and quinone, are arranged about a pyrrolo[1,2-a]indole nucleus (Fig. 1) (Webb *et al.*, 1962). This arrangement is such that, as the molecule occurs naturally, these functions are relatively inactive. However, upon chemical or biochemical reduction of the quinone (**1**) to a hydroquinone (**2**) (Scheme 1), the indolic nitrogen becomes much more electron-rich and it is able to participate in a number of interactions which activate the aziridine and carbamate functions as alkylating groups. The biological consequences of these interactions will be discussed in Section III. Among the important chemical consequences of the quinone reduction are the facility with which the 9a-methoxy group (hydroxy group in mitomycin B) can be replaced by other nucleophiles or lost along with the 9-hydrogen as methanol (or water). Thus, Hornemann *et al.* (1976) have shown that reduction of mitomycin C with sodium bisulfite followed by oxidation gave, among a complex mixture, a product (**7**) in which the 9a-methoxy group was replaced by sodium bisulfite. This product could be converted back into mitomycin C by reduction in methanol. When sodium borohydride was the reducing agent the product was a "demethoxymitomycin" (**8**) in which the 9a-methoxy group had been replaced by hydrogen (Kinoshita *et al.*, 1970a). Patrick and co-workers (1964) observed that the catalytic reduction of mitomycin B or *N*-methylmitomycin A followed by air oxidation gave 7-methoxy-1,1-(*N*-methylaziridino)mitosene (**6**), whose structure corresponds to the loss of water or methanol from starting material. Mitomycin B gave better yields in this reaction, which might reflect either the different leaving group or the difference in stereochemistry at position 9 in mitomycin B (Fig. 1). The structures and numbering of mitomycins are shown in Fig. 1. Compounds with the aziridine ring and full substitution at positions 9 and 9a are called mitosanes. Those with a 9,9a-double bond are called mitosenes. The name mitosene does not require an intact aziridine ring.

The chemical transformations described above can be rationalized in terms of a process in which the hydroquinone (**2**) reversibly loses methoxide or hydroxide to give a stabilized carbonium ion (**3**) capable of interacting with nucleophiles such as bisulfite ion or hydride to give the observed products, (**7**) and (**8**), after reoxidation to the quinone (Scheme 1). In the absence of a good nucleophile, carbonium ion (**3**) could lose a proton to give the hydroquinone of an

TABLE I

Structure–Activity Relationships for Mitomycins

Substituent (synonym)				Reduction[a] potential E½ (V)	Toxicity[a] LD$_{50}$ (mg/kg) in mice	Antibacterial activity[b] composite MIC (μg/mg)		Antitumor activity			
								L-1210[c]		P-388[d]	
X	Y	Z	R			Gram (+)	Gram (−)	Dose (mg/kg)	ILS	Dose (mg/kg)	ILS
CH$_3$O (Mitomycin A)	OCH$_3$	H	H	−0.19	2.0	0.01	0.57	—[e]	—	3.2	80
C$_6$H$_5$NH	OCH$_3$	H	H	−0.23	9.3	0.09	7.55	—	—	—	—
C$_2$H$_5$NH	OCH$_3$	H	H	−0.31	9.4	0.11	17.1	—	—	—	—
C$_3$H$_7$NH	OCH$_3$	H	H	—	—	—	—	2.7	45	—	—
HO(CH$_2$)$_2$NH	OCH$_3$	H	H	—	—	—	—	2.7	42	—	—
H$_2$N	OCH$_3$	H	H	−0.40	9.0	0.21	0.48	1.5	60	3.2	129–170

R₁	R₂	R₃	R₄	Compound	π							
H_2N	OCH_3	CH_3	H	(Mitomycin C)	—	55	0.27	11.4	10	50	12.8	125
				(Porfiromycin)	—	—	—	—	—	—	3.2	80
CH_3O	OCH_3	CH_3	H		−0.21	3.0	0.65	50	—	—	—	—
CH_3O	OH	CH_3	H	(Mitomycin B)	—	—	—	—	—	—	—	—
CH_3O	H		H		—	>200	>50	>50	—	—	—	—
C_6H_5NH	OH	CH_3	H		−0.23	5.6	0.47	>50	—	—	—	—
C_2H_5NH	OH	CH_3	H		−0.32	>100	4.4	>50	—	—	—	—
H_2N	OH	CH_3	H		−0.37	180	2.4	23	35	25	—	—
N	OH	CH_3	H		—	—	—	—	5.3	51	—	—
N	OCH_3	CH_3	H		—	—	—	—	0.5	53	—	—
H_2N	OCH_3	H	CH_3		—	—	—	—	27.5	49	—	—
H_2N	OCH_3	CH_3	C_2H_5		—	—	—	—	200	32	—	—
H_2N	OCH_3	H	CH_2CH_2OH		—	—	—	—	148	49	—	—
				7-Methoxy-1,2-(N-methyl-aziridino)mitosene (10)	−0.39	18.7	1.1	2.5	1.0	22	—	—
				7-Methoxy-1-hydroxy-2-methylaminomitosene (11)	−0.39	37.5	1.1	28	naᵉ	na	—	—
				7-Methoxy-1-acetoxymitosene (32)	—	—	—	—	—	—	6.4	62
				N-Methylmitomycin B (19)	—	9.4	—	—	7	32	—	—

[a] Taken from the data of Kinoshita and co-workers (Kinoshita et al., 1970b; Matsui et al., 1968).

[b] Taken from the data of Kinoshita and co-workers (Matsui et al., 1968). Six species each of Gram (+) and Gram (−) bacteria were used, and MICs (minimum inhibitory concentrations) were measured by the plate dilution method with heart infusion agar.

[c] Taken from the data of Kojima and co-workers (Kojima et al., 1972). Percent increase in length of survivals (ILS) of treated mice (single dose) relative to controls. Only the optimal dose is shown in the table.

[d] Unpublished data determined by W. T. Bradner at Bristol Laboratories. Percent increase in length of survivals (ILS) of treated mice (single dose) relative to controls. Only the optimal dose is shown in the table.

[e] na, Not active; —, not determined.

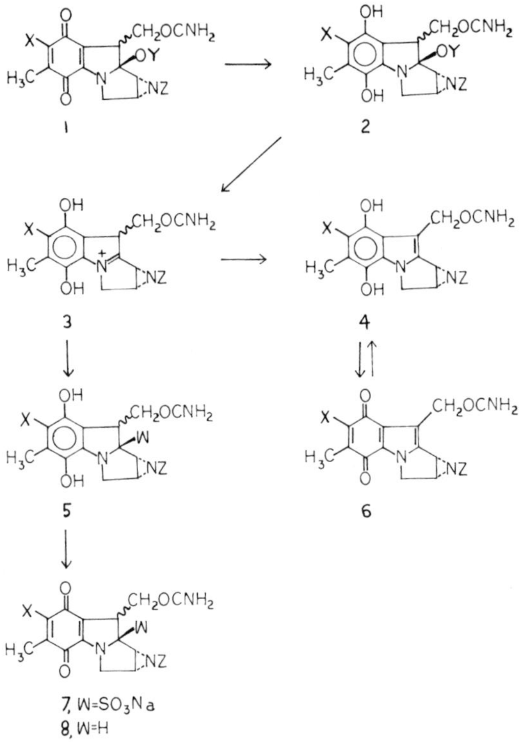

Scheme 1.   Transformations of mitomycin hydroquinones.

aziridinomitosene (**4**). Although this process accounts for the observed products, there is at present little direct supporting evidence for it. Additional implications for the process will be described for the mode-of-action studies in Section III.

When mitomycins such as N-methylmitomycin A (**9**) are treated with acidic reagents, they undergo transformations involving opening of the aziridine rings and the loss of methanol or water from the 9 and 9a positions (Matsui *et al.*, 1968; Webb *et al.*, 1962). The stereochemistry of substitutents in the resulting 1,2-disubstituted mitosenes (e.g., **11** and **12**) was examined in detail by Remers and co-workers. They found that the amino group and the hydroxyl group, or derivatives containing these functions, were invariably at positions 2 and 1, respectively (Taylor and Remers, 1975). The stereochemistry was predominantly *cis,* which is contrary to expectations based upon the ring opening of simple aziridines (Cheng and Remers, 1977). Aziridinomitosenes such as (**10**) also gave mostly the *cis*-aminohydrins (Scheme 2) (Cheng and Remers, 1977).

Mild alkaline hydrolysis of a 7-aminomitosane such as mitomycin C (**13**) gives the corresponding 7-hydroxy compound (**14**) (Garrett, 1963). This compound

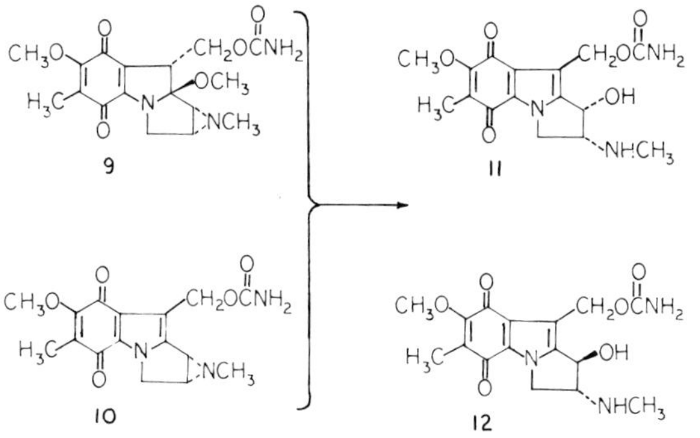

Scheme 2. Mild acid hydrolysis of mitomycins.

can be methylated with diazomethane to furnish mitomycin A (**15**) (Cheng and Remers, 1977; Matsui *et al.*, 1968). Conversely, mitomycin A can be converted into mitomycin C by treatment with ammonia (Webb *et al.*, 1962). Treatment of mitomycins with sodium methylate in methanol results in hydrolysis of the carbamate group (Kinoshita *et al.*, 1970a, 1971a). The resulting "decarbamoyl-mitomycins" (**16**) can be converted into a variety of acyl or carbamoyl derivatives (Kinoshita *et al.*, 1971a). Mitomycins such as A and C which have unsubstituted aziridine nitrogens can be alkalyted, sulfonylated, or acylated (Kinoshita *et al.*, 1971b; Uzu *et al.*, 1964; Wagner and Gitterman, 1962). For example, methyl iodide and potassium carbonate convert mitomycin C into porfiromycin (**17**) (Scheme 3) (Uzu *et al.*, 1964; Wagner and Gitterman, 1962; Webb *et al.*, 1962). The foregoing types of transformations have been used to interrelate the mitomycins. They also have afforded a large number of mitomycin analogs for antitumor testing (Section IV).

One rather novel type of mitomycin analog is *N*-methylmitomycin B, which is formed by treatment of mitomycin B with methyl iodide and strong base (Scheme

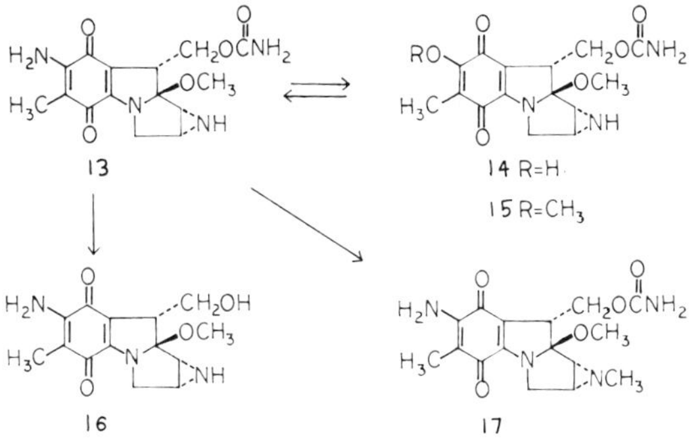

Scheme 3. Mitomycin interconversions.

4) (Kinoshita *et al.*, 1970a). There is some doubt about the reported structure of this compound, and confirmative evidence for this structure never has appeared. However, the compound and certain of its derivatives showed antileukemia activity in mice (Section IV).

A variety of synthetic routes to mitomycins and their simpler analogues have been reported. They are not discussed in detail in this chapter. However, they have been especially important for the preparation of numerous mitosene and indoloquinone analogs that are not available by chemical transformation of the natural mitomycins.

III. MODE OF ACTION

Mitomycins bind to DNA in a complex process that involves reductive activation of the mitomycin and alkylation of the DNA. The lethal event in cancer cells is believed to be cross-linking of strands in double helical DNA (Iyer and Szybalski, 1964). However, these cross-links involve only 10% of the total covalent interactions. Monofunctional alkylation of DNA is thought to be lethal in certain organisms that have deficient repair mechanisms (Szybalski and Iyer, 1967). Inhibition of DNA replication is an important result of covalent bonding by mitomycins. This result can lead to inhibition of RNA synthesis and a chain of events that produce cell morphology and death (Iyer and Szybalski, 1964).

Although the cross-linking of mitomycins to DNA in living cells was readily observed and NADPH-dependent cell extracts capable of activating mitomycins were prepared, it was difficult to reproduce the process in model *in vitro* systems. Reduction of the mitomycins could be readily accomplished enzymically or chemically, but covalent binding to DNA was only marginal and cross-linking was not observed. Tomasz *et al.* (1974) solved this problem by reducing the mitomycin under conditions in which an excess of unreduced mitomycin was always present. These conditions are optimum for the maintainance of the semiquinone radical, and it is this species which is thought to bind initially to the DNA. The existence of semiquinone radicals under these binding conditions has not been verified, but these species have been observed by electron spin resonance studies on mitomycins reduced electrochemically or by borohydride. Pat-

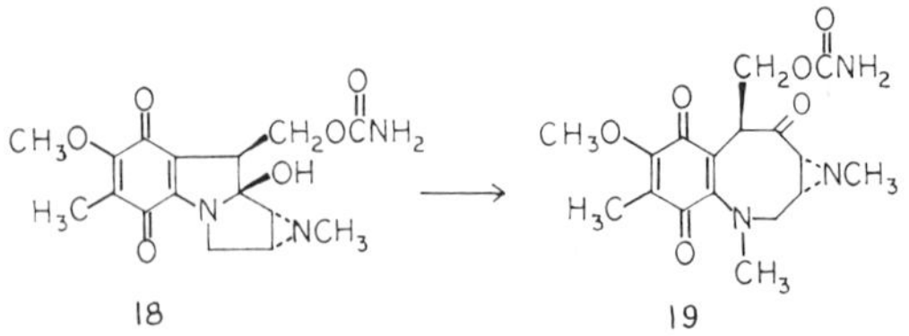

Scheme 4. Preparation of *N*-methylmitomycin B.

rick *et al*. (1964) reported that the semiquinone radical retained its 9a-hydroxyl group, whereas the corresponding hydroquinone readily eliminated it.

The foregoing observations can be combined with the pathway suggested by Iyer and Szybalski (1964) to provide the hypothetical mechanism for mitomycin activation and binding shown in Scheme 5. In this scheme the radical anion (21) binds to double-helical DNA. It is reduced further to hydroquinone (23), which readily loses methanol to give the aziridinomitosene hydroquinone (22). The structure of (22) is such that the indolic nitrogen can readily stabilize carbonium ions formed at position 1 by opening of the aziridine ring and at position 10 by loss of the carbamoyloxy group. Thus covalent bonds can be formed between these two positions and the DNA as in structure (24). The distances between these two positions are presumably appropriate for cross-linking. Since the covalent binding of mitomycins to DNA is pH-dependent, Lown *et al*. (1976) concluded that the aziridine ring is involved in the first binding interaction.

Tomasz (1976) reported that reduction of mitomycin C in the presence of a cell extract, followed by exposure to air resulted in the formation of hydrogen peroxide. This process occurs even more readily with DNA-bound mitomycin, and the redox cycle (e.g., 24 and 25 and return) may be repeated with hydrogen peroxide generation at each oxidation step. Thus the mitomycin can produce additional cell damage beyond DNA cross-linking. The observation that mitomycin C caused single-strand cleavage in covalently closed circular DNA of PM2 supports this concept (Lown *et al*., 1976).

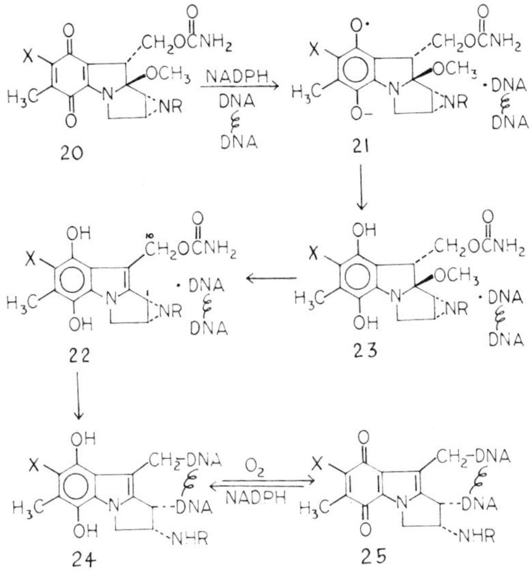

Scheme 5. Mitomycin activation and binding.

The specific sites at which mitomycins alkylate DNA are not known with certainty. Guanine residues appear to have a specific role in the binding, but the great heat stability of the binding products was said to rule out the N-7 of guanine (Tomasz *et al.*, 1974). Acid sensitivity of mitomycin-DNA linkages led Szybalski to suggest alkylation at O-6 of guanine. Further studies are needed to evaluate this possibility (Iyer and Szybalski, 1964).

The requirement that mitomycins and other quinone-containing antitumor agents must be reduced to the hydroquinone form before their functional groups are fully active as alkylating centers led Sartorelli to formalize the concept of "bioreductive alkylation" (see Lin *et al.*, 1974a). This concept has become important in understanding the mode of action of many antitumor agents, and it has led to the design of simpler benzoquinone analogues of the mitomycins (Section V).

IV. STRUCTURE–ACTIVITY RELATIONSHIPS

Because of their need for bioreductive activation and complex mode of action involving bifunctional alkylation, the mitomycins present a fascinating but difficult problem in defining structure–activity relationships. The quinone reduction potential is obviously of major importance to the bioreductive alkylation process. Kinoshita and colleagues (1971b; Matsui *et al.*, 1968) recognized this importance by comparing the half-wave polarographic reduction potentials of 7-substituted mitosanes against their toxicities, antibacterial activities, and antitumor activities (Table I). It can be generalized that the most readily reduced quinones are the most active against cultures of bacteria, especially the Gram-positive species. Antitumor activity in mice presents a more complex relationship to quinone reduction potential because selectivity in the lethal effect is a prime consideration. It is anticipated that readily reduced compounds such as mitomycin A would be active at low doses, assuming that they reach the cancer cells efficiently, but that their therapeutic indices might not be as high as those of less readily reduced compounds, since the latter might be more selective between cancer and normal cells. Some support for this anticipation can be found in Table I. For example, mitomycins A and B are highly toxic, but less efficaceous in treating experimental tumors than mitomycin C and porfiromycin. Mitomycin A shows some activity against P-388 murine leukemia at doses as low as 0.025 mg/kg/day, whereas mitomycin C is not active at this dose (Bradner, personal communication). However, we must note that factors other than quinone reduction potential can contribute to the activity of mitomycin analogs (see below).

The optimum quinone reduction potential is not known, although it probably is close to that of mitomycin C (about -0.4 volts). Mitosene analogues with 7-methoxy groups (such as compound **6**; $X = CH_3O$, $Y = H$) have reduction potentials in this range (-0.39 V), and they show significant antitumor activity in mice. However, 7-aminomitosenes, which are more difficult to reduce (-0.5 V) have poor antitumor and antibacterial activity. One type of substituent at position 7 that confers high potency to mitomycin analogs is the aziridino group, as found in compound (**26**) (Fig. 2) (Kojima *et al.*, 1972). It is not known whether the most significant feature of this group is its contribution to quinone reduction potential, its potential ability to provide a third alkylation site, or some other property such as an effect on lipophilicity. A large number of mitomycin analogs with variants at position 7 have been prepared and tested against tumors in mice. The 7-hydroxy analog (**14**) has received considerable attention in Japan because it causes less leukopenia that mitomycin C (Imai, 1976).

A variety of substituents have been introduced onto the aziridino nitrogen of mitomycins. The methyl group decreased both potency and toxicity, as indicated by the properties of mitomycin C and porfiromycin in Table I. Other substituents generally reduced the antitumor activity. This effect is expected for substituents that stabilize the aziridine ring; however, the methanesulfonyl derivative of mitomycin C was active against the Hirosaki ascites sarcoma at high doses. Cleavage of the aziridine ring, as in mitosenes (**11**) and (**12**), resulted in the loss of significant antitumor activity (W. T. Bradner, personal comunication; Kojima *et al.*, 1972).

At the 9a-position, a methoxyl group appears to confer the best antitumor activity (Table 1). Mitomycin B and derivatives which have 9a-hydroxyl groups are less active. However, these compounds also have stereochemistry at position

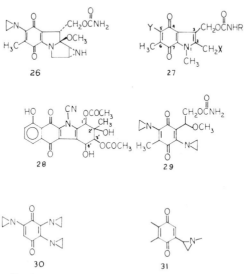

Figure 2.   Some important mitomycins and related compounds.

9 different from that of the other mitomycins. The 9a-hydrogen analogue of mitomycin C (**8**; X = NH$_2$, Z = H) was inactive. This result was attributed to its inability to form an indolohydroquinone such as (**22**) (Scheme 5), in which the aziridine and carbamoyloxy functions are activated as alkylation centers (Kinoshita *et al.*, 1970a).

Hydrolysis of the mitomycin carbamoyl group leads to the loss of antitumor activity. If the resulting 9-hydroxymethyl compound (e.g., **16**) is reconverted into a carbamate, the activity is restored. Certain substituted carbamates show antitumor activity in mice, but at dose levels higher than those of the corresponding unsubstituted carbamates (Table I) (Kojima *et al.*, 1972).

As mentioned previously, *N*-methylmitomycin B (**19**) and certain derivatives, which are reported to have an 8-membered ring structure, show activity against L-1210 leukemia in mice.

Weiss and co-workers (1968) prepared a large series of indoloquinone analogs of the mitomycins and tested them as antibacterial agents. This series was based on the systematic variation of substituents at each position on the indoloquinone nucleus. They made the following observations about structure–activity relationships (see structure **27**): a methyl or ethyl group was best at position 1; a methyl group was best at position 2; at position 3, the carbamoyloxymethyl group was necessary, with best activities obtained for the hydroxyethyl, methyl, and unsubstituted carbamates; at position 5 the ethyleneimino group was best, followed by the methoxyl group; and at position 6 the methyl group was best (Weiss *et al.*, 1968). It does not necessarily follow that these relationships will apply equally well to the antitumor activity of mitosanes, but Table I shows that at least the 7-aziridino analogue and the hydroxyethyl carbamate analogue of mitomycin C are highly active antitumor agents.

V. DESIGN AND SYNTHESIS OF NEW ANALOGUES

The most obvious starting point for the preparation of new mitomycin analogues was the transformation of reactive functional groups in the natural mitomycins. As discussed in Section IV, there were four different functionalities that could be manipulated without drastic loss of activity: the 7-methoxy group of mitomycin A or mitomycin B, the aziridine ring of mitomycin A or mitomycin C, the carbamoyloxymethyl sidechain, and elimination of the elements of methanol or water from positions 9 and 9a to give aziridinomitosenes. Retention of the capability for bifunctional alkylation was a requirement for antitumor activity in any mitomycin derivative. Also important was a quinone reduction potential optimum for selective action against cancer cells. This effect can be related to the fact that some cancer cells have greater reducing capabilities than normal cells (Cater and Phillips, 1954). Within the natural mitomycins, an octanol/water partition coefficient less than unity was found to correlate with the best antileukemia activity (Remers and Schepman, 1974). This correlation needs to be examined with additional analogues.

New substituents at the mitomycin 7-position offer the easiest way in which to modify the lipophilicity and quinone reduction potential without disturbing the alkylation sites, assuming that the mode of mitomycin action as outlined in Scheme 5 is substantially correct. A large number of amine substituents have been prepared by displacement of the 7-methoxy group (Matsui *et al.*, 1968). Although none of the analogues prepared in this manner was superior in anti-tumor effect to mitomycin C, it would appear that a systematic attempt to optimize activity by variation in the 7-substituent still is a viable goal. As men-tioned previously, the 7-hydroxy analogue (14) is under intensive investigation in Japan because of its lower toxicity. The aziridinomitosenes have been studied very little due to the difficulty in preparing these compounds. They, too, would appear to be good targets for further study, especially since they have different reduction potentials and lipophilicities from similarly substituted mitosanes. New substituents on the aziridine nitrogen serve mainly to deactivate this func-tion. The main hope in this type of analogue would be reduced toxicity. Since the carbamate group is lost during alkylation, it would not be expected that com-pounds with new variants on this group would have greatly enhanced intrinsic activity. However, it is possible that their properties such as absorption, distri-bution, binding, and excretion would be improved.

A few examples of 1-substituted mitosenes have been provided by Remers and co-workers (Leadbetter *et al.*, 1974; Taylor *et al.*, 1977). These compounds were designed according to the hypothesis that a mitosene with a good leaving group at position 1 would afford a species capable of bifunctional alkylation upon reduction of the quinone ring (Scheme 6). Thus far, one compound of this type, 1-acetoxy-7-methoxymitosene (32), has shown activity against P-388 leukemia in mice. Further examples of this type should be prepared; however, the length of synthetic routes to these compounds limits their development. It should be noted that kinamycin C (28) bears a structural resemblance to mitosene (32) in that the 1'-carbon atom can readily afford an alkylation site upon loss of its acetoxyl substituent (Fig. 2) (Omura *et al.*, 1970).

Indoloquinone mitomycin analogues with leaving groups on the 2-methyl sub-stituent (27; X = Cl, CH_3SO_3) should be capable of bifunctional alkylation in the

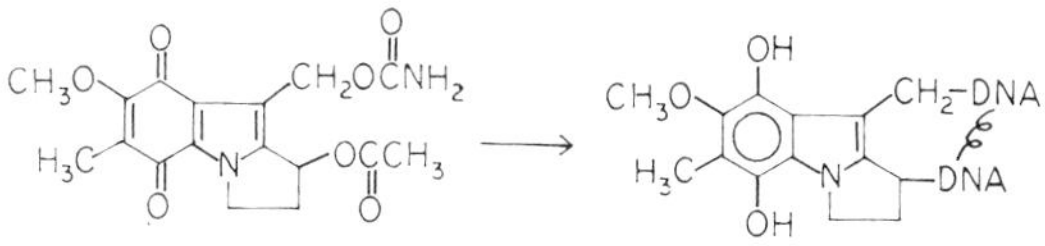

Scheme 6.   Bifunctional alkylation with a 1-acetoxymitosene.

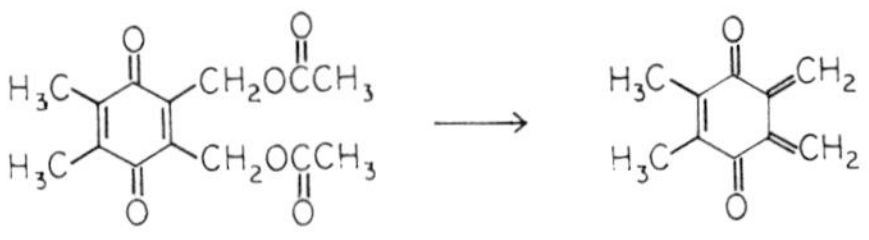

Scheme 7. Quinone methide formation.

same manner as mitosene (**32**). Some compounds of this type were prepared and found active against bacteria in culture. However, they were inactive against *Staphylococcus* infections in mice. No antitumor activity was reported for them (Weiss *et al.*, 1968; Allen *et al.*, 1967).

Sartorelli, Lin, and co-workers applied the principle of bioreductive alkylation to the design of benzoquinones that would form quinone methides (Scheme 7) (Lin *et al.*, 1973, 1974b; Lin and Sartorelli, 1973). These compounds, exemplified by (**34**), bear one or more CH$_2$X groups on the quinone ring, and they have been shown to add nucleophiles to the quinone methide system *in vitro*. Compound (**34**) and certain analogues showed activity against adenocarcinoma 755 ascites cells. Carbazilquinone (**29**) was specifically designed as a mitomycin analogue (Arakawa *et al.*, 1970). It showed the greatest activity of a large number of quinones tested against L-1210 murine leukemia (Driscoll *et al.*, 1974). However, there is some question whether it acts by a bioreductive alkylation mechanism or by simple nucleophilic attack on the aziridine ring as in the case of aziridinobenzoquinones such as (**30**). Moore (1977) has suggested that 2-quinonylaziridines with the general structure **31** might offer interesting possibilities for bioreductive alkylation.

The bioreductive alkylation concept, derived in part from knowledge of the mitomycin mode of action, has been used by Moore to rationalize the antitumor activity of many other classes of compounds including anthracyclines, camptothecin, α-methylene lactones, and various alkaloids (Moore, 1977). It has led Lin and Sartorelli to prepare some active benzoquinones, naphthoquinones, quinolinequinones and naphthazirins with CH$_2$X substituents (Lin *et al.*, 1973, 1974b,c, 1975). Further stimulating research based on this concept is anticipated in the future.

REFERENCES

Allen, G. R., Jr., Poletto, J. F., and Weiss, M. J. (1967). *J. Med. Chem.* **10**, 14.

Arakawa, M., Aoki, T., and Nakano, H. (1970). *Gann* **61**, 485.

Bristol Laboratories (1974). "Mutamycin." Syracuse, New York.

Carter, S. K., and Kershner, L. M. (1975). *Pharm. Times* Aug., p. 56.

Cater, D. B., and Phillips, A. F. (1954). *Nature (London)* **174**, 121.

Cheng, L., and Remers, W. A. (1977). *J. Med. Chem.* **20**, 767.

De Boer, C., Dietz, A., Lummus, N. E., and Savage, G. M. (1961). *Antimicrob. Agents Annu. 1960*, 17.

Driscoll, J. S., Hazard, G. F., Wood, H. B., and Goldin, A. (1974). *Cancer Chemother. Rep.* **4**, 1.

Frank, W., and Osterberg, A. E. (1960). *Cancer Chemother. Rep.* **9**, 114.

Garrett, E. R. (1963). *J. Med. Chem.* **6**, 488.

Hata, T., Sano, Y., Sugawara, R., Matsume, A., Kanamori, K., Shima, T., Hoshi, T. (1956). *J. Antibiot., Ser. A* **9**, 141.

Hornemann, U., Ho, Y. K., Mackey, J. K., Jr., and Srivastava, S. C. (1976). *J. Am. Chem. Soc.* **98**, 7096.

Imai, R. (1976). *Jpn. Chemother. Soc., 24th Meet.*, Tokyo.

Iyer, V. N., and Szybalski, W. (1964). *Science* **145**, 55.

Izbicki, R., Al-Sarraf, M., Reed, M. L., Vaughn, C. B., and Vaitkevicius, V. K. (1972). *Cancer Chemother. Rep.* **56**, 615.

Jones, R., Jr. (1959). *Cancer Chemother. Rep.* **2**, 3.

Kinoshita, S., Uzu, K., Nakano, K., and Takahashi, T. (1970a). *Prog. Antimicrob. Anticancer Chemother.*, Vol. 2, p. 112. Univ. Park Press, Baltimore, Maryland.

Kinoshita, S., Uzu, K., Nakano, K., Shimizu, M., Takahashi, T., Wakaki, S., and Matsui, M. (1970b). *Prog. Antimicrob. Anticancer Chemother.*, Vol. 2, p. 1058. Univ. Park Press, Baltimore, Maryland.

Kinoshita, S., Uzu, K., Nakano, K., Shimizu, M., and Takahashi, T. (1971a). *J. Med. Chem.* **14**, 109.

Kojima, R., Driscoll, J., Mantel, N., and Goldin, A. (1972). *Cancer Chemother. Rep., Part 2* **3**, 121.

Leadbetter, G., Fost, D. L., Ekwuribe, N. N., and Remers, W. A. (1974). *J. Org. Chem.* **39**, 3508.

Lefemine, D. V., Dann, M., Barbatschi, F., Hausmann, W. K., Zbinovsky, V., Monnikendam, P., Adam, J., and Bohonos, N. (1962). *J. Am. Chem. Soc.* **84**, 3184.

Lin, A. J., and Sartorelli, A. C. (1973). *J. Org. Chem.* **38**, 813.

Lin, A. J., Pardini, R. S., Crosby, L. A., Lillis, B. J., Shansky, C. W., and Sartorelli, A. C. (1973). *J. Med. Chem.* **16**, 1268.

Lin, A. J., Cosby, L. A., and Sartorelli, A. C. (1974a). *Cancer Chemother. Rep.* **4**, 23.

Lin, A. J., Shansky, C. W., and Sartorelli, A. C. (1974b). *J. Med. Chem.* **17**, 558.

Lin, A. J., Pardini, R. S., Lillis, B. J., and Sartorelli, A. C. (1974c). *J. Med. Chem.* **17**, 668.

Lin, A. J., Lillis, B. J., and Sartorelli, A. C. (1975). *J. Med. Chem.* **19**, 917.

Lown, J. W., Begleiter, A., Johnson, D., and Morgan, A. R. (1976). *Can. J. Biochem.* **54**, 110.

Matsui, M., Yamada, Y., Uzu, K., and Hirata, T. (1968). *J. Antibiot.* **21**, 189.

Moore, H. W. (1977). *Science* **197**, 527.

Omura, S., Nakagawa, A., Yamada, H., Hata, T., Furusaki, A., and Watanabe, T. (1970). *Chem. Pharm. Bull.* **21**, 931.

Patrick, J. B., Williams, R. P., Meyer, W. E., Fulmor, W., Cosulich, D. B., Broschard, R. W., and Webb, J. S. (1964). *J. Am. Chem. Soc.* **86**, 1889.

Remers, W. A., and Schepman, C. S. (1974). *J. Med. Chem.* **17**, 729.

Stevens, C. L., Taylor, K. G., Munk, M. E., Marshall, W. S., Noll, K., Shah, G. D., Shah, L. G., and Uzu, K. (1964). *J. Med. Chem.* **8**, 1.

Szybalski, W., and Iyer, V. N. (1967). *In* "Antibiotics I, Mechanism of Action" (D. Gottlieb and P. D. Shaw, eds.), pp. 221–245. Springer-Verlag, Berlin and New York.

Taylor, W. G., and Remers, W. A. (1975). *J. Med. Chem.* **18**, 307.

Taylor, W. G., Leadbetter, G., Fost, D. L., and Remers, W. A. (1977). *J. Med. Chem.* **20**, 138.

Tomasz, M. (1976). *Chem.-Biol. Interact.* **13**, 89.

Tomasz, M., Mercado, C. M., Olson, J., and Chatterjie, N. (1974). *Biochemistry* **13**, 4878.

Tulinsky, A. (1962). *J. Am. Chem. Soc.* **84**, 3188.

Uzu, K., Harada, Y., and Wakaki, S. (1964). *Agric. Biol. Chem.* **28**, 388.

Wagner, A. F., and Gitterman, C. O. (1962). *Antibiot. Chemother. (Washington, D.C.)* **12**, 464.

Wakaki, S., Marumo, H., Tomioka, K., Shimizu, G., Kato, E., Kamada, H., Kudo, S., and
    Fujimoto, Y. (1958). *Antibiot. Chemother. (Washington, D.C.)* **8,** 228.
Webb, J. S., Cosulich, D. B., Mowat, J. H., Patrick, J. B., Broschard, R. W., Meyer, W. E.,
    Williams, R. P., Wolf, C. F., Fulmor, W., Pidacks, C., and Lancaster, J. E. (1962). *J. Am.
    Chem. Soc.* **84,** 3185, 3186.
Weiss, M. J., Redin, G. S., Allen, G. R., Jr., Dornbush, A. C., Lindsay, H. L., Poletto, J. F.,
    Remers, W. A., Roth, R. H., and Sloboda, A. E. (1968). *J. Med. Chem.* **11,** 742.

CHAPTER 5

Recent Progress in Bleomycin Studies

HAMAO UMEZAWA

Bleomycin has exhibited therapeutic effect against squamous-cell carcinoma, Hodgkin's disease, and testis tumors, as reviewed in a book edited by Carter *et al.* (1976). About 80% of fresh cases of well differentiated squamous-cell carcinoma respond to bleomycin treatment, and complete remission has been observed in about 25%. Bleomycin in combination with radiation has exhibited a curative effect against this type of tumors. Therapeutic effect on squamous-cell carcinoma has been shown to be due to a low content of bleomycin hydrolase and a high uptake of this antibiotic in this tumor. About 80% of Hodgkin's tumor cases respond to bleomycin treatment, which has shown a curative effect against this tumor in combination with other cytotoxic antitumor agents. Even a small dose, such as 0.5 or 1.0 mg daily, can exhibit therapeutic effect on Hodgkin's disease. The treatment with bleomycin in combination with a vinca alkaloid or a platinum compound or with both of them causes the complete regression of testis tumors in a high frequency.

Mechanism of action of bleomycin has been studied from various aspects. The structure of bleomycin–metal complex was recently elucidated, and the action of bleomycin to cause DNA strand scission has been shown in relation to the structure of bleomycin–ferrous-ion complex.

147

Anticancer Agents Based on Natural Product Models

The side effect which limits the total dose of bleomycin is the pulmonary toxicity. Therefore, new bleomycins which have a lower pulmonary toxicity than the present one used clinically have been studied.

In this chapter, the author reviews the recent studies on chemistry and action of bleomycin, together with those on new bleomycins and their analogues.

I. CHEMISTRY AND BIOSYNTHESIS

Structure studies up to 1975 have been reviewed by the author in detail (Umezawa, 1976). When bleomycin was discovered in a culture filtrate of a strain classified as *Streptomyces verticillus* (Umezawa *et al.*, 1966a), bleomycin extracted from culture filtrates was separated into each bleomycin–copper complex by CM-Sephadex chromatography (Umezawa *et al.*, 1966a,b). Although the reason was not clear, each individual bleomycin–copper complex was inferior to their mixture in the degree of therapeutic effect against Ehrlich carcinoma. Among individual bleomycin–copper complexes, the complex of A5 showed the best therapeutic activity but was still inferior to the mixture (Umezawa *et al.*, 1968). Therefore, the mixture which contained A2 at 55–70%, B2 at 25–32% and others in very small proportions was applied to the clinical study. Bleomycins B4 and B6, which had a strong renal toxicity, were controlled to be less than 1.0% in the mixture. In the early clinical study in 1966, the copper complex was found to damage the vein where bleomycin was injected. Therefore, copper-free bleomycin was used clinically. As is well known, copper-free bleomycin hydrochloride or sulfate can be applied not only intravenously but also subcutaneously or intramuscularly.

As reported by the author (Umezawa, 1971), before the end of 1970, structures of all partial hydrolysis products of bleomycins were elucidated as follows:

1 = β-amino-β-(4-amino-6-carboxy-5-methylpyrimidin-2-yl)propionic acid; 2 = L-β-aminoalanine; 3 = β-hydroxyhistidine; 4 = 4-amino-3-hydroxy-2-methyl-*n*-pentanoic acid; 5 = L-threonine; 6 = 2'-(2-aminoethyl)-2,4'-bithiazole-4-carboxylic acid; 7 (of bleomycin A2) = 3-aminopropyldimethyl-sulphonium; sugar moiety = 2-O-(3-O-carbamoyl-α-D-mannopyranosyl)-L-gulopyranose.

As seen from the formulas, the structure of tripeptide S was easily elucidated. Tripeptide A is called also "pseudotetrapeptide" because its total hydrolysis gives four amino acids. In concentrated hydrochloric acid at 37°C overnight, the carboxyl group of the β-hydroxyhistidine moiety in tripeptide A migrates to the hydroxyl group of the 4-amino-3-hydroxymethylpentanoic acid moiety, and the hydrogenolysis of the product of this migration gives 4-amino-3-hydroxymethyl-pentanoic acid and a reduced dipeptide, which on further hydrolysis gives β-hydroxyhistidinol and a tricarboxylic amino acid containing a pyrimidine

Tripeptide S:

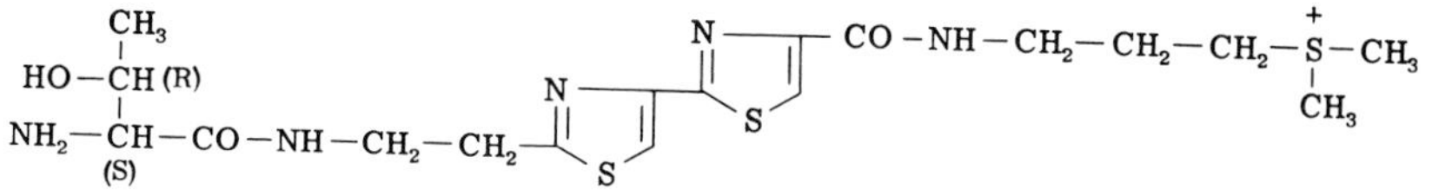

Tripeptide A:

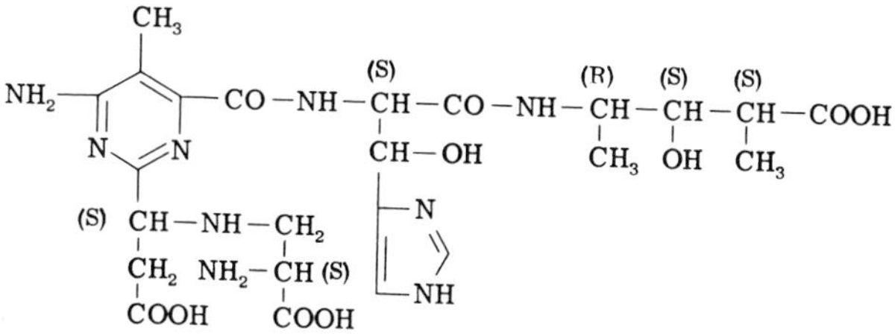

Sugar moiety:

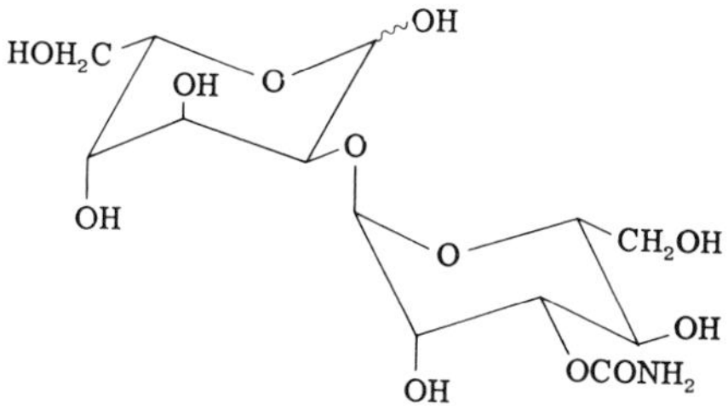

chromophore. Treatment of bleomycin with N-bromosuccinimide gives a tetrapeptide, 4-amino-3-hydroxy-2-methylpentanoyl tripeptide S. Therefore, in the bleomycin molecule the carboxyl group of the 4-amino-3-hydroxy-methylpentanoic acid moiety of tripeptide A should link to the amino group of the threonine moiety of tripeptide S.

The disaccharide in bleomycin was isolated by testing about 100 hydrolysis conditions. Hydrolysis for 3 hours in 0.3 N H_2SO_4 at 82–83°C gives the disaccharide containing a carbamoyl group. Moreover, before 1971, it was found that various bleomycins were different from one another in the terminal amine as follows, where RCOOH is bleomycinic acid:

$$\text{O}$$
Bleomycin A1: $\text{RCO}-\text{NH}-\text{CH}_2-\text{CH}_2-\text{CH}_2-\overset{\overset{\displaystyle \parallel}{}}{\text{S}}-\text{CH}_3$
Demethyl-A2: $\text{RCO}-\text{NH}-\text{CH}_2-\text{CH}_2-\text{CH}_2-\text{S}-\text{CH}_3$

Hamao Umezawa

$$CH_3$$
$$|$$

A2: RCO—NH—CH$_2$—CH$_2$—CH$_2$—S—CH$_3$
A2'-a: RCO—NH—CH$_2$—CH$_2$—CH$_2$—CH$_2$—NH$_2$
A2'-b: RCO—NH—CH$_2$—CH$_2$—CH$_2$—NH$_2$
A5: RCO—NH—CH$_2$—CH$_2$—CH$_2$—NH—CH$_2$—CH$_2$—CH$_2$—CH$_2$—NH$_2$
A6: RCO—NH—CH$_2$—CH$_2$—CH$_2$—NH—CH$_2$—CH$_2$—CH$_2$—CH$_2$—NH—CH$_2$—CH$_2$—
CH$_2$—NH$_2$
B2: RCO—NH—CH$_2$—CH$_2$—CH$_2$—CH$_2$—NH—C—NH$_2$
                                        ‖
                                        NH
B4: RCO—NH—CH$_2$—CH$_2$—CH$_2$—CH$_2$—NH—C—NH—CH$_2$—CH$_2$—CH$_2$—NH—C—NH$_2$;
                                         ‖   ‖
                                         NH   NH

Later, the following bleomycins containing histamine or ammonia were isolated (Fujii *et al.*, 1973):

A2'-c: RCO—NH—CH$_2$—CH$_2$

B1' :   RCO—NH$_2$

These differences in the terminal amine moieties of various bleomycins suggest that an amine added to a fermentation medium may be incorporated into bleomycins. In fact, the one methyl carbon labeled 3-aminopropyl dimethylsulfonium is incorporated into bleomycin A2, and the addition of this amino acid suppressed the production of other bleomycins (Umezawa, 1971). About 300 artificial bleomycins have been produced by addition of various amines to fermentation media (Umezawa, 1971, 1976).

Structures containing a β-lactam ring in the N-terminal amino acid moiety was proposed for bleomycins in 1972 (Takita *et al.*, 1972b; Umezawa, 1976) on the basis of the following experimental results.

The treatment of bleomycin in 0.1 N NaOH at room temperature liberated its disaccharide, with appearance of a one-proton signal at $\delta 8.52$, suggesting production of a double bond. The hydrogenation of bleomycin thus treated with palladium carbon, followed by hydrolysis, gave D,L-histidine. Thus, the disaccharide was shown to link to the hydroxyl group of the β-hydroxyhistidine moiety. The α-glycosidic linkage was determined by application of ^{1}H-nmr analysis. A mild hydrolysis of bleomycin methylated with methyliodide and triethylamine gave β-aminoalanine betaine amide indicating the presence of the α-aminocarboxamide moiety.

Potentiometric titration of bleomycin demonstrated the presence of three measurable functional groups in bleomycin molecule except for the terminal amine: pK 7.4, pK 4.7, and pK 2.7. The pK 7.4 was assigned to the amino group of the α-aminocarboxamide moiety; pK 4.7 was assigned to the imidazole by pH dependence of the ^{1}H and ^{13}C-nmr shifts. At that time, pK 2.7 was assigned to the 4-aminopyrimidine moiety, because the study of the pK of the 4-amino-pyrimidine moiety of (4-amino-6-carboxy-5-methylpyrimidin-2-yl)-β-aminop-ropionic acid and its derivatives indicated that the 4-aminopyrimidine ring gave a measurable pK of 2.7 when its β-amino group was masked by acetylation. At that time, it was thought that if the β-amino group of the β-aminoalanine moiety was not masked and existed as a secondary amino group in bleomycin, it should have a higher pK value than 2.7. Moreover, the methylation of this nitrogen was not successful. Therefore, pK 2.7 was assigned to the 4-aminopyrimidine moiety, and the structure in which this secondary amino group of the β-aminoalaninamide moiety was masked by formation of the β-lactam ring was proposed. The number of carbon atoms in bleomycin molecule was definitely determined by ^{13}C-nmr.

In this structure determination, the infrared spectrum of bleomycin was also carefully studied. One model compound, 4-phenyl-2-azetidinone containing a β-lactam ring, indicated the absorbance at 1710 cm^{-1}. The β-lactam structure of this model compound was supported by x-ray crystal analysis.

Stereochemistry of all asymmetric carbon atoms was elucidated mainly by x-ray crystal analysis of the hydrolysis products (Umezawa, 1976). However, the β-lactam structure part was recently revised and the structures shown in Fig. 1 were proposed (Takita *et al.*, 1978a).

From the general concept, it can be imagined that the peptide part of bleomy-cin is biosynthesized on a multienzyme in bleomycin-producing cells, starting from the synthesis of the dipeptide containing the first and the second N-terminal amino acids. The labeled 2'-(2-aminoethyl)-2,4'-bithiazole-4-carboxylic acid (**6** described above) added to the fermentation medium is not incorporated into bleomycin, and this amino acid residue in bleomycin has been confirmed to be biosynthesized from β-alanine and cysteine.

If we call the first amino acid in bleomycin structure pyrimidoblamic acid, then the following biosynthetic intermediate peptides were recently isolated (Fujii *et al.*, 1979):

(1) Demethylpyrimidoblamylhistidine
(2) Demethylpyrimidoblamylhistidylalanine
(3) a. Demethylpyrimidoblamylhistidyl-(4-amino-3-hydroxy-2-methyl)
      pentanoic acid
    b. Demethylpyrimidoblamylhistidyl-(2-amino-3-oxo)pentane
(4) Demethylpyrimidoblamylhistidyl-(4-amino-3-hydroxy-2-methyl)pen-
    tanoylthreonine

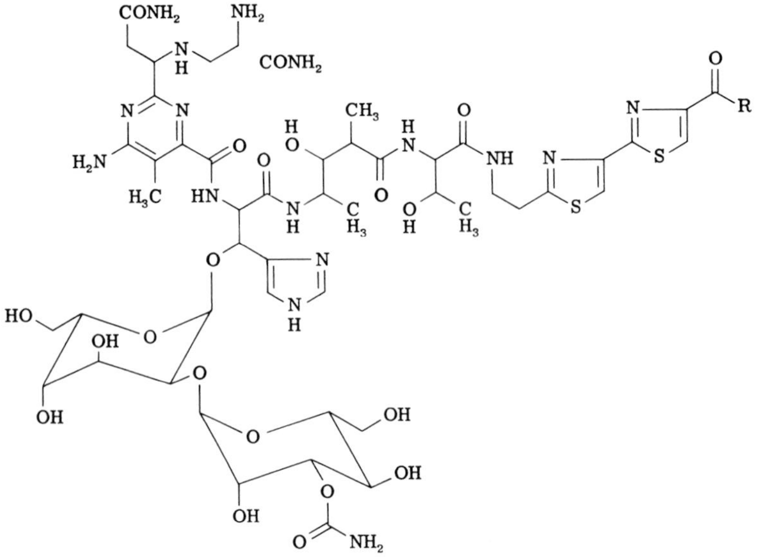

R = terminal amine

Figure 1.   Revised structure of bleomycin.

(5) Pyrimidoblamylhistidyl- (4- amino- 3- hydroxy- 2- methyl)pentanoylthreo-
nine

(6) Pyrimidoblamylhistidyl - (4 - amino - 3 - hydroxy - 2 - methyl)pentanoyl-
threonyl-2'-(2-aminoethyl)-2,4'-bithiazole-4-carboxylic acid

(7) Pyrimidoblamyl-β-hydroxyhistidyl-(4-amino-3-hydroxy-2-methyl)pen-
tanoylthreonyl-2'-(2-aminoehtyl)-2,4'-bithiazole-4-carboxylic acid

(8) Deglycobleomycin

One of the unexpected things found in this study is that the methyl group in the
pyrimidoblamic acid moiety was introduced after the peptide was elongated up to
the demethylpyrimidoblamylhistidyl- (4- amino- 3- hydroxy- 2- methyl)pentanoyl-
threonine. This methyl group of bleomycin is highly labeled by addition of
CH_3-labeled methionine to the fermentation medium. The other unexpected thing
is that the β-hydroxyhistidine moiety of bleomycin is produced by oxidation of
the histidine moiety of peptides at the step of the synthesis of the peptide part of
bleomycinic acid. Bleomycinic acid is the main structural part common to all
bleomycins. These may be easily understood, if we assume that these enzymes
involved in the methylation of the pyrimidine ring or in the β-hydroxylation of
the histidine moiety can react with the pyrimidine ring or the histidine moiety,
when these parts are separated far distant enough from the multienzyme.

The intermediate 3b described above can be imagined to be produced by decarboxylation. During biosynthesis, the carboxyl group of 3a may link to the SH group of the enzyme, as is well known in cases of the biosynthesis of bacterial peptide antibiotics. From the structure, it can be imagined that the 4-amino-3-hydroxy-2-methylpentanoyl moiety may be synthesized from alanine and propionic acid. However, the 2-methyl group of this amino acid was confirmed to be derived from the methyl group of methionine.

This biosynthesis study presented very important information on bleomycin biosynthesis and also presented the isolation of demethylpyrimidoblamylhistidylalanine, which when crystallized as its copper complex provided very useful information to the structure of bleomycin. The x-ray crystal analysis of the copper complex of this biosynthesis intermediate gave its complete structure, except for the exact assignment of two electron-dense molecules (X and Y in Fig. 2) in the sidechain of demethylpyrimidoblamic acid moiety as shown in Fig. 2. The part which was not exactly shown by the x-ray crystallography was elucidated by chemical studies (Iitaka *et al.*, 1978), and this study led to the revised open structure of bleomycin.

^{13}C-nmr signals of all carbon atoms in bleomycin molecule have been identified (Naganawa *et al.*, 1977). The comparison of the ^{13}C-nmr spectrum of this biosynthetic intermediate with that of bleomycin indicates that both of them should have the same structure in the sidechain on the aminopyrimidine ring. The presence of the β-lactam ring in the biosynthetic intermediate (peptide 3 described above) was denied by the bond angle between the two bonds from the methin carbon atom linking to the pyrimidine ring. Free carboxylic acid group in bleomycin cannot be shown, and the carboxylic acid structure for the carbon atom a in Fig. 3 is denied by the chemical shifts of the ^{13}C-nmr: the signal of the α-carbon atom existing at b in Fig. 3 should be seen at $\delta 36.9-38.7$ in the case of the carboxylic acid structure but was seen at $\delta 40.9-41.0$ in bleomycin and the crystallized intermediate. Pyrimidoblamide (A in Fig. 3), which contains three amide groups, was synthesized from deamidopyrimidoblamic acid obtained by

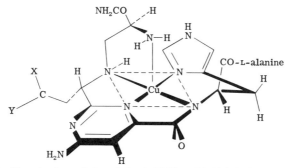

Figure 2. The structure of demethylpyrimidoblamylhistidyl-alanine, a biosynthetic intermediate, shown by x-ray crystal analysis.

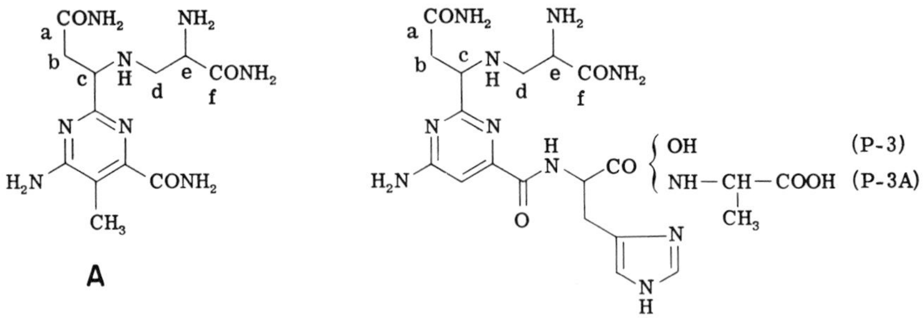

	a	b	c	d	e	f
A	176.9	41.0	53.2	47.8	60.8	171.9
P-3A	177.0	40.9	53.3	47.9	60.8	171.5
P-3	177.0	41.0	53.3	47.9	60.8	171.5
Bleomycin A2'-c	177.0	41.0	53.3	47.8	60.5	171.9

Figure 3. Chemical shifts of six carbon atoms in the sidechain of pyrimidoblamide (A), and in the sidechain of demethylpyrimidoblamyl or pyrimidoblamyl moiety in demethylpyrimidoblamylhistidyl-alanine (P-3), demethylpyrimidoblamylhistodylalanylpropionic acid (P-3A), and bleomycin.

hydrolysis of bleomycin. The ^{13}C-nmr spectra of the sidechain of the pyrimidoblamide (A in Fig. 3), of the biosynthetic intermediates (P-3 and P-3A), and of bleomycin are completely the same, as shown in Fig. 3 (Iitaka et al., 1978).

As already described in 1972, the pK of 2.7 in bleomycin was assigned to the 4-aminopyrimidine ring, because the following (4-amino-6-carbamoyl-5-methyl-pyrimidine-2-yl)-2-acetylaminopropionamide showed pK 2.7:

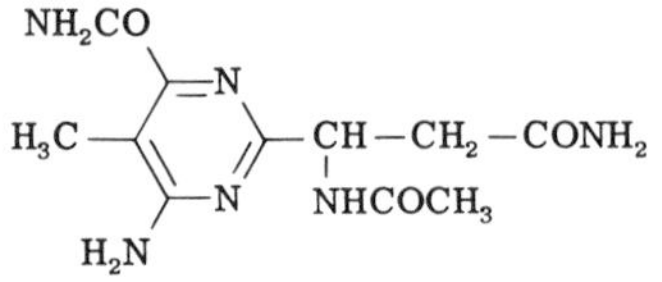

Potentiometric titration of pyrimidoblamide and a model compound containing the same sidechain on a benzene ring indicated extraordinarily low pK values such as 3.4 or 2.7. Thus, the pK of 2.7 can be assigned to the secondary amine in the sidechain of the pyrimidoblamic acid moiety of bleomycin. As already described, the pK value of the 4-aminopyrimidine moiety in the pyrimidoblamic acid moiety where the secondary amine of the sidechain is not protected should be smaller than 1.0.

Phleomycin, which was discovered by the author in 1956, is closely related to bleomycin. Phleomycin showed a high therapeutic index against Ehrlich carcinoma, but it showed a strong renal toxicity in dogs. Therefore, the study of phleomycin-like antibiotics was continued by the author, and bleomycin was discovered. The structure of phleomycin is different from bleomycin in the hydrogenation of one double bond in the bithiazole acid moiety as follows (Takita *et al.*, 1972b):

Bleomycin:

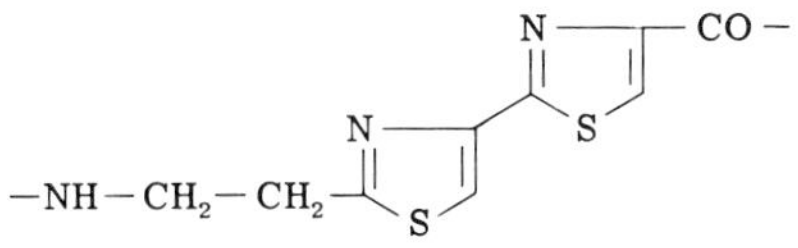

Phleomycin:

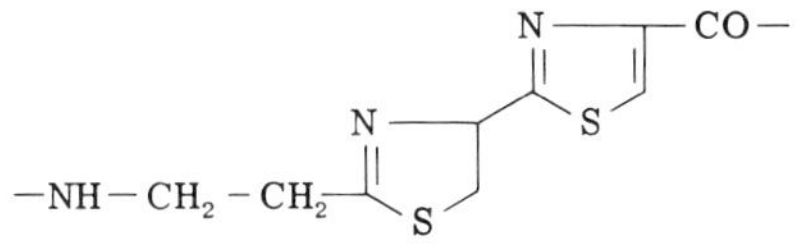

Various phleomycins are produced by a phleomycin-producing strain, as in the case of bleomycins, and they also are different from each other in their terminal amine moiety. Main components of phleomycins produced by fermentation contain the same terminal amines as bleomycins B2, B4, and B6 (Takita *et al.*, 1972a). Phleomycins B4 and B6 which contained two or more guanidine groups produce a strong renal toxicity. Bleomycin–phleomycin-group compounds which contain two or more guanidine groups or an amidine group in the terminal amine moiety cause a strong damage in dog kidney. Phleomycins which have no renal toxicity can be produced by addition of proper amines to the fermentation medium (Umezawa, 1976).

II. MECHANISM OF ACTION OF BLEOMYCIN TO INHIBIT GROWTH OF CELLS AND STRUCTURES OF BLEOMYCIN–METAL COMPLEXES

In 1964, phleomycin was found to inhibit DNA synthesis in *E. coli* and HeLa cells (Tanaka *et al.*, 1963) and to increase the Tm of DNA (Falaschi and Kornberg, 1964). The phleomycin used in these experiments was a mixture of more than three phleomycin–copper complexes containing one, two, or more guanidine groups in their terminal amine moiety. Bleomycins and phleomycins closely resemble each other in their chemical properties, and therefore at the beginning of the study of the mechanism of bleomycin action it was expected that bleomycin–copper complex might increase the Tm of DNA, as did phleomycin. However, contrary to expectation, being different from the old phleomycin–copper complex, copper complexes of all bleomycins did not increase the Tm of DNA. Moreover, the copper complex of a freshly prepared phleomycin containing (3-aminopropyl)dimethylsulfonium (the amine of bleomycin A2) did not increase the Tm of DNA. Bleomycin B4 containing two guanidine groups in the terminal amine also did not increase the Tm of DNA. The reason why phleomycin–copper complex prepared in its early study increased the Tm of DNA remains as an interesting problem which shall be solved in the future.

As first reported by the author in 1969, copper-free bleomycin causes DNA strand scission in the presence of a sulfhydryl compound, peroxide, or ascorbic acid, and lowers the Tm of DNA in the same reaction mixture (Nagai et al., 1969a,b; Umezawa, 1976). Thereafter, the reaction mechanism of bleomycin in DNA strand scission was further studied (Umezawa et al., 1973a; Asakura et al., 1975; Haidle et al., 1971, 1972a,b; Kuo and Haidle, 1973a,b; Müller et al., 1972, 1973; Yamazaki et al., 1973).

The results may be summarized as follows:

Bleomycin first binds to DNA. There are base sequences to which bleomycin can bind selectively. After the binding, the reaction which releases thymine occurs. If the concentration of bleomycin and the concentration of a sulfhydryl compound are high, other bases are also released. Strand scission occurs at the same time with the release of a base; or after the release of a base, strand scission is caused by alkaline reaction. It was shown by Müller and Zahn (1976) that after causing strand scission, bleomycin remains intact.

The action of bleomycin in causing strand scission of SV40 virus DNA has been studied in detail. This system can be used to test the binding ability of nucleotides to bleomycin. If a nucleotide which can bind to bleomycin is added to this system, SV40 DNA is protected from the action of bleomycin. All double-stranded DNA tested protected: that is, they were shown to bind with bleomycin. RNA does not bind with bleomycin. Double-stranded oligonucleotides also bind to bleomycin. For instance, single strands of $d(pTpG)_{6-9}$ or $d(pCpA)_{6-9}$ does not protect SV40 DNA from bleomycin action: that is, these single strands to not bind with bleomycin. If they are mixed and added, then the double strand thus formed binds to bleomycin. If $d(pTpG)_{6-9}$ is mixed with $d(pCpA)_2$ and added, then SV40 DNA is protected from the action of bleomycin: that is, the double strands formed from these deoxynucleotides bind with bleomycin. However, if instead of $d(pCpA)_2$, $d(pCpA)$ is mixed with $d(pTpG)_{6-9}$ and added, then SV40 DNA is not protected from bleomycin action. It suggests that bleomycin binds strongly to the double-strand structure of the tetradeoxynucleotide pair (Asakura et al., 1978).

The binding of bleomycin to DNA can be shown by measuring the quenching of the fluorescence due to the bithiazole moiety of bleomycin. If bleomycin solution is excited at 290 nm, which is the adsorbance peak due to the bithiazole moiety, the maximum is observed at 353 nm in the emission spectrum. This fluorescence is quenched by calf thymus DNA (Chien et al., 1977).

As is well known, copper-free bleomycin causes the strand scission of DNA, but bleomycin–copper complex does not. But the quenching of the bithiazole fluorescence indicates that bleomycin–copper complex also binds with DNA. In cases of copper-free and copper-chelated bleomycin A2, the calculated association constants with calf thymus DNA were 3.2×10^5 and 2.3×10^5. The association constant of bleomycin A5 to calf thymus DNA was 3.0×10^5. Both monodeoxynucleotides and monomucleotides do not quench the bithiazole fluorescence. Poly G and poly dG cause the strongest quenching compared with poly A, poly U, poly C, poly dA, poly dT and poly UC. Association constants of bleomycin A5 with poly C, poly A, and poly U were 5.7×10^6, 1.0×10^6 and 1.0×10^5, respectively (Kasai et al., 1978). This indicates that the bithiazole moiety of bleomycin selects the guanine in DNA for its binding.

Bleomycin is inactivated in an aqueous solution by a sulfhydryl compound. This inactivation is caused by ferrous ion and oxygen (Umezawa, 1976). In the presence of a reducing agent such as a sulfhydryl compound, this inactivation occurs in the presence of a trace amount of ferrous ion. The most sensitive part of the bleomycin molecule to this inactivation reaction is the pyrimidoblamic acid moiety: by this reaction, the sidechain is split from the pyrimidine ring. Bleomycin–copper complex is not inactivated by this reaction. The conditions of this inactivation reaction are the same as those which cause the strand scission of DNA. As reported by Sausville et al. (1976, 1978), the reaction of copper-free bleomycin with DNA to cause strand scission requires ferrous ion and oxygen, and is promoted by a sulfhydryl compound. Bleomycin–copper complex binds to DNA as shown by quenching of the fluorescence, as described above, but bleomycin–copper complex does not cause DNA strand scission. Therefore, in order to elucidate the reaction mechanism of bleomycin in causing DNA strand scission, it is necessary to elucidate the metal ligands in the bleomycin molecule and the structure of bleomycin–metal complexes.

Before the x-ray crystal analysis of the copper complex of a biosynthetic intermediate described in the previous section, the α-amino group of the α-aminocarboxamide group of the pyrimidoblamic acid moiety, the N^1 of the pyrimidine ring of the pyrimidoblamic acid moiety, the N^π of the imidazole ring of the β-hydroxyhistidine moiety and the carbamoyl group of the sugar moiety were shown to be the cupric ligands of bleomycin molecule as follows (Muraoka et al., 1976):

(1) In phleomycin, as described in the previous section, one of the double bonds in the bithiazole group of bleomycin is saturated. The difference in UV absorption spectrum between copper-chelated and metal-free bleomycins is the same as the difference in spectrum between copper-chelated and metal-free phleomycins. This indicates that the bithiazole moiety of bleomycin is not involved in the metal binding. Phleomycins produce equimolar copper complexes as do the bleomycins.

(2) The pK value of 7.4 of the α-amino group of the α-aminocarboxamide moiety is not seen in the bleomycin–copper complex. This amino group of

copper-free bleomycin is easily acetylated in a weakly base condition, but that in the bleomycin–copper complex is not acetylated.

(3) The pK of 4.7 due to the imidazole ring of the β-hydroxyhistidine moiety is not seen in the bleomycin–copper complex. Pauli color reaction is positive with copper-free bleomycin but negative with bleomycin copper complex.

(4) Epi-bleomycin, in which the carbon atom c in Fig. 3 has the R configuration, is obtained by alkaline treatment of bleomycin–copper complex but is not obtained from copper-free bleomycin.

(5) The migration of the carbamoyl group from the 3-OH of the mannose moiety to the adjacent 2-OH group occurs in the copper-free bleomycin in alkaline solution but not in the copper-chelated bleomycin.

Electron spin resonance studies by Sugiura (1978) supported the structure for the bleomycin–copper complex shown in Fig. 4 (Takita *et al.*, 1978b), which is the same as that of the copper complex (Fig. 2) of demethylpyrimidoblamyl-histidyl-alanine shown by x-ray crystal analysis.

If CM-Sephadex chromatography is developed with pH 6.0 buffer, the retention time of bleomycin–copper complex is the same as that of copper-free bleomycin. This indicates that both copper-free and copper-chelating bleomycins should have the same positive charge. At this pH, metal-free bleomycin has one positive charge, except for the charge in the terminal amine. Therefore, in the copper complex, there must be a deprotonated functional group as one of the coordination sites. The amide nitrogen of the β-hydroxylhistidylamide moiety is involved in the binding with Cu^{2+}, and it can therefore be inferred that the amide nitrogen is deprotonated and binds to the cupric ion. Esr studies by Sugiura indicated that bleomycin–Co^{2+} complex should have the same structure as the bleomycin–cupric complex, and this Co^{2+} in the complex was shown to bind to the oxygen.

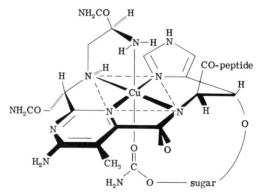

Figure 4. The structure of Cu(II) complex of bleomycin.

Bleomycin also forms a complex with ferrous ion at about neutral pH. Thus, the structure of the bleomycin–ferrous-ion complex can be shown as in Fig. 5 (Takita *et al.*, 1978b). The structure of the bleomycin–ferrous-ion complex provides important information to clarify the reaction mechanism for the inactivation of copper-free bleomycin in the presence of ferrous ion and oxygen, and also explains the reaction to cause DNA strand scission in the presence of ferrous ion and oxygen.

The oxygen molecule which binds to the ferrous ion in the bleomycin complex is activated. This activated oxygen can react with the sidechain on the pyrimidine ring of the pyrimidoblamic acid moiety. As already mentioned, copper-free bleomycin is inactivated in an aqueous solution containing a trace amount of ferrous ion, oxygen, and a reducing agent. The cupric ion in the bleomycin–copper complex cannot be replaced by ferrous ion, and the bleomycin–copper complex is not inactivated. If the oxygen on the ferrous ion in the bleomycin–ferrous-ion complex which has bound to DNA is brought close to a sensitive group of the deoxyribose moiety of DNA, the reaction which results in strand scission can occur. Sugiura and Kikuchi (1978) have confirmed that oxygen bubbling to the bleomycin Fe^{2+} complex at 1.0 mM generates hydroxy radical, and with bleomycin–Fe^{2+} complex at 0.02 mM it generates superoxide ion or its protonated form.

As first reported by Müller *et al.* (1972), as the result of the reaction of bleomycin with DNA, thymine is released. In the case of a high concentration of bleomycin and a sulfhydryl compound, other bases are also released (Haidle *et al.*, 1972b). As found by Haidle *et al.* (1972b) and Takeshita *et al.* (1978), the deoxyribose moiety is degraded and malondialdehyde is released. The base sequence most sensitive to bleomycin action has been reported to contain GC or GT (Takeshita *et al.*, 1978).

Bleomycin–^{57}Co complex does not inhibit the growth of animal and bacterial

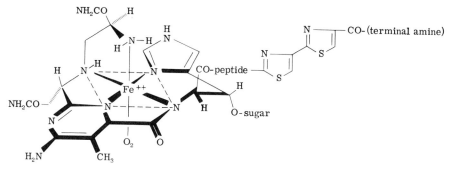

Figure 5. The structure of bleomycin–ferrous-ion complex, which can bind to oxygen molecule.

cells, but intact bleomycin can be separated from this complex. Injected bleomycin-^{57}Co complex has been found bound to DNA (Kono, 1977). This bleomycin-Co^{3+} complex does not cause DNA strand scission. Therefore, the binding of bleomycin to DNA alone does not show cytotoxic action. It has been reported that bleomycin inhibits DNA ligase, DNA synthesis, and cell division (Umezawa, 1976), probably due to the strand scission of DNA. Exposure of cells to bleomycin causes G2 block (Carter *et al.*, 1976).

III. MECHANISM OF THERAPEUTIC ACTION OF BLEOMYCIN AGAINST SQUAMOUS-CELL CARCINOMA

In the study of cancer chemotherapy, the mechanism of therapeutic action should be studied parallel to the mechanism of cytotoxic action. If there are cells which lack the ability to repair the damage caused by bleomycin, these cells should be most sensitive to bleomycin treatment. As described at the beginning of this chapter, Yagoda and Krakoff (1976) have emphasized that Hodgkin's lymphoma responds to a very low daily dose of bleomycin. It must be very interesting to study the mechanism of the therapeutic action against this tumor.

The mechanism of therapeutic action against squamous-cell carcinoma has been studied (Umezawa, 1976). Immediately after the effect on penile cancer was first observed by Ichikawa in 1966 (Carter *et al.*, 1976), the author confirmed that bleomycin was distributed in the skin of mice at a high concentration compared with other organs. Bleomycin was also confirmed to be effective against squamous-cell carcinoma in mouse skin induced by methylcholanthrene, but not effective against sarcoma in mouse skin induced by the same agent. If the radioactivity indicating the total amount of bleomycin per gram of skin and the antibacterial activity indicating the amount of the active form of bleomycin are measured one hour after the subcutaneous injection of copper-free ^{3}H-bleomycin, the total uptake of bleomycin to a tissue and the degree of inactivation in this tissue can be measured. In three such experiments, the radioactivity showed a total concentration of 15.9–22.7 μg per gram of bleomycin in carcinoma and a lower total concentration such as 4.5–14.4 μg/gm in sarcoma. Moreover, a marked difference was recognized in the concentrations of bleomycin which was not inactivated: 11.0–17 μg/gm in the carcinoma and 0–2.8 μg/gm in the sarcoma. Thus, therapeutic effect of bleomycin on squamous-cell carcinoma has been shown to be due to a high uptake of bleomycin and a low inactivation of the antibiotic in this tumor (Umezawa *et al.*, 1972; Umezawa, 1976).

It has been found by the author that all tissues contain an enzyme which inactivates bleomycin. This enzyme has been extracted and called bleomycin hydrolase. During inactivation of one mole of copper-free bleomycin by bleomycin hydrolase, one mole of ammonia is released (Umezawa, 1971). The inactivated bleomycin has a slightly lower basicity than the intact bleomycin, suggesting the hydrolysis of a carboxamide bond during inactivation. Bleomycin hydrolase has been proved to hydrolyze the α-aminocarboxamide group of the pyrimidoblamic acid moiety. The pK value of 7.3 of bleomycin shifts to pK 9.4 after the hydrolysis. This enzyme is an aminopeptidase-like enzyme, and hydrolyzes arginine β-naphthylamide and lysine β-naphthylamide, but not leucine β-naphthylamide. The hydrolysis of arginine or lysine β-naphthylamide is competitively inhibited by bleomycin (Umezawa et al., 1974).

The content of bleomycin hydrolase in squamous-cell carcinoma in mouse skin, which is sensitive to bleomycin treatment, has been confirmed to be significantly smaller than in sarcoma (Umezawa et al., 1972). It has also been confirmed that the content of this enzyme is significantly smaller in a rat hepatoma sensitive to bleomycin treatment than in another rat hepatoma resistant to bleomycin treatment (Umezawa, 1976).

Copper-free bleomycin is hydrolyzed by bleomycin hydrolase, but the bleomycin–copper complex is resistant to this enzyme.

Information on the behavior of bleomycin in animals can be summarized as follows (Umezawa, 1976):

A portion or all of injected bleomycin binds with cupric ion in blood (Kanao et al., 1973). After penetration into cells, the cupric ion in the bleomycin–copper complex is reduced by sulfhydryl compounds, which can be found in a small molecular fraction of a 105,000-g supernatant of cell homogenates: the treatment of bleomycin–copper complex with this small molecular fraction or a sulfhydryl compound like cysteine in the presence of neocuproine (a cuprous ion-binding reagent) gives copper-free bleomycin and cuprous ion in equimolar ratio, and this reaction is inhibited by N-ethylmaleimide. Cuprous ion is easily converted by oxygen to cupric ion, which readily binds to copper-free bleomycin. Therefore, in removing cupric ion from bleomycin–copper complex, cuprous ion produced from the cupric ion must be captured by a cellular material (Takahashi et al., 1977). The ability of this protein to bind with cuprous ion is eliminated by preincubation with N-ethylmaleimide, and the active group of this protein was suggested to be a sulfhydryl group.

As described above, the cupric ion of bleomycin–copper complex is reduced in cells, and the cuprous ion thus produced is captured by a cellular protein. Copper-free bleomycin thus formed undergoes the action of bleomycin hydrolase and is inactivated. Copper-free bleomycin which was not inactivated reaches the nucleus and binds with DNA. Ferrous ion is taken into bleomycin binding with

DNA, an oxygen molecule on the ferrous ion is activated, and DNA strand scission occurs. The incubation of rat hepatoma cells with copper-free bleomycin at 0°C causes DNA fragmentation, but the incubation with bleomycin–copper complex at 0°C does not (Takahashi *et al.*, 1977). In some cells, it is possible that bleomycin is inactivated also by a trace amount of ferrous ion and oxygen in the presence of small molecular reducing compounds. There is also a possibility that bleomycin–copper complex from which the cupric ion has not been removed binds to DNA, and the cupric ion in the bleomycin bound to DNA is reduced and replaced by ferrous ion.

As already described, the therapeutic effect of bleomycin on squamous-cell carcinoma is due to a low content of bleomycin hydrolase and a high distribution of the antibiotic in this tumor. The side effect in skin is due to the same reason (Umezawa *et al.*, 1972; Umezawa, 1976). Bleomycin does not cause bone-marrow depression. This has been suggested to be due to a high content of high-molecular-weight bleomycin-inactivating material, probably bleomycin hydrolase.

IV. NEW USEFUL BLEOMYCINS

As already described, various bleomycins differ from each other in their terminal amine moiety. If an amine is added to a fermentation medium and a bleomycin-producing strain is cultured, a bleomycin containing the amine added is produced, and the production of other bleomycins are suppressed. By this fermentation method, about 250 bleomycins have been prepared. The hydrolysis of bleomycin B2 by a *Fusarium* enzyme (Umezawa *et al.*, 1973b) which was named acylagmatine amidohydrolase, or the application of the cyanogen bromide method to bleomycin demethyl-A2 (Takita *et al.*, 1973), gives bleomycinic acid.

Bleomycinic acid (R=OH in Fig. 1) is the main molecular part common to all bleomycins. Various bleomycins can be synthesized from bleomycinic acid. Aminoalkyl esters of bleomycinic acid also causes a single-strand scission of SV40 DNA (Umezawa, 1976). Moreover, various bleomycins are different in the degree of the renal and pulmonary toxicity.

Bleomycin does not show pulmonary toxicity in young mice, because this antibiotic is rapidly inactivated in lungs of young animals. This inactivation is much slower in old mice. A method to test the pulmonary toxicity using mice older than 15 weeks of age has been established by Matsuda *et al.*, of Nippon Kayaku Co. Recently, a much simpler method was presented by S. Cohen. Among bleomycins which have lower pulmonary toxicity than the present bleomycin, bleomycin PEP, which is called pepleomycin and contains *N*-(3-aminopropyl)-α-phenethylamine as the terminal amine, has been studied clinically. It inhibits Ehrlich carcinoma and squamous-cell carcinoma induced by methylcholanthrene in mouse skin in the same or stronger degree than the present bleomycin. Its high dose also inhibits adenocarcinoma in rat stomach induced by N-methyl-N'-nitro-N-nitrosoguanidine. The present bleomycin does not inhibit this adenocarcinoma in rat stomach. In a test for pulmonary toxicity using aged mice and dogs, pepleomycin has been shown to have a significantly lower pulmonary toxicity than the present bleomycin. This bleomycin may be more useful in treatment of bleomycin-sensitive tumors.

As reviewed by Svanberg (1976), bleomycin is useful in treatment of squamous-cell carcinoma in the lungs. Therefore, it is thought that a bleomycin which has a lower pulmonary toxicity than the present bleomycin is more useful in treatment of lung cancer.

Miyamoto et al. (1977) reported that daily infusion of 5 mg of bleomycin for 7 days and 10 mg of mitomycin after the last injection of bleomycin with two to five repetitions of this treatment with 7-day intervals, exhibited a strong therapeutic effect on metastasis of cervix cancer. A daily does of 5 mg of bleomycin PEP for 7 days and the repetition of the treatment with 7-day intervals may be an interesting administration schedule in treatment of tumors sensitive to bleomycin treatment. Bleomycin BAPP, which contains N-(3-aminopropyl)-N'-butyldiaminopropane in the terminal amine, was shown by the National Cancer Institute (U.S.A.) to exhibit an inhibition against B16 melanoma. It was about five times lower pulmonary toxicity than the present bleomycin. It may be also one of new bleomycins worth clinical study.

Tallisomycins A and B (Konishi et al., 1977) are bleomycin-group antibiotics. Tallysomycin A is weaker than bleomycin in causing DNA strand scission, but in a smaller dose than bleomycin it can exhibit antitumor effect against experimental animal tumors (Strong and Crooke, 1978). Tallysomycin A, which has strong basic groups, may have a strong renal toxicity. It may also have a strong pulmonary toxicity. It is quite possible that the study of new bleomycins and new bleomycin-group antibiotics will provide effective agents which will increase the rate of cure of tumors sensitive to bleomycin treatment.

Up to the present, various bleomycins which are different in the terminal amine from known bleomycins have been studied in detail. As already mentioned, the carboxamide of the α-aminocarboxamide moiety is hydrolyzed by bleomycin hydrolase. Therefore, starting from this hydrolysis product, new bleomycins have been synthesized. These bleomycins which are resistant to bleomycin hydrolase should have a wider antitumor spectrum. In the revised structure, there is another carboxamide, and it is possible to develop derivatives based on this amide group. Thus, it may be said that the study of new bleomycins and new bleomycin-group antibiotics will lead to the development of effective drugs useful in treatment of various types of human cancer.

V. COMBINATION TREATMENT OF BLEOMYCIN WITH AN IMMUNE-ENHANCING AGENT

As one of the characteristics, bleomycin does not suppress the bone marrow and immune response. The bone stem cells are rich in bleomycin hydrolase and resistant to the action of bleomycin. On the basis of this characteristic, it can be imagined that therapeutic effect of bleomycin may be enhanced by an immune-enhancing agent.

Recently the author found that microorganisms produced small molecular compounds enhancing immune response (Umezawa, 1977). Aminopeptidases are located on the surface of all kinds of animal cells, and bestatin [$(2S,3R)$-3-amino-2-hydroxy-4-phenylbutanoyl-L-leucine], which inhibits aminopeptidase B, binds to macrophages, lymphocytes, and other animal cells. Very low doses of bestatin (Umezawa, 1977), such as 1–100 μg per mouse, enhance delayed-type hypersensitivity, and the administration to mice bearing slowly growing tumors exhibits a therapeutic effect.

The cancer reduces the cellular immune activity of the host. Therefore, it is thought that effective cancer chemotherapy decreases the size of a tumor and may restore partially the reduced immunity. The additional treatment with an immune-enhancing agent should increase the rate of the complete cure by chemotherapy.

The combination of bleomycin treatment with a low-molecular-weight immune-enhancing agent will increase the rate of cure of tumors sensitive to bleomycin treatment.

VI. BLEOMYCIN AS MODEL FOR CANCER CHEMOTHERAPEUTIC AGENTS

Bleomycin radioactive metal complexes have been reported to distribute in malignant tumors and are used for their diagnosis. The study up to 1976 has been reviewed by Nouel (1976). It suggests that the bleomycin molecule is taken by tumors at a higher concentration than by normal cells and inflammatory lesions. Therefore, the utilization of bleomycin as a carrier to tumor cells has been studied. If it is necessary, it is possible to attach a metal-binding group to bleomycinic acid or to the terminal amine. The study on a receptor of bleomycin may give a useful information on why bleomycin molecule is selectively taken by tumor cells.

As described in a previous section, the bithiazole moiety selects a guanine base in polynucleotides for the binding of bleomycin. It is also possible that other molecular parts are also involved in the binding with DNA double strands. After the binding, DNA strand scission is caused by the activated oxygen molecule which has bound to the ferrous ion in bleomycin–ferrous-ion complex or by radicals produced by bleomycin–ferrous-ion complex binding to oxygen molecule. Therefore, the structural part involved in the metal-complex formation may be an important structure model which activates oxygen molecule. It also indicates that compounds which can form a ferrous-ion complex and which have the DNA-binding groups are worth study in the aspect of cancer chemotherapeutic agents.

REFERENCES

Asakura, H., Hori, M., and Umezawa, H. (1975). *J. Antibiot.* **28**, 537–542.
Asakura, H., Umezawa, H., and Hori, M. (1978). *J. Antibiot.* **31**, 156–158.
Carter, S. K., Ichikawa, T., Mathé, G., and Umezawa, H., eds. (1976). *Gann Monogr. Cancer Res.* No. 19.
Chien, M., Grollman, A. P., and Horwitz, S. B. (1977). *Biochemistry* **16**, 3641–3647.
Falaschi, A., and Kornberg, A. (1964). *Fed. Proc., Fed. Am. Soc. Exp. Biol.* **23**, 940–945.
Fujii, A., Takita, T., Maeda, K., and Umezawa, H. (1973). *J. Antibiot.* **26**, 398–399.
Haidle, C. W. (1971). *Mol. Pharmacol.* **7**, 645–652.
Haidle, C. W., Kuo, M. T., and Weiss, K. K. (1972a). *Biochem. Pharmacol.* **21**, 3308–3312.
Haidle, C. W., Weiss, K. K., and Kuo, M. T. (1972b). *Mol. Pharmacol.* **8**, 531–537.
Iitaka, Y., Nakamura, H., Nakatani, T., Muraoka, Y., Fujii, A., Takita, T., and Umezawa, H. (1978). *J. Antibiot.* **31**, 1070–1072.
Kanao, M., Tomita, S., Ishihara, S., Murakami, A., and Okada, H. (1973). *Kagaku Ryoho* **21**, 1305–1310.
Kasai, H., Naganawa, H., Takita, T., and Umezawa, H. (1978). *J. Antibiot.* **31**, 1316.
Konishi, M., Saito, K., Numata, K., Tsuno, T., Asama, K., Tsukiura, H., Naito, T., and Kawaguchi, H. (1977). *J. Antibiot.* **30**, 789–805.
Kono, A. (1977). *Chem. Pharm. Bull.* **25**, 2882–2886.
Kuo, M. T., and Haidle, C. W. (1973a). *Biochim. Biophys. Acta* **335**, 109–114.
Kuo, M. T., and Haidle, C. W. (1973b). *Biochem. J.* **13**, 1296–1306.
Miyamoto, T., Takabe, Y., Watanabe, M., and Terashima, T. (1977). *Gann; Kagaku Ryoho* **4**, 273–291.
Müller, W. E. G., and Zahn, R. K. (1976). *Gann Monogr. Cancer Res.* No. 19, 51–62.
Müller, W. E. G., Yamazaki, Z., Breter, H., and Zahn, R. K. (1972). *Eur. J. Biochem.* **31**, 518–525.
Müller, W. E. G., Yamazaki, Z., Zoellner, J. E., and Zahn, R. K. (1973). *FEBS Lett.* **31**, 217–221.
Muraoka, Y., Kobayashi, H., Fujii, A., Kunishima, M., Fujii, T., Nakayama, Y., Takita, T., and Umezawa, H. (1976). *J. Antibiot.* **29**, 853–856.
Nagai, K., Suzuki, H., Tanaka, N., and Umezawa, H. (1969a). *J. Antibiot.* **22**, 569–573.
Nagai, K., Suzuki, H., Tanaka, N., and Umezawa, H. (1969b). *J. Antibiot.* **22**, 624–628.
Naganawa, H., Muraoka, Y., Takita, T., and Umezawa, H. (1977). *J. Antibiot.* **30**, 388–396.
Nakatani, T., Muraoka, Y., Fujii, A., Takita, T., and Umezawa, H. (1980) *J. Antibiot.* (in press).
Nouel, J. P. (1976). *Gann Monogr. Cancer Res.* No. 19, 301–319.
Sausville, E. A., Reisach, J., and Horwitz, S. B. (1976). *Biochem. Biophys. Res. Commun.* **73**, 814–822.
Sausville, E. A., Stein, R. W., Reisach, J., and Horwitz, S. B. (1978). *Biochemistry* **17**, 2746–2754.

Strong, J. E., and Crooke, S. T. (1978). *Cancer Res.* **38**, 3322–3326.

Sugiura, Y. (1978). *J. Antibiot.* **31**, 1206.

Sugiura, Y., and Kikuchi, T. (1978). *J. Antibiot.* **31**, 1310.

Svanberg, L. E. (1976). *Gann Monogr. Cancer Res.* No. 19, 193–220.

Takahashi, K., Yoshioka, O., Matsuda, A., and Umezawa, H. (1977). *J. Antibiot.* **30**, 861–869.

Takeshita, M., Grollman, A. P., Ohtsubo, E., and Ohtsubo, H. (1978). *Proc. Natl. Acad. Sci. USA* **75**, 5983.

Takita, T., Muraoka, Y., Fujii, A., Itoh, H., Maeda, K., and Umezawa, H. (1972a). *J. Antibiot.* **25**, 197–199.

Takita, T., Muraoka, Y., Yoshioka, T., Fujii, A., Maeda, K., and Umezawa, H. (1972b). *J. Antibiot.* **25**, 755–758.

Takita, T., Fujii, A., Fukuoka, T., and Umezawa, H. (1973). *J. Antibiot.* **26**, 252–254.

Takita, T., Muraoka, Y., Nakatani, T., Fujii, A., Umezawa, Y., Naganawa, H., and Umezawa, H. (1978a). *J. Antibiot.* **31**, 801–804.

Takita, T., Muraoka, Y., Nakatani, T., Fujii, A., Iitaka, Y., and Umezawa, H. (1978b). *J. Antibiot.* **31**, 1073–1077.

Tanaka, N., Yamaguchi, H., and Umezawa, H. (1963). *J. Antibiot., Ser. A* **16**, 86–91.

Umezawa, H. (1971). *Pure Appl. Chem.* **28**, 665–680.

Umezawa, H. (1976). *Gann Monogr. Cancer Res.* No. 19, 3–37.

Umezawa, H. (1977). *Jpn. J. Antibiot.* Suppl., 138–163.

Umezawa, H., Maeda, K., Takeuchi, T., and Okami, Y. (1966a). *J. Antibiot., Ser. A* **19**, 200–209.

Umezawa, H., Suhara, Y., Takita, T., and Maeda, K. (1966b). *J. Antibiot., Ser. A* **19**, 210–215.

Umezawa, H., Ishizuka, M., Kimura, K., Iwanaga, J., and Takeuchi, T. (1968). *J. Antibiot.* **21**, 592–602.

Umezawa, H., Takeuchi, T., Hori, S., Sawa, T., Ishizuka, M., Ichikawa, T., and Komai, T. (1972). *J. Antibiot.* **25**, 409–420.

Umezawa, H., Asakura, H., and Hori, M. (1973a). *J. Antibiot.* **26**, 521–527.

Umezawa, H., Takahashi, Y., Fujii, A., Saino, T., Shirai, T., and Takita, T. (1973b). *J. Antibiot.* **26**, 117–119.

Umezawa, H., Hori, S., Sawa, T., Yoshioka, T., and Takeuchi, T. (1974). *J. Antibiot.* **27**, 419–424.

Yagoda, A., and Krakoff, I. H. (1976). *Gann Monogr. Cancer Res.* No. 19, 255–268.

Yamazaki, Z., Müller, W. E. G., and Zahn, R. K. (1973). *Biochim. Biophys. Acta* **308**, 412–421.

CHAPTER 6

Streptozocin

PAUL F. WILEY

I. INTRODUCTION

The antibiotic streptozocin (**1**) was isolated approximately 20 years ago as an antibacterial agent (Herr *et al.*, 1960), and was originally named streptozotocin. Subsequent investigations (Evans *et al.*, 1965) established its antitumor activity, and led eventually to clinical investigation showing its usefulness as an antitumor agent (Murray-Lyon *et al.*, 1968). The shortcomings of streptozocin, such as a very limited spectrum and its diabetogenic effect, coupled with ease of chemical modification led to extensive efforts to synthesize analogous new compounds having improved properties. It appears at present that these programs have achieved considerable success in chlorozotocin, and, as they are still in progress, greater improvement may be expected in the future.

II. STREPTOZOCIN

A. Fermentation and Isolation

Streptozocin is a metabolite of *Streptomyces achromogenes* var. *streptozoticus* (Vavra *et al.*, 1960). This organism was isolated from a soil sample

167

Anticancer Agents Based on Natural Product Models
Copyright © 1980 by Academic Press, Inc.
All rights of reproduction in any form reserved.
ISBN 0-12-163150-8

$$\overset{6}{C}H_2OH$$

1

collected at Blue Rapids, Kansas, and is a variant of *S. achromogenes* Waksman 3700. The streptozocin-producing organism differs from the Waksman organism in producing a brown pigment absent in the latter. In addition to streptozocin, *S. achromogenes* var. *streptozoticus* produces enteromycin (**2**) and *trans*-β-(oximinoacetylamino)acrylamide (**3**), a closely related compound, although neither is chemically related to streptozocin (Wiley *et al.*, 1965).

2   3

Fermentation of *S. achromogenes* var. *streptozoticus* in a medium consisting of cerelose (3 gm/liter), starch (15 gm/liter), refined corn meal (40 gm/liter), and peptone (3 gm/liter), with utilization of a temperature cycle of 32°C for four days and 28°C for two days, gave optimal yields. Addition of 2–4 gm of ammonium sulfate assists in maintaining the pH at 4–6, which is required because of the instability of streptozocin at neutral or alkaline pH. Yields of about 0.5 gm/liter can be obtained under optimum conditions (Vavra *et al.*, 1960).

Isolation of streptozocin is a relatively simple process. After filtration of the culture broth, the filtrate is concentrated, and acetone is added to precipitate the impurities. After filtration streptozocin is purified further by partition chromatography using a solvent system consisting of methylethylketone–cyclohexane–MacIlvaine's pH 4.0 buffer (9:1:1.43). The product from the chromatography is purified further by recrystallization from ethanol (Herr *et al.*, 1960). The recrystallization must be carried out with care, or decomposition occurs to give glucosamine (Wiley, 1968). At the present time, production by fermentation has been superseded by synthesis.

B. Chemistry

1. Structure

Streptozocin was shown to have the molecular formula $C_8H_{15}N_3O_7$ by analysis and molecular weight determinations. Alkaline treatment gave diazomethane and an amorphous solid (**8**). The latter was converted to glucosamine and carbon dioxide by acidic hydrolysis. The formation of diazomethane suggested the presence of an N-nitroso-N-methylureido moiety. A band in the infrared spectrum of streptozocin at 1705 cm^{-1} was consistent with the presence of a urea carbonyl, and a maximum in the ultraviolet spectrum at 228 nm probably arose from a nitroso group. The proton magnetic resonance (PMR) spectrum of streptozocin had a three-proton signal as a singlet at $\delta 3.5$, indicating the presence of a methyl group on nitrogen. The carbon magnetic resonance (CMR) spectrum showed two sets of lines which were very similar to those in the spectrum of a mixture of α- and β-D-2-deoxy-2-acetamidoglucopyranose. This spectrum was consistent with the isolation of glucosamine, and its presence as a moiety in streptozocin in α and β forms. The CMR spectrum indicated that the α form of streptozocin predominates (Wiley, 1978). As streptozocin is neutral, the moiety which gives rise to diazomethane must be attached to the nitrogen at C-2 of the glucose residue, particularly in view of the CMR spectrum. Furthermore, streptozocin forms a tetraacetyl derivative whose PMR spectrum has peaks for acetyl with very similar chemical shifts suggesting that they are all attached to oxygen. These data established the structure of streptozocin to be that shown in structure (**1**) (Wiley *et al.*, 1979). This structure was confirmed by synthesis.

2. Synthesis

The first reported synthesis of streptozocin was by Herr *et al.* (1967). The starting material [see Eq. (1)] was tetra-O-acetyl-β-D-glucosamine hydrochloride (**4**), which reacted with methyl isocyanate to give a urea (**5**). The urea was nitrosated with nitrosyl chloride, and the acetyl groups were then removed with ammonia in methanol. The yield was rather poor in this conversion, as the removal of the acetyl groups resulted in several side reactions. The necessity to provide supplies of streptozocin for clinical use and the desire to prepare analogues has resulted in greatly improved syntheses. Hardegger *et al.* (1969) used D-glucosamine as their starting material and introduced the acyl sidechain *in toto* at one step [Eq. (2)] using an acyl azide. The yield was 31%. The process

$$
\begin{array}{c}
\text{CH}_2\text{OH} \\
\end{array}
\qquad
\xrightarrow[\text{CH}_3\text{NCON}_3]{\overset{\text{NO}}{|}}
\qquad 1
\qquad (2)
$$

was a very simple one, but it required preparation of the acyl azide. Hessler and Jahnke (1970) a little later reported a two-step process which gave much better yields. Their starting material was also glucosamine, which was converted to a urea **(7)** by reaction with methyl isocyanate. The urea was then nitrosated with nitrous acid [Eq. (3)]. The overall yield of 77–80% was excellent. It was necessary to maintain the temperature at $0 \pm 2°\text{C}$ for the best yield in Step 1 and to use nitrous acid generated from N_2O_3 rather than sodium nitrite–sulfuric acid for a good yield in Step 2.

$$
\xrightarrow{\text{CH}_3\text{NCO}} \qquad \xrightarrow{\text{HNO}_2} \qquad 1 \qquad (3)
$$

7

3. Reactions

The instability of streptozocin in most solvents results in ready conversion to other products. In aqueous solutions it is most stable at pH 4 (Herr *et al.*, 1960), being rapidly decomposed at pH above 7 and in strong acid. It can be recrystallized from ethanol, but if care is not exercised, decomposition can also occur in this solvent, giving as the only identified product glucosamine (Wiley, 1968).

Treatment of streptozocin with 2 N sodium hydroxide solution gives diazomethane (Herr *et al.*, 1960) and a bicyclic product **(8)** arising from rearrangement of the pyranose ring to a furanose ring and attack of the urea carbonyl at the hydroxyl oxygen attached to C-1 [Eq. (4)] (Wiley *et al.*, 1979). The structure originally proposed for **(8)** was that shown as **(10)** which retains the pyranose

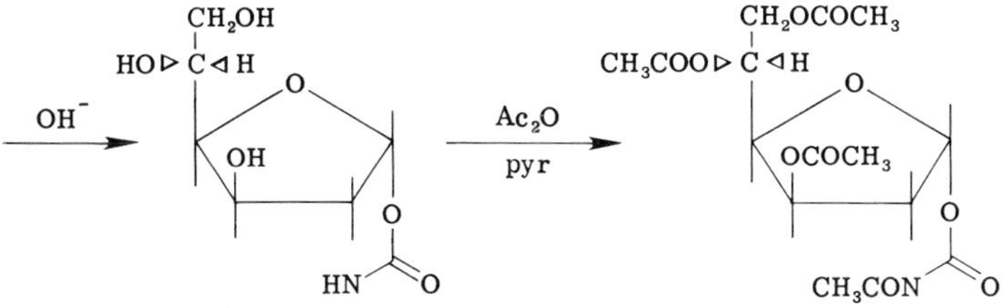

$$\text{(4)}$$

8   9

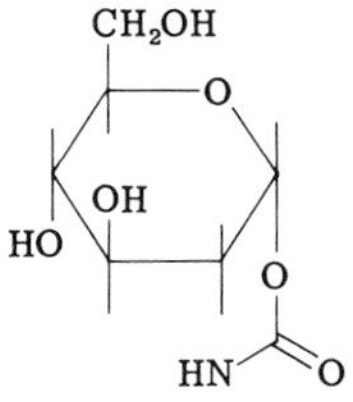

10

ring (Herr *et al.*, 1967). However, the PMR spectrum and rotation of (**8**) and its acetate (**9**) did not fit such a structure well. Consequently, the structure of the tetraacetyl derivative (**9**) was established by x-ray crystallography and found to be as indicated in structure (**9**) (Wiley *et al.*, 1979).

Streptozocin has been found to have two modes of decomposition in the presence of acid. Reaction with 0.1 *M* sulfamic acid gave a product having a structure analogous to that of (**8**), while stronger acid (1 *M* sulfamic acid) caused much more deep-seated decomposition. The milder acid causes a cyclization which results in substitution at C-1 by the nitrogen atom substituted by the nitroso group with loss of the elements of nitrous acid. The structure first proposed (Herr *et al.*, 1967) for the product was that depicted as (**11**). This structure implied the α configuration at C-1 although there was no specific mention of chirality. In a subsequent publication (Wiley *et al.*, 1976), the catalytic reduction of streptozocin to give an isomer to the mild acid degradation product was reported, and it was proposed that this product was the α isomer (**11**). It was suggested that the previously mentioned isomer was the β form (**12**). After it

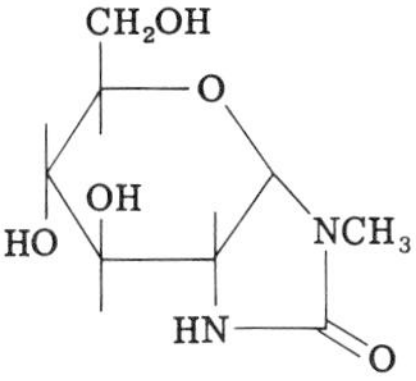

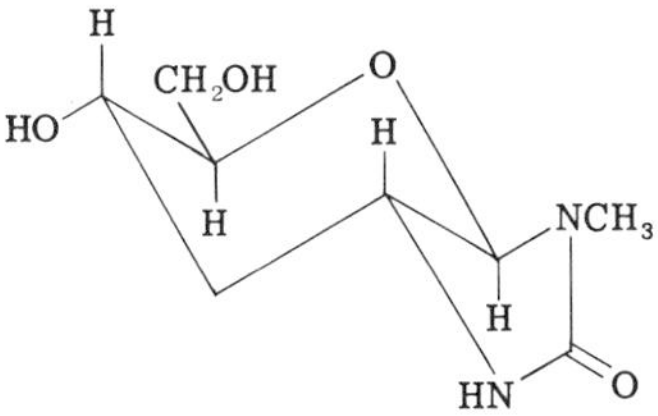

11   12

was found that the base cyclization product (**8**) was a furo-oxazolone, the structure of the mild acid degradation product was reconsidered. On the basis of PMR spectra of the product and of its triacetyl derivative, the structures were assigned as (**13**) and (**14**) [Eq. (5)].

13   14

$$\text{(5)}$$

Argoudelis *et al.* (1974) have discussed the treatment of streptozocin with 1 M sulfamic acid. The same product is obtained by the action of 1 M sulfamic acid on (**13**), and it is considered probable that the streptozocin is first converted to (**13**) followed by more extensive degradation. It was at once apparent that the product (**15**) was highly unsaturated as it was colored, and its electronic spectrum confirmed this observation. The product was found to have the molecular formula $C_8H_{10}N_2O_3$. Alkaline hydrolysis of (**15**) gave carbon dioxide, methylamine, acetaldehyde and glyoxal. The PMR spectrum showed the presence of a moiety $HOCH_2CH{=}CHCH{=}$, and the infrared spectrum suggested that this moiety was attached to the hydantoin ring. Such data could be accommodated in the structure (**15**) at least in the crystal form. PMR spectra indicated the carbonyl between the nitrogen atoms was enolized in solution, and the spectra would be more consistent with structure (**16**). Degradation of (**15**) could occur by hydration of the diene system followed by retroaldol reactions. Protonation and subsequent elimi-

15   16

$$\text{(6)}$$

nation and rearrangement, as indicated in Eq. (7), could be the route followed in the conversion of (**13**) to (**15**). Acidic degradation of streptozocin and (**13**) does not give glucosamine which must be the result, in the case of streptozocin, of the initial formation of (**13**) under the influence of acid. The dehydration of (**13**) to (**15**) then occurs much more readily than hydrolysis of the imidazolone ring to give glucosamine.

Although streptozocin is most stable in aqueous solutions at pH 4, it decomposes in saline at pH 4.6 with evolution of nitrogen and formation of at least three products and perhaps more (Wiley *et al.*, 1979). The product was separable into

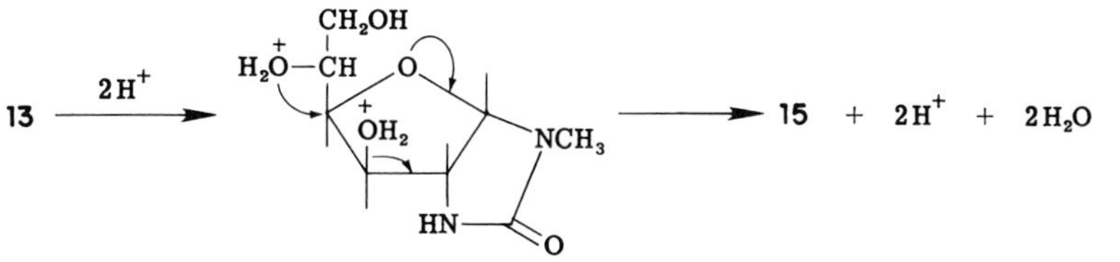

$$13 \xrightarrow{2H^+} \quad \longrightarrow \quad 15 \;+\; 2H^+ \;+\; 2H_2O \quad (7)$$

three components only after acetylation; only one component has been identified. The mass spectrum of the identified crystalline component (**17**) suggested that it resulted from the loss of the CH_3NNO group, followed by cyclization to the oxygen atom at C-3 and dimerization of the product formed by cyclization. Signals in the PMR spectrum indicated only 19 hydrogen atoms requiring strict symmetry in the dimer, as did the CMR spectrum, which showed signals for only 15 carbon atoms. Such properties could only be present in structure (**17**).

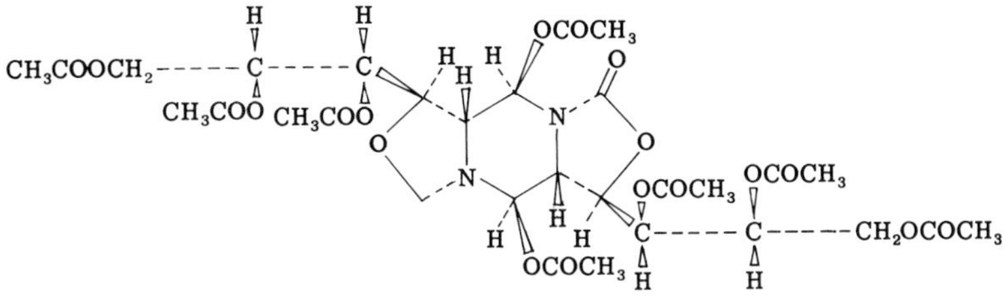

17

One of the more unusual reactions of streptozocin is its decomposition in DMSO to form (**18**), Eq. (8) (Wiley *et al.*, 1979). This reaction does not occur if a small amount of acetic acid is added to the DMSO. As a result of failure to crystallize (**18**), it was converted to a crystalline tetraacetyl derivative (**19**). Analysis and mass spectra established that reaction had occurred with loss of the

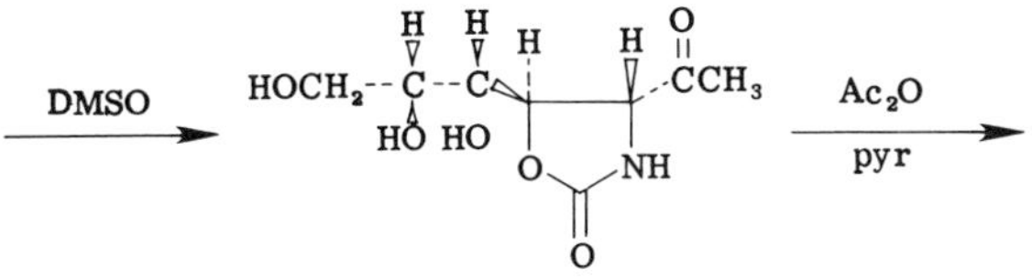

18

(8)

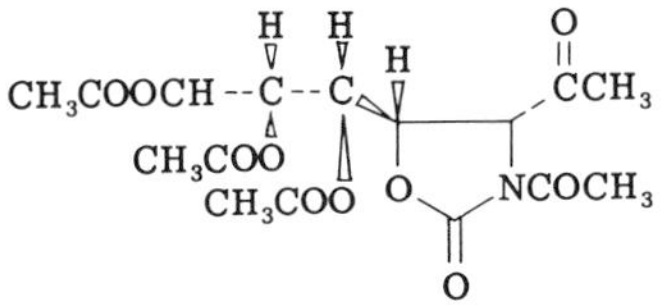

19

elements of water and nitrogen only. The usual spectral data showed the retention of the methyl group. These data, combined with the formation of a 2,4-dinitrophenylhydrazone, indicated structure (**18**). In some fashion the methyl group, originally attached to nitrogen, is transferred to the original C-1 in streptozocin.

C. Biological Activity

1. Pharmacology and Toxicology

In lower animals, streptozocin is rapidly excreted with 72% being excreted in 4 hours (Bhuyan *et al.*, 1974). Some accumulation in organs occurs with the highest levels in the liver and the next highest in the kidney, but none reaches the brain. Apparently there is no concentration in the pancreas (Ryo *et al.*, 1974). However, it has been reported (Karunanayake *et al.*, 1974, 1975), on the basis of excretion studies with streptozocin labeled with ^{14}C at C-1, C-2′, and N-3′CH_3, that the methyl on nitrogen is retained much more than is the rest of the molecule. It was suggested that this is caused by degradation of streptozocin with transfer of the methyl group on nitrogen to other compounds present in the body.

The most serious toxicity of streptozocin in lower animals is the effect on pancreatic cells resulting in diabetes. Rats and hamsters become diabetic after doses of 75–80 mg/kg i.v., but mice are somewhat more resistant requiring higher doses (Losert *et al.*, 1971). Streptozocin also causes diabetes in rabbits and guinea pigs but with more difficulty (Lazarus and Shapiro, 1972; Kushner *et al.*, 1969). The diabetogenic effect of streptozocin is prevented by simultaneous administration of nicotinamide (Dulin and Wyse, 1969). The diabetogenic effect involves damage to the islets of Langerhans and granules in β cells of the pancreas (Rakieten *et al.*, 1963). Cataract quite frequently occurs in animals given streptozocin, and is believed to be related to the diabetogenic effect (White and Cinotti, 1972). The same effect has been found by others in rats (Arison *et al.*, 1967) and in mice (Verheyden, 1967). Single doses of 65 mg/kg caused the appearance of cataracts in 10–18 weeks in these animals. The cataract-producing effect of streptozocin has also been seen in hamsters in which one dose gave cataracts in 14% of the animals after several months (Sibay *et al.*, 1971). A far larger percentage suffered retinal vascular pathology. Hepatotoxicity occurs in dogs and monkeys after administration of streptozocin and is not prevented by nicotinamide (Rakieten *et al.*, 1969a,b). Rats, after a single dose of 65 mg/kg, exhibited renal tubule damage suggesting renal toxicity (Arison *et al.*, 1967).

2. Tumorigenic and Mutagenic Effects

The known mutagenic activity of nitrosoureas suggested that streptozocin was mutagenic, and this property was confirmed by the studies of several investigators. Kolbye and Legator (1968) compared the mutagenic actions of streptozocin and N-methyl-N-nitroso-N'-nitroguanidine on a histidine-requiring mutant of *Salmonella typhimurium*. It was found that the mutagenic activities of the two compounds were about the same. Ficsor *et al.* (1974) also found that streptozocin was highly mutagenic with respect to the same organism. About the same time, Gichner *et al.* (1968) established that streptozocin was mutagenic to plant cells, but much less so than N-methyl-N-nitroso-N'-nitroguanidine. The mutational effect of streptozocin against a human cell line (L-132) *in vitro*, as measured by inhibition of DNA synthesis and by chromosome aberrations, was far less than that of the guanidine (Kelly and Legator, 1971). It appears that the mutagenic effect is enhanced by some effect of mammalian metabolism. The mutagenic effect of streptozocin against V79 Chinese hamster cells *in vitro* has been studied by Bhuyan *et al.* (1976). It was found that the mutation paralleled DNA damage, and it was concluded that the two effects are related.

The tumorigenic activity of streptozocin has been studied for the most part in male rats, as it seems to be much more pronounced in males than in females. However, this appears to differ somewhat, depending on the strain of rats. Using male Holtzman rats and a dose of 50 mg/kg, Arison and Feudale (1967) found a 52% incidence of tumors, mostly in the renal cortex. A later study (Rakieten *et al.*, 1968) using both male and female rats reported an incidence of 16% renal tumors. The same dosage combined with massive doses of nicotinamide caused 64% of male Holtzman rats to develop pancreatic islet cell tumors (Rakieten *et al.*, 1971). A study conducted with hamsters resulted in an incidence of 87% of biliary tumors which were for the most part benign.

3. Antibacterial and Antitumor Activity

The earliest interest in streptozocin was because of its excellent antibacterial activity rather than its antitumor activity. Vavra *et al.* (1960) and Lewis and Barbiers (1960) have published extensive data describing in detail the *in vitro* antibacterial activity of streptozocin. Table I taken from the work of Vavra *et al.* (1960) shows the broad spectrum activity of streptozocin. As indicated in the table, streptozocin is inactive against *P. aeruginosa*. It is also inactive against *Shigella dysenteriae* and various strains of neisseria, clostridia, and mycobacteria (Olitzki *et al.*, 1967).

The *in vitro* activity of streptozocin against various infections in mice was discussed by Lewis and Barbiers (1960). When streptozocin was given subcutaneously or orally in doses of 1–20 mg/kg/day, mice were protected against *S. aureus*, *P. multocidia*, and *P. vulgaris*. Oral administration required somewhat higher doses than did subcutaneous dosage. No effect was seen against *P. aeruginosa*, *Salmonella gallinarum*, and *Mycobacterium tuberculosis*.

Streptozocin has been reported to be toxic *in vitro* to V79 Chinese hamster cells (Bhuyan *et al.*, 1972, 1976), the ID_{50} against these cells being 370 μg/ml.

TABLE I

Antibacterial Activities of Streptozocin

Microorganism	Minimal inhibitory concentration (μg/ml)
Salmonella pullorum	0.15
Proteus vulgaris	0.20
Escherichia coli	0.50
Salmonella typhi	0.90
Streptococcus faecalis	2.0
Klebsiella pneumoniae	3.0
Staphylococcus aureus	0.75
Salmonella schotmuelleri	0.35
Pasturella multocidia	<0.25
Aerobacter aerogenes	10.0
Pseudomonas aeruoginosa	>50

The effect of streptozocin on the pentose phosphate pathway intermediates in Krebs 2 ascites cells from mice has been studied (Gumaa and McLean, 1969). In these cells, 6-phosphogluconate concentration was lowered, while that of sedoheptulose-7-phosphate was increased. The oxidation of glucose at C-1 was lowered markedly. It was hypothesized that the effect on 6-phosphogluconate concentration was due to reduced NADP$^+$ effects on glucose-6-phosphate dehydrogenase.

The original observation of streptozocin antitumor activity was made by Evans *et al.* (1965). Activity of a streptozocin–zedalan (85:15) mixture against the ascitic form of the lymphoblast L-5178Y, S-180, Ehrlich carcinoma, and Walker 256 carcinosarcoma in mice was reported. Both components were considered to be necessary for activity against the L-5178Y murine lymphoblast. However, it has since been found that zedalan is unneeded. White (1963) reported an ILS of 65% in mice having L-1210 leukemia when they were treated with 90 mg/kg i.p. Somewhat higher activity has been reported by Schein *et al.* (1973). A dose of 50 mg/kg inhibited the growth of Walker carcinoma implanted in dogs.

4. Mode of Action

Heinemann and Howard (1965) have shown that streptozocin inhibits DNA synthesis, decreases RNA synthesis by 25%, and decreases protein synthesis by 20% in *Escherichia coli*. These results were confirmed and extended by Bhuyan (1970), who found that incorporation of thymidine into DNA in L-1210 cells was totally inhibited while inhibition of uridine and adenosine incorporation into DNA was much greater than into RNA. The inhibition of lysine incorporation was substantially less than was that of nucleosides into DNA and RNA. The effect of streptozocin on the enzymes regulating these processes was insignificant.

A number of publications suggested that the streptozocin effects were a result of methylation of nucleoside bases by formation of diazomethane which acted as the alkylating agent (Gabridge *et al.*, 1969; Kolbye and Legator, 1968; Rakieten *et al.*, 1968; Schein and Loftus, 1968). These suggestions arose from the idea that nitrosomethylureas acted on tumor cells by formation of diazomethane followed by methylation and the knowledge that streptozocin contains the $CH_3N(NO)$-CO group which forms diazomethane under appropriate conditions. Gabridge *et al.* (1969), however, found that in mutagenesis streptozocin behaves quite differently than does *N*-methyl-*N*-nitroso-*N* '-nitroguanidine, whose action presumably involves diazomethane formation. It has been shown that *N*-methyl-*N*-nitrosourea is not cross resistant with streptozocin, differs from streptozocin in its pattern of macromolecular synthesis inhibition, and the two differ in other biological parameters (Rosenkrantz and Carr, 1970). These differences would suggest that streptozocin does not act through formation of diazomethane. There is evidence in the form of isolated methylated DNA bases that streptozocin can alkylate, but it probably does not do so through diazomethane formation. It is at least theoretically possible that streptozocin can carbamoylate various components of biological systems, and it may be that this is how it exerts its action.

C. Clinical Studies

The first clinical use of streptozocin was by Murray-Lyon *et al.* (1968), who treated one patient with a malignant islet cell carcinoma of the pancreas. The patient was given a total of 8.5 gm of streptozocin in three doses over a period of 6 weeks. Considerable improvement was seen and was maintained over a period of some months.

This preliminary finding led to extensive clinical investigation of the effect of streptozocin on various tumors. It was found that streptozocin was most effective against malignant islet cell carcinoma of the pancreas. Broder and Carter (1973) have reported results of clinical trials involving 52 patients which are typical of results with streptozocin against islet cell carcinoma. Doses of 1–3 gm/m^2 of body surface were given i.v. weekly for a total of 8.1–10 gm/m^2 of body surface. Complete or partial remissions were obtained in 64% of the cases, with median survival of responders being 1268 days as opposed to 518 days for nonresponders. Toxic symptoms were seen in virtually all of the patients. The most prevalent forms of toxicity were gastrointestinal (94%), renal (65%), and hepatic (67%). Diabetogenesis was not seen in humans. Schein)1972) reported much the same results in terms of response of islet cell carcinomas to streptozocin. No impairment of glucose tolerance was observed (Sadoff, 1970, 1972).

A series of publications has reported the effectiveness of streptozocin in carcinoid tumors of the small intestine (Carter and Broder, 1974), bronchi (Feldman et al., 1972) and the gall bladder (Iweze et al., 1972). The usual dose was 8–10 gm over a period of months. Davis et al. (1973) have reported that a combination of streptozocin, given at a dosage of 50 mg/m²/d × 5, and 5-fluorouridine given at a dosage of 12 mg/kg/d × 5, and repeated at 6-week intervals, is more effective against malignant carcinoid tumors than streptozocin alone. Since carcinoid tumors secrete serotonin and pancreatic tumors secrete insulin, it has been proposed that streptozocin would be beneficial against all tumors which secrete hormones (Feldman et al., 1972; Carter and Broder, 1974).

Stolinsky et al. (1972) have found streptozocin to be beneficial in carcinoma of the lung and in squamous carcinoma of the oral cavity. Activity has also been reported against Hodgkin's disease (7/16), lymphocytic lymphoma (3/11), Burkitt's lymphoma (1/12), and acute lymphocytic leukemia (Schein et al., 1974). However, at present streptozocin is recommended only for malignant islet cell carcinoma of the pancreas and carcinoid carcinoma.

III. ANALOGUES

A. Chemistry

A discussion of chemically derived analogues of streptozocin makes necessary a rather arbitrary decision as to what should be included. In the present discussion, compounds included are those derived from hexoses and pentoses, reduced hexoses and pentoses, and polyhydroxylated cycloalkane analogues of pyranose and furanose sugars. The 105 such compounds reported in the literature are listed in Table II. For the most part they are derivatives of a limited number of hexoses and pentoses, which in all cases have the D configuration. These have been modified by introducing ureido groups at various positions, variation in the ureido groups, and formation of a limited number of glycosides. Although emphasis has been on preparing sugar analogues with free hydroxyl groups, the acetates have also been made in most cases.

Synthesis of the analogues of streptozocin was by the two general procedures already exemplified in Eqs. (2) and (3). Gassmann et al. (1975) and Hardegger and Meier (1970) have prepared a series of analogues of streptozocin derived from glucosamine, 2-deoxy-2-aminogalactose, arabitol, and various hexitols by their original procedure [Eq. (2)] involving introduction of the R-N(NO)CO group as a unit by way of its azide. However, all other workers in this field have used the Hessler and Jahnke (1970) procedure [Eq. (3)] with minor variations. For the most part the starting material contained a primary amino group in its basic form. However, in a few instances hydrochlorides were used necessitating neutralization. In some cases this was done with an organic base (Ichikawa et al., 1976), but in others silver carbonate was used (Suami, 1974; Suami and Machinami, 1971). The aminosugars used as starting materials could be and frequently were used without any derivatization to prevent reaction of hydroxyl groups with the isocyanate (Bannister, 1972). The hydroxyl groups could be partially covered as by starting with an alkyl glycoside (Iwasaki *et al.*, 1976) or a glycoside which had been mesylated (Machinami *et al.*, 1975a). When ribofuranoses were starting materials, the acetonides were used (Montéro and Imbach, 1974). In many cases, however, completely O-acylated aminosugars were the starting materials. The usual nitrosating agent was nitrous acid prepared *in situ* by reaction of sodium nitrite with a mineral acid or an organic acid. Fujiwara *et al.* (1974) used N_2O_3 as preferred by Hessler and Jahnke (1970), and Suami and Machinami (1970a) nitrosated with nitrosyl chloride. Occasionally the ureas prepared by reaction of aminosugars with isocyanates were acylated before nitrosation (Montero and Imbach, 1974). In some instances in which an O-acylated urea was nitrosated, it was desired to remove the acyl groups. This was done with methanolic ammonia or methanolic sodium hydroxide (Machinami *et al.*, 1975b). More commonly the nitrosoureido compounds were prepared first with free hydroxyl groups, and, if acylated analogues were desired, acylation was carried out by the usual methods (Burns and Heindel, 1974; Montéro *et al.*, 1977).

As a rule, stereochemistry was established, except for that at C-1, by the starting material; and in some compounds such as the cycloalkane analogues, the stereochemistry was completely determined by the configuration of the starting material. The sugar analogues having a free hydroxyl group at C-1 were always a mixture of α and β forms. However, in those compounds whose configuration at C-1 was determined, the α form predominated to the extent of about 80% (Johnston *et al.*, 1975; Meier *et al.*, 1974). A number of streptozocin analogues were α- and β-glycosides. The starting materials for the preparation of such compounds were mixtures of sugar glycosides which were separated into their isomers as their *N*-carbobenzoxy derivatives. The carbobenzoxy groups were removed and the α and β isomers were converted to nitrosoureas in the usual way (Bannister, 1972; Iwasaki *et al.*, 1976). The physical properties of melting point and rotation for the α-methyl and β-methyl glycosides of streptozocin have been reported by Bannister (1972), Suami and Machinami (1970b), and Iwasaki *et al.* (1976). While there was reasonable agreement as to rotations, there was some as yet unresolved inconsistency in the melting points. Those compounds having the 3-methyl-3-nitrosoureido substituent at C-1 of a sugar nucleus also have the possibility of α and β isomers. Bannister (1972) prepared 3-β-D-glucopyranosyl-1-methyl-1-nitrosourea (**20**) and its acetate, in which the β configuration at C-1 was assumed on the basis of preparation from tetra-O-acetyl-β-D-1-deoxy-1-aminoglucopyranose of known β configuration. The analogue from galactose was reported to have the same configuration. Montéro *et al.* (1977) have pre-

TABLE II

Streptozocin Analogues

A. Glucosamine Analogues

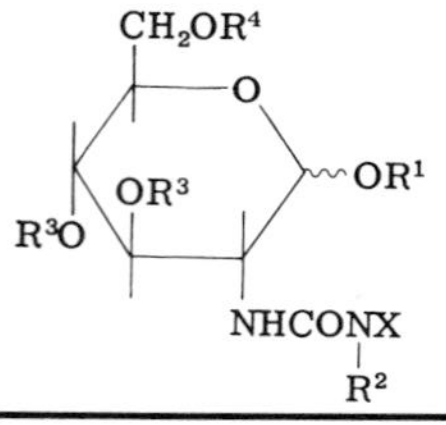

CH_2OR^4, O, OR^1, OR^3, NHCONX, R^2

No.	C-1	R^1	R^2	R^3	R^4	X	References
30	β	CH_3CO	CH_3	CH_3CO	CH_3CO	NO	Herr et al. (1967)
31	α,β	H	C_2H_5	H	H	NO	Meier et al. (1974)
32	α,β	H	$n\text{-}C_3H_7$	H	H	NO	Meier et al. (1974)
33	α,β	H	$n\text{-}C_4H_9$	H	H	NO	Meier et al. (1974)
34	α,β	H	$C_6H_5CH_2$	H	H	NO	Meier et al. (1974)
24	α,β	H	$ClCH_2CH_2$	H	H	NO	Burns and Heindel (1974); Johnston et al. (1975)
35	α	CH_3CO	C_2H_5	CH_3CO	CH_3CO	NO	Meier et al. (1974)
36	α	CH_3CO	$n\text{-}C_3H_7$	CH_3CO	CH_3CO	NO	Meier et al. (1974)
37	α	CH_3CO	$n\text{-}C_4H_9$	CH_3CO	CH_3CO	NO	Meier et al. (1974)
38	α	CH_3CO	$C_6H_5CH_2$	CH_3CO	CH_3CO	NO	Meier et al. (1974)
39	β	CH_3CO	C_2H_5	CH_3CO	CH_3CO	NO	Meier et al. (1974)
40	β	CH_3CO	$n\text{-}C_3H_7$	CH_3CO	CH_3CO	NO	Meier et al. (1974)
41	β	CH_3CO	$n\text{-}C_4H_9$	CH_3CO	CH_3CO	NO	Meier et al. (1974)
42	α,β	CH_3CO	$ClCH_2CH_2$	CH_3CO	CH_3CO	NO	Burns and Heindel (1974); Johnston et al. (1975); Montéro and Imbach (1974)

No.		R	R	R	R	R	Reference
43	α	CH_3	CH_3	H	H	NO	Bannister (1972); Iwasaki *et al.* (1976); Suami and Machinami (1970b)
44	β	CH_3	CH_3	H	H	NO	Bannister (1972); Iwasaki *et al.* (1976); Suami and Machinami (1970b)
45	α	CH_3	$n\text{-}C_3H_7$	H	H	NO	Iwasaki *et al.* (1976)
46	α	CH_3	$n\text{-}C_4H_9$	H	H	NO	Iwasaki *et al.* (1976)
47	β	CH_3	C_2H_5	H	H	NO	Iwasaki *et al.* (1976)
48	β	CH_3	$n\text{-}C_3H_7$	H	H	NO	Iwasaki *et al.* (1976)
49	β	CH_3	$n\text{-}C_4H_9$	H	H	NO	Iwasaki *et al.* (1976)
50	α	C_2H	CH_3	H	H	NO	Iwasaki *et al.* (1976)
51	α	C_2H_5	C_2H_5	H	H	NO	Iwasaki *et al.* (1976)
52	β	C_2H_5	CH_3	H	H	NO	Iwasaki *et al.* (1976)
53	β	C_2H_5	C_2H_5	H	H	NO	Iwasaki *et al.* (1976)
54	α	$n\text{-}C_3H_7$	CH_3	H	H	NO	Iwasaki *et al.* (1976)
55	β	$n\text{-}C_3H_7$	CH_3	H	H	NO	Iwasaki *et al.* (1976)
56	α	$n\text{-}C_4H_9$	CH_3	H	H	NO	Iwasaki *et al.* (1976)
57	α	$n\text{-}C_4H_9$	$n\text{-}C_4H_9$	H	H	NO	Iwasaki *et al.* (1976)
58	β	$n\text{-}C_4H_9$	CH_3	H	CH_3SO_2	NO	Iwasaki *et al.* (1976)
59	β	CH_3	CH_3	H	CH_3SO_2	NO	Machinami *et al.* (1975a); Suami *et al.* (1975)
60	β	CH_3	CH_3	CH_3CO	CH_3SO_2	NO	Machinami *et al.* (1975a)
61	α,β	H	CH_3	H	H	C_6H_5	Wiley *et al.* (1976)
62	β	CH_3CO	CH_3	CH_3CO	CH_3CO	CH_3	Wiley *et al.* (1976)
63	β	CH_3CO	CH_3	CH_3CO	CH_3CO	HCO	Wiley *et al.* (1976)
64	β	CH_3CO	CH_3	CH_3CO	CH_3CO	CH_3CO	Wiley *et al.* (1976)
65	β	CH_3CO	CH_3	CH_3CO	CH_3CO	C_6H_5	Wiley *et al.* (1976)

(Continued)

TABLE II (*Continued*)

B. Other Aminohexopyranoses

1. Derived from D-Glucose

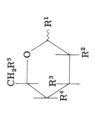

No.	C-1	R^1	R^2	R^3	R^4	R^5	References
20	β	NHCONCH₃ —NO	OH	OH	OH	OH	Machinami et al. (1975a); Suami et al. (1974); Bannister (1972)
66	β	NHCONC₂H₅ —NO	OH	OH	OH	OH	Machinami et al. (1975a)
67	β	NHCONC₃H₇ —NO	OH	OH	OH	OH	Machinami et al. (1975a)
68	β	NHCONC₄H₉ —NO	OH	OH	OH	OH	Machinami et al. (1975a)
69	β	NHCONCH₂CH₂Cl —NO	OH	OH	OH	OH	Machinami et al. (1975b)
70	β	NHCONCH₃ —NO	CH₃COO	CH₃COO	CH₃COO	CH₃COO	Bannister (1972); Machinami et al. (1975a)

No.		R_1	R_2	R_3	R_4	R_5	Reference
71	β	$NHCONC_2H_5$ —NO	CH_3COO	CH_3COO	CH_3COO	CH_3COO	Machinami et al. (1975a)
72	β	$NHCONC_3H_7$ —NO	CH_3COO	CH_3COO	CH_3COO	CH_3COO	Machinami et al. (1975a)
73	β	$NHCONC_4H_9$ —HO	CH_3COO	CH_3COO	CH_3COO	CH_3COO	Machinami et al. (1975a)
74	β	$NYCONCH_2CH_2Cl$ —NO	CH_3COO	CH_3COO	CH_3COO	CH_3COO	Machinami et al. (1975b)
75	α	CH_3O	OH	$NHCONCH_3$ —NO	OH	OH	Fujiwara et al. (1974)
76	α,β	OH	OH	OH	OH	$NHCONCH_3$ —NO	Gassman et al. (1975); Fujiwara et al. (1974)
77	α,β	CH_3COO	CH_3COO	CH_3COO	CH_3COO	$NHCONCH_3$ —NO	Gassman et al. (1975)
78	α	CH_3O	OH	OH	OH	$NHCONCH_3$ —NO	Fujiwara et al. (1974); Machinami and Suami (1973)
79	α	CH_3O	OH		OH	$NHCONCH_3$ —NO	Machinami and Suami (1973)
80a	β	CH_3O	$NHCONCH_3$ —NO	$NHCONCH_3$ —NO	OH	$NHCONCH_3$ —NO	Machinami et al. (1975a)
80b	β	CH_3O	$NHCONCH_3$ —NO	$NHCONCH_3$ —NO	CH_3COO	$NHCONCH_3$ —NO	Machinami et al. (1975a)

(Continued)

TABLE II (*Continued*)

2. Derived from D-Mannose

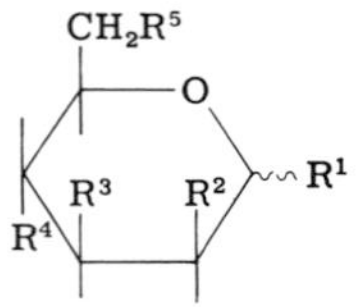

No.	C-1	R^1	R^2	R^3	R^4	R^5	References
81	β	NHCONCH$_3$ $\overset{\textstyle\mid}{\text{NO}}$	OH	OH	OH	OH	Machinami *et al.* (1975a)
82	β	NHCONCH$_2$CH$_2$Cl $\overset{\textstyle\mid}{\text{NO}}$	OH	OH	OH	OH	Machinami *et al.* (1975b)
83	β	NHCONCH$_2$CH$_2$Cl $\overset{\textstyle\mid}{\text{NO}}$	CH$_3$COO	CH$_3$COO	CH$_3$COO	CH$_3$COO	Machinami *et al.* (1975b)
84	α,β	OH	NHCONCH$_3$ $\overset{\textstyle\mid}{\text{NO}}$	OH	OH	OH	Bannister (1972)

3. Derived from D-Galactose

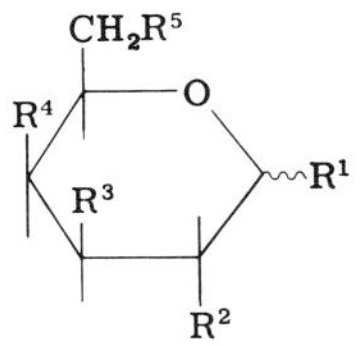

No.	C-1	R¹	R²	R³	R⁴	R⁵	References
85	β	$NHCON(NO)CH_3$	OH	OH	OH	OH	Bannister (1972)
86	β	$NHCON(NO)CH_3$	CH_3COO	CH_3COO	CH_3COO	CH_3COO	Bannister (1972)
87	β	$NHCON(NO)CH_2CH_2Cl$	OH	OH	OH	OH	Machinami et al. (1975b)
88	β	$NHCON(NO)CH_2CH_2Cl$	CH_3COO	CH_3COO	CH_3COO	CH_3COO	Machinami et al. (1975b)
89	α,β	OH	$NHCON(NO)CH_3$	OH	OH	OH	Bannister (1972); Gassmann et al. (1975)
90	α,β	OH	$NHCON(NO)CH_2CH_2Cl$	OH	OH	OH	Ichikawa et al. (1976)

(*Continued*)

TABLE II (*Continued*)

4. Miscellaneous

	No.	R	References
	91	R = H	Machinami *et al.* (1975a); Suami *et al.* (1975)
	92	R = CH$_3$O	Machinami *et al.* (1975a)
	93		Fujiwara *et al.* (1974)

C. Pentoses

1. Pyranoses

No.		R^1	R^2	R^3	R^4	R^5	R^6	References
94	β	NHCONCH$_3$ | NO	OH	H OH		H	OH	Machinami *et al.* (1975a)

No.								Reference
95	α,β	$NHCON(NO)CH_2CH_2Cl$	OH	H	OH	H	OH	Montéro and Imbach (1974)
96	α,β	$NHCON(NO)CH_2CH_2Cl$	CH_3COO	H	CH_3COO	H	CH_3COO	Imbach et al. (1976); Montéro and Imbach (1974); Montéro et al. (1977)
97	β	$NHCON(NO)CH_2CH_2Cl$	C_6H_5COO	H	C_6H_5COO	H	C_6H_5COO	Montéro et al. (1977)
98	β	$NHCON(NO)CH_2CH_2Cl$	$p\text{-}NO_2C_6H_4COO$	H	$p\text{-}NO_2C_6H_4COO$	H	$p\text{-}NO_2C_6H_4COO$	Montéro et al. (1977)
99	β	$NHCON(NO)CH_3$	OH	H	H	OH	OH	Machinami et al. (1975a)
100	β	$NHCON(NO)CH_3$	CH_3COO	H	H	CH_3COO	CH_3COO	Machinami et al. (1975a)
101	β	$NHCON(NO)CH_2CH_2Cl$	OH	H	H	OH	OH	Machinami et al. (1975b)
102	β	$NHCON(NO)CH_2CH_2Cl$	CH_3COO	H	H	CH_3COO	CH_3COO	Machinami et al. (1975b); Imbach et al. (1976)
103	β	CH_3O	OH	H	H	$NHCON(NO)CH_3$	OH	Fujiwara et al. (1974)
104	β	CH_3O	H	H	$NHCON(NO)CH_3$	H	OH	Fujiwara et al. (1974)

(Continued)

TABLE II (*Continued*)

2. Furanoses

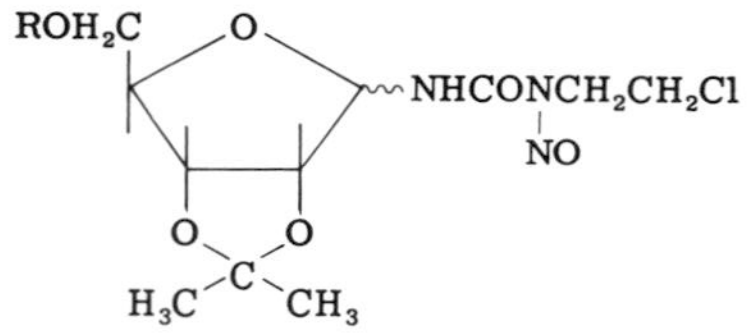

No.	C-1	R	References
105	β	H	Montéro and Imbach (1974);
106	α,β	CH_3CO	Montéro and Imbach (1974); Montéro et al. (1977)
107	α,β	C_6H_5CO	Montéro et al. (1977)
108	α,β	o-$NO_2C_6H_4CO$	Montéro et al. (1977)
109	α,β	m-$NO_2C_6H_4CO$	Montéro et al. (1977); Montéro and Imbach (1974)
110	α,β	p-$NO_2C_6H_4CO$	
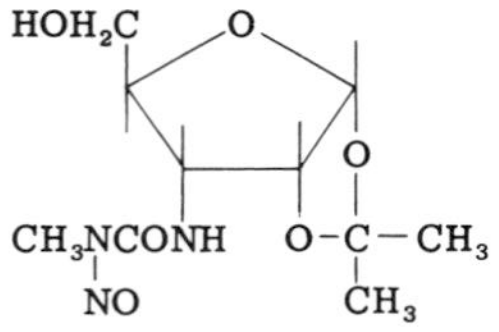 **111**			Fujiwara et al. (1974)

D. Cycloalkanes

No.			References
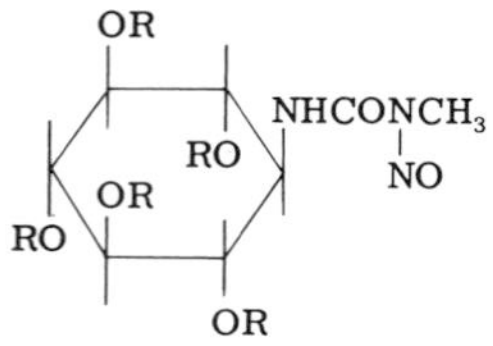 **112**		R = H	Suami and Machinami (1970a)
113		R = CH_3CO	Suami and Machinami (1970a)

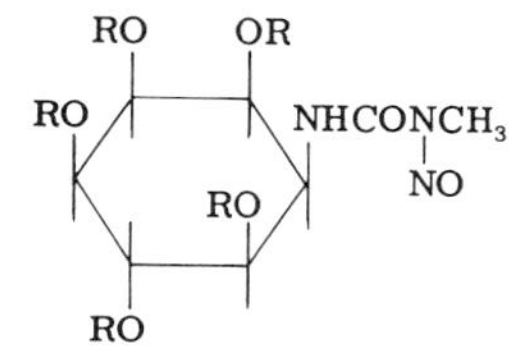

114	R = H		Suami and Machinami (1970a)
115	R = CH$_3$CO		Suami and Machinami (1970a)
116			Suami and Machinami (1970a)
117	R = H		Machinami et al. (1975b); Suami (1976)
118	R = CH$_3$CO		Machinami et al. (1975b); Suami (1976)
119	R = H		Machinami et al. (1975b); Suami (1976)
120	R = CH$_3$CO		Machinami et al. (1975b); Suami (1976)
121			Panasci et al. (1977)

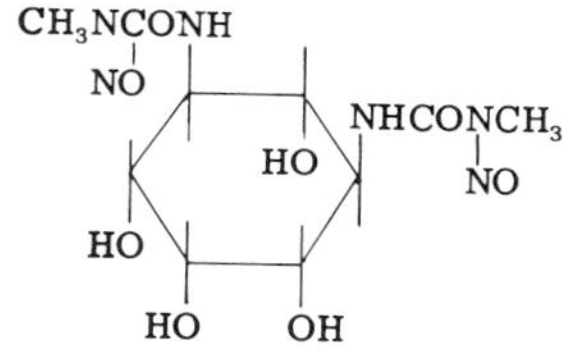

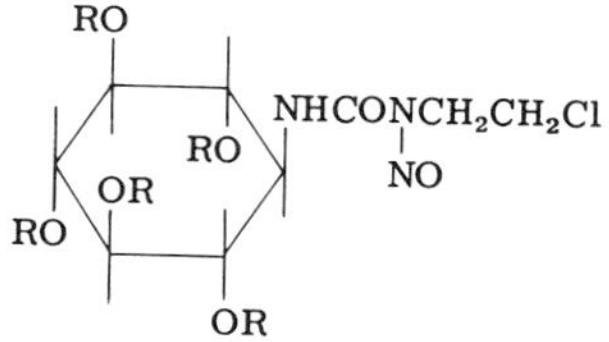

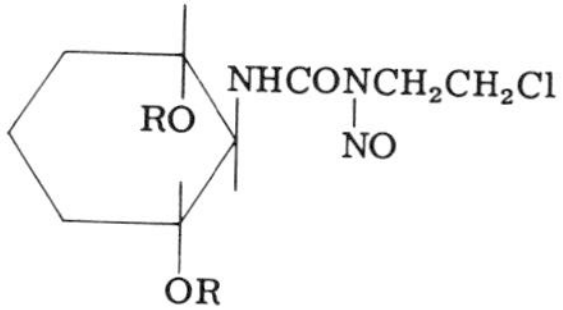

NHCONCH$_3$
NO
OR
RO
RO
RO
RO

NHCONCH$_3$
NO
HO
OH
CH$_3$NCONH
NO
HO
HO

NHCONCH$_2$CH$_2$Cl
NO
RO
RO
OR
OR
RO

NHCONCH$_2$CH$_2$Cl
NO
RO
OR

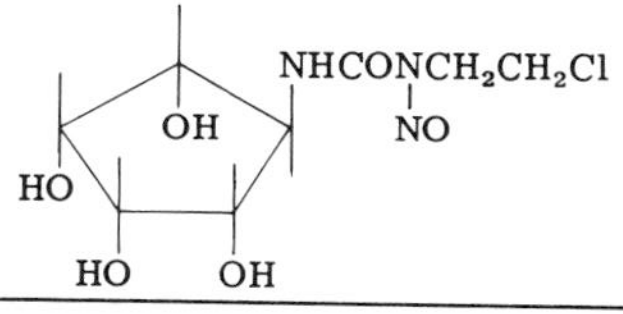

NHCONCH$_2$CH$_2$Cl
NO
OH
OH
HO
HO

(Continued)

189

TABLE II (*Continued*)

E. Reduced Sugar Analogs

No.	Compound	References
122	1-Deoxy-1-(3-methyl-3-nitrosoureido)-D-arabitol	Gassman *et al.* (1975)
123	2,3,4,5-Tetra-*O*-acetyl-1-deoxy-1-(3-methyl-3-nitrosoureido)-D-arabitol	Ichikawa *et al.* (1976)
124	1-Deoxy-1-(3-methyl-3-nitrosoureido)-D-sorbitol	Bannister (1972); Gassman *et al.* (1975); Ichikawa *et al.* (1976)
125	2,3,4,5,6-Penta-*O*-acetyl-1-deoxy-1-(3-methyl-3-nitrosoureido)-D-sorbitol	Bannister (1972); Gassman *et al.* (1975); Ichikawa *et al.* (1976)
126	2-Deoxy-2-3-methyl-3-nitrosoureido)-D-sorbitol	Bannister (1972); Gassmann *et al.* (1975)
127	1,3,4,5,6-Penta-*O*-acetyl-2-deoxy-2-(3-methyl-3-nitrosoureido)-D-sorbitol	Bannister (1972); Gassman *et al.* (1975)
128	1-Deoxy-1-(3-methyl-3-nitrosoureido)-D-mannitol	Gassmann *et al.* (1975)
129	2,3,4,5,6-Penta-*O*-acetyl)-1-deoxy-1-(3-methyl-3-nitrosoureido)-D-mannitol	Gassmann *et al.* (1975)
130	1-Deoxy-1-[3-(2-chloroethyl)-3-nitrosoureido]-D-galactitol	Gassman *et al.* (1975); Ichikawa *et al.* (1976)
131	2,3,4,5,6-Penta-*O*-acetyl-1-deoxy-1-[3-(2-chloroethyl)-3-nitrosoureido]-D-galactitol	Gassmann *et al.* (1975); Ichikawa *et al.* (1976)

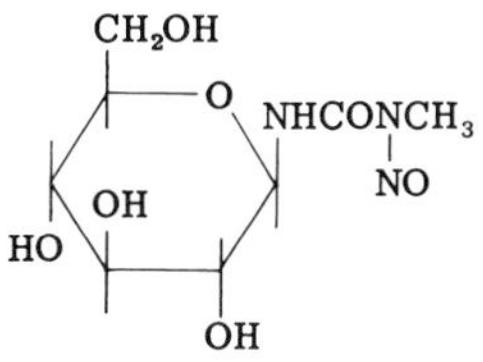

20

pared derivatives of ribose in both its pyranose and furanose forms in which the 3-methyl-3-nitrosoureido substituent is at C-1. Configuration at C-1 was determined by PMR spectra, and it was found that most of the compounds were α/β mixtures with the β form predominating. In some cases, pure β isomers were obtained. Suami and Machinami and their collaborators (Suami *et al.*, 1974; Machinami *et al.*, 1975a,b) have prepared similar compounds and indicated that they all had the β configuration at C-1. However, no mention was made of how this was determined.

A few analogues of streptozocin have been prepared having the nitroso group replaced by other groups (Wiley *et al.*, 1976). The starting material for these analogues was 2-deoxy-2-isocyanato-β-D-glucopyranose (**21**), which was treated with an amine or an amide, as in Eq. (9), resulting in the formation of substituted ureas.

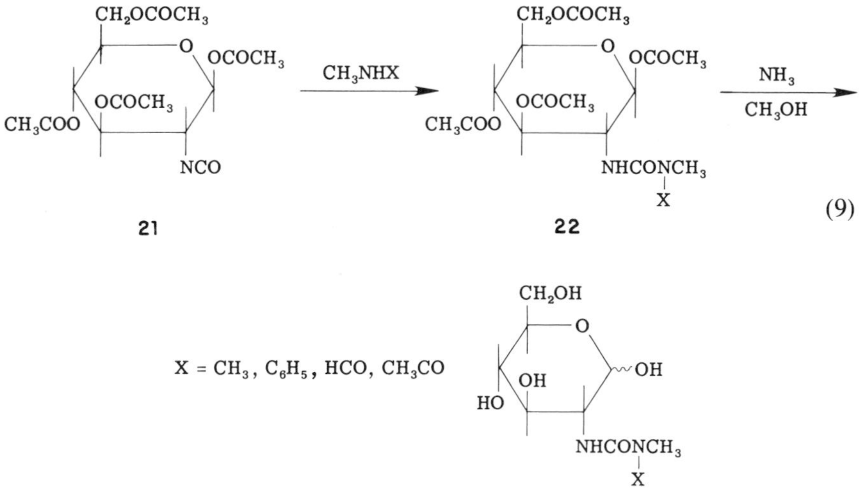

$$\tag{9}$$

21   **22**

$X = CH_3, C_6H_5, HCO, CH_3CO$

23

The chemistry of streptozocin analogues has been studied very little, except for various acylations and deacylations. Montgomery *et al.* (1975) have studied the decomposition of chlorozotocin (**24**) in aqueous solutions. It was suggested that the $ClCH_2CH_2N(NO)$ portion of the molecule undergoes decomposition as indicated in Eq. (10).

$$\underset{\text{24}}{\underset{\substack{\text{NO}}}{\underset{\text{NHCONCH}_2\text{CH}_2\text{Cl}}{\text{(sugar ring with CH}_2\text{OH, OH, HO, OH)}}}}$$

(10)

$$\text{CH}_2\text{=CHN=NOH} + \text{H}^+ + \text{Cl}^- \qquad\qquad \underset{\text{25}}{\text{ClCH}_2\text{CH}_2\text{N=NOH}}$$

$$\text{CH}_2\text{=CH}^+ + \text{N}_2 + \text{OH}^- \qquad \underset{\text{26}}{\overset{+}{\text{ClCHCH}_3} + \text{ClCH}_2\text{CH}_2^+ + \text{N}_2 + \text{OH}^-}$$

$$\text{CH}_3\text{CHO} \longleftarrow \text{CH}_3\text{CHClOH} \quad \text{ClCH}_2\text{CH}_2\text{OH}$$

B. Structure-Activity Relationships

Comparison of antitumor activities within the group of analogues of streptozocin will be almost entirely on the basis of effects on murine leukemias and largely L-1210, although some studies with P-388 have been reported. Other systems used to a minor extent have been the ascites forms of Sarcoma 180 and Ehrlich carcinoma. Very little has been published regarding activities of these compounds against bacteria. Iwasaki *et al.* (1976) have reported that a series of alkyl glycoside analogues having different small alkyl groups on nitrogen are inactive against several microorganisms. Bannister (1972) has found that in a series of seven compounds only one, the mannosamine analogue of streptozocin (**84**), had any activity against *P. vulgaris*. In this one case, activity was far less than that of streptozocin.

Modification of streptozocin has been carried out in only six different ways with, in many instances, combinations of two or more. The modifications are (1) different sugars as starting materials or the same sugar in its furanose or pyranose form and also having the possibility of α and β forms; (2) location of ureido groups on the ring which, if at C-1, offers the possibility of α and β forms; (3) acylation of hydroxyls in which α and β forms are possible; (4) glycoside formation with simple alkyl groups, which also involves the possibility of α and

β forms; (5) variation of the alkyl group on nitrogen; and (6) replacement of NO with other groups. The various compounds listed in Table II have been numbered to facilitate comparison in the discussion.

One of the means most often used for making analogues of streptozocin has been the use of various aminosugars, reduced aminosugars (aminoarabitol, aminosorbitol, etc.), and even aminopolyhydroxycycloalkanes as starting materials for the introduction of various RN(NO)—CO groups. It appears that alterations in this part of the molecule have rather small effects on activity. For example, Bhuyan *et al.* (1972) compared the activities of glucose and galactose substituted at C-1 with 3-methyl-3-nitrosureido groups (**20** and **85**, respectively) against L-1210 leukemia in mice, and found they were identical. The same authors report that the sorbitol analogue (**124**) has only slightly less activity, with all three being somewhat less active than streptozocin. Machinami *et al.* (1975b) have reported a derivative of mannose (**81**) of the same type, but a quantitative value for its activity was not given, although it was stated to be a remarkably active antitumor agent. In the same publication, compounds derived from the pyranose forms of ribose and xylose with 3-methyl-3-nitrosoureido substituents at C-1 (**94** and **99**, respectively) were also reported to be remarkably active antitumor agents.

A comparison of the activities of methyl- α- D- 3- deoxy- 3- (3- methyl- 3- nitrosoureido) altroside (**93**) and its glucose analogue (**75**) was made by Fujiwara *et al.* (1974). The activities were tested against L-1210 leukemia in mice using two regimens, and it was found that there was very little difference between the two compounds. A further comparison of the ribose and xylose derivatives can be made from the results of Imbach *et al.* (1976). In this case the compounds were (**96**) and (**102**), in which the five-carbon sugars were acetylated and had the ureido group at C-1, with $ClCH_2CH_2$ being the alkyl group on N. Again it was found that there was little difference in activity against murine L-1210 leukemia, both giving greater than 50% cures but with a small difference in the doses. A similar comparison of the same derivative of a sugar in its furanose form and in its pyranose form is not possible with the present published data. However, the compounds (**95**) (pyranose form) and (**105**) (furanose form but as the acetonide) would suggest very little effect of ring form, as both gave ILS values of over 100% against murine L-1210 leukemia (Montéro *et al.*, 1977).

A number of ureido polyhydroxycyclohexanes have been reported by Suami and Machinami (Suami and Machinami, 1970a; Suami, 1976; Machinami *et al.*, 1975b). Very little quantitative data concerning antitumor activities have been forthcoming. However, it has been claimed that 3- β- D- mannopyranosyl- 1- (2-chloroethyl)- 1- nitrosourea (**82**) and the corresponding derivative of *scyllo*-inosamine (**117**) are both highly active against L-1210 leukemia in mice. The latter compound at a dose of 20 mg/kg cured two out of two mice. The analogous diol (**119**) cured three out of three mice at a dose of 4 mg/kg. A tetrahydroxycyclopentane having as a substituent the 3-(2-chloroethyl)-3-nitrosoureido group (**121**) has been reported to be extremely active against L-1210 leukemia in mice showing an ILS of 429% (Panasci *et al.*, 1977).

Although quite a large number of streptozocin analogues having 3-alkyl-3-nitrosoureido substituents at any one of the six carbon atoms of hexoses or at any of the five possible positions in pentoses are possible, only a small number have been prepared having such a substituent at positions other than C-1 or C-2. Furthermore, direct comparisons of compounds substituted at different positions are few, and the data are somewhat equivocal. In a comparison of streptozocin with the analogue (**29**) having the ureido substituent at C-1, Bhuyan *et al.* (1972) have found that the two compounds have quite comparable activities against murine L-1210 leukemia. In tests with the methyl glycoside of the C-3 isomer (**75**), comparable to those of Bhuyan *et al.*, it was found that its activity was rather similar to those reported for streptozocin and 3-β-D-glucopyranosyl-1-methyl-1-nitrosourea (**20**) (Fujiwara *et al.*, 1974). These latter authors also found that methyl α-D-6-deoxy-6-(3-methyl-3-nitrosoureido) glucopyranoside (**78**) had essentially the same activity as its 3-isomer. However, in the cases of 1-deoxy-1-(3-methyl-3-nitrosoureido)-D-sorbitol (**124**) and its 2-isomer (**126**), the 1-isomer was found to give an ILS of 33% when given to mice infected with L-1210 leukemia at a dose of 100 mg/kg/day for 7-days, but the 2-isomer had no activity at the same dose and under the same regimen. Chlorozotocin (**24**) and its isomer (**69**) having the ureido group at C-1 have been directly compared by Aoshima and Sakurai (1977) and by Panasci and colleagues (1977), in both cases against murine L-1210 leukemia using 10–20 mg/kg doses given i.p. Panasci *et al.* (1977) report an ILS of 332% for chlorozotocin and 90% for its isomer. Aoshima and Sakurai found that both gave an ILS of greater than 200%, although these authors did report that the 1-isomer was highly active orally (ILS 187% at 20 mg/kg), while chlorozotocin was essentially inactive orally. 2,4-Dideoxy-2,6-bis(3-methyl-3-nitrosoureido)-D-glucose (**79**) has been reported to be active against Ehrlich ascites tumors in mice (Machinami and Suami, 1973).

As can be seen by an inspection of Table II, most of the streptozocin analogues which have been prepared were either prepared first as acetates or acetylated. Only a few other esters have been prepared. Quite early in the work on streptozocin, it was found that the β form of the tetraacetate (**30**) was even more active against L-1210 *in vitro* than was streptozocin. Bhuyan *et al.* (1972) have compared the two compounds against murine L-1210 leukemia giving approximately the same dosage over a 7-day period. Streptozocin gave an ILS of 85%, while the acetate gave 69%. In the case of the glucose compound with the ureido substituent at C-1 (**20**) and its acetate (**70**) in the same type of test, the ILS of the mice given the acetate was 70% as compared with a result of 56% from (**20**). The C-1 galactose analogue (**85**) and its acetate (**86**) gave similar results. However, in all of these cases the acetates were substantially more active *in vitro* than was the parent compound. Montéro *et al.* (1977) have reported that both 3-D-ribopyranosyl-1-(2-chloroethyl)-1-nitrosourea as an α/β mixture (**95**) and its acetate (**96**) have ILS values of greater than 100% when tested against murine L-1210 leukemia. These authors also report that the benzoate ester (**97**) has similar activity but that of the *p*-nitrobenzoate (**98**) was less. A series of esters of 3-β-D-ribofuranosyl-1-(2-chloroethyl)-1-nitrosourea as its acetonide (**105**) were tested in the same system and were all as active as the parent compound, having ILS values of greater than 100% (Montéro *et al.*, 1977). The esters were acetyl (**106**), benzoyl (**107**), *o*-nitrobenzoyl (**108**), *m*-nitrobenzoyl (**109**), and *p*-nitrobenzoyl (**110**). Johnston and associates (1975) have found that chlorozotocin (**24**) is substantially more active *in vivo* against murine L-1210 leukemia than is its acetate (**42**). Using a dosage regimen of Days 1–9 and giving 2.5 mg/kg/day, chlorozotocin gave an ILS of 63%, and with a single dose of 30 mg/kg cured 90% of the mice. The ester, on the other hand, prolonged life by only 36% at 7.5 mg/kg/day, and in a single dose of 20 mg/kg cured 40% of the mice. It appears that, while acetates and other esters are quite active, they are usually less active than is the parent compound.

Both Bannister (1972) and Iwasaki *et al.* (1976) report that conversion of streptozocin to its methyl glycoside, either α or β, destroys its antibacterial activity and also its diabetogenic effect (Machinami *et al.*, 1975a). The *in vitro* effect against L-1210 cells remains much the same, although the α isomer is slightly less active. Rather extensive testing data for these two glycosides (α, **43** and β, **44**) have been reported by Iwasaki *et al.* (1976). These authors have compared activities against ascites Sarcoma 180, Ehrlich ascites carcinoma, and L-1210 leukemia, all in mice. In two of the three sets of tests, streptozocin was included. The overall results were that the methyl glycosides were approximately as active as streptozocin against ascites Sarcoma 180 but were less active against the other tumors. Both the α- and β-glycosides of streptozocin, in which the alkyl groups were ethyl (**50, 52**), *n*-propyl (**54, 55**), and *n*-butyl (**56, 58**), were tested in the same series of tests. These were all active against the ascites tumors, although somewhat less active than streptozocin, but in the L-1210 tests they had the same order of activity as did streptozocin. Comparisons were also made with glycosides of streptozocin analogues in which the methyl group at N-3′ in streptozocin was replaced with ethyl and butyl. For example, methyl-β-D-2-deoxy-2-(3-ethyl-3-nitrosoureido) glucopyranoside (**47**) was tested, as were its ethyl α- and β-glycoside analogues (**50, 52**). Similarly, ureido analogues having butyl on N-3′ (**46, 49, 57**) were tested. None of these compounds had significant activity, which is as expected, as the analogues with free hydroxyl at C-1 were also inactive. Only a few other glycosides have been prepared, and test results on even fewer have been reported. Fujiwara *et al.* (1974) have tested five methyl glycosides, some of which were α and some were β. Methyl-α-D-3-deoxy-3-(3-methyl-3-nitrosoureido) glucopyranoside (**75**) increased the lifespan of mice with

L-1210 leukemia by 66% at fairly high doses. Its altroside isomer (**93**) and the 6-ureido glucose analogue (**78**) had about the same activity. Methyl α- D- 3- deoxy- 3- (3- methyl- 3- nitrosoureido) xylopyranoside (**103**) was astonishingly active, showing an ILS of 124% at a dose of 100 mg/kg/day over 9 days against murine L-1210 leukemia. Thus, it is evident that alkyl glycoside analogues of streptozocin are highly active as antitumor agents, although in general somewhat less active than the free aldopyranose forms. However, formation of glycosides affects other properties, such as diabetogenicity, much more significantly.

In the general formula (**27**) for streptozocin and its analogues in which R¹ is a

R¹NHCONR²
|
NO
27

sugar or sugarlike moiety, R² has usually been methyl; but other alkyl groups, although a surprisingly small number, have been used to replace methyl. The other groups have been ethyl, *n*-propyl, *n*-butyl, benzyl, and 2-chloroethyl. Excluding 2-chloroethyl analogues, most of the other analogues of this type have been either acetates or alkyl glycosides. Meier *et al.* (1974) have prepared a series of analogues in which the *N*-methyl group of streptozocin (**27**, $R^2 = CH_3$) was replaced by C_2H_5, *n*-C_3H_7, *n*-C_4H_9, and $C_6H_5CH_2$ (**31, 32, 33**, and **34**). However, no activities were reported. The acetates of the ethyl, *n*-propyl, and *n*-butyl analogues were reported by the same authors as well as by Machinami *et al.* (1975a), but again essentially no activities were reported. Bannister (1972) and Bhuyan *et al.* (1972) have reported that the N-3' ethyl and butyl analogues of streptozocin and their acetates are essentially inactive *in vitro* against L-1210 leukemia. The latter authors tested the N-3' ethyl analogue and its acetate (**21** and **35**) and the N-3' butyl analogue acetate (**41**) against murine L-1210 leukemia and found no activity. Iwasaki *et al.* (1976) have prepared a number of glycosides of streptozocin in which the N-3' R groups were ethyl, *n*-propyl, and *n*-butyl alkyl groups. These compounds are those numbered **65, 46, 47, 48, 49, 51, 53,** and **57.** Tests against three systems, ascites Sarcoma 180, Ehrlich ascites carcinoma, and L-1210 leukemia, all in mice, indicated that none of them were active. Thus it appears that replacement of N-3' methyl by other hydrocarbon moieties destroys activity.

However, replacement of N-3' methyl in streptozocin by 2-chloroethyl results in a compound, chlorozotocin (**24**), which has the greatest probability of any of the streptozocin analogues of becoming a clinically useful drug. Johnston *et al.* (1975) first reported the remarkable activity of this compound. It was found that dosages of 7.5 mg/kg/day over a nine-day period starting on Day 1 after infection of mice with L-1210 leukemia prolonged life by 63%. A somewhat heavier dose (30 mg/kg) given only once on Day 1 resulted in 90% cures. Similar results have been reported by others. Anderson *et al.* (1975), Aoshima and Sakurai (1977), Schein *et al.* (1976), and Panasci *et al.* (1977) all report increased lifespans of over 300% with high percentages of cures in the same infection using similar dosages but varying regimens. The acetate was also active, but less so (Johnston *et al.*, 1975). Chlorozotocin causes very slight bone marrow depression and has little effect on peripheral neutrophil count (Anderson *et al.*, 1975). Schein *et al.* (1976) have compared chlorozotocin with simpler nitrosourea analogues and found considerably less toxicity. It has been shown that the 2-chloroethyl group of chlorozotocin and its simpler analogues such as N,N^1-*bis*(2-chloroethyl)-*N*-nitrosourea (BCNU) decompose in aqueous solutions to give (**25**) and (**26**) [Eq. (10)], which presumably alkylate DNA (Montgomery *et al.*, 1975). Carbamoylation can occur by the action of the isocyanate obtained by elimination of the elements of nitrosamines. Panasci *et al.* (1977) attempted to correlate the carbamoylating and alkylating properties of chlorozotocin and its simpler nitrosourea analogues with activity against L-1210 leukemia, toxicity, and myelosuppression. Carbamoylating activity did not correlate with any of the three. Alkylating activity correlated with toxicity but not with L-1210 activity. Chlorozotocin is now being tested in phase I and phase II clinical trials and is showing considerable activity.

A number of compounds have been prepared in which sugars and reduced sugars have been substituted at C-1 with the 3-(2-chloro-ethyl)-3-nitrosoureido group. The activity of 3-D-glucopyranosyl-1-(2-chloroethyl)-1-nitrosourea (**69**) has been reported to be quite good (Aoshima and Sakurai, 1977; Panasci *et al.*, 1977) with substantial oral activity (ILS 187%). Montéro *et al.* (1977) and Imbach *et al.* (1976) have also found high activity with similar pentose analogues.

A series of streptozocin analogues has been reported by Wiley *et al.* (1976) in which the nitroso group of streptozocin has been replaced by other groups as indicated in Eq. (9). Most of these were prepared only as their acetates (**22**), but in the case of the phenyl analog (**41**) the acetyl groups were removed. None of these compounds showed activity against P-388 leukemia in mice, and none were highly active against L-1210 *in vitro*, although their order of activity was approximately that of streptozocin tetraacetate.

REFERENCES

Anderson, T., McMenamin, M. G., and Schein, P. S. (1975). *Cancer Res.* **35**, 761–765.
Aoshima, M., and Sakurai, Y. (1977). *Gann* **68**, 247–250.
Argoudelis, A. D., Mizsak, S. A., and Meulman, P. A. (1974). *J. Antibiot.* **27**, 564–566.
Arison, R. N., and Feudale, E. L. (1967). *Nature (London)* **214**, 1254–1255.
Arison, R. N., Ciaccio, E. I., Glitzer, M. S., Cassaro, J. A., and Pruss, M. B. (1967). *Diabetes* **16**, 51–56.
Bannister, B. (1972). *J. Antibiot.* **25**, 377–386.

Bhuyan, B. K. (1970). *Cancer Res.* **30**, 2017–2023.
Bhuyan, B. K., Scheidt, L. G., and Fraser, T. J. (1972). *Cancer Res.* **32**, 398–407.
Bhuyan, B. K., Kuentzel, S. L., Gray, L. G., Fraser, T. J., Wallach, D., and Neil, G. L. (1974). *Cancer Chemother. Rep., Part 1* **58**, 157–165.
Bhuyan, B. K., Peterson, A. R., and Heidelberger, C. (1976). *Chem. Biol. Interact.* **13**, 173–179.
Broder, L. E., and Carter, S. K. (1973). *Ann. Intern. Med.* **79**, 108–118.
Burns, H. D., and Heindel, N. D. (1974). *Org. Prep. Proced. Int.* **6**, 259–263.
Carter, S. K., and Broder, L. E. (1974). *Clin. Gastroenterol.* **3**, 733–745.
Davis, Z., Moertel, C. G., and McIlrath, D. C. (1973). *Surg., Gynecol. Obstet.* **137**, 637–644.
Dulin, W. E., and Wyse, B. M. (1969). *Diabetes* **18**, 459–466.
Evans, J. S., Gerritsen, G. C., Mann, K. M., and Owen, S. P. (1965). *Cancer Chemother. Rep.* **48**, 1–6.
Feldman, J. M., Quickel, K. E. G., Jr., Maracek, R. L., and Lebowitz, H. E. (1972). *South. Med. J.* **65**, 1325–1327.
Ficsor, G., Zuberi, R. I., Suami, T., and Machinami, T. (1974). *Chem. Biol. Interact.* **8**, 395–402.
Fujiwara, A. N., Acton, E. M., and Henry, D. W. (1974). *J. Med. Chem.* **17**, 392–396.
Gabridge, M. G., Denunzio, A., and Legator, M. S. (1969). *Nature (London)* **221**, 68–70.
Gassmann, N., Stoos, F., Meier, A., Büyük, G., Helali, S. E., and Hardegger, E. (1975). *Helv. Chim. Acta* **58**, 182–185.
Gichner, T., Veleminsky, J., and Krepinsky, J. (1968). *Mol. Gen. Genet.* **102**, 184–186.
Gumaa, K. A., and McLean, P. (1969). *Biochem. Biophys. Res. Commun.* **35**, 86–93.
Hardegger, E., and Meier, A. (1970). Ger. Patent 2,008,578 [*C.A.* **73**, 131267x (1970)].
Hardegger, E., Meier, A., and Stoos, A. (1969). *Helv. Chim. Acta* **52**, 2555–2560.
Heinemann, B., and Howard, A. J. (1965). *Antimicrob. Agents Chemother.* 488–492.
Herr, R. R., Eble, T. E., Bergy, M. E., and Jahnke, H. K. (1960). *Antibiot. Annu. 1959–1960* 236–240.
Herr, R. R., Jahnke, H. K., and Argoudelis, A. D. (1967). *J. Am. Chem. Soc.* **89**, 4808–4809.
Hessler, E. J., and Jahnke, H. K. (1970). *J. Org. Chem.* **35**, 245–6.
Ichikawa, K., Murakami, M., Sato, N., and Kawamura, T. (1976). Jpn. Patent 76 39,629 [*C.A.* **85**, 124298c (1976)].
Imbach, J. L., Hayat, M., Chenu, E., Serrow, B., and Mathe, G. (1976). In "Chemotherapy" (K. Hellmann and T. A. Connors, eds.), Vol. 8, pp. 221–227. Plenum, New York.
Iwasaki, M., Ueno, M., Hinomuya, K., Sekine, J., Nogamatsu, Y., and Kimura, G. (1976). *J. Med. Chem.* **19**, 918–923.
Iweze, F. J., Owen-Smith, M., and Pulak, J. (1972). *Proc. R. Soc. Med.* **65**, 164–165.
Johnston, T. P., McCaleb, G. S., and Montgomery, J. A. (1975). *J. Med. Chem.* **18**, 104–106.
Karunanayake, E. H., Hearse, D. J., and Mellows, G. (1974). *Biochem. J.* **142**, 673–683.
Karunanayake, E. H., Baker, J. R. J., Christian, R., Hearse, D. J., and Mellows, G. (1975). *Biochem. Soc. Trans.* **3**, 414–417.
Kelly, F., and Legator, M. (1971). *Mutat. Res.* **12**, 183–190.
Kolbye, S. M., and Legator, M. S. (1968). *Mutat. Res.* **6**, 387–389.
Kushner, B., Lazar, M., Furman, M., Lieberman, T. W., and Leopold, I. H. (1969). *Diabetes* **18**, 542–544.
Lazarus, S. S., and Shapiro, S. H. (1972). *Diabetes* **21**, 129–137.
Lewis, C., and Barbiers, A. R. (1960). *Antibiot. Annu. 1959–1960* 247–254.
Losert, W., Rilke, A., Loge, O., and Richter, K. D. (1971). *Arzneim.-Forsch.* **21**, 1643–1653.
Machinami, T., and Suami, T. (1973). *Bull. Soc. Chem. Jpn.* **46**, 1013–1014.
Machinami, T., Kobayashi, K., Hayakawa, Y., and Suami, T. (1975a). *Bull. Chem. Soc. Jpn.* **48**, 3761–3762.

Machinami, T., Nishiyama, S., Kikuchi, K., and Suami, T. (1975b). *Bull. Chem. Soc. Jpn.* **48**, 3763-3764.

Meier, A., Stoos, F., Martin, D., Büyük, G., and Hardegger, E. (1974). *Helv. Chim. Acta* **57**, 2622-2626.

Montéro, J. L., and Imbach, J. L. (1974). *C. R. Acad. Sci., Ser. C* **279**, 809-811.

Montéro, J. L., Maruzzi, A., Oiry, J., and Imbach, J. L. (1977). *Eur. J. Med. Chem.* **12**, 397-407.

Montgomery, J. A., James, R., McCaleb, G. S., Kirk, M. C., and Johnston, T. P. (1975). *J. Med. Chem.* **18**, 568-571.

Murray-Lyon, I. M., Eddleston, A. L. W. F., Williams, R., Brown, M., Hogbin, B. M., Edwards, J. C., Bennett, A., and Taylor, K. W. (1968). *Lancet ii*, 895-898.

Olitzki, A. L., Godinger, D., Israeli, M., and Honigman, A. (1967). *Appl. Microbiol.* **15**, 994-1001.

Panasci, L. C., Green, D., Nagourney, R., Fox, P., and Schein, P. S. (1977). *Cancer Res.* **37**, Part 1, 2615-2618.

Rakieten, N., Rakieten, M. L., and Nadkarni, M. V. (1963). *Cancer Chemother. Rep.* **29**, 91-98.

Rakieten, N., Gordon, B. S., Cooney, D. A., Davis, R. D., and Schein, P. S. (1968). *Cancer Chemother. Rep.* **52**, 563-567.

Rakieten, N., Cooney, D. A., and Davis, R. D. (1969a). *U.S. Gov. Res. Dev. Rep.* **69**, 48.

Rakieten, N., Cooney, D. A., and Davis, R. D. (1969b). *U.S. Gov. Res. Dev. Rep.* **69**, 56.

Rakieten, N., Gordon, B. S., Beaty, A., Cooney, D. A., Davis, R. D., and Schein, P. S. (1971). *Proc. Soc. Exp. Biol. Med.* **137**, 280-283.

Rosenkrantz, H. S., and Carr, H. S. (1970). *Cancer Res.* **30**, 112-117.

Ryo, U. Y., Bierwaltes, W. H., Feehan, P., and Ice, R. D. (1974). *J. Nucl. Med.* **15**, 572-576.

Sadoff, L. (1970). *Cancer Chemother. Rep., Part 1* **54**, 457-459.

Sadoff, L. (1972). *Cancer Chemother. Rep., Part 1* **56**, 61-69.

Schein, P. S. (1972). *Cancer (Philadelphia)* **30**, 1616-1626.

Schein, P. S., and Loftus, S. (1968). *Cancer Res.* **28**, 1501-1506.

Schein, P. S., McMenamin, M. G., and Anderson, T. (1973). *Cancer Res.* **33**, 2005-2009.

Schein, P. S., O'Connell, M. J., Blom, I., Hubbard, S., Magrath, I. T., Bergevin, P., Wiernik, P. H., Ziegler, J. L., and Devita, V. J. (1974). *Cancer (Philadelphia)* **34**, 993-1000.

Schein, P. S., Panasci, L., Woolley, P. V., and Anderson, T. (1976). *Cancer Treat. Rep.* **60**, 801-805.

Sibay, T. M., Hausler, H. R., and Hayes, J. A. (1971). *Ann. Ophthalmol.* **3**, 596-901.

Stolinsky, D. C., Sadoff, L., Braunwald, J., and Bateman, J. R. (1972). *Cancer (Philadelphia)* **30**, 61-67.

Suami, T. (1974). Br. Patent 1,353,513 [*C.A.* **81**, 63923w (1974)].

Suami, T. (1976). Jpn. Patent 76 52160 [*C.A.* **86**, 30032e (1977)].

Suami, T., and Machinami, T. (1970a). *Bull. Chem. Soc. Jpn.* **43**, 2953-2956.

Suami, T., and Machinami, T. (1970b). *Bull. Chem. Soc. Jpn.* **43**, 3013-3015.

Suami, T., and Machinami, T. (1971). Ger. Patent 2 119964 [*C.A.* **76**, 59979y (1972)].

Suami, T., Machinami, T., and Hisamatsu, M. (1974). Jpn. Patent 74 36679 [*C.A.* **81**, 120944e (1974)].

Suami, T., Machinami, T., and Hisamatsu, T. (1975). Jpn. Patent 75 101326 [*C.A.* **84**, 31365m (1976)].

Vavra, J. J., DeBoer, C., Dietz, A., Hanka, L. J., and Sokolski, W. T. (1960). *Antibiot. Annu. 1959-1960* 230-235.

Verheyden, C. (1967). *Bull. Soc. Belge Ophthalmol.* No. 147, 479-485.

White, F. R. (1963). *Cancer Chemother. Rep.* **30**, 49-53.

White, J. H., and Cinotti, A. A. (1972). *Invest. Ophthalmol.* **11**, 56-57.

Wiley, P. F. (1968). Unpublished observations.

Wiley, P. F. (1978). Unpublished observations.

Wiley, P. F., Herr, R. R., MacKellar, F. A., and Argoudelis, A. D. (1965). *J. Org. Chem.* **30,** 2330-2334.

Wiley, P. F., McMichael, D. L., Koert, J. M., and Wiley, V. H. (1976). *J. Antibiot.* **29,** 1218-1225.

Wiley, P. F., Herr, R. R., Jahnke, H. K., Chidester, C. G., Mizsak, S. A., Spaulding, L. B., and Argoudelis, A. D. (1979). *J. Org. Chem.* **44,** 9-16.

CHAPTER 7

Terpenoid Antitumor Agents

JOHN M. CASSADY AND MATTHEW SUFFNESS

I. INTRODUCTION

The terpenoids which are discussed in this chapter represent a wide variety of structural types which have differing potentials for development as antitumor agents. Before the specific classes of terpenoids which have shown activity are taken up, it is necessary to give the reader some background on the tumor systems which have been used for screening, especially with regard to the significance of positive results in these systems as related to the potential for further development of the active compounds.

All of the screening data on the compounds discussed are from National Cancer Institute (NCI) files, and no attempt has been made to include data from other sources, since the NCI program is the largest of its kind and the large number of compounds screened at NCI is adequate to give a comprehensive picture of each of the classes discussed. Also, since NCI has well-established

201

Anticancer Agents Based on Natural Product Models
Copyright © 1980 by Academic Press, Inc.

uniform screening protocols for each tumor system, NCI data are internally consistent and it is possible to compare the activity of the compounds discussed. It could be extremely misleading to compare NCI screening data with data from the literature, since different sublines of tumors and different testing protocols used in other laboratories or programs could markedly influence the screening data. Due to the large number of individual compounds discussed in this chapter, it is not possible to individually cite the literature references to the isolation of the compounds or to individually name the suppliers of each compound, but we do wish to acknowledge with thanks all of the scientists who were kind enough to donate their compounds, often obtained with great difficulty, to the NCI program.

Table I lists those tumor systems which have been used to screen the compounds discussed. Many of these tumor systems are no longer part of the current NCI screening program and have been dropped either because they are excessively sensitive and give a large number of false positive leads or because there is a poor correlation between activity in these tumor systems and known clinically effective drugs. Thus activity in such systems as CA, DL, EA, SA, or WA must be viewed with some skepticism unless activity is also demonstrated in tumors of current interest. Activity in the P-388 leukemia (PS) must also be viewed with the fact in mind that this system has been deliberately chosen as the initial *in vivo* tumor system in the NCI current screen because it is quite sensitive and will pick up a large number of initial actives for more rigorous screening in other tumor systems. The PS system is therefore a prescreen, and low level activity in PS is not of much significance unless activity is also demonstrated in other systems. The KB system measures toxicity to human cancer cells grown in cell culture and has no necessary relationship to *in vivo* activity. Many compounds which are simply toxic and have no *in vivo* antitumor activity are active in this system, and KB activity alone is therefore not meaningful as a criterion for antitumor activity. Within a series of compounds with *in vivo* activity there may be a fairly good correlation between the cytotoxicity of these compounds in the KB system and their *in vivo* activity, and the KB can therefore be a useful screen for analogues since the testing is rapid and requires only small amounts of material. Unfortunately, there is considerable misuse of the KB data in the literature and some investigators have used the terms "anticancer activity" or "antitumor activity" to describe compounds active in the KB system, for which the term "cytotoxicity" is correct and appropriate.

We have included in this chapter those groups of compounds in which sufficient examples of analogues or derivatives are available to discuss what the requirements for activity may be. Other terpenoids with antitumor or cytotoxic activity have been reported, but sufficient data are not available for analysis of the structural requirements for activity.

TABLE I

National Cancer Institute Tumor Systems

Tumor code	Name	Parameter measured	Activity criteria[a] T/C[b]	Use in current screen
B1	B16 melanosarcoma	Survival	>125	Yes
CA	Adenocarcinoma 755	Tumor inhibition	<42	No
CD	CD8F$_1$ mammary tumor	Tumor inhibition	<42	No
C6	Colon 26	Survival	>140	Yes
C8	Colon 38	Tumor inhibition	<42	Yes
DL	Dunning leukemia	Survival	>125	No
EA	Ehrlich ascites	Tumor inhibition	<42	No
KB	Carcinoma of nasopharynx (cell culture)	Cell growth inhibition	ED$_{50}$ <4 μg/ml	Yes
LE	L-1210 lymphoid leukemia	Survival	>125	Yes
LL	Lewis lung carcinoma	Survival (or inhibition)	>140 (or <42)	Yes
P4	P-1534 leukemia	Survival	>125	No
PS	P-388 lymphocytic leukemia	Survival	>120	Yes
8P	P-1534 leukemia	Survival	>125	No
SA	Sarcoma 180	Tumor inhibition	<42	No
WA	Walker carcinosarcoma 256	Tumor inhibition	<42	No

[a] Criteria for minimal activity; criteria for further development are more rigorous.

[b] T/C = ratio of test to control tumored animals expressed as a percentage.

II. MONO- AND SESQUITERPENES

A. Iridoids

Only a few members of this group have been evaluated as antitumor agents, and all of these contain an unsaturated lactone group. Lack of material has prevented extensive evaluation; however, allamandin and isoplumericin have shown moderate and preliminary P-388 activity as reported by Kupchan *et al.* (1974), who isolated allamandin and related compounds from *Allamanda cathartica*. Cytotoxicity and probably *in vivo* antitumor activity rely on the presence of the α,β-unsaturated γ-lactone group. In support of this view, compounds **1**, **3**, and **4** in Table II are cytotoxic (Table III), but the saturated lactone **2** is inactive. Compounds **1**, **3**, and **4** are currently undergoing further evaluation in the NCI tumor panel. Penstemide (**5**), an iridoid glucoside, was also reported to possess significant P-388 activity (Jolad *et al.*, 1976) at higher doses.

B. Sesquiterpene Lactones

This structurally diverse and interesting group of compounds, primarily isolated from plants in the family Compositae (Asteraceae), has produced a large number of the active agents isolated in the program of antineoplastic agents from plants. A majority of the hundreds of compounds evaluated are cytotoxic, and a small number have shown activity *in vivo* against P-388 leukemia and other tumor systems. Only a few of these compounds have been evaluated in other tumor systems of current interest to NCI, and additional testing of selected compounds in the animal tumor panel is underway.

The majority of this group can be classified structurally into four major groups based on their carbon skeleton. These classes include the germacranolides, eudesmanolides, guaianolides, and pseudoguaianolides (Fig. 1). All of these compounds and additional miscellaneous groups are derived biosynthetically from *trans,trans*-farnesyl pyrophosphate. The interesting physiological activity exhibited by these compounds has led to numerous reviews of their biosynthesis, distribution, chemotaxonomy, and chemistry (Geissman, 1973; Herout, 1971; Herz, 1973; Mabry and Bohlmann, 1978; Rodriguez and Mabry, 1978; Yoshioka *et al.*, 1973). In addition, several authors (Fujita and Nagao, 1977; Hartwell, 1976; Kupchan, 1974; Lee, 1978; Rodriguez *et al.*, 1976; Suffness and Douros, 1979) have discussed the cytotoxic and antitumor activity of this group.

1. Chemical Reactivity

The activity of these compounds derives from the presence of one or more groups which are subject to nucleophilic attack, presumably resulting in alkylation of essential biomolecules and ultimately cell death. The essential

TABLE II

Iridoid Lactones

Compound no.	NSC no.	Name	
1	251690	Allamandin	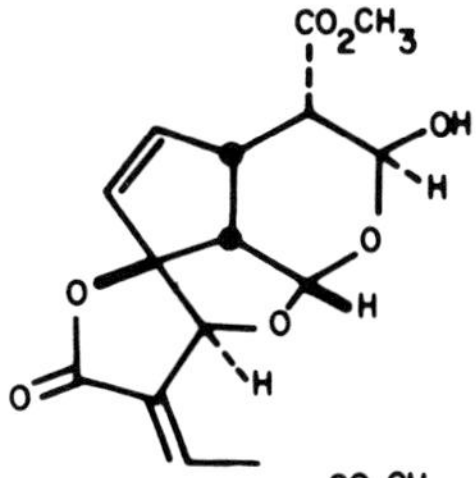
2	251691	Allamandicin	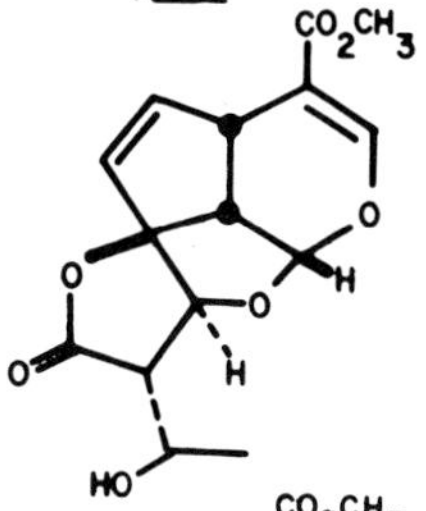
3	112152	Plumericin	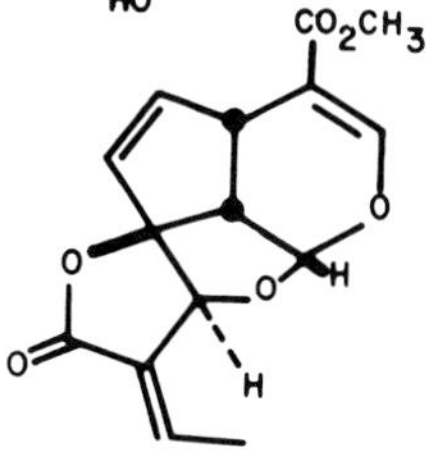
4	112153	Isoplumericin	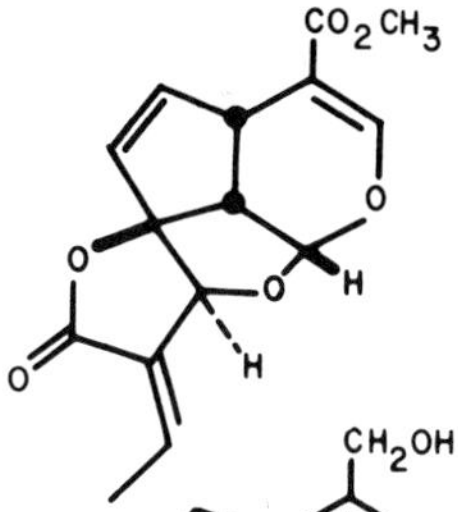
5		Penstemide	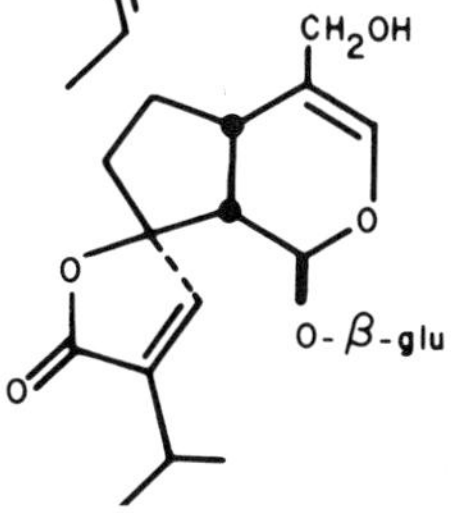

TABLE III

Iridoid Lactones, Screening Data

Compound no.	KB (ED_{50}, μg/ml)	*In vivo* system[a]
1	2.1×10^0	145 at 2 in PS
2	$>1 \times 10^1$	Inactive in PS
3	2.7×10^0	Inactive in LE
		126 at 6×10^{-4} in PS
4	2.6×10^0	145 at 1.25 in PS
		Inactive in LE
5		184 at 50 in PS

[a] Optimal T/C (%), dose (mg/kg).

reactive groups are usually enones, including α,β-unsaturated lactones, α,β-unsaturated ketones and α,β-unsaturated esters or epoxides. One of the most common groups found is the α-methylene-γ-lactone (**6**) which can undergo conjugate addition with a nucleophile such as a thiol resulting in alkylation of the thiol group. Kupchan (1971) argued that the repeated isolation of α-methylene-γ-lactones, and not the corresponding endocyclic analogues (butenolides), was significant. In general, the endocyclic compounds react more slowly and reversibly with cysteine, while the α-methylene-γ-lactones react rapidly to form stable adducts.

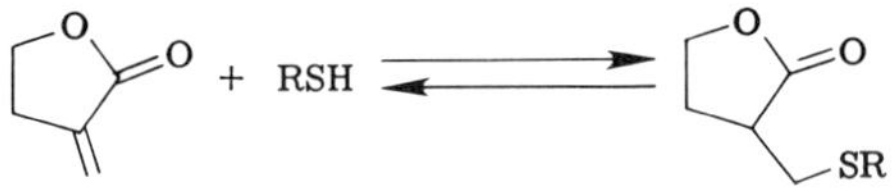

| 6 | 6a |

The rate constants for reaction of α-methylene-γ-lactones with cysteine at pH 7.4 range from about 100 LM^{-1} min^{-1} to 15,000 LM^{-1} min^{-1} (Kupchan *et al.*, 1971). Substitution of alkyl, alkylamino, or alkoxy groups directly at the β position (replacing one hydrogen on the α-methylene group) resulted in a decrease in reactivity with cysteine and loss of cytotoxic activity (Howie *et al.*, 1974, 1976; Stamos *et al.*, 1977; Cassady *et al.*, 1978). In general, the natural products with significant *in vivo* activity showed rates of reaction greater than 1,000 LM^{-1} min^{-1}, and part of this large difference in rates was attributed to neighboring group effects by Kupchan *et al.* (1971), since no lactone containing

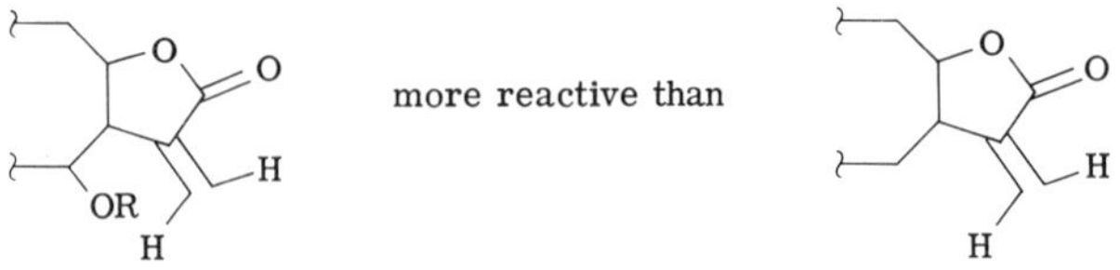

more reactive than

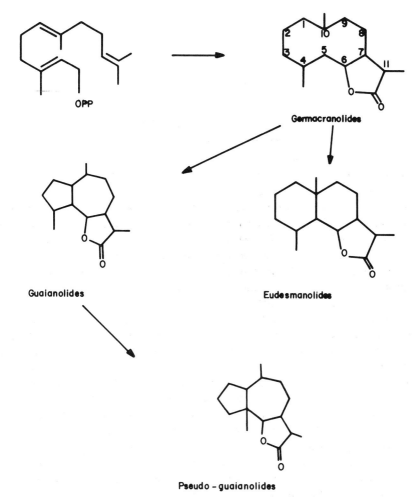

Figure 1. Classes of sesquiterpenoids.

an adjacent —OH or —OCOR group had a rate of reaction less than 720 LM^{-1} min^{-1}. Lactones lacking an adjacent group were consistently less reactive. This trend does not hold in comparison of the rates of reaction of model compounds with a neighboring —OH group, where no significant increase in reactivity is observed even though the —OH group is very close to the α-methylene hydrogens (Cassady *et al.*, 1978). The α-methylene-γ-lactone group will react with amines in a fashion analogous to thiol addition, however the reaction is much slower and no reaction occurs between guanine and elephantopin (Kawanata and Inayama, 1971; Lantz *et al.*, 1976).

$78 \pm 16 \ LM^{-1} \ min^{-1}$ $284 \pm 8 \ LM^{-1} \ min^{-1}$

2. Reaction with Other Biological Nucleophiles and Possible Mechanism of Action

Kupchan and co-workers tested the reactivity of α-methylene-γ-lactones and other electrophilic tumor inhibitors with cysteine, other model biological nucleophiles (Kupchan et al., 1970a), and sulfhydryl-containing enzymes in order to test the hypothesis that tumor growth inhibition resulted from selective alkylation of sulfhydryl groups in key enzymes controlling cell division (Kupchan, 1974). A series of compounds were studied on phosphofructokinase, and vernolepin, elephantopin, eupacunin, and euparotin were found to inhibit this enzyme as a result of reaction with sulfhydryl groups (Hanson et al., 1970). In addition, vernolepin was found to inactivate glycogen synthetase (Smith et al., 1972).

Lee et al. (1977) and Hall et al. (1977) extended this view in their studies on helenalin and tenulin, and in the process refined the probable mechanism of antitumor activity for these compounds. In these compounds, a cyclopentenone appears to be essential to antitumor activity, and in fact tenulin lacks an α-methylene-γ-lactone group. Reaction of helenalin and tenulin with the model nucleophiles reduced glutathione and cysteine, led to alkylation by conjugate addition to the enone system(s). These compounds also inhibited the glycolytic enzymes phosphofructokinase and hexokinase, the respiratory process, and DNA synthesis (Hall et al., 1977; Lee et al., 1977). The generally accepted mechanism of the classical antitumor alkylating agents is direct interaction with DNA. Lee and co-workers, in agreement with Kupchan, found no evidence of any interaction between tenulin and helenalin and DNA bases or DNA itself. The action of these compounds on DNA synthesis was presumably due to an effect on DNA polymerase. Hall and co-workers (1978a,b) further refined the picture by showing that helenalin supresses anaerobic glycolytic enzymes at a number of sites, with hexokinase showing the maximum inhibition. The compound also inhibited aerobic energy processes in tumor cells.

Hladon and co-workers (1977) studied the cytotoxic germacranolide alatolide in HeLa cells and, in contrast to earlier workers, found an inhibition of protein synthesis and RNA synthesis which they ascribed to a process of complex formation with DNA and not exclusively as a result of DNA synthesis inhibition.

Although a considerable amount of research has been carried out in attempts to elucidate the mechanism of action for this group (Fujita and Nagao, 1977), there is still no clear indication of the critical biochemical lesion leading to cytotoxicity. Based on the current evidence, it could be one of a number of effects. The final answer will require further research, along the lines of recent studies by Hartmann and co-workers (1978) which led to refinement of the mechanism of action of the reactive terpenes ovalicin and the trichothecenes betainyl-anguidine and verrucarin A. This research may uncover a diversity of mechanisms depending on the specific sesquiterpene lactone subtype under investigation.

3. Structure–Activity Relationships

Several studies have been undertaken to relate the structural features of sesquiterpene lactones to cytotoxicity (Cassady *et al.*, 1978; Kupchan *et al.*, 1971; Lee *et al.*, 1971). In general, an α-methylene-γ-butyrolactone or cyclopentenone group is a necessary and usually sufficient condition. It should be noted, however, that there are exceptions to this structural requirement. The compound bakkenolide A (**7**), which contains a β-methylene-γ-lactone group, has been reported to be cytotoxic (Jamieson *et al.*, 1976). Quadrone (**8**), a sesquiterpene isolated from *Aspergillus terreus*, is also cytotoxic (Ranieri and Calton, 1978).

7   8

The cytotoxicity of the eremantholides (**9a,b,c;** Fig. 2) is apparently due to the reactivity of the γ,δ double bond conjugated to the furanone ring, based on reaction with a model nucleophile (Le Quesne *et al.*, 1978).

The piptocarphins (**10a–f**), a series of novel cytotoxic germacranolides, contain an unusual enol lactone group in addition to a cyclic ketal and α,β-unsaturated ester (Cassady *et al.*, 1979). Preliminary analysis indicates that the conjugated enol lactone is necessary for activity; however, the most reactive site in the molecule has not been established. Piptocarphin C shows borderline activity in P-388 in addition to its cytotoxicity.

The critical question to be answered is the relationship between structure and *in vivo* antitumor activity, and here the results are incomplete and less conclusive. This type of correlation has been hampered by the limited number of sesquiterpene lactones which have been tested *in vivo*, and by a lack of sufficient analogues within a structural class. This latter situation is especially critical, since comparisons of compounds with different carbon skeletons (i.e., germac-

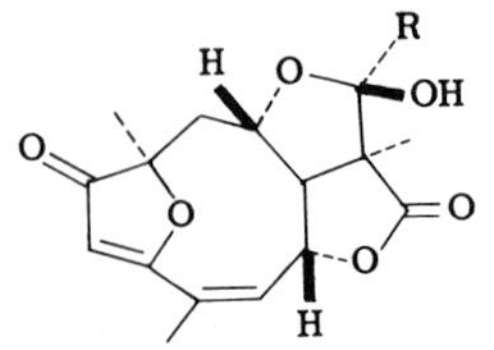

9a , R =

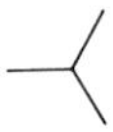

9b , R =

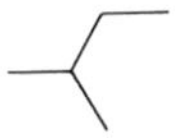

9c , R =

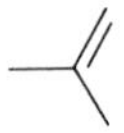

Figure 2. Cytotoxic eremantholides.

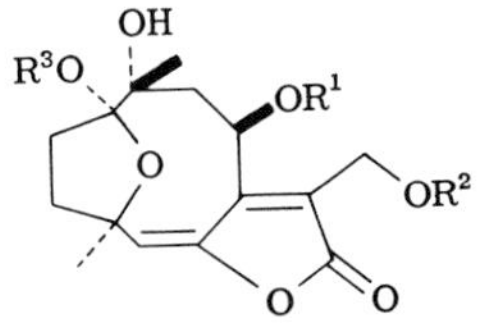

		R^1	R^2	R^3
10a	Piptocarphin A	Methacrylate	Acetate	H
10b	Piptocarphin B	Tiglate	Acetate	H
10c	Piptocarphin C	Methcrylate	H	H
10d	Piptocarphin D	H	Acetate	H
10e	Piptocarphin E	Methacrylate	Acetate	Et
10e	Piptocarphin F	Methacrylate	Et	H

ranolides and guaianolides) or with diverse functionalities may result in comparisons of compounds with different modes of action.

The two structural classes which have sufficient data for some tentative conclusions to be drawn are the pseudoguaianolides, where a detailed study of helenalin and derivatives has been reported by Lee *et al.* (1973, 1978a,b) and Lee (1978) and the germacranolides (Kupchan *et al.*, 1978b). This chapter will attempt to update the analysis of the germacranolides based on NCI data files and recent results from our laboratory.

The largest group of germacranolides with demonstrated activity are the *trans*, *trans*-cyclodecadiene type which are designated type A. The structures of these compounds are shown in Table IV, and P-388 activity data are given in Table V. Some general conclusions can be derived by examination of this data. First of all, in general, a single reactive group such as the α-methylene-γ-lactone group is not sufficient to impart significant P-388 activity. Compounds (11), (12), (13), and numerous other cytotoxic sesquiterpene lactones are examples of this. It should be noted, however, that (14) and (15) show moderate P-388 activity. In general, the presence of one other reactive group in addition to the unsaturated lactone, such as epoxide (16) or an α,β-unsaturated ester (19, 20), do not substantially influence the potency or level of activity. The most significantly active compounds are multifunctional, and placement of hydroxyl groups at various positions on the ring appears to substantially increase both T/C and potency. Substitution of OH in the vicinity of the 1,10 double bond appears to be especially favorable and much more important, for example, than epoxide at either the 4,5 (26) or 1,10 position (21, 22). Substitution of two hydroxyl groups at positions adjacent to the 1,10 double bond produces the most active compound in this series, eriofertopin (27). Activity is decreased by acetylation of these hydroxyl groups (28, 29) and apparently by substitution at the double bond of the unsaturated ester (compare 30), although this is a subtle effect.

Elephantopin (33) and related compounds are shown in Table VI, and the importance of multifunctionality is graphically illustrated here even though the series is very small. Maximum activity requires epoxide, unsaturated ester, and lactone. Loss of epoxide, as in (35), or reduction of the unsaturated ester, as in (36), leads to derivatives with significantly reduced activity.

Based on this analysis, both eriofertopin (27) and elephantopin (33) have been selected for further tumor panel evaluation. Other compounds of interest based on P-388 activity would include liatrin (37 in Table VII), and helenalin (52 in Table X). Helenalin and related compounds have been described extensively by Lee *et al.* (1978a), and here a cyclopentenone function combined with the α,β-unsaturated lactone group is the prominent feature. It is of interest to note the activity of the bis-helenalin analogue (60), as this opens a route to modify other active prototypes in other series. It should be productive to attempt to combine the best structural features of active prototypes such as elephantopin and eriofertopin in an attempt to further increase activity in this series. Other compounds of this type that show P-388 activity are listed in Tables VIII, IX, XI, and XII; however, there are insufficient numbers to draw any relationships between structure and *in vivo* activity.

III. DITERPENES

This group has produced several compounds with high activity in mouse leukemia systems and two compounds which are in preclinical development.

TABLE IV

Structures of Type A Germacranolides

Compound (NSC no.)	R¹	R²	R³	R⁴	R⁵	R⁶	Other
				Substituents			
11 (106404)	H	H	H	H	H	H	—
12 (135019)	H	H	H	H	α-OH	H	—
13 (152861)	β-OH	H	H	H	H	H	—

14 (177853)	OCOCH(CH$_3$)CH$_2$OH	H	H	H	H	OH	—
15 (251668)	OAc	H	OAc	H	H	H	—
16 (203821)	H	α-OH	H	H	H	H	4,5-epoxy
17 (290497)	H	H	OH	H	H	H	4,5-epoxy
18 (295426)	OCOCH(CH$_3$)$_2$	H	OH	H	H	OH	—
19 (295427)	OCOC(CH$_2$OH)=CHCH$_2$OH	H	H	H	H	H	—
20 (141349)	OCOC(CH$_2$OH(=CHCH$_2$OH	H	H	H	H	OH	1,10-dihydro
21 (144151)	OCOC(CH$_3$)=CH$_2$	H	H	H	β-OH	H	1,10-epoxy
22 (251667)	OCOC(CH$_3$)=CH$_2$	H	H	H	β-OAc	H	1,10-epoxy
23 (135023)	OCOC(CH$_2$OAc)=CHCH$_3$	H	H	α-OH	H	H	—
24	OCOC(CH$_2$OH)=CHCH$_3$	H	H	α-OH	H	H	—
25	OCOC(CH$_2$OAc)=CHCH$_3$	H	H	α-OAc	H	H	4,5-epoxy
26 (292662)	OCOC(CH$_3$)=CHCH$_2$OH	H	H	H	H	OH	—
27 (283439)	OCOC(CH$_3$)=CH$_2$	H	OH	α-OH	H	H	—
28 (283784)	OCOC(CH$_3$)=CH$_2$	H	OH	α-OAc	H	H	—
29	OCOC(CH$_3$)=CH$_2$	H	OAc	α-OAc	H	H	—
30 (283440)	OCOC(CH$_3$)=CHCH$_3$	H	OH	α-OH	H	H	—
31 (283441)	OCOC(CH$_3$)=CHCH$_3$	H	OH	α-OAc	H	H	—
32	OCOC(CH$_3$)=CHCH$_3$	H	OAc	α-OAc	H	H	—

TABLE V

Antitumor (P-388) Activity and Cytotoxicity of Type A Germacranolides

Compound no.	NSC[a] no.	Common name	Dose range (mg/kg)	Optimal T/C	Optimal dose (mg/kg)	Lowest toxic dose (mg/kg)	KB, ED$_{50}$ (µg/ml)
11	106404	Costunolide	12.5–100.0	113	100	—	3×10^{-1}
12	135019	Taumaulipin B	Inactive in LE, not tested in P-388				N.T.[b]
13	152861	Eupatolide	5.00–20.00	114	10	—	N.T.
14	177853	Arctiopicrin	2.75–22.00	140	11	—	2×10^{0}
15	251668	Ovalifolin acetate	9.00–105.00	143	70	—	2×10^{0}
16	203821	Parthenolide	8.75–35.00	136	35	—	1×10^{-1}
17	290497	9-α-Hydroxyparthenolide	60.00–90.00	150	80	200	N.T.
18	295426	Alatolide	3.75–30.00	113	7.5	—	N.T.
19	295427	Eupatoriopicrin	10.00–80.00	140	20	—	3×10^{0}
20	141349		3.50–14.00	125	14	—	N.T.
21	144151	Erioflorin	6.00–48.00	127	24	—	N.T.
22	251667	Acetylerioflorin	11.25–90.00	131	45	—	1.8×10^{0}
23	135023	Eupaserrin	20.00–44.00	150	44	—	N.T.
24		Desacetyleupaserrin	8.00–100.00	160	8	—	2×10^{-1}
25		Eupaserrin acetate	7.5–30.00	123	30	—	N.T.
26	292662	Eupahyssopin	2.00–16.00	125	2	—	N.T.
27	283439	Eriofertopin	1.25–40.00	167	20	40	1×10^{-1}
28	283784	2-O-Acetyleriofertopin	3.75–30.00	130	30	—	1×10^{0}
29		Diacetyleriofertopin	7.5–30.00	Inactive	—	—	N.T.
30	283440	Eriofertin	10.00–40.00	148	20	—	2×10^{0}
31	283441	2-O-Acetyleriofertin	3.75–30.00	Inactive	—	—	1.5×10^{0}
32		Diacetyleriofertin	20.00	126	20	—	N.T.

[a] NSC number = National Cancer Institute identification number assigned to these compounds.

[b] N.T. = not tested.

TABLE VI

Elephantopin and Related Compounds

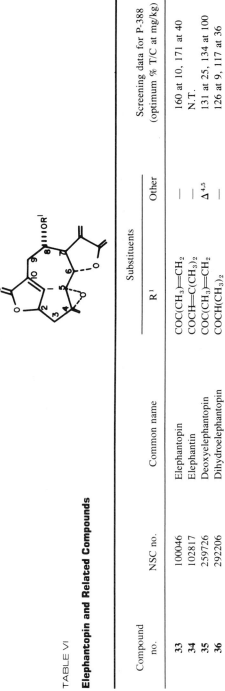

Compound no.	NSC no.	Common name	Substituents		Screening data for P-388 (optimum % T/C at mg/kg)
			R^1	Other	
33	100046	Elephantopin	$COC(CH_3)=CH_2$	—	160 at 10, 171 at 40
34	102817	Elephantin	$COCH=C(CH_3)_2$	—	N.T.
35	259726	Deoxyelephantopin	$COC(CH_3)=CH_2$	$\Delta^{4,5}$	131 at 25, 134 at 100
36	292206	Dihydroelephantopin	$COCH(CH_3)_2$	—	126 at 9, 117 at 36

TABLE VII

Liatrin and Related Compounds

Compound no.	NSC no.	Common name	Substituents			Screening data for P-388 (optimum % T/C at mg/kg)
			R^1	R^2	Other	
37	135034	Liatrin	$COC(CH_2OAc){=}CHCH_3$	H	—	157 at 8
38	249959	Tirotundin	$COCH(CH_3)_2$	H	1,2,4,5-tetrahydro	129 at 200
39	291860	Tagitinin-F	$COCH(CH_3)_2$	H	—	161 at 50
40	249956	Chapliatrin	$COC(CH_2OAc){=}CHCH_3$	OAc	1,2,4,5-tetrahydro 5,10-epoxy	132 at 25

TABLE VIII

Other Germacranolides with P-388 Activity

Compound	No.	NSC no.	Common name	Screening data P-388 (optimum % T/C at mg/kg)	Other
	41	135020	Eupacunin (R^1 = H; R^2 = OCOC(CH$_3$)=CHCH$_3$)	135 at 60	LE, WA (active)
	42	292663	Eupacurvin (R^1 = OH; R^2 = OCOCH(CH$_3$)CH$_2$CH$_3$)	119 at 25	B1 (borderline active)
	43	277282	Acantholide (R^1 = H; R^2 = COCH(CH$_3$)$_2$; R^3 = CHO; R^4, R^5 = H)	Inactive	LE (borderline active)
	44	277283	19-Hydroxyglabratolide (R^1 = COCH(CH$_3$)$_2$; R^2 = H; R^3 = CHO; R^4, R^5 = H)	144 at 10	N.T.
	45	282701	Dihydroanthospermal A (R^1 = COCH(CH$_3$)$_2$; R^2 = COC(OH)(CH$_3$)$_2$; R^3 = CH$_2$OH; R^4, R^5 = H)	136 at 20	N.T.
	46	294602	Melampodin A acetate (R^1 = CO-C-CHCH$_3$ epoxide, CH$_3$, OH; R^2 = COCH$_3$; R^3 = COOCH$_3$; R^4, R^5 = O)	133 at 6	N.T.
	47	294600	Melampodinin (R^1 = COC(CH$_3$)CH(OAc)CH$_3$; R^2 = COCH$_3$; R^3 = COOCH$_3$; R^4, R^5 = O)	140 at 12	N.T.

TABLE IX

Eudesmanolides and Seco-Eudesmanolides with P-388 Activity

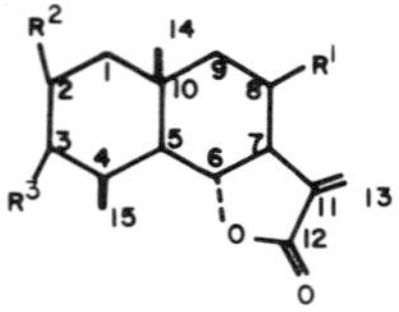

Type A (C-6)

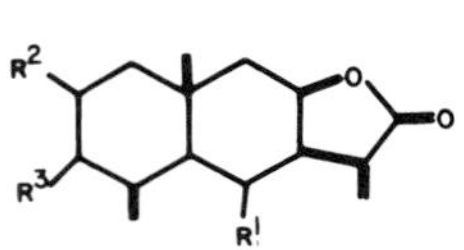

Type B (C-8)

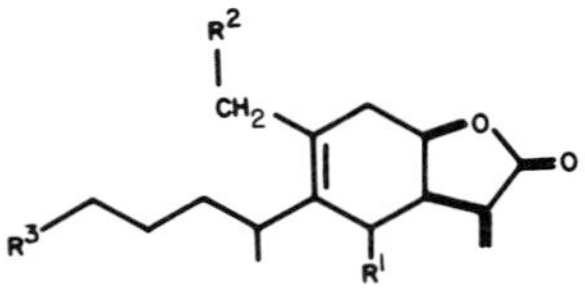

Type C (seco)

| Compound no. | NSC no. | Common name | Type | Substituent | | | | Screening data for P-388 (optimum % T/C at mg/kg) | Other |
				R^1	R^2	R^3	Other		
48	293065	Tuberiferine	A	H	$\Delta^{1,2}$	$=O$	—	Inactive 12.00–400	N.T.
49	306896		A	H	Br	$=O$	—	133 at 200	N.T.
50	233032	Pulchellin E	B	H	OH	OAc	$\Delta^{4,15}$	133 at 100	B1
51	144152	Eriolanin	C	—OCOC(CH$_3$)=CH$_2$	OH	OH	—	152 at 32	KB active

TABLE X

Pseudoguaianolides with P-388 Activity

Compound no.	NSC no.	Structure	Name	Screening data (optimum % T/C at mg/kg)	Other screening
52	085236	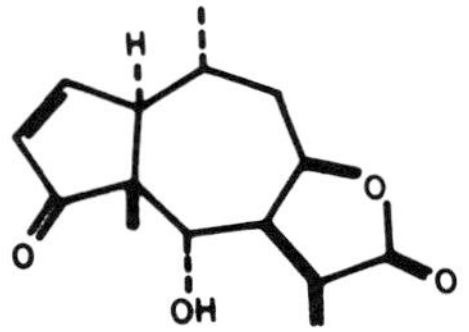	Helenalin	Toxic at 16, 168 at 8 220 at 3	B1, LE, LL, SA, WA (active)
53	280414	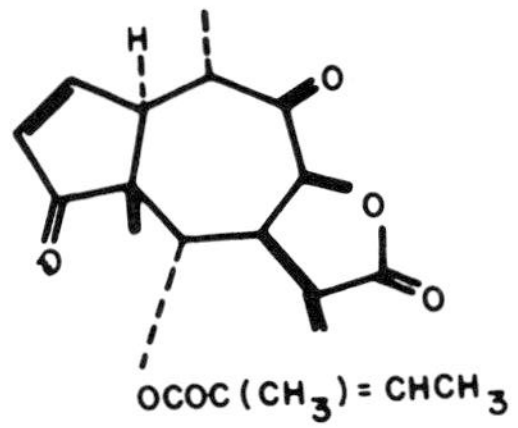 OCOC(CH₃) = CHCH₃	Multistatin	131 at 32	—
54	265203	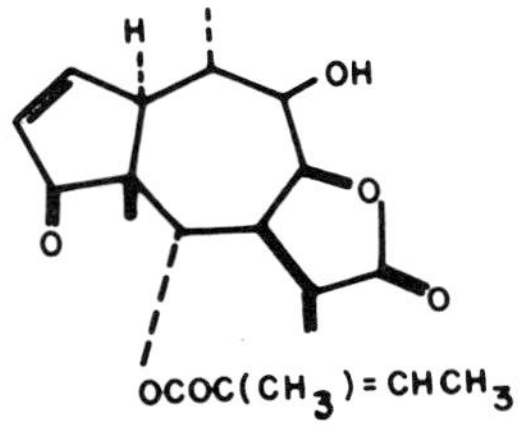 OCOC(CH₃) = CHCH₃	Multigilin	164 at 12.5	B1, C6

(Continued)

219

TABLE X (*Continued*)

Compound no.	NSC no.	Structure	Name	Screening data (optimum % T/C at mg/kg)	Other screening
55	176507		Fastigilin C	153 at 3.12	
56	085242		Coronopilin	140 at 100	
57	249995			160 at 75	

58   136772   Paucin   138 at 35

59   Dihydrohelenalin
(K. H. Lee, 1978)   138 at 25
(Non-NCI data)

60   Bis-succinyl helenalin
(K. H. Lee, 1978)   168 at 25
(Non-NCI data)

TABLE XI

Guaianolides with P-388 Activity

Compound no.	NSC no.	Structure	Name	Screening data (optimum % T/C at mg/kg)
61	114556		Eupatundin	133 at 40
62	187686		Liatrigramin	140 at 50 173 at 50

TABLE XII

Elemanolides with P-388 Activity

Compound no.	NSC no.	Structure	Name	Screening data (optimum % T/C at mg/kg)	Other
63	106398		Vernolepin	145 at 4	WA active
64	116070		Vernomenin	136 at 1.88	WA active
65	124459		Vernodalin	Not tested	WA active

Most of the representatives in this group are members of small series of compounds which have not been extensively developed. In general, these compounds are chemically reactive and are in that way similar to the sesquiterpene lactones; however, it is likely that each group represents a unique class in terms of structure–activity relationships and mode of action.

Jatrophone (66) is a complex macrocyclic diterpene which shows only moderate activity in P-388 (Kupchan et al., 1970a, 1976a).

Triptolide (67) and tripdiolide (68) are potent antileukemic agents which were isolated from Tripterygium wilfordii by Kupchan and co-workers (1972a) (see Fig. 3). Tripdiolide is scheduled for preclinical pharmacology based on its activity in the L-1210 system; however, neither compound shows significant activity in solid tumors. No research is reported which utilizes these compounds as prototypes for the development of synthetic analogues. The unique structure and reactivity of the triepoxide system would appear to be a useful model for further development.

Another diterpene, taxol (69), was discovered by Wani and co-workers (1971). This unique, complex taxane derivative has a relatively broad spectrum of activity which includes activity in B1 in addition to the mouse leukemias. No systematic modification of the basic structure has been accomplished to date.

The quinone methides, taxodone (70) and taxodione (71), were reported by Kupchan and co-workers (1968) and are further examples of reactive terpenes. The reactivity in this case presumably stems from the susceptibility of the quinone methide system to nucleophilic attack with consequent aromatization.

Taxodone and taxodione have not stimulated much interest in analogue development, and in fact these compounds have not received extensive evaluation in mouse tumor systems.

A. Norditerpene Dilactones

A number of nor- and bisnorditerpene dilactones have been isolated from plants in the genus Podocarpus (family Podocarpaceae), and this interesting group of compounds is the subject of a recent review (Ito and Kodama, 1976). Insufficient numbers of compounds of this type have been evaluated for antitumor activity for any firm structure–activity correlations to be derived; however, members of this group including podolide (Kupchan et al., 1975a) and nagilactone C (Hayashi and Sakan, 1975) have shown marginal P-388 activity as shown in Table XIII.

B. Phorbol Esters and Related Compounds

Several compounds in the phorbol and daphnetoxin classes of diterpenes have been shown to possess antitumor activity. The activity of this group appears

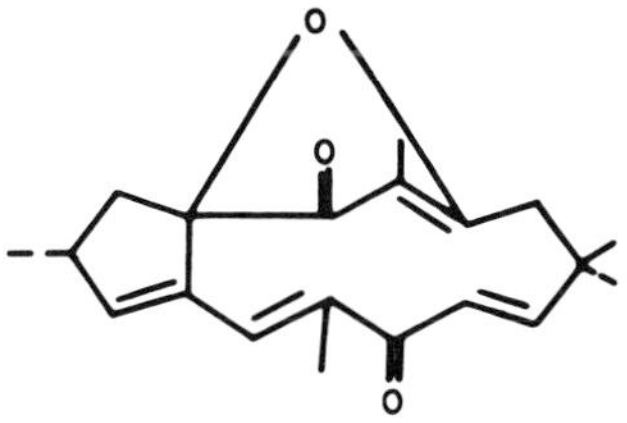

Jatrophone, 66

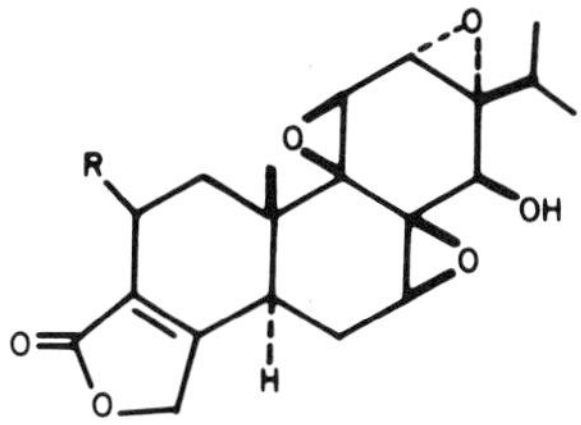

Triptolide, 67, R = H
Tripdiolide, 68, R = OH

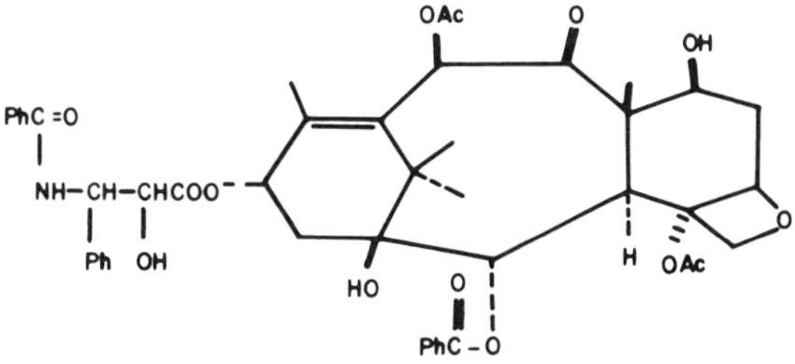

Taxol, 69

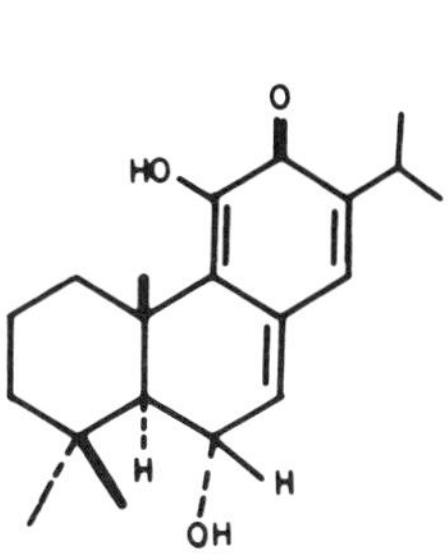

Taxodone, 70

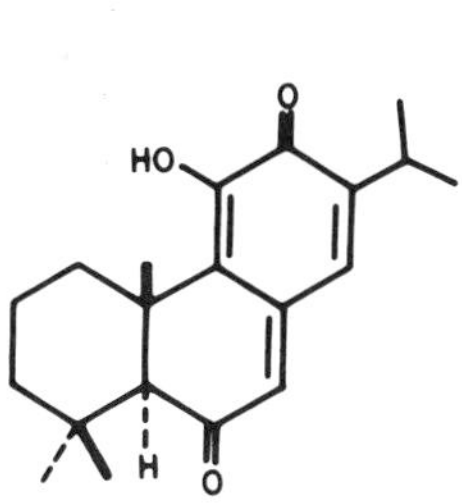

Taxodione, 71

Figure 3. Antitumor diterpenes.

TABLE XIII

Norditerpene Dilactones, Structure and Antitumor Activity

Compound no.	NSC no.	Structure	Name	Dose range (mg/kg)	T/C at optimal dose (mg/kg)	Lowest toxic dose (mg/kg)	KB ED_{50}
72			Nagilactone E	Active vs. Yoshida sarcoma			
73	238978		Podolide	(Hayashi et al., 1975) 12.5–50	127 at 50	—	2×10^{-1}

74	245351	Podolactone B		1-40	129 at 10	—	—
75	117884	Nagilactone C		10-60	145 at 40	60	—

to be limited to PS *in vivo,* with no substantial activity in other leukemias or any solid tumor systems. The P-388 leukemia cell appears to have a very marked sensitivity to these compounds since high cytotoxicity is seen against the PS *in vitro* cell line in culture as well as P-388 *in vivo,* but little cytotoxicity towards the KB cell line is observed. Based on this and the potential tumor-promoting properties of this group (Evans and Soper, 1978), there appears to be little utility in further development as potential clinical agents. The structures and activity of the phorbol series are given in Tables XIV and XV. The daphnetoxin series in Tables XVI and XVII. The phorbol series is short, but tentative conclusions can be drawn. It is apparent that the esters are significantly more active than the corresponding alcohols and that the conjugated ketone is essential to significant P-388 activity.

IV. BUFADIENOLIDES, CARDENOLIDES, AND WITHANOLIDES

All of these groups have in common a steroidal nucleus and a modified sidechain at position 17 or position 20 in the form of a five- or six-membered unsaturated lactone. These groups all have shown some biological activity assumed to be related to cancer and have been of some interest in that regard. The antitumor activities of these groups of compounds are summarized, and structure-activity relationships in the KB cell cytoxicity system are developed and discussed through use of the data on compounds screened at the National Cancer Institute (NCI). The NSC numbers in the tables are the NCI identification codes for the compounds.

A. Bufadienolides

These compounds occur in both plants and animals and are well known as constituents of squill (*Urginia maritima*—Liliaceae) and related plants and of toads of the genus *Bufo.* The well-known cardiac activity of these compounds will not be discussed in this presentation.

The basic structure and stereochemistry of the bufadienolides is indicated in Fig. 4. The key stereochemical features which are assumed important for activity are the *cis* relationship between the C–D rings, the *beta* sidechain at C-17, and the A–B ring juncture which is generally *cis* due to the presence of a 5β hydrogen or hydroxyl. In the naturally occurring compounds, the most frequent sites of functionalization are: (1) C-3, where a beta hydroxyl group invariably occurs, often as a glycoside; (2) C-5, which sometimes bears a beta hydroxyl; (3) C-14, which is usually hydroxylated or is the terminus for a C-14,15 epoxide; and (4) C-19, which may be oxidized to an aldehyde.

TABLE XIV

Phorbol Derivatives

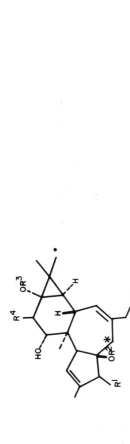

Compound no.	NSC no.	Name	R^1	R^2	R^3	R^4	Other
76	154778	Phorbol	$=O$	H	H	OH	—
77	262643	Phorbol 12,13-dibenzoate	$=O$	H	COPh	OCOPh	—
78	262644	Phorbol 12-tetradecanoyl-13-Acetate	$=O$	H	$COCH_3$	$OCO(CH_2)_{12}CH_3$	—
79	262645	4-O-Methyl-12-tetradecanoyl-phorbol-13-acetate	$=O$	CH_3	$COCH_3$	$OCO(CH_2)_{12}CH_3$	—
80	278618	Phorbolol myristate acetate	$—OH$	H	$COCH_3$	$OCO(CH_2)_{12}CH_3$	—
81	262642	Phorbol-12-tiglate-13-decanoate	$=O$	H	$CO(CH_2)_8CH_3$	$OCOC(CH_3)=CHCH_3$	—
82	281268	12-Deoxy phorbol-13-palmitate	$=O$	H	$CO(CH_2)_{14}CH_3$	H	—
83	281269	12-Deoxy-5β-hydroxy phorbol-13-myristate	$=O$	H	$CO(CH_2)_{12}CH_3$	H	*—OH
84	281270	12-Deoxy-16-hydroxy phorbol-13-palmitate	$=O$	H	$CO(CH_2)_{14}CH_3$	H	—OH

TABLE XV

Antileukemic Activity of Phorbol Derivatives

Compound no.	NSC no.	Name	Dose range (mg/kg)	Optimal T/C and dose (mg/kg)	Lowest toxic dose (mg/kg)
76	154778	Phorbol	6.25–50.0	109 at 25	—
77	262643	Phorbol 12,13-dibenzoate	3–800 (μg/kg)	172 at 3 (μg/kg)	—
78	262644	Phorbol 12-tetradecanoyl-13-acetate	3–800 (μg/kg)	140 at 400 (μg/kg)	—
79	262645	4-O-Methyl-12-tetradecanoyl-phorbol-13-acetate	3–800 (μg/kg)	169 at 800 (μg/kg)	—
80	278618	Phorbolol myristate acetate	0.12–2.0	111 at 0.5	—
81	262642	Phorbol-12-tiglate-13-decanoate	0.03–2.00	163 at .25	.50
82	281268	12-Deoxy phorbol-13-palmitate	0.06–1.00	131 at .13	1.0
83	281269	12-Deoxy-5β-hydroxy phorbol-13-myristate	0.13–1.00	115 at .25	—
84	281270	12-Deoxy-16-hydroxy phorbol-13-palmitate	0.03–.12	119 at .03	—

TABLE XVI

Daphentoxin Derivatives

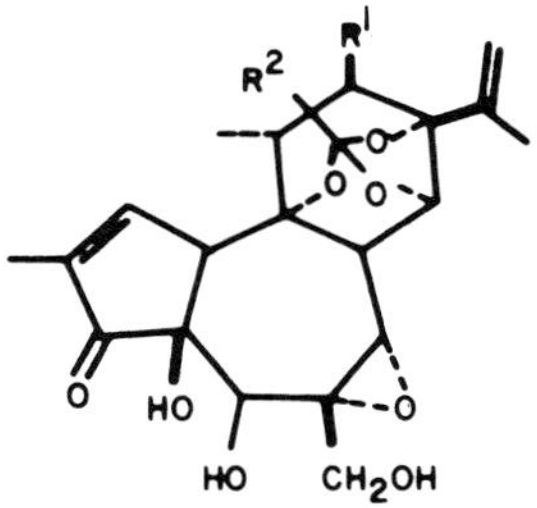

Compound no.	Name	NSC no.	R^1	R^2
85	12-Hydroxy-daphnetoxin	239073	OH	Ph
86	Gnidiglaucin	270917	OCOCH$_3$	—(CH$_2$)$_8$CH$_3$
87	12-Benzoyl-daphnetoxin	239074	OCOPh	Ph
88	Gnidicin	238941	OCOCH=CHPh	Ph
89	Gnidilatidin	270919	OCOPh	—(CH=CH)$_2$(CH$_2$)$_4$CH$_3$
90	Mezerein	239072	OCO(CH=CH)$_2$Ph	Ph
91	Gnididin	238942	OCO(CH=CH)$_2$(CH$_2$)$_4$CH$_3$	Ph
92	Gniditrin	238943	OCO(CH=CH)$_2$(CH$_2$)$_2$CH$_3$	Ph
93	Gnidilatin	266489	OCOPh	—(CH$_2$)$_8$CH$_3$

TABLE XVII

Antileukemic Activity of Daphnetoxin Derivatives

Compound no.	NSC no.	Dose range (μg/kg)	Optimum T/C and dose (μg/kg)	Lowest toxic dose
85	239073	25–1600	131 at 400	—
86	270917	12.5–400	123 at 200	—
87	239074	6.3–200	150 at 200	1.60 mg/kg
88	238941	6.3–400	173 at 200	—
89	270919 (261422)	0.8–200	145 at 25	200 μg/kg
90	239072	6.3–800	200 at 25	200 μg/kg
91	238942	6.3–400	127 at 50	—
92	238943	6.3–400	168 at 100	—
93	266489	12.5–400	174 at 400	—

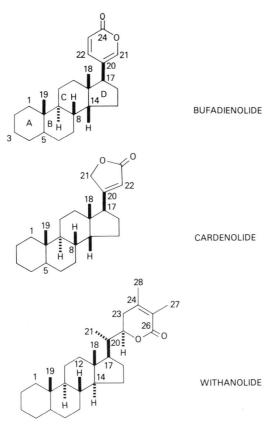

BUFADIENOLIDE

CARDENOLIDE

WITHANOLIDE

Figure 4. General structures.

The major property which distinguishes these compounds from other steroids, in addition to the obvious lactone sidechain, are the cis fusions of the A–B and C–D rings, so the compounds screened have been divided into groups based on the substituents at C-14 (epoxide, hydroxyl, olefin, or hydrogen) and the substituents at C-5 (hydrogen, hydroxyl, or Δ4-5 unsaturation).

Table XVIII contains all those compounds with 5β-H and 14β,15β-epoxides (Group A, Fig. 5). There are two main subclasses, those with an oxygen substituent at C-16 (compounds **94–107**) and those without (compounds **108–119**). For the cases where the C-16 substituent is β and small (hydroxyl, acetyl, formyl), there is little apparent difference in KB cytotoxicity from the compounds where there is no C-16 substituent. However, when the C-16 substituent is of the β orientation and is larger (compounds **100–106**), activity is lost, and likewise where the C-16 substituent is α, as in compound (**98**), total inactivity results. With regard to C-5, note that compounds (**107**) and (**114**), which have

TABLE XVIII

Bufadienolides with 3-OR; 5β-H; 14β,15β-Epoxy Substituents

| Compound no. | NSC no. | Substituents[a] | | | Screening data | |
		C-3[b]	C-16[b]	Other	KB[c,d]	In vivo systems[d,e]
94	90325	OH	OAc		3×10^{-2}	B1, LE, LL, PS, WA
95	234203	OH	OH		$\sim 5 \times 10^{0}$	B1, LE, PS
96	234204	OSu	OAc		2×10^{-1}	LE, PS
97	237024	OAc	OH		9×10^{0}	N.T.
98	248530	OAc	α-OH		>100	N.T.
99	267899	OAc	O-CHO		2×10^{-1}	N.T.
100	267900	OAc	O(CO)Et		>100	N.T.
101	267901	OAc	O(CO)Pr		>100	N.T.
102	267902	OAc	O(CO)iPr		>100	N.T.
103	267903	OAc	O(CO)nBu		7×10^{0}	N.T.
104	267904	OAc	O(CO)sBu		>100	N.T.
105	267905	OAc	O(CO)nPe		5×10^{1}	N.T.
106	267906	OAc	O(CO)Ph		9×10^{1}	N.T.
107	90326	OH	OAc	5β-OH	2×10^{-2}	LE
108	90783	OH	H		4×10^{0}	B1, LE, PS
109	135688	OAc	H		N.T.	LE
110	234206	OH	H	19-oxo	2×10^{-1}	LE, PS
111	267712	OMA	H		N.T.	B1, PS
112	267908	OPO$_2$ClH	H		9×10^{0}	N.T.
113	237022	OAc	H	14β-Cl,15α-OH	2×10^{0}	N.T.
114	237023	OAc	H	14β-OH,15α-Cl	6×10^{0}	N.T.
115	135687	OAc	H	5β-OH	N.T.	LE
116	234205	OH	H	5β-OH	$\sim 10^{-1}$	EA(MA), LE, PS, WA, 8P
117	237769	OH	H	14α,15α-epoxy	>100	N.T.
118	248525	α-OH	H	14α,15α-epoxy	>100	N.T.
119	248526	oxo	H	14α,15α-epoxy	>100	N.T.

[a] Su = succinyl, Ac = acetyl, MA = methacrylyl.

[b] All substituents at this position are β unless otherwise indicated.

[c] ED$_{50}$ in μg/ml.

[d] N.T. = not tested.

[e] Test data in the systems indicated are negative unless followed by MA (marginal activity).

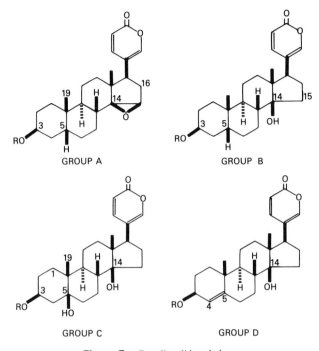

Figure 5. Butadienolide subclasses.

the β hydroxyl instead of hydrogen retain cytotoxicity. The most important conclusion to be drawn from this table is the effect of having a C-14,C-15 α epoxide (compounds **117–119**), which results in complete loss of cytotoxicity.

Table XIX includes those compounds with C-5 β hydrogens and C-14 β hydroxyl groups (Group B, Fig. 5). Here note compounds (**126**) and (**127**), which have an α hydroxyl and a ketone at C-3, respectively. These are considerably less active than the 3β series. Those compounds (**128–129**), which have C-15 substituents adjacent to the C-14 hydroxyl, are inactive so the β oxygen at C-14 may play an important role in cytotoxicity. This point will become clearer when those compounds in Table XXII (Group E compounds) which do not contain oxygen at C-14 are discussed.

Table XX shows those bufadienolides which have both C-5 β and C-14 β hydroxyls (Group C, Fig. 5). The cytotoxicity of these compounds is generally in the 10^{-2} to 10^{-3} μg/ml range, which is an order of magnitude more toxic than the compounds in Table XVIII and Table XIX which have an oxygen function at C-14 but not at C-5. This could also be attributed to the aldehyde at C-19, however. Compounds (**141–142**) which contain the 16 β hydroxyl group are less active than compound (**140**) which does not contain this group, but more examples would be necessary to develop this point.

TABLE XIX

Bufadienolides with 3-OR; 5β-H; 14β-OH Substituents

Compound no.	NSC no.	Substituents[a]		Screening data	
		C-3[b]	Other	KB[c,d]	In vivo systems[g]
120	89595	OH		$\sim 10^{-1}$	B1, PS
121	90384	OH	11α-OH	2×10^{-2}	N.T.
122	7534	OH	Δ8,9; 12-OH; 23-OH	N.T.	EA, LE, SA, WA
123	135689	OAc	16β-OAc	N.T.	LE
124	234668	OSu		2×10^{-1}	N.T.
125	248517	OAc		2×10^{-2}	N.T.
126	248518	α-OH		2×10^{0}	N.T.
127	248519	oxo		1×10^{0}	N.T.
128	248527	OAc	15α-OH	>100	N.T.
129	248528	OAc	15-oxo	3×10^{1}	N.T.
130	268347	OGlu		4×10^{-3}	PS
131	270462	OMA		N.T.	PS
132	270463	e		>100	N.T.
133	270464	OMs		3×10^{1}	N.T.
134	270465	f		5×10^{-1}	N.T.

[a] Su = succinyl, Ac = acetyl, Glu = glucosyl, MA = methacrylyl, Ms = mesyl.

[b] All substituents are β unless noted.

[c] ED_{50} μg/ml.

[d] N.T. = not tested.

[e] $-O-PO(OCH_3)-N\underset{O}{\overset{\frown}{\smile}}$

[f] $-O-PO(OH)-N\underset{O}{\overset{\frown}{\smile}}$

[g] All data are negative in the system tested.

TABLE XX

Bufadienolides with 3-OR; 5β-OH; 14β-OH Substituents

		Substituents[a]			Screening data	
Compound no.	NSC. no.	C-3[b]	C-19	Other	KB[c]	In vivo systems[d]
135	89594	OH	CHO		1×10^{-2}	LE
136	90782	OH	Me		3×10^{-2}	BI, LE, PS
137	93134	O-Glu-Rh	CHO		2×10^{-2}	BI, LE, LL, PS, SA, WA
138	106676	OAc	CHO		6×10^{-3}	WA(A)
139	109330	OAc	CHO	5β-OAc	5×10^{-3}	WA
140	135076	OAc	CHO	1β-OH	$\sim 10^{-4}$	N.T.[e]
141	135079	OAc	CHO	1β-OH,16β-OH	2×10^{-1}	N.T.
142	135080	OH	CHO	1β-OAc, 16β-OH	5×10^{-1}	N.T.

[a] Rh = rhamnose, Glu = glucose.

[b] All substituents are β.

[c] ED_{50} μg/ml.

[d] Systems listed described in Table I. Test data are negative unless indicated by the following codes: (MA) = marginal activity of statistical significance but not of biological interest; (A) = activity of both statistical significance and biological interest.

[e] N.T. = not tested.

The derivatives in Table XXI (**143–150**), all of which contain a Δ4,5 unsaturation and the 14β-OH (Group D, Fig. 5) are quite cytotoxic, whereas those in Table XXII which have a Δ14,15 unsaturation (compounds **151–155**) are inactive. This again points up the critical nature of C-14. Compound (**157**) has a β hydrogen at C-14 and has marginal activity, while a similar compound with C-14 α hydrogen (**158**) is totally inactive.

One may summarize the roles of substituents and stereochemistry with regard to cytotoxicity among the bufadienolides as follows:

1. One must have a C-14β substituent for maximal activity. Hydroxyl or epoxy are both suitable, but there are insufficient data on a C-14β hydrogen to draw a conclusion as to whether this group might also be associated with cytoxicity.

2. At C-5, either a β hydroxyl group or an olefin (Δ4,5) show high cytotoxicity, while a β hydrogen appears somewhat less active.

3. The substituent at C-3 should be a β hydroxyl or derivative. Free hydroxyl, acetyl, or sugar units all have similar activity.

4. The C-19 aldehyde is present in active compounds, but there is insufficient data to determine its role as a possible enhancer of cytotoxicity.

B. Cardenolides

The cardenolides have been isolated from numerous plants, mainly in the families Apocynaceae, Asclepiadaceae, Fabaceae, Melianthaceae, and Scrophulariaceae, and are steroidal compounds characterized by the presence of a five-membered unsaturated lactone ring at C-17 and cis ring fusions at the AB and CD ring junctions. The basic structure, numbering, and stereochemistry was given in Section IV,A in Fig. 4. Since the cardiotonic activity of these molecules has been shown to be related to the stereochemistry of the A–B and C–D ring junctions, these compounds were divided into groups based on the substituents at C-5 and C-14 in order to determine if the same relationship would hold there for cytotoxicity in the KB cell culture system.

Group A consists of those cardenolides having β hydroxyl groups at C-5 and C-14 (Fig. 6; Table XXIII; compounds **160–174**). The parent compound in this series can be considered as the well-known genin strophanthidin (**160**). Compounds (**160–167**) in Table XXIII clearly demonstrate consistent cytotoxicity irrespective of the presence of a sugar residue at C-3β, although the glycosides appear to be somewhat more cytotoxic than the genins. Compound (**167**) (dihydrostrophanthidin) is interesting since it retains cytotoxicity in spite of the saturated lactone ring. Compounds (**168**) and (**169**) have C-19 derivatized as a carboxylic acid and its methyl ester, respectively, and are not cytotoxic. Compounds (**170–173**) all have C-19 hydroxyl groups and retain activity but are somewhat less active than the C-19 aldehydes. The presence of 1-β or 11-α hydroxyl groups is consistent with cytotoxicity.

TABLE XXI

Bufadienolides with 3β-OR, Δ4,5 and 14β-OH Substituents

Compound no.	NSC no.	Substituents		Screening data	
		C-3β	Other	KB[b]	In vivo systems[c]
143	7521	ORh		~10⁻³	B1, EA, LE, LL(MA), PS, SA, WA
144	7523	O-Glu	6β-OAc; 8β-OH	2 × 10⁻²	EA, LE, PS, SA, WA(MA)
145	7525	ORh-Glu		2 × 10⁻²	CA(MA), LE, LL, PS, SA, WA
146	117180	OH	6β-OAc; 8β-OH; 1α,2α-epoxy	N.T.	LE, PS
147	135036	OH	19-oxo	2 × 10⁻²	PS
148	135037	OH	16β-OH; 19-oxo	7 × 10⁻²	N.T.[d]
149	194683	O-TADT[a]		N.T.	LE
150	234669	OH		~10⁻³	PS

[a] TADT = 2,3,4-tri-O-acetyl-6-deoxytalopyranosyl.

[b] ED₅₀ μg/ml.

[c] Systems listed are described in Table I. Test data are negative unless indicated by the following codes: (MA) = marginal activity of statistical significance but not of biological interest; (A) = activity of both statistical significance and biological interest.

[d] N.T. = not tested.

TABLE XXII

Other Bufadienolides Not Covered in Tables XVIII–XXI

| Compound no. | NSC no. | Substituents | | | | | Screening data | |
		$C-3^a$	$C-5^a$	$C-14^a$	Other		KB^b	In vivo system^c
151	248520	OH	H	Δ14			3×10^1	N.T.^{d}
152	248521	OAc	H	Δ14			5×10^1	N.T.
153	248522	α-OH	H	Δ14			>100	N.T.
154	248523	oxo	H	Δ14			>100	N.T.
155	248529	OAc	H	Δ14	16-oxo		3×10^0	N.T.
156	136989	OAc	H	H			N.T.	LE
157	248531	OAc	H	H	15-oxo		4×10^0	N.T.
158	237021	OH	H	α-H	15-oxo		3×10^1	N.T.
159	134654	OAc	αH	α-H			3×10^0	N.T.

^{a} Substituents at these positions are β unless otherwise indicated.

^{b} ED_{50} $\mu g/ml$.

^{c} Systems listed are described in Table I. Test data are negative unless indicated by the following codes: (MA) = marginal activity of statistical significance but not of biological interest; (A) = activity of both statistical significance and biological interest.

^{d} N.T. = not tested.

TABLE XXIII

Cardenolides with 5β-OH and 14β-OH Substituents

Compound no.	NSC no.	Substituents[a] C-3β	C-19	Other	KB[b]	Screening data In vivo systems[c]
160	86078	OH	oxo		2×10^{-1}	B1, LE, LL, PS
161	93373	OH	oxime		1×10^{-1}	LE, LL(MA), PS, WA
162	92954	OAc	oxo		4×10^{-2}	CA, LE, SA
163	7670	O-Cy-Gl	oxo		3×10^{-2}	CA, LE, SA
164	7530	O-Cy-Gl-Gl	oxo		4×10^{-2}	EA, LE, PS, SA, WA
165	407808	O-Rh	oxo		3×10^{-1}	LE, WA
166	7522	O-Cy	oxo		3×10^{-2}	CA, LE, LL, PS, SA(MA), WA
167	87319	OH	oxo	21,22-Dihydro	4×10^{-2}	N.T.[d]
168	87320	OAc	COOH		3×10^{-1}	N.T.
169	87321	OAc	COOMe		3×10^{-1}	N.T.
170	87314	OH	OH		4×10^{-1}	LE
171	152149	OTa	OH	1β-OH	2×10^{-1}	LE
172	97088	OH	OH	1β-OH; 11α-OH	3×10^{0}	N.T.
173	25485	ORh	OH	1β-OH; 11α-OH	$<10^{-1}$	CA, LE, PS, SA, WA
174	173716	OTm	OH	1β-OH; 11α-OH	5×10^{-2}	N.T.

[a] Cy = cymarose, Gl = glucose, Rh = rhamnose, Ta = Talose, Tm = talomethylose.
[b] ED_{50} μg/ml.
[c] All data are negative unless followed by (MA) = marginal activity.
[d] N.T. = not tested.

Table XXIV contains structural information and bioassay data on the second group of cardenolides (Group B in Fig. 6) which possess 5-β hydrogen and 14-β hydroxyl groups (compounds **175–192**). The parent for these compounds is digitoxigenin (**175**). The cytotoxicity of the Group B cardenolides is approximately equivalent to the Group A compounds, showing that there is no substantial difference between a hydrogen and hydroxyl at C-5β with regard to cytotoxicity. Those compounds with β-hydroxy or β-acetyl substituents at C-16 (**180–186**) seem slightly less cytotoxic than those compounds without these substituents, and the one example of a compound with an unsaturation at C-16 (**188**) is inactive. Compounds (**189–191**) show that the 12β-OH group does not markedly affect activity.

Table XXV shows the structural information and screening data for other cardenolides which contain the 14β-OH group but do not contain either 5β-H or 5β-OH substitutents (Group C cardenolides, Fig. 6). Compounds (**193–198**) in Table XXV have $\Delta4,5$ unsaturations, and therefore the substituents at C-5 should all be in a plane. This apparently has no negative effect on cytotoxicity. Compound (**198**) is sufficiently different from the other compounds that the inactivity in the KB system cannot be considered to be related to C-5. Since compounds (**199–201**) are not active in the KB system, it can be concluded that substituents involving C-6 have a negative effect on cytotoxicity. Those compounds with α hydrogens at C-5 (**203–207**) retain activity but are notably less cytotoxic (10° range versus 10^{-1} to 10^{-2} range) than the Group A and Group B compounds.

Table XXVI contains those compounds without 14β-OH substituents (compounds **208–214**). There is a large loss in activity when a 14,15 unsaturation is introduced in the molecule (compare **208** with compound **162** in Table XXIII).

When all the cytotoxicity data on the cardenolides are compared, the conclusions about structural requirements for cytotoxicity in the KB system are:

1. There is some influence of the C-3 substituent on cytotoxicity with glycosides being slightly more active than aglycones.

2. C-5 should be a β hydrogen for best activity. A 4,5 unsaturation retains activity, while a 5,6 unsaturation or epoxide does not. A compound with a 5-α hydrogen is considerably less cytotoxic.

3. C-14 requires a β hydroxyl for maximum cytotoxicity, as 14,15 unsaturated compounds are inactive.

4. C-16 β hydroxyl or acetyl groups are tolerated.

5. C-19 can be an alcohol or aldehyde and retain cytotoxicity, but a C-19 acid is inactive.

6. Other substituents which are tolerated without large changes in cytotoxicity are 1β-OH, 11α-OH, and 12β-OH groups.

Review of the biological screening data in Tables XXIII–XXVI reveals that, like the bufadienolides, the cardenolides are devoid of useful activity in *in vivo*

TABLE XXIV

Cardenolides with 14β-OH and 5β-H Substituents

Compound no.	NSC no.	Substituents[a] C-3β	C-16	Other	KB[b]	In vivo systems[c]
175	407806	OH			2×10^1	CA, LE, SA, WA
176	254671	O-Th			2×10^{-2}	N.T.[d]
177	251674	O-AcTh			N.T.	PS(MA)
178	7529	O-Di-Di-Di			9×10^{-2}	SA, WA
179	7532	O-Di-Di-AcDi-Gl			4×10^{-1}	CA(MA), LE, LL, PS, SA, WA
180	407807	OH	β-OH		2×10^1	CA, LE, LL, SA, WA
181	148790	OH	β-OAc		2×10^{-1}	LE
182	152153	OAf	β-OAc		1×10^{-1}	N.T.
183	152154	O-Af-Gl-Gl	β-OAc		5×10^0	N.T.
184	254670	O-Cy	β-OAc		3×10^{-2}	N.T.
185	95099	O-Di-Di-Di	β-OH		5×10^0	N.T.
186	7535	O-Di-Di-AcDi-Gl	β-OH		4×10^{-1}	B1, CA, LE, PS, SA, WA
187	160843	OH	β-O(CO)Et		4×10^0	N.T.
188	160844	OH	$\Delta^{16,17}$		2×10^1	N.T.
189	95100	O-Di-Di-Di		12β-OH	$<1 \times 10^0$	CA, LE, SA, WA
190	7531	O-Di-Di-Di-Gl		12β-OH	2×10^{-1}	CA, LE, SA, WA
191	7533	O-Di-Di-AcDi-Gl		12β-OH	1×10^{-1}	CA(MA), LE, SA, WA
192	83216	O-Cy		19-oxo	1×10^{-1}	LE, LL, PS, SA, WA

[a] Af = acofriose, Cy = cymarose, Di = digitoxose, Gl = glucose, Th = thevetose.
[b] ED_{50} in μg/ml.
[c] All data are negative unless followed by (MA) = marginal activity.
[d] N.T. = not tested.

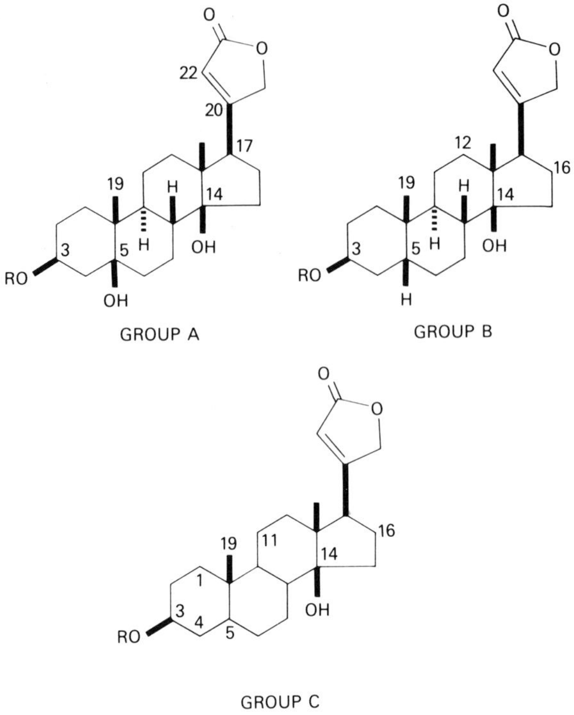

Figure 6. Cardenolide subclasses.

systems. The activity when seen is close to the cut-off point for activity in the tumor system, is not easily reproduced, and is at doses approximating the toxic dose so that there is no therapeutic implication to the activity. From the volume of negative or extremely marginal *in vivo* data on cardenolides and bufadienolides and the large number of compounds tested, it is apparent that there is little if any prospect of developing therapeutically useful anticancer compounds from these series.

Thus, the premise that the stereochemistry of the ring fusions of the A–B and C–D rings is critical to the biological activity of the bufadienolides is borne out. The structural requirements for cytotoxicity among the bufadienolides as herein described are analogous to the requirements for cardiotonic activity in the cardenolide series as reviewed by Boutagy and Gelbart (1974) and extend the conclusions of Kamano *et al.* (1977) on a much smaller series of bufadienolides.

Review of the *in vivo* testing data in the tables shows it to be almost entirely negative. Negative data have been obtained in a wide variety of tumor systems and no useful biological results have been forthcoming. It is quite clear that the cell cytotoxicity data have no correlation whatsoever with *in vivo* antitumor activity

TABLE XXV

Other Cardenolides with 14β-OH Substituents

| | | Substituents[a] | | | | | Screening data |
| | | | | | | | |
Compound no.	NSC no.	C-3[b]	C-5	C-19	Other	KB[c]	In vivo system[d]
193	237527	OH	$\Delta^{4,5}$			6×10^{-1}	PS
194	152150	O-Af	$\Delta^{4,5}$			2×10^{-1}	N.T.[e]
195	152148	O-Av	$\Delta^{4,5}$		1β-OH	6×10^{-1}	N.T.
196	281266	O-Xy	$\Delta^{4,5}$	oxo		2×10^{-2}	N.T.
197	256926	O-Xy-Gl	$\Delta^{4,5}$	oxo		7×10^{-1}	B1, C8, LE, LL, PS(MA)
198	65944	oxo	$\Delta^{4,5}$	11β-19 epoxy		>100	CA, LE, PS, SA, WA
199	87315	OAc	$\Delta^{5,6}$	oxo	11α-OMe	2×10^{1}	N.T.
200	87318	OAc	$5\alpha,6\alpha$-epoxy	oxo		2×10^{1}	N.T.
201	93374	OAc	$5\beta,6\beta$-epoxy	oxo		>100	N.T.
202	93070	OAc	α-Cl	α-OH	6b-19b-epoxy	1×10^{1}	N.T.
203	119993	OH	αH			3×10^{0}	B1, LE, PS, WA
204	251698	O-Cy	αH			N.T.	B1, LE, PS
205	277288	O-Gl	αH			3×10^{0}	LE, PS
206	277289	O-Gl-Gl	αH			4×10^{0}	N.T.
207	144150	OH	αH	OH		2×10^{0}	N.T.

[a] Af = acofriose, Av = acevenose, Cy = cymarose, Gl = glucose, Xy = xylose.
[b] Substituents at this position are β where two configurations are possible.
[c] ED_{50} in $\mu g/ml$.
[d] Data are negative unless followed by (MA) = marginal activity.
[e] N.T. = not tested.

TABLE XXVI

Cardenolides without 14-OH Substituents

		Substituents					Screening data	
Compound no.	NSC no.	C3-β	C-5	C-14	C-19	Other	KB[a]	In vivo systems[b]
208	87316	OAc	β-OH	Δ14	oxo		100	N.T.[c]
209	87317		Δ5	Δ14	OEt	3β,19-epoxy	100	N.T.
210	93446	OAc	Δ5	Δ14	oxo		5×10^1	N.T.
211	137447	OAc	β-H	Δ14			N.T.	LE
212	107129	OH	α-H	β-H			N.T.	LE, PS(MA)
213	92955	OAc	5β,6β-epoxy	14α,15α-epoxy			2×10^1	N.T.
214	251673	O-O1[d]	β-H	8β,14β-epoxy			N.T.	B1, LL

[a] ED_{50} in μg/ml.
[b] Data are negative unless followed by (MA) = marginal activity.
[c] N.T. = not tested.
[d] O1 = oleandrose.

and that the bufadienolides are a dead end for development of useful anticancer agents.

C. Withanolides

The withanolides differ from the cardenolides and bufadienolides in two major respects: first, the lactone ring is not attached directly to the steroid ring system and has carbon substituents at positions 24 and 25, and second, the C–D ring fusion is trans, instead of the cis ring fusion seen in the cardenolides and bufadienolides.

The basic structure and numbering system for the withanolides was shown in Fig. 4 (Section IV,A). Table XXVII shows the structures and screening data for the withanolides screened at NCI (compounds **215–227**). It is not possible to develop any structure–activity relationships for this series because of the limited number of compounds screened and because the screening data available are from a number of different tumor systems which cannot be compared to each other. However, some general remarks can be made based on the data available. There are hints of activity in the PS and WA systems which may or may not be significant, since these are among the most sensitive systems used in the NCI screen and by themselves are not predictive for clinical activity. The toxicity of the withanolides in *in vivo* systems is much lower than that of the cardenolides and bufadienolides, which indicates that if a good active compound were discovered it might have a reasonable therapeutic index. The preponderance of data thus far shows no high-level activity, but since a limited amount of screening has been done, particularly in the newer tumor systems, there is a chance that this group might produce a useful compound.

Recently, the outlook has become more promising, since two compounds, NSC 179834 (withanolide E, **220**) and NSC 212509 (4β-hydroxy withanolide E, **221**), have shown good activity in the B16 melanoma system (T/C ~ 170), and NSC 212509 has been selected by the NCI for further development. It is noteworthy that in both of these cases the C-17 side chain has the α orientation instead of the normal β orientation found in most steroids including the bufadienolides, cardenolides, and other withanolides.

V. CUCURBITACINS

Many plants belonging to the family Cucurbitaceae have been known to contain toxic principles since ancient times. More recently there has been some interest in the toxic principles (cucurbitacins) since there have been a number of reports of the toxicity of these compounds to tumor cells in *in vitro* systems and also some indications of antitumor activity *in vivo*. The chemistry of the cucur-

TABLE XXVII

Structures and Screening Data of the Withanolides

		Substituents				
Compound no.	NSC no.	C-4	C-5	C-6	Other	Screening data[b] test systems
215	10108	β-OH	5β,6β-epoxy		27-OH	B1, LE, PS, SA(MA), WA(A), KB(A)
216	135073	β-OH	5β,6β-epoxy		18-OAc	KB(A)
217	179841	β-OH	5β,6β-epoxy		20β-OH	LE, PS
218	179882	β-OH	5β,6β-epoxy		14α-OH	KB
219	273755	β-OH	5β,6β-epoxy		14α-OH; 27-OH	KB
220	179834	H	5β,6β-epoxy		14α-OH; 17β-OH[a]; 20β-OH	B1(A), LE, PS(A)
221	212509	OH	5β,6β-epoxy		14α-OH; 17β-OH[a]; 20β-OH	PS(A), B1(A)
222	109437	H	5β,6β-epoxy		27-OH	WA
223	109439	H	α-OH	β-OH	27-OH	WA
224	285115	H	α-OH 6α,7α-epoxy		12α-OH; 17α-OH, 22[S][a]	PS
225	109438	H	Δ[4,5]	β-OH	27-OH	WA
226	179835	H	Δ[5,6]		14α-OH; 20β-OH	PS(MA)
227	179840	H	Δ[5,6]		14α-OH; 17α-OH; 20β-OH	PS(MA)

[a] Reversed configuration from normal at this position.

[b] Antitumor or cytotoxicity data in the systems tested are negative unless followed by (MA) = marginal activity or (A) = activity.

bitacins as well as the history and pharmacologic activity has been reviewed by Lavie and Glotter (1971). Since the time of that review, there have been several other reports of cytotoxicity of the cucurbitacins to tumor cells in culture (Kupchan *et al.*, 1972b, 1973a, 1978a), and the purpose of this section is to review the cytotoxicity and antitumor data of the cucurbitacins screened at NCI, to comment on structure–activity relationships among them and to evaluate the potential of this class of compounds as antitumor agents.

The characteristic feature of the cucurbitane ring system which differentiates it from other steroids and triterpenes is the presence of the C-19 methyl group at position 9 instead of the usual position 10. The cucurbitane ring system is therefore equivalent to 19 (10→9β)-*abeo*-lanostane. Structure (**228**) shows the numbering and stereochemistry of the cucurbitane nucleus. Commonly found substituents among the cucurbitacins include: (1) oxygenation at positions 2 and 3 in ring A either as diosphenols, α ketols, or diols; (2) a 5,6 unsaturation which is a consequence of the biosynthesis of this group; (3) a ketone at C-11; (4) an α-hydroxyl group either free or esterified at C-16; (5) a β-hydroxyl group at C-20; (6) a ketone at C-22, most usually an α,β-unsaturated ketone with a 23,24 unsaturation; and (7) a hydroxyl group at C-25, either free or acetylated.

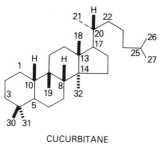

CUCURBITANE

228

Structure–activity correlations with regard to cell cytotoxicity in the KB system have been made on small series of cucurbitacins by Kupchan and co-workers (1967, 1970b) and Kupchan and Tsou (1972).

It was concluded that the α,β-unsaturated ketone in the sidechain was important for maximum cytotoxicity and that, further, a free 16α-hydroxyl could possibly hydrogen-bond with the C-24 ketone to increase the susceptibility of the α,β-unsaturated ketone to nucleophilic attack. This was based on the idea that the biological activity of the cucurbitacins could be at least partly accounted for by alkylation of biological nucleophiles through Michael-type addition to the unsaturated ketone.

The cucurbitacins screened at the NCI have been arranged into several groups based on structural similarities, particularly with regard to the A ring substituents. Group A contains those cucurbitacins having a diosphenol group in ring A (Fig. 7). The cytotoxicity and *in vivo* antitumor data for all the cucurbitacins

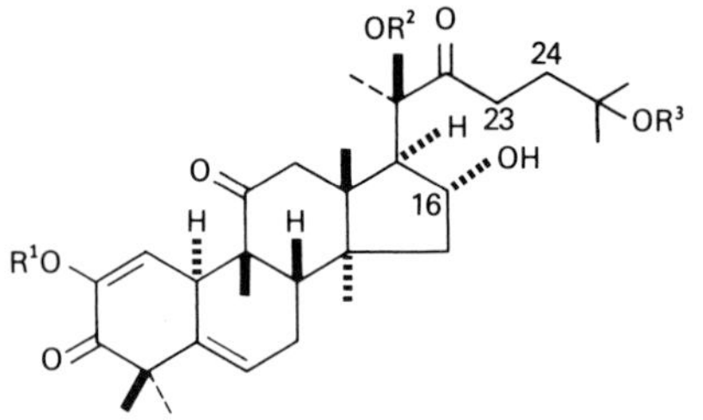

Compound no.	Name	R^1	R^2	R^3	Other
229	Cucurbitacin I	H	H	H	Δ^{23-24}
230	2-O-Isopropyl cucurbitacin I	i-C_3H_7	H	H	Δ^{23-24}
231	2-O-Butyl cucurbitacin I	n-C_4H_9	H	H	Δ^{23-24}
232	2-O-Glucosyl cucurbitacin I	$C_6H_{11}O_5$	H	H	Δ^{23-24}
233	Datiscacin	H	Ac	H	Δ^{23-24}
234	Cucurbitacin E	H	H	Ac	Δ^{23-24}
235	Cucurbitacin L	H	H	H	
236	Cucurbitacin J[a]	H	H	H	24-OH
237	Cucurbitacin K[a]	H	H	H	24-OH

[a] Stereochemistry at C-24 unknown.

Figure 7. Cucurbitacin derivatives.

studied are found in Table XXVIII. The extremely high cytotoxicity of cucurbitacin E (**234**) (10^{-8} μg/ml) indicates that this compound contains all the correct functionalities for optimal *in vitro* activity. Note that cucurbitacin I (**229**), which is identical to cucurbitacin E except for the absence of the C-25 acetate, is much less cytotoxic ($\sim 10^{-3}$ μg/ml), so the C-25 acetate is needed for maximum cytotoxicity. Cucurbitacins J, K, and L (**235–237**), which lack the α,β-unsaturated ketone in the sidechain, are 100-fold less cytotoxic than cucurbitacin I. Datiscacin (**233**), which is the 20-acetate of cucurbitacin I (**229**), is less cytotoxic, bearing out Kupchan's observation that the free hydroxyl at 16 is needed for maximum activity. The comparison of the cytotoxicity of datiscacin (10^{-2} μg/ml) with that of cucurbitacin E (10^{-8} μg/ml) is instructive, since both are monoacetates. The three compounds in Group A which have substitutions on the 2-hydroxyl group (**230–232**) have not been tested for cytotoxicity, but the *in vivo* toxicity data show that these compounds are essentially devoid of toxicity up to high doses, and therefore one can predict that these compounds would not be cytotoxic. This implies that the free diosphenol system is needed for maximum cytotoxicity.

Group B contains those compounds having a 2β-hydroxy-3-ketone in ring A (Fig. 8). The most cytotoxic compound in this group is cucurbitacin B (**241**)

TABLE XXVIII

Cytotoxicity and In Vivo Anticumor Activity of the Cucurbitacins

Compound no.	NSC no.	Name	KB, ED_{50} (μg/ml)	*In vivo* system — System tested[a]	Toxic dose (mg/kg)[b]
229	521777	Cucurbitacin I	6×10^{-3}	LE, PS, WA	2.0
230	177855	2-O-Isopropylcucurbitacin I	N.T.[c]	LE	>150.0
231	177857	2-O-Butylcucurbitacin I	N.T.	LE	>200.0
232	177857	2-O-Glucosylcucurbitacin I	N.T.	LE	200.0
233	251679	Datiscacin	2×10^{-2}	PS	>4.0
234	106399	Cucurbitacin E	5×10^{-8}	B1, CD, DL, LE, LL(MA), PS, SA, WA, 8P	10.0
235	112167	Cucurbitacin L	3×10^{-1}	PS, WA	12.5
236	112165	Cucurbitacin J	2×10^{1}	PS(MA), WA	4.5
237	112166	Cucurbitacin K	5×10^{-1}	PS, WA	18.0
238	521776	Cucurbitacin D	2×10^{-3}	LE, PS, WA, 8P	2.0
239	144154	*Datisca* principle B	3×10^{-2}	PS, WA(MA)	3.0
240	144153	Datiscoside	2×10^{-1}	B1, LE, LL, PS, WA(MA)	2.5
241	49451	Cucurbitacin B	2×10^{-6}	B1, CA, CD, C6, DL, LE, LL, P4, SA, WA, 8P, PS(MA)	2.0
242	106401	Dihydrocucurbitacin B	2×10^{-3}	WA	>4.0
243	49452	Fabacein	1×10^{-2}	LE, SA, WA, PS(MA)	50.0
244	94743	Cucurbitacin A	1×10^{-3}	DL, LE, LL, P4, SA, WA, 8P	0.5
245	106400	Isocucurbitacin B	4×10^{-1}	N.T.	
246	305982	3-Epi-isocucurbitacin D	2×10^{-1}	PS(MA)	>2.0
247	305981	Isocucurbitacin D	3×10^{-2}	PS	>2.0
248	251680	Cucurbitacin F	2×10^{-1}	PS, WA(MA)	4.0
249	135074	Cucurbitacin P	5×10^{-1}	N.T.	
250	135075	Cucurbitacin Q	3×10^{-2}	N.T.	
251	212564	Bryogenin	N.T.	PS	>400.0
252	94744	Cucurbitacin C	1×10^{-3}	DL, LE, LL, PS, P4, SA 8P, WA(MA)	4.0

[a] Activity is negative for all systems named unless followed by (MA) indicating marginal activity.

[b] Typical toxic dose in tumored animals. The symbol > indicates that this was the highest dose tested and was nontoxic.

[c] N.T. = not tested.

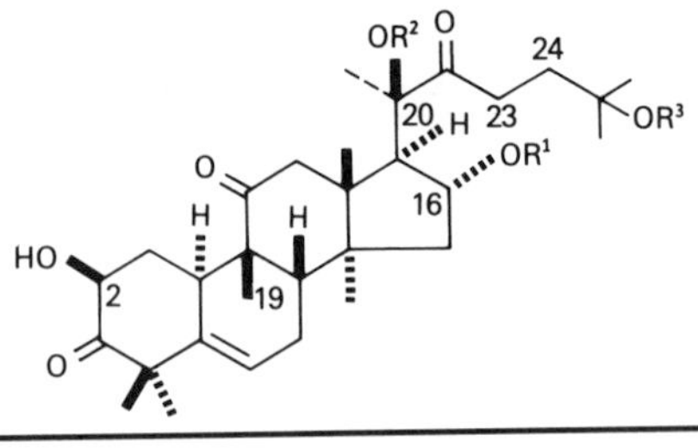

Compound no.	Name	R^1	R^2	R^3	Other
238	Cucurbitacin D	H	H	H	$\Delta^{23,24}$
239	Datisca principle B	H	Ac	H	$\Delta^{23,24}$
240	Datiscoside	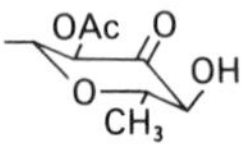 H	H	H	$\Delta^{23,24}$
241	Cucurbitacin B	H	H	Ac	$\Delta^{23,24}$
242	Dihydrocucurbitacin B	H	H	Ac	
243	Fabacein	Ac	H	Ac	$\Delta^{23,24}$
244	Cucurbitacin A	H	H	Ac	$\Delta^{23,24}$; 19-OH

Figure 8. Cucurbitacin derivatives.

$(10^{-6}\ \mu g/ml)$, which again has an acetate at C-25 and is approximately 1000 times as cytotoxic as cucurbitacin D (**238**), which lacks this group. Reduction of the unsaturation on the sidechain of cucurbitacin B leads to dihydrocucurbitacin B (**242**), which is much less cytotoxic than its parent. Fabacein (**243**), which is the 16-acetate of cucurbitacin B, is four orders of magnitude less cytotoxic, pointing up again the need for a free hydroxyl group at position 16. This is also noted in the relative cytotoxicities of datiscoside (**240**) $(10^{-1}\ \mu g/ml)$ and cucurbitacin D (**238**) $(10^{-3}\ \mu g/ml)$. *Datisca* Principle B (**239**), the 20-acetate of cucurbitacin D, is less active than its parent, and this is also observed in the Group A compounds comparing datiscacin with cucurbitacin I (**229**). Therefore the presence of acetates at C-20 reduce cytotoxicity. Cucurbitacin A (**244**) is identical to cucurbitacin B (**241**), except that C-19 is a hydroxymethyl instead of a methyl group and there is a large difference in cytotoxicity, indicating that this modification reduces activity.

Group C (Fig. 9) contains those compounds (**245–247**) which have a 2-keto-3-hydroxy system (isocucurbitacins), and these are clearly much less cytotoxic than their corresponding parents.

Group D (Fig. 9) consists of cucurbitacins F, P, and Q (**248–250**), which all contain 2,3-diols in ring A. Again these are relatively less cytotoxic than the corresponding compounds with either diosphenol or 2β-hydroxy-3-ketone groups in ring A.

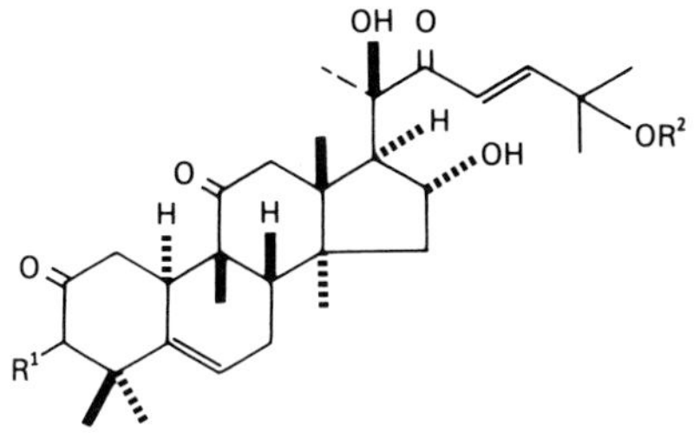

Compound no.	Name	R^1	R^2
245	Isocucurbitacin B	α-OH	Ac
246	3-epi-isocucurbitacin D	β-OH	H
247	Isocucurbitacin D	α-OH	H

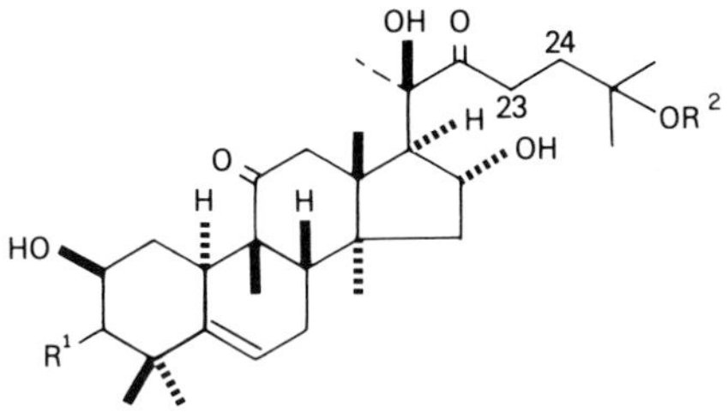

Compound no.	Name	R^1	R^2	Other
248	Cucurbitacin F	α-OH	H	Δ^{23-24}
249	Cucurbitacin P	β-OH	H	
250	Cucurbitacin Q	β-OH	Ac	Δ^{23-24}

Figure 9.   Curcurbitacin derivatives.

The last two compounds to be discussed, bryogenin (**251**) and cucurbitacin C (**252**) (Fig. 10), both have only one oxygen substituent in ring A, a 3β-hydroxyl group. Bryogenin was not tested in cell culture but is nontoxic at 400 mg/kg *in vivo*. This compound lacks so many of the typical active features of the other cucurbitacins that it cannot be effectively compared. Cucurbitacin C (**252**) is fairly cytotoxic ($\sim 10^{-3}$ μg/ml) and has the same order of cytotoxicity as cucurbitacin A which is identical except that the latter has an α ketol in Ring A. The presence of a compound with only one ring A oxygen function with such high cytotoxicity is difficult to correlate with the idea that a diosphenol or α ketol is an absolute requirement for cytotoxicity.

The summary of these data is that the following substituents are needed for

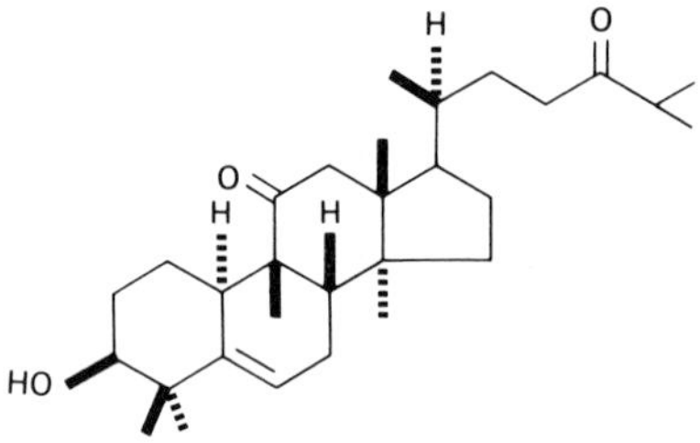

251   BRYOGENIN

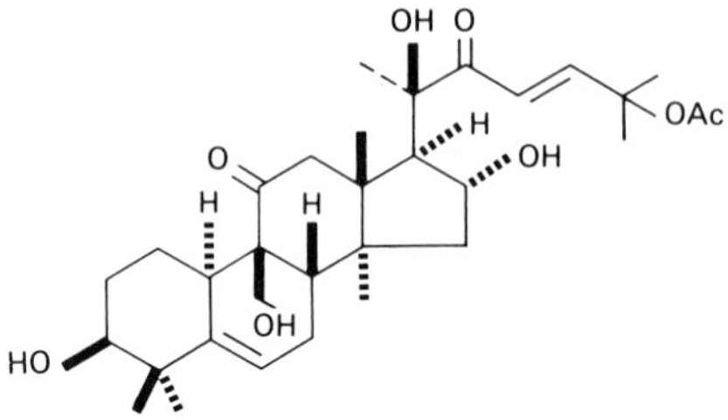

252   CUCURBITACIN C

Figure 10.   Cucurbitacin derivatives.

maximum cell cytotoxicity in the KB system: (1) Ring A—either a diosphenol or a 2β-hydroxy-3-ketone; (2) C-16α—a free hydroxyl group is needed; (3) C-20β—a free hydroxyl group is needed; (4) C-22 to C-24—an α,β-unsaturated ketone is required; (5) C-25—an acetate is much more active than a free hydroxyl group. When the *in vivo* data on the cucurbitacins (Table XXVIII) are examined, it can be seen that they are inactive in a wide variety of tumor systems with only occasional indications of marginal activity. The activities when seen are not dramatic and are at doses very close to the toxic dose of the compound. Since a fairly large number of compounds representing the common structural types in this series have been tested and found to present no possibilities for further development as antitumor agents, it must be concluded that there is no correlation between the high cytotoxicity of these compounds and their potential as antitumor agents. Unless new and very different data come to light, it must be considered that cucurbitacins are a dead end for further antitumor studies.

VI. QUASSINOIDS

The quassinoids, sometimes called simaroubolides, are a group of lactonic compounds biogenetically derived by degradation of triterpenoids and found in the family Simaroubaceae. These were originally discovered as the bitter principles of this family and have been shown to possess antiamoebic activity (Geissman, 1964) and more recently antitumor activity (Wall and Wani, 1970; Kupchan *et al.*, 1973b; Kupchan and Lacadie, 1975; Kupchan and Streelman, 1976; Wani *et al.*, 1977, 1978; Seida *et al.*, 1978). The chemistry of the quassinoids has been reviewed in an excellent article by Polonsky (1973). There are three types of quassinoid skeletons as shown in Fig. 11, but only compounds with type I skeletons have demonstrated antileukemic activity and types II and III will not be considered further. The compounds to be discussed are all named according to the current rules of nomenclature for the picrasane nucleus. The numbering and standard stereochemistry of this system are shown in Fig. 12, and all structures shown have the indicated stereochemistry unless otherwise noted. The previously used numbering system regarded the quassinoids of type I as degraded triterpenes, with the result that the methyl group at C-8 was numbered as C-30 instead of C-20 as in the present usage.

The type I quassinoids can be further divided into three groups (A, B, and C) as in Fig. 12, where the division is made on the basis of the presence or absence of an additional ring in the form of a hemiketal (Group A), an ether (Group B), or no additional ring (Group C). This is a somewhat arbitrary division, but the substituent patterns tend to be fairly consistent within each group, which makes the use of this classification advantageous for structure–activity analysis.

A. Group A Quassinoids

The quassinoids in Group A (Fig. 13) are all characterized structurally by the presence of an oxide bridge between positions 11 and 20. The major structural differences apparent in this group relate to the presence or absence of

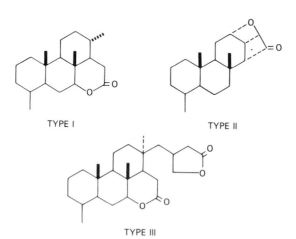

TYPE I   TYPE II

TYPE III

Figure 11.   Types of quassinoid skeletons.

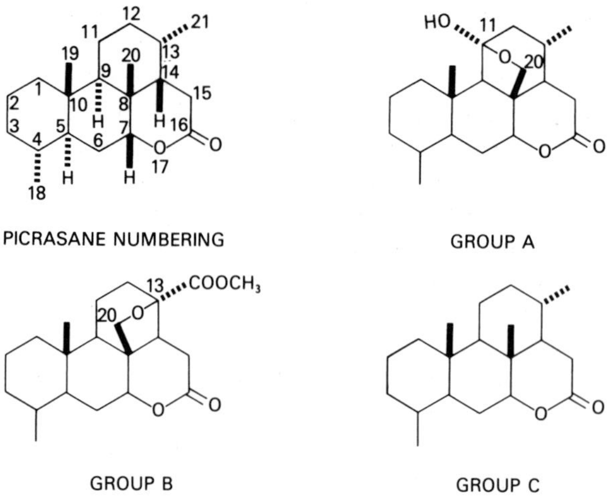

PICRASANE NUMBERING   GROUP A

GROUP B   GROUP C

Figure 12. Picrasane numbering and structural groups.

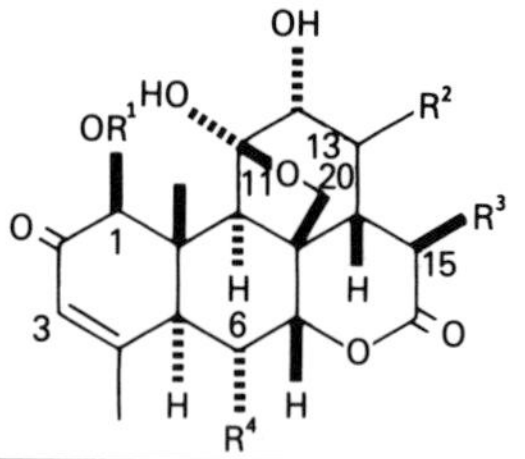

Compound no.	R^1	R^2	R^3	R^4
253	H	α-Me	H	H
256	H	α-Me	OH	H
257	H	$=CH_2$	OH	H
260	H	α-Me	OAc	H
261	H	α-Me	O(CO)C(Me)(OH)Et	H
262	H	α-Me	O(CO)C(Me)(OAc)Et	H
263	H	$=CH_2$	O(CO)C(Me)(OH)Et	H
264	H	α-Me	O(CO)C$_6$H$_5$	H
265	H	α-Me	O(CO)C(Me)=CH Me	H
270	H	α-Me	O(CO)CH(Me)Et	H
271	H	$=CH_2$	O(CO)CH(Me)Et	H
272	Me	$=CH_2$	O(CO)CH(Me)Et	H
273	H	α-Me	OAc	O(CO)C(Me)=CH Me
274	H	α-Me	H	O(CO)CH=C(Me)$_2$
275	H	α-Me	H	O(CO)C(Me)=CH Me
276	Ac	α-Me	H	O(CO)CH=C(Me)$_2$

Figure 13. Group A quassinoids.

an ester substitutent (either at position 6 or 15), the presence of an α-methyl or a methylene group at 13, and the oxidation state of positions 1, 2, and 3.

The screening data on these compounds are presented in Table XXIX. The data represent a composite of the available data for any given compound. The screening was performed over a period of more than 10 years under NCI auspices in a number of different laboratories, and the data are not as reliable as if the screening had all been performed by a single laboratory in the same control run. Also, limitations of the quantities of compounds available prevented complete testing in many cases, so where there is no toxic dose indicated in the table it may be assumed that the compound might have had more activity if tested at higher doses. Nonetheless, the screening data represent a much larger sampling of this series of compounds that has been reported previously, and the data on closely related compounds within this series show good internal consistency.

Those compounds which do not contain an ester sidechain (compounds **253, 256, 257**) all are negative in the P-388 leukemia (PS) or show weak activity at high doses (>8 mg/kg). When the compounds with C-15 esters are examined (**260–265; 270–273**), several differences become apparent. Compounds (**260–265**), (**270**), (**271**), and (**273**) all show strong or moderate activity (T/C ≥ 150) at doses in the .25–5 mg/kg range, and are both more potent and more active than the compounds without ester groups. Interestingly enough, compounds (**266**) and (**267**) (Fig. 14), which have a hydroxyl group at C-2 instead of the ketone found in the other members of this group, are inactive at these doses, which indicates that the 3-ketone is important for activity. Compounds (**268**) and (**269**), which do not possess the 3,4 unsaturation in ring A, are active but only at much higher doses, suggesting that the unsaturated ketone in ring A is required for optimal potency. Compound (**272**) is inactive at the doses tested, suggesting that a free 1-hydroxyl is important but the doses tested are too low to draw firm conclusions on this point. Compounds (**273–276**) are moderately to highly active; (**277**) and **278**) are inactive as tested. Note that 1-acetoxy-6α-senecioyloxychaparrinone (**276**) has comparable activity to its unacetylated parent (**274**) but at a 10-fold higher dose. It thus appears that substitution on the C-1 alcohol reduces potency in this series. From a comparison of compounds (**276**) and (**277**), it appears that the free hydroxyl at C-12 may be important, but this one example is insufficient to prove this point. Compound (**278**), which lacks the oxide bridge altogether, is inactive, but whether this is due to the opening of the cyclic hemiketal or to the presence of the 1,12-diacetate is unknown.

From examination of the data in Table XXIX, there is no apparent difference in activity between those Group A compounds having C-13 α-methyl groups and those having methylene groups at this position.

When the cytotoxicity to KB cells is examined, it can be seen that there is at least some correlation between cytotoxicity and potency *in vivo* since all of those compounds with cytotoxicity in the 10^{-2}–10^{-3} range have optimal doses in the

TABLE XXIX

Antileukemic Activity (P-388 Leukemia) and Cytotoxicity (KB) of Group A Quassinoids

Compound no.	NSC no.	Common name	Dose range (mg/kg)	Optimal T/C	Optimal dose (mg/kg)	Lowest toxic dose (mg/kg)	KB ED$_{50}$ (μg/kg)
253	288754	Chaparrinone	40–5.0	145	40.0	—	4×10^{-1}
254	079404	Chaparrin	10–1.25	92	5.0	—	N.T.[a]
255	132792	Ailanthone	10–1.25	111	2.5	10.0	N.T.
256	126764	Glaucarubolone	12–1.25	150	8.0	—	4.3×10^{-1}
257	238189	Δ13-Dehydroglaucarubolone	1–0.06	106	0.25	—	2×10^{0}
258	305726	Dihydroglaucarubolone	N.T.				6.5×10^{1}
259	014974	Glaucarubol	200–25.0	100	25	100.0	N.T.
260	126765	Holacanthone	13–0.13	190	4.0	16.0	2×10^{-1}
261	132791	Glaucarubinone	10–0.07	177	0.25	1.0	4×10^{-2}
262	194251	Glaucarubinone-2'-acetate	8–0.25	157	1.25	8.0	5×10^{-3}
263	279504	Δ13-Dehydroglaucarubinone	16–0.25	154	2.0	4.0	N.T.
264	283453	Glaucarubolone-15-benzoate	32–0.06	201	2.0	4.0	2×10^{-1}
265	305727	Glaucarubolone-15-tiglate	8–0.03	147	2.0	—	2×10^{-1}
266	297898	Glaucarubine-2'-acetate	4–0.25	103	0.25	—	N.T.
267	014975	Glaucarubine	30–1.0	127	2.0	15.0	N.T.
268	305725	Dihydroglaucarubolone-15-benzoate	32–0.12	171	32.0	—	5.7×10^{0}
269	305724	Dihydroglaucarubinone	32–0.12	152	16.0	—	4.6×10^{1}
270	238187	Ailanthinone	8–0.25	148	2.0	—	3×10^{-2}
271	238188	Δ13-dehydroailanthinone	8–0.06	184	0.5	4.0	1×10^{-2}
272	239070	Δ13-dehydroailanthinone methyl ether	2–0.06	120	1.0	—	$>1 \times 10^{1}$
273	267709	Undulatone	5–0.5	163	5.0	—	3×10^{-1}
274	290494	6α-Senecioyloxychaparrinone	8–0.25	198	1.0	4.0	7×10^{-3}
275	281271	6α-Tigloyloxychaparrinone	1.2–0.08	163	0.6	—	2×10^{-2}
276	295151	1-Acetoxy-6α-senecioyloxy-chaparrinone	10–0.5	177	10.0	—	N.T.
277	301732	6α-Senecioyloxychaparrinone 1,12-diacetate	10–1.25	107	10.0	—	N.T.
278	295152	6α-Senecioyloxychaparrinone 1,12,20-triacetate	8–0.05	102	8.0	—	N.T.

[a] N.T. = not tested.

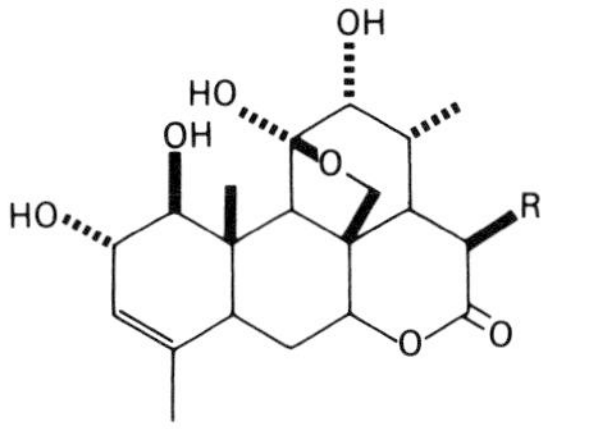

254 R=H
259 R=OH
266 R=[S]O(CO)C(Me)(OAc)Et
267 R=[S]O(CO)C(Me)(OH)Et

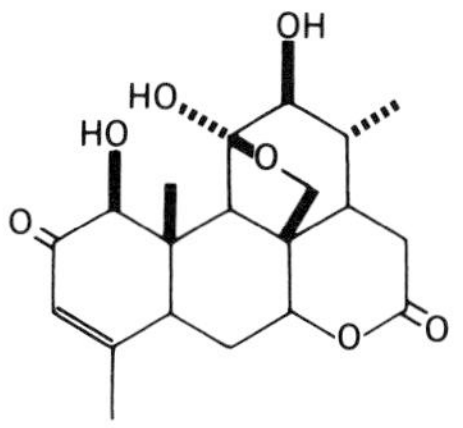

255

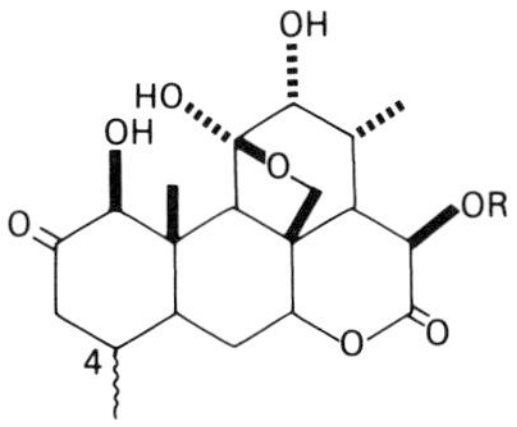

258 R=H
268 R=COC₆H₅
269 R=[S]COC(Me)(OH)Et

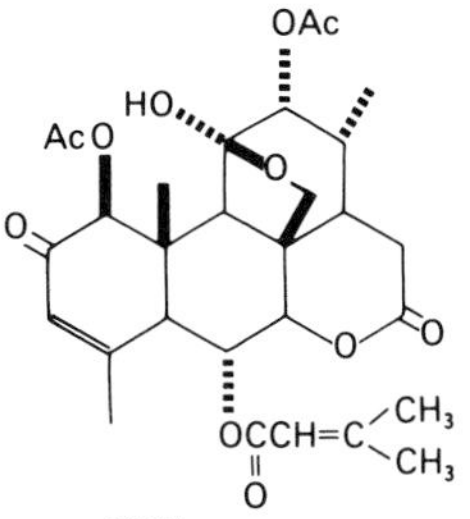

277

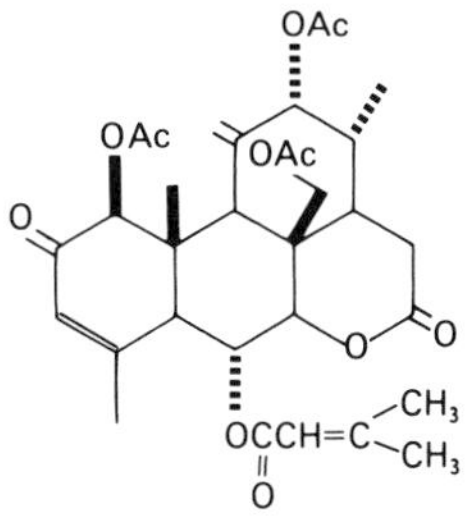

278

Figure 14.   Group A quassinoids.

PS system below 2.0 mg/kg. Those compounds having a modified A ring, either without the 3,4 unsaturation (**258, 268, 269**) or having the C-1 hydroxyl blocked (**272**) are either much less cytotoxic or show no significant cytotoxicity. Since some of these compounds are active *in vivo* at higher doses (**268, 269**), cytotoxicity is probably not a good screen for absolute *in vivo* activity, but is rather related more to potency.

The most interesting information to be gathered from the data is that compounds with esters at C-15β or C-6α both show potent activity against the P-388 leukemia. When one examines models of these compounds, it is apparent that esters at these two positions are both equatorial and would not sterically obstruct binding to either the α or β face of the molecule. It would be highly interesting to prepare esters of reversed configuration at these positions, as such esters would obstruct the β face of the molecule (6β-esters) or the α face of the molecule (15α-esters) and might provide some insight into which groups in the molecule are involved in receptor binding.

One possible role of the esters is in transport, since lipophilic esters have been seen to greatly affect *in vivo* antileukemic activity in other natural products (e.g., maytansine).

Based on the structure–activity relationships developed thus far, several potential modifications of this series suggest themselves. These include:

a. Substitution of longer chain fatty acid esters such as myristate, palmitate, stearate, etc. for the esters at 15β or 6α.

b. Preparation of compounds with esters at both 6α and 15. Thus far undulatone (**273**) is the only example of this group.

c. Preparation of compounds with esters at C-6α or C-15β, which are resistant to hydrolysis either from steric or electronic considerations.

d. Further investigation of the role of the functional groups in ring A. Since the dihydro derivatives of glaucarubinone (**269**) and glaucarubolone-15-benzoate (**268**) are active (although at higher doses), the question of whether the α-ketol present in these molecules is converted to a diosphenol is of interest. This could be answered by preparation of compounds without the hydroxyl group at C-1. Likewise, preparation of C-1 ketones or α-hydroxyl compounds should shed light on the role of the C-1β hydroxyl group. Both the Group A quassinoids in this section and the Group B quassinoids discussed in the next section have two adjacent oxygen functionalities in ring A, and an obvious question is: What would be the activity of a compound with only one oxygen function such as a $\Delta4,5$-3-ketone?

e. The presence of the 12α-OH moiety in both Group A and Group B quassinoids is noteworthy. This is a tempting position to explore. Based on the difference in reactivity of the C-1 and C-12 hydroxyl groups, it should be possible to start with a compound such as chapparinone (**253**) and selectively prepare small-, medium-, and long-chain esters at either C-1 or C-12 to determine if compounds with esters at either of these positions would have comparable activity to those compounds with esters at C-15 or C-6.

Kupchan and co-workers (1976b) examined a more limited group of glaucarubolone esters than considered here and concluded that: (a) an ester group was needed for maximal *in vivo* activity but the nature of the ester had little effect on activity; (b) the probable role of the ester was in biological transport; and (c) the presence of an unsaturation at the 3,4 position is advantageous for optimal activity.

Wall and Wani (1977) concurred in these observations as a result of data on related compounds.

The data presented in Table XXIX lead to the same conclusions but also allow further extension of the structure–activity relationships to point out that: (a) the 3-ketone is necessary for maximal activity; (b) derivatization of the hydroxyl group at position 1 decreases potency; and (c) activity is unaffected by change of the C-13 methyl to a methylene group. Further studies of structure–activity relationships in this series should be concerned with the role of the two adjacent oxygen atoms in ring A, the need for the 12α-hydroxyl group, and extension of the role of the esters to longer-chain esters, further variance of the position of the esters, and preparation of esters resistant to hydrolysis.

B. Group B Quassinoids

The Group B quassinoids (Figs. 15 and 16), which possess a 13–20 oxide linkage, can be considered for the most part as variants of bruceolide (**279**), which possess oxygen functions at positions 2 and 3 and a 3,4 unsaturation. The exceptions are isobruceine A (**291**), isobruceine B (**292**), samaderine E (**296**), quassamarin (**294**), and simalikalactone D (**295**), which contain the same 1-hydroxy-2-keto-Δ3,4 system in ring A as the Group A quassinoids.

Compounds (**279–285**) in Table XXX show the effect of variation of the ester side chain at C-15 on antileukemic activity. Bruceolide (**279**) is much less active than bruceantin (**282**), bruceantinol (**283**), bruceantarin (**284**), or dihydrobruceantin (**285**), and this can be attributed to lack of the ester sidechain. Likewise, bruceine B (**280**), which has an acetyl sidechain, is inactive at the doses tested. Brusatol (**281**) is inactive at doses where bruceantin is highly active and this is rather surprising, since the sidechains in these compounds are both highly substituted unsaturated acids. The P-388 data on brusatol have been repeated and there is no reason to doubt its accuracy, so one may surmise that, while the Group A quassinoids, the nature of the sidechain is more critical in the brusatol–bruceantin series. This is further supported by the somewhat decreased activity seen in bruceantarin (**284**), which is a benzoate ester, and dihydrobruceantin (**285**), where the sidechain is saturated when compared to bruceantin (**282**). The importance of the nature of the sidechain ester is demonstrated in the work of Liao et al. (1976), in which a number of compounds in this series were studied with regard to their ability to inhibit protein synthesis in *in vitro* systems. These workers established that bruceantin acts through inhibition of the initiation phase of protein synthesis and found that the order of inhibition in the compounds tested was bruceantin ≃ dihydrobruceantin > bruceantarin >> bruceine B >>> bruceolide > tetrahydrobruceantin. Bruceolide and tetrahydrobruceantin were essentially without useful effect and bruceine B was only somewhat effec-

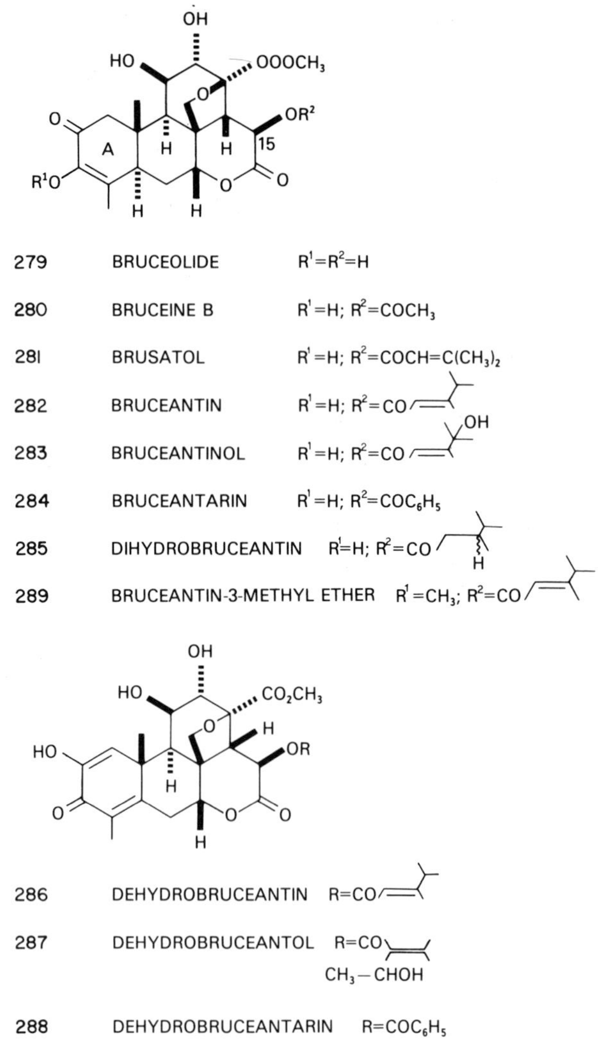

279	BRUCEOLIDE	$R^1=R^2=H$
280	BRUCEINE B	$R^1=H; R^2=COCH_3$
281	BRUSATOL	$R^1=H; R^2=COCH=C(CH_3)_2$
282	BRUCEANTIN	$R^1=H; R^2=CO$
283	BRUCEANTINOL	$R^1=H; R^2=CO$
284	BRUCEANTARIN	$R^1=H; R^2=COC_6H_5$
285	DIHYDROBRUCEANTIN	$R^1=H; R^2=CO$
289	BRUCEANTIN-3-METHYL ETHER	$R^1=CH_3; R^2=CO$

286	DEHYDROBRUCEANTIN	$R=CO$
287	DEHYDROBRUCEANTOL	$R=CO$ CH_3-CHOH
288	DEHYDROBRUCEANTARIN	$R=COC_6H_5$

Figure 15. Group B quassinoids.

tive at high molar concentrations relative to bruceantin, since it required 16 times the molar concentration of bruceantin for 50% inhibition of protein synthesis. The protein synthesis inhibition data appear to correlate much better with the antileukemic activity of this series than does the KB cell cytotoxicity assay.

The ring A dehydro compounds (**286–288**) show only weak antileukemic activity, indicating that this modification of ring A is undesirable. Likewise, bruceantin-3-methyl ether (**289**) is only marginally active at much higher doses

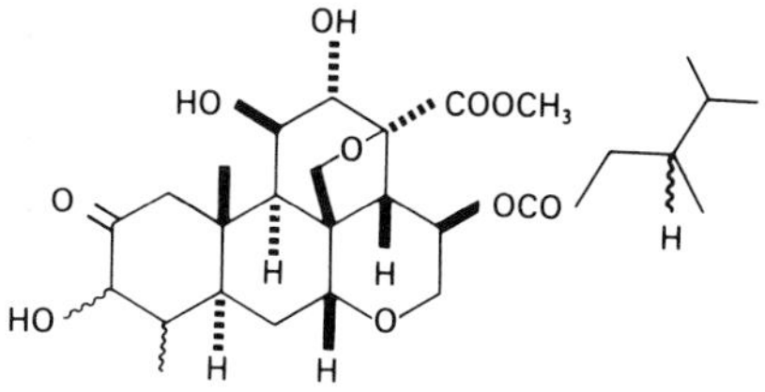

290 TETRAHYDROBRUCEANTIN

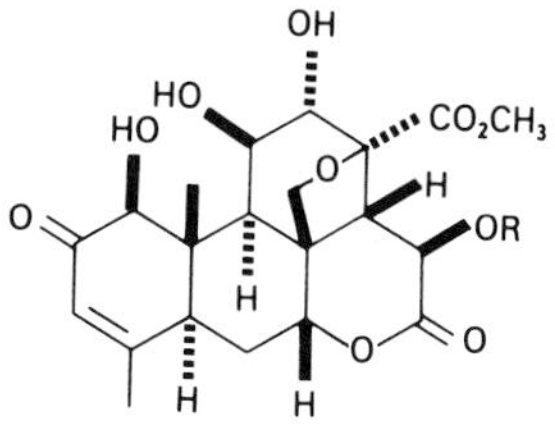

291 ISOBRUCEINE A R=$COCH_2CH(CH_3)_2$
292 ISOBRUCEINE B R=$COCH_3$

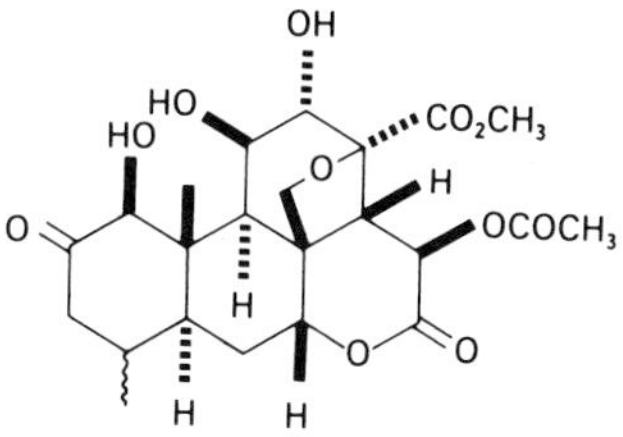

293 DIHYDROISOBRUCEINE B

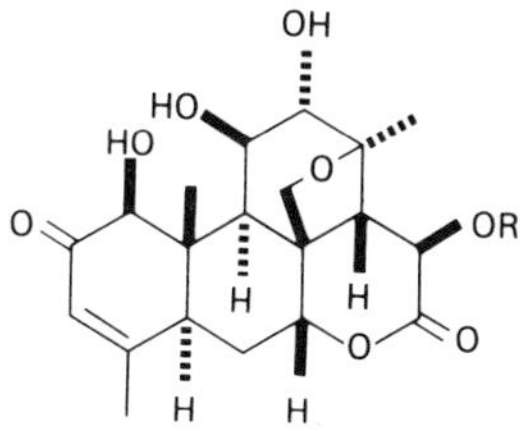

294 QUASSIMARIN R=$COC(OAc)(CH_3)C_2H_5$
295 SIMILIKALACTONE D R=$COCH(CH_3)C_2H_5$

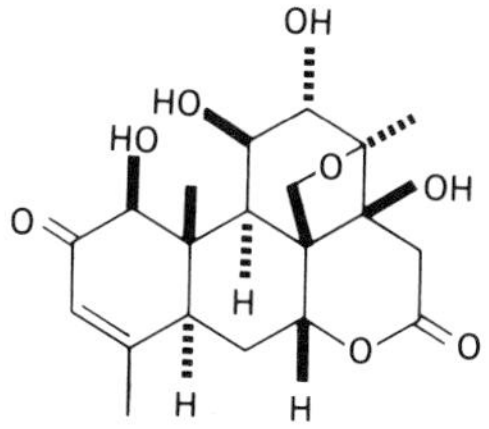

296 SAMADERINE E

Figure 16. Group B quassinoids.

than bruceantin, leading to the conclusion that derivatization at this position causes loss of activity. It appears thus far that the intact A ring without modifications is needed for best activity in the brusatol series. Further support for this comes from tetrahydrobruceantin (**290**), which shows only minimal activity at high doses against the P-388 leukemia.

The remaining compounds in Group B (**291**–**296**) have the same arrangement of functional groups in ring A as in the Group A quassinoids. Isobruceine A (**291**) is more active than isobruceine B (**292**), and this is expected from the more lipophilic sidechain present. Dihydroisobruceine B (**293**), in which the 3,4 unsaturation is not present, shows the same order of cytotoxicity as its parent, but unfortunately no *in vivo* data are available for this compound.

Antileukemic Activity and Cytotoxicity of Group B Quassinoids

Compound no.	NSC no.	Common name	Dose range (mg/kg)	Optimal T/C	Optimal dose (mg/kg)	Lowest toxic dose (mg/kg)	KB, ED$_{50}$ (µg/kg)
279	238185	Bruceolide	4–0.25	147	4.0	10.0	3×10^{-2}
280	132793	Bruceine B	10–1.25	115	5.0		N.T.
281	172924	Brusatol	4–0.125	104	0.5	2.0	N.T.
282	165563	Bruceantin	4–0.007	220	0.5		6×10^{-1}
283	238177	Bruceantinol	2–0.015	238	1.0		6×10^{-2}
284	175399	Bruceantarin	2–0.03	150	2.0		4×10^{-2}
285	238186	Dihydrobruceantin	8–2.0	176	4.0	8.0	3×15^{-3}
286	238178	Dehydrobruceantin	2–0.03	135	0.125		$>1 \times 10^{0}$
287	238180	Dehydrobruceantol	N.T.[a]				3×10^{-1}
288	238179	Dehydrobruceantarin	4–0.04	133	0.4		2×10^{-1}
289	305729	Bruceantin-3-methyl ether	16–0.06	139	16.0		2×10^{1}
290	305728	Tetrahydrobruceantin	16–0.06	128	16.0		4×10^{-1}
291	279503	Isobrucein A	4.0–0.5	163	2.0	4.0	N.T.
292	238181	Isobruceine B	4–0.015	140	1.0	2.0	4×10^{-2}
293	238182	Dihydroisobruceine B	N.T.				2×10^{-2}
294	266493	Quassimarin	4.0–0.06	180	0.06	4.0	3×10^{-3}
295	266494	Simalikalactone D	4.0–0.06	198	1.0		3×10^{-3}
296	269442	Samaderine E	4.0–0.5	156	0.5		N.T.

[a] N.T. = not tested.

Quassimarin (**294**) and simalikalactone D (**295**) are essentially similar to glaucarubinone-2'-acetate (**262**) and ailanthinone (**270**) from Group A, except that the oxide bridge is from C-13 to C-20 instead of from C-11 to C-20. This may be a favorable modification since the activity of the former two compounds (**294, 295**) is somewhat greater than in the latter two (**262, 270**).

The last compound in the Group B quassinoids is samaderine E (**296**), in which the hydroxyl group is at 14 instead of 15. This compound is active but not highly so although since there is no ester group present, strong activity would not be expected. The mere fact that (**296**) is active indicates that there is potential in C-14 esters of this compound.

The conclusions to be drawn from the Group B quassinoid data are that in the brusatol–bruceantin series, changes in ring A are not tolerated and the nature of the sidechain ester is much more critical than in the Group A compounds. In those compounds which have the Group A type of A ring (**291–296**), antileukemic activity is equivalent to or slightly better than comparable Group A compounds, demonstrating that the orientation of the oxide bridge is not critical for activity.

C. Group C Quassinoids

The quassinoids of Group C (Fig. 17) which do not possess an oxide bridge are all inactive in the tumor systems tested. This can not be attributed to the lack of an oxide bridge, however, since there are other structural features in the examples tested in the NCI program which would result in the prediction of inactivity. Quassin (**297**) and simalikalactone C (**298**) have methoxyl groups in ring A. Picrasin B (**299**) and 6-hydroxy picrasin B (**300**) lack an unsaturation in ring A. Glaucarubol pentaacetate (**301**) and chaparrin triacetate (**302**) both lack a ring A carbonyl group. Since none of these compounds have either of the two ring A systems associated with antileukemic activity which were discussed in the Group A and B compounds, no judgment can be made on the need for an oxide bridge in the quassinoid series at this time.

D. Summary

From the structure–activity relationships discussed among the Group A and Group B quassinoids, it is clear that there are a number of variants of the basic picrasane system which retain activity and certain variations which are incompatible with antileukemic activity. Those features that appear essential are found in Fig. 18 and should be taken as the starting point for development of analogues.

Bruceantin, which thus far is the most consistently active quassinoid over the broadest dose range, has shown activity in the Colon 38, L-1210 leukemia, and

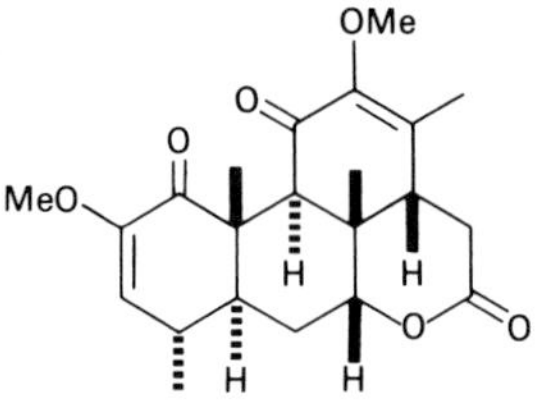

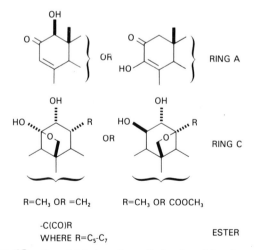

297 QUASSIN

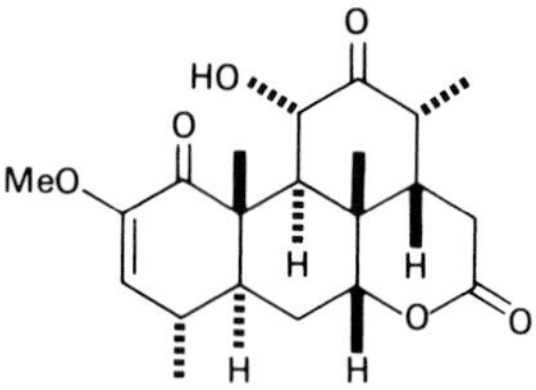

298 SIMALIKALACTONE C

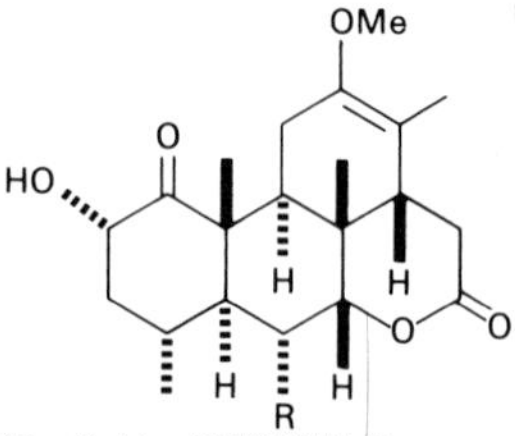

299 R=H PICRASIN B
300 R=OH 6-HYDROXY-PICRASIN B

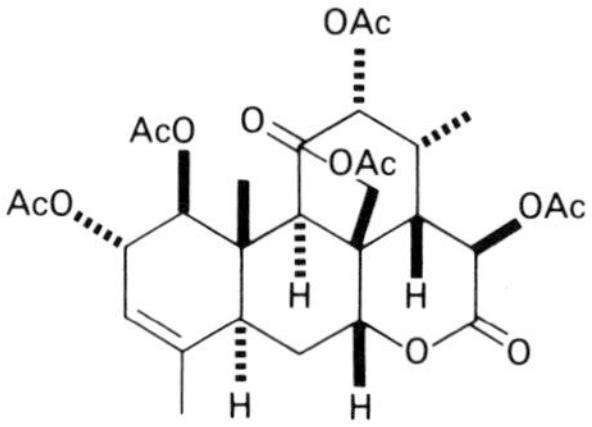

301 GLAUCARUBOL PENTAACETATE

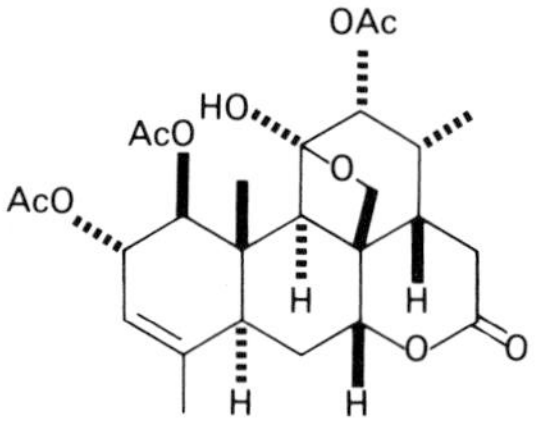

302 CHAPPARRIN TRIACETATE

Figure 17. Group C quassinoids.

OH

OR   RING A

OH   OH

OR   RING C

R=CH₃ OR =CH₂   R=CH₃ OR COOCH₃

-C(CO)R
WHERE R=C₅-C₇   ESTER

Figure 18. Essential features for antileukemic activity of quassinoids.

B16 melanoma systems in mice in addition to the P-388 leukemia, and is now being studied in phase II clinical trials in man. No definitive results are available as yet, but it is clear that the quassinoids are a group of compounds deserving further study as potential anticancer agents.

REFERENCES

Boutagy, J., and Gelbart, A. (1974). *J. Pharm. Sci.* **63**, 1699.

Cassady, J. M., Byrn, S. R., Stamos, I. K., Evans, S. M., and McKenzie, A. (1978). *J. Med. Chem.* **21**, 815.

Cassady, J. M., Cowall, P., and Chang, C.-J. (1979). Unpublished data from P. Cowall, Ph.D. Thesis, Purdue Univ., Lafayette, Indiana.

Evans, F. J., and Soper, C. J. (1978). *Lloydia* **43**, 193.

Fujita, E., and Nagao, Y. (1977). *Bioorg. Chem.* **6**, 287.

Geissman, T. A. (1964). *Annu. Rev. Pharmacol.* **4**, 305.

Geissman, T. A. (1973). *Recent Adv. Phytochem.* **6**, 65.

Hall, I. H., Lee, K.-H., Mar, E. C., Starnes, C. O., and Waddell, T. G. (1977). *J. Med. Chem.* **20**, 333.

Hall, I. H., Lee, K.-H., Starnes, C. O., El Gebaly, S. A., Ibuka, T., Wu, Y.-S., Kimura, T., and Haruna, M. (1978a). *J. Pharm. Sci.* **67**, 1235.

Hall, I. H., Lee, K.-H., and El Gebaly, S. A. (1978b). *J. Pharm. Sci.* **67**, 552.

Hanson, R. L., Lardy, H. A., and Kupchan, S. M. (1970). *Science* **168**, 378.

Hartmann, G. R., Richter, H., Weiner, E. M., and Zimmerman, W. (1978). *Planta Med.* **34**, 231.

Hartwell, J. L. (1976). *Cancer Treat. Rep.* **60**, 1031.

Hayashi, Y., and Sakan, T. (1975). *Gann* **66**, 587.

Herout, V. (1971). *In* "Pharmacognosy and Phytochemistry" (H. Wagner and L. Horhammer, eds.), pp. 93-111. Springer Publ., New York.

Herz, W. (1973). *In* "Chemistry in Botanical Classification" (G. Bendz and J. Santesson, eds.), Nobel Symposium 25, pp. 153-173. Academic Press, New York.

Hladon, B., Bobkiewicz, T., and Drozdz, B. (1977). *Arch. Immunol. Ther. Exp. (Engl. Transl.)* **25**(2), 243.

Howie, G. A., Manni, P. E., and Cassady, J. M. (1974). *J. Med. Chem.* **17**, 840.

Howie, G. A., Stamos, I. K., and Cassady, J. M. (1976). *J. Med. Chem.* **19**, 309.

Ito, S., and Kodama, M. (1976). *Heterocycles* **4**, 595.

Jamieson, F. R., Reid, E. H., Turner, B. P., and Jamieson, A. T. (1976). *Phytochemistry* **15**, 1713.

Jolad, S., Hoffmann, J. J., Wiedhopf, R. M., Cole, J. R., Bates, R. B., and Kriek, G. R. (1976). *Tetrahedron Lett.*, 4119.

Kamano, Y., Pettit, G. R., Inoue, M., Tozawa, M., and Komeichi, Y. (1977). *J. Chem. Res.* 840.

Kawanata, T., and Inayama, S. (1971). *Chem. Pharm. Bull.* **19**, 643.

Kupchan, S. M. (1974). *Fed. Proc., Fed. Am. Soc. Exp. Biol.* **33**, 2288.

Kupchan, S. M., and Lacadie, J. A. (1975). *J. Org. Chem.* **40**, 654.

Kupchan, S. M., and Streelman, D. R. (1976). *J. Org. Chem.* **41**, 3481.

Kupchan, S. M., and Tsou, G. (1972). *J. Org. Chem.* **38**, 1055.

Kupchan, S. M., Gray, A. H., and Grove, M. D. (1967). *J. Med. Chem.* **10**, 337.

Kupchan, S. M., Karim, A., and Marcks, C. (1968). *J. Am. Chem. Soc.* **90**, 5923.

Kupchan, S. M., Sigel, C. W., Matz, M. J., Saenz Renauld, J. A., Haltiwanger, R. C., and Bryan, R. F. (1970a). *J. Am. Chem. Soc.* **92**, 4476.

Kupchan, S. M., Smith, R. M., Aynechi, Y., and Maruyama, M. (1970b). *J. Org. Chem.* **35**, 2891.

Kupchan, S. M., Fessler, D. C., Eakin, M. A., and Giacobbe, T. J. (1970c). *Science* **168**, 376.

Kupchan, S. M., Eakin, M. A., and Thomas, A. M. (1971). *J. Med. Chem.* **14**, 1147.

Kupchan, S. M., Court, W. A., Dailey, R. F., Jr., Gilmore, C. J., and Bryan, R. F. (1972a). *J. Am. Chem. Soc.* **92**, 7194.

Kupchan, S. M., Sigel, C. W., Guttman, L. J., Restivo, R. J., and Bryan, R. F. (1972b). *J. Am. Chem. Soc.* **94**, 1353.

Kupchan, S. M., Tsou, G., and Sigel, C. W. (1973a). *J. Org. Chem.* **38**, 1420.

Kupchan, S. M., Britton, R. W., Ziegler, M. F., and Sigel, C. W. (1973b). *J. Org. Chem.* **38**, 178.

Kupchan, S. M., Dessertine, A. L., Blaylock, B. T., and Bryan, R. F. (1974). *J. Org. Chem.* **39**, 2477.

Kupchan, S. M., Baxter, R. L., Zeigler, M. G., Smith, P. M., and Bryan, R. F. (1975a). *Experientia* **31**, 137.

Kupchan, S. M., Britton, R. W., Lacadie, J. A., Zeigler, M. F., and Sigel, C. W. (1975b). *J. Org. Chem.* **40**, 648.

Kupchan, S. M., Jarvis, B. B., Dailey, R. G., Jr., Bright, W., Bryan, R. F., and Shizuri, Y. (1976a). *J. Am. Chem. Soc.* **98**, 7092.

Kupchan, S. M., Lacadie, J. A., Howie, G. A., and Sickles, B. R. (1976b). *J. Med. Chem.* **19**, 1130.

Kupchan, S. M., Meshulam, H., and Sneden, A. T. (1978a). *Phytochemistry* **17**, 767.

Kupchan, S. M., Ashmore, J. W., and Sneden, A. T. (1978b). *J. Pharm. Sci.* **67**, 868.

Lantz, C. H., Larner, J., Schubert, R. M., Howie, G. A., and Kupchan, S. M. (1976). *Cancer Biochem. Biophys.* **1**, 229.

Lavie, D., and Glotter, E. (1971). *Fortschr. Chem. Org. Naturst.* **29**, 307.

Lee, K.-H. (1978). *Abstr., Natl. Med. Chem. Symp., 16th, Kalamazoo, Mich.* p. 44.

Lee, K.-H., Huang, E.-S., Piantadosi, C., Pagano, J. S., and Geissman, T. A. (1971). *Cancer Res.* **31**, 1649.

Lee, K.-H., Meck, R., Piantadosi, C., and Huang, E.-S. (1973). *J. Med. Chem.* **16**, 299.

Lee, K.-H., Hall, I. H., Mar, E. C., Starnes, C. O., El Gebaly, S. A., Waddell, T. G., Hadgraft, R. I., Ruffer, C. G., and Weidner, I. (1977). *Science* **196**, 533.

Lee, K.-H., Ibuka, T., Mar, E.-C., and Hall, I. H. (1978a). *J. Med. Chem.* **21**, 698.

Lee, K.-H., Mar, E.-C., Okamoto, M., and Hall, I. H. (1978b). *J. Med. Chem.* **21**, 819.

Le Quesne, P. W., Levery, S. B., Menachery, M. D., Brennan, T. R., and Raffauf, R. F. (1978). *J.C.S. Perkin I*, 1572.

Liao, L. L., Kupchan, S. M., and Horwitz, S. B. (1976). *Mol. Pharmacol.* **12**, 167.

Mabry, T. J., and Bohlmann, F. (1977). In ''The Biology and Chemistry of the Compositae'' (V. H. Heywood, J. B. Harborne, and B. L. Turner, eds.), Vol. 2, pp. 1097–1104. Academic Press, New York.

Polonsky, J. (1973). *Fortschr. Chem. Org. Naturst.* **30**, 101.

Ranieri, R. L., and Calton, G. J. (1978). *Tetrahedron Lett.*, 499.

Rodriguez, E., and Mabry, T. J. (1977). In ''The Biology and Chemistry of the Compositae'' (V. H. Heywood, J. B. Harborne, and B. L. Turner, eds.), Vol. 2, pp. 788–797. Academic Press, New York.

Rodriguez, E., Towers, G. H. N., and Mitchell, J. C. (1976). *Phytochemistry* **15**, 1573.

Seida, A. A., Kinghorn, A. D., Cordell, G. A., and Farnsworth, N. R. (1978). *Lloydia* **41**, 584.

Smith, C. H., Larner, J., Thomas, A. H., and Kupchan, S. M. (1972). *Biochim. Biophys. Acta* **276**, 94.

Stamos, I. K., Howie, G. A., Manni, P. E., Haws, W. J., Byrn, S. R., and Cassady, J. M. (1977). *J. Org. Chem.* **42**, 1703.

Suffness, M., and Douros, J. (1979). *Methods Cancer Res.* **16A**, 73.

Wall, M. E., and Wani, M. C. (1970). *Int. Symp. Chem. Nat. Prod., 7th, Riga* Abstr., 614.

Wall, M. E., and Wani, M. C. (1977). *Annu. Rev. Pharmacol. Toxicol.* **17,** 117.

Wani, M. C., Taylor, H. L., Wall, M. E., Coggon, P., and MacPhail, A. T. (1971). *J. Am. Chem. Soc.* **93,** 2325.

Wani, M. C., Taylor, H. L., and Wall, M. E. (1977). *J.C.S. Chem. Commun.* 295.

Wani, M. C., Taylor, H. L., Thompson, J. B., and Wall, M. E. (1978). *Lloydia* **41,** 578.

Yoshioka, H., Mabry, T. J., and Timmerman, B. N. (1973). "Sesquiterpene Lactones: Chemistry, NMR and Plant Distribution." Univ. of Tokyo Press, Tokyo.

CHAPTER 8

Dimeric Catharanthus Alkaloids

KOERT GERZON

I. INTRODUCTION

Studies of the chemical modification of vinblastine reported in this chapter are described from the point of view of the medicinal chemist, desirous to gain a deepened understanding of the chemical, biochemical, and biological qualities of the important, clinically useful *Catharanthus* ("Vinca") alkaloids. This understanding is needed to fathom the mode of action by which these alkaloid agents express their chemotherapeutic benefit. It is also needed to guide

Anticancer Agents Based on Natural Product Models
Copyright © 1980 by Academic Press, Inc.

efforts of the medicinal chemist in the careful design of new modification products with improved oncolytic activity.

In addition to the need for basic understanding at the experimental level, there is a need for feedback information from clinical experience with the established agents, vinblastine and vincristine, and agents in trial, presently vindesine. Therefore, the present discussion, though not immediately involved with clinical aspects, does cite preliminary reports of clinical experience which furnish feedback information for aid in drug design.

In the same vein, additional topics, dealing with the role of the alkaloids as inhibitors of cell mitosis, their ability to bind to microtubular protein units (tubulin dimers), and other qualities, are included for their impact on understanding and design of drug action.

II. HISTORICAL BACKGROUND

A. Discovery: Experimental Antitumor Activity

The discovery of vinblastine, an antiproliferative factor in leaf extracts of the periwinkle plant (*catharanthus roseus* G. Don., *Vinca Rosea* Linn.) at the University of Western Ontario (Cutts *et al.*, 1960), and, concurrently, at the Lilly Research Laboratories (Johnson *et al.*, 1959) was soon followed by the further discovery in these extracts of the alkaloids vincristine and leurosidine (Svoboda, 1961).

Vincristine at 0.15–0.3 mg/kg and leurosidine at 7.5–10.0 mg/kg given i.p. for 10 days to mice bearing the P-1534 leukemia produced long-term survivors Vinblastine in a dose range of 0.1–0.5 mg/kg produced no survivors, though survival time of treated mice was increased appreciably over that of the control mice (Johnson *et al.*, 1963). Leurosine, the fourth active alkaloid isolated (Svoboda *et al.*, 1962), produced significant prolonging of survival time at 10–20 mg/kg × 10.

In addition to P-1534 activity, vincristine was shown to have activity against the Ridgway osteogenic sarcoma, a tumor system insensitive to vinblastine. No significant activity was noted when the four alkaloids were assayed against the L-1210 leukemia, a model system used widely in other laboratories engaged in screening potential antitumor agents. Additional results for the four alkaloids in a total of 20 tumor model systems were reported in 1963 (Johnson *et al.*, 1963). The inactivity of vincristine against the Gardner lymphosarcoma reported at that time is now considered in error.

The inactivity reported for desacetyl vinblastine against the P-1534 leukemia is curious in view of the significant activity observed recently for this agent against the Gardner lymphosarcoma and Ridgway osteogenic sarcoma as well as against the P-388 and the newer P-1534(J) leukemias (Barnett *et al.*, 1978). Desacetyl vinblastine has been recognized as a vinblastine metabolite in the dog (Creasey *et al.*, 1975; Creasey, 1977) and also in man (Owellen, 1975; Owellen *et al.*, 1977a).

Only a few years after serving in the discovery of vinblastine, the P-1534 leukemia became less responsive to the *Vinca* alkaloids, especially to vinblastine, and at the time of the current modification program (1972) the P-1534 leukemia implanted i.p. was essentially unresponsive to *Catharanthus* alkaloids.

Historically, it is worthy of note that the discovery of vinblastine's chemotherapeutic potential stemmed from the propitious use of the P-1534 leukemia model, a tumor system exquisitely sensitive to the *Vinca* alkaloids (Johnson *et al.*, 1963) and sensitive also to other types of antitumor agents (Gerzon *et al.*, 1959).

Mitotic arrest of mammalian cells by *Vinca* alkaloids presents another facet of their biological activity found to be useful for quantitative assay purposes (Sweeney *et al.*, 1978). Mitotic accumulation of Chinese hamster ovary cells in culture reveals a greater potency for vincristine than vinblastine. A similar potency relationship for these alkaloids is apparent in the avidity for binding to turbulin, an interaction deemed basic to their antimitotic action (Sartorelli and Creasey, 1969).

B. Vinblastine and Vincristine: Clinical Experience

Following the first report of the clinical use of vinblastine (Hodes *et al.*, 1960), numerous reports and reviews have appeared (Calabresi and Parks, 1975; DeConti and Creasey, 1975; Johnson *et al.*, 1963; Mathé *et al.*, 1966; Zubrod, 1974) describing clinical experience with vinblastine and vincristine as single agent (Livingston and Carter, 1970) and in combination with other anticancer agents (DeVita *et al.*, 1970; Moxley *et al.*, 1967). Zubrod (1974) has tabulated the drugs of choice for 13 human tumor types that are highly responsive to chemotherapy resulting in substantial increases in life expectancy. Vincristine is recommended for seven of these tumors, including acute lymphocytic leukemia, advanced stages of Hodgkin's disease ("MOPP regimen"), lymphosarcoma, and others. Among the drugs reported to have a "substantial" effect, vinblastine is mentioned for the treatment of choriocarcinoma and Hodgkin's disease, vincristine for myelocytic leukemia and retinoblastoma.

The agents vinblastine and vincristine differ substantially in clinical utility and clinical toxicity. The major use of vinblastine is in the treatment of patients with Hodgkin's disease, while vincristine, often in combination, is widely used in the treatment of acute lymphocytic leukemia in childhood. In the latter disease, vincristine in combination with amethopterin, 6-mercaptopurine, and prednisone

("VAMP") has produced induction remission and survival results not achieved in single-drug treatment (Freireich *et al.*, 1968; Henderson and Samaha, 1969; Holland, 1969).

Vinblastine and vincristine differ substantially also in the nature of dose-limiting toxicity in patients receiving these agents in therapeutically effective dosage. Bone-marrow toxicity is the dose-limiting factor in clinical therapy with vinblastine, whereas vincristine in children and in adults frequently produces neuropathy requiring reduction of dosage or discontinuation of therapy (Gröbe and Palm, 1972; Rosenthal and Kaufman, 1974; Weiss *et al.*, 1974).

A special quality of Vinca alkaloid therapy relates to the biochemical mode of action, thought to involve the interaction of the alkaloid with tubulin (Sartorelli and Creasey, 1969). While a number of clinically useful antitumor agents express their activity by interacting—covalently or noncovalently—with DNA function and carry a potential risk of carcinogenesis for the patient, the Vinca alkaloids, interacting noncovalently with tubulin, do not carry this risk. Lack of carcinogenic activity has been demonstrated for vinblastine in an animal model (Schmähl and Osswald, 1970). In a bacterial assay utilizing *Salmonella typhimurium* mutants, vinblastine and vincristine have been reported to be not mutagenic (Seineo *et al.*, 1978).

Reporting on a bioassay program for carcinogenic hazards of cancer chemotherapeutic agents, Weisburger (1977) ranks vinblastine and vincristine as agents that "fairly consistently, depending on the animal model, fall into the group which increases tumor incidence only slightly over that in the controls or even show an effect similar to that in the controls" (cf. Section V).

C. Structure–Activity Relationship Observations (1960–1966)

Structural investigations leading to the formulation of vinblastine (**1**) and vincristine (**2**) (see Fig. 1) as dimeric indole–dihydroindole alkaloids (Neuss *et al.*, 1964) and further chemical aspects of these and related dimeric *Catharanthus* alkaloids—leurosidine, leurosine (Fig. 2), and others—have been reviewed in detail (Abraham, 1975; Johnson *et al.*, 1963).

A corrected structure of leurosidine (see Fig. 3) as the C-4' epimer of vinblastine has been reported recently (Wenkert *et al.*, 1975). The detailed stereochemical configuration of vincristine methiodide has been elucidated by x-ray crystallographic analysis (Moncrief and Lipscomb, 1965).

It is readily appreciated that modification studies reported in this chapter are based on the elegant structural studies of Neuss and associates (1964).

Experimental antitumor activities of leurosidine, leurosine, and leurosine methiodide observed in the P-1534 leukemia model have been described as the basis for selection of these three alkaloids for clinical evaluation (Johnson *et al.*,

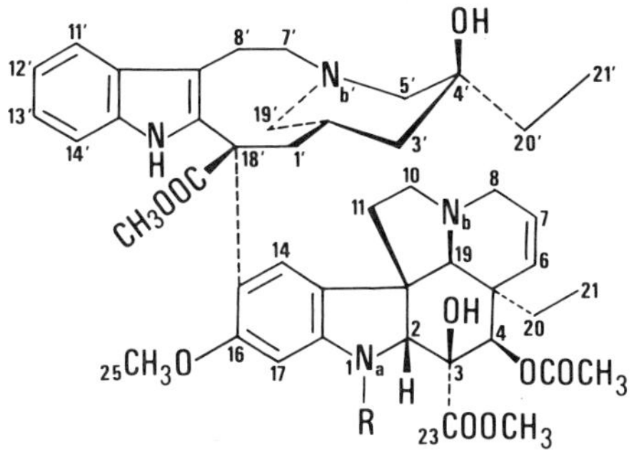

Figure 1. Molecular structures of the dimeric *Catharanthus* alkaloids vinblastine (VLB, **1**, R = CH$_3$) and vincristine (VCR, **2**, R = CHO).

1963). Leurosidine and leurosine methiodide were shown to produce survivors in this model, while vinblastine does not. This apparent experimental therapeutic advantage is procured at the price of a considerable increase in dose required of these two agents.

According to information available at this time (Johnson *et al.*, 1963; DeConti and Creasey, 1975), clinical experience with the three alkaloids can be summarized as follows:

In a brief clinical trial, restricted because of limited supply, leurosidine sulfate produced no therapeutic benefit in patients receiving doses high enough to elicit side effects (Gailani *et al.*, 1966).

Clinical trial of leurosine sulfate was discontinued because of lack of convincing therapeutic benefit and/or unacceptable side effects (Gailani *et al.*, 1966; Mathé *et al.*, 1965). Lastly, the antitumor benefits in patients receiving leurosine methiodide were inferior to those noted with vinblastine and vincristine, while side effects were found to be more severe (Hodes *et al.*, 1963).

Vinglycinate, a chemical modification product of vinblastine (Hargrove,

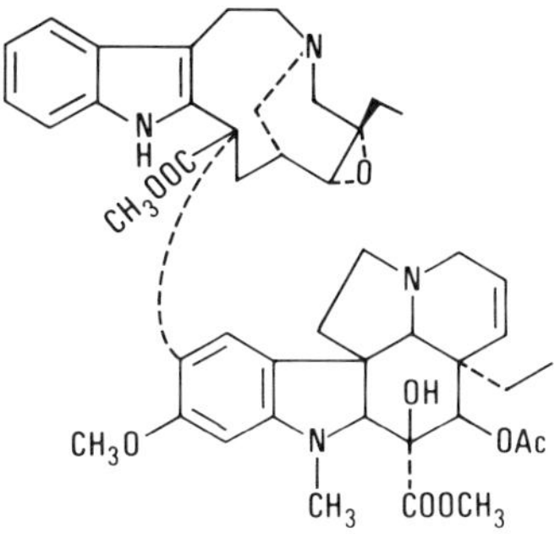

Figure 2. Molecular structure of leurosine (Abraham and Farnsworth, 1969).

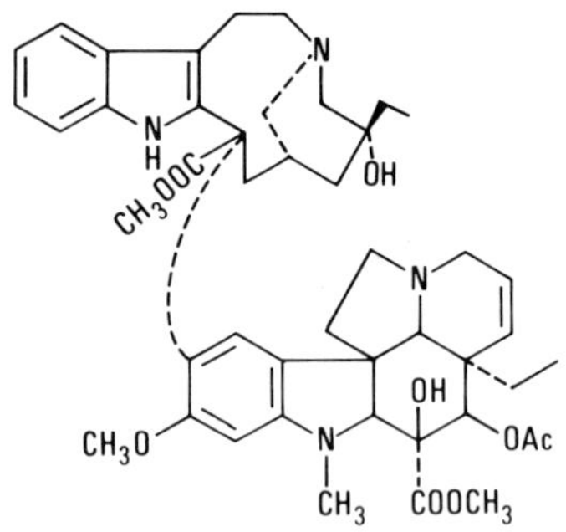

Figure 3. Molecular structure of leurosidine (Wenkert *et al.*, 1975).

1964), produced a high percentage of survivors in mice bearing the P-1534 leukemia. The dose required to produce this effect (5.6 mg/kg × 10) was 10–15 times larger than the dose of vinblastine or vincristine required for optimum effect. Clinical trial of vinglycinate was halted because of chemical instability problems encountered before conclusive results were obtained.

In summarizing the experience with leurosidine, leurosine, leurosine methiodide, and vinglycinate, the results of these clinical evaluations are seen as negative or inconclusive. The question presenting itself with considerable urgency is: Does this disappointing experience with four alkaloids of considerably lesser potency, relative to that of vinblastine (against the P-1534 leukemia), indicate that in the course of modification studies (with VLB) only such products as have retained the potency (of VLB) can be considered worthy candidates for clinical evaluation?

This question concerning clinical promise of agents of reduced potency (relative to the parent alkaloid) returns in the sequel with reference to *N*-formyl-*N*-desmethyl leurosine (Somfai-Relle *et al.*, 1975) in Section IV.

Since 1972, separate publications reporting work reviewed in this chapter have appeared. These concern the chemistry and biological assay of vindesine (VDS, Fig. 4) (Barnett *et al.*, 1978; Conrad *et al.*, 1979), antitumor activity and tissue culture assay (Sweeney *et al.*, 1978), tissue distribution and disposition of vindesine in the rat (Culp *et al.*, 1977), pharmacokinetics and preliminary clinical observations with vindesine (Dyke and Nelson, 1977), and other topics.

III. CHEMICAL MODIFICATION OF VINBLASTINE; VINDOLINE MOIETY

A. Approach and Objectives

In a review of the auspicious clinical performance of vinblastine and vincristine 10 years after their discovery, Johnson (1968) stresses the desirability of further structural modification of the dimeric "Vinca" alkaloids.

Significant differences in clinical activity and toxicity of these alkaloids, linked with the relatively minor difference in their chemical structure, prompted the effort to probe what effect other such "minor" structural alterations might have on antitumor efficacy.

The difference between the structures of vinblastine (VLB, 1) and vincristine (VCR, 2) resides in the substituent, a methyl or a formyl group, respectively, on the N_a-atom of the vindoline ("lower") moiety (see Fig. 1). The structures of leurosine (Fig. 2) and leurosidine (Fig. 3) differ from that of VLB because of changes in the carbomethoxyvelbanamine ("upper") moiety. Leurosidine is shown to be epimeric at C-4' (Wenkert *et al.*, 1975). Leurosine, also epimeric at C-4', has a α-3',4'-epoxy function (Abraham and Farnsworth, 1969).

Because it was realized that "minor" changes in the vindoline moiety of VLB do not cause loss of clinical activity whereas such changes in the velbanamine ("upper") moiety appear to lower activity, vinblastine was favored as the parent alkaloid for modification. This choice, made in 1972 at the beginning of the program, was reinforced by the relative availability of vinblastine.

Considerations of chemical stability also favored vinblastine over vincristine. The latter alkaloid is readily de-formylated, for example, in the course of reactions involving aminolysis, hydrazinolysis, etc., leading to loss of activity. Most relevant, however, the choice of vinblastine over vincristine was made under pressure of the uncertainty concerning the structural basis for the dose-limiting neurotoxicity seen with vincristine. True, no evidence existed to implicate the N-formyl group of VCR—the distinguishing feature—as causing or contributing to neurotoxicity, but neither was evidence at hand to deny such connection.

Therefore, a vinblastine analogue, which—hopefully—would emulate the superior antitumor activity of vincristine but lack dose-limiting neurotoxicity, was viewed as a major objective of modification research. Hardly less important, increased understanding of the mechanism underlying the antitumor activity of the dimeric alkaloids was a consistent aim of the program.

B. Vindesine and N-Methyl Vindesine: Chemistry and Experimental Antitumor Activity

Desacetyl vinblastine amide (vindesine, VDS, 3, see Fig. 4) first was obtained in moderate yield from vinblastine (VLB, 1) or desacetyl VLB (4) by the preferential ammonolysis of the vindoline-$COOCH_3$ function and concurrent deacetylation at C-4 (see Scheme I). Preferential hydrazinolysis and deacetylation, followed by Raney nickel hydrogenolysis of the resulting desacetyl VLB monohydrazide (5), furnished a fair yield of vindesine. The purification of vindesine free base by high-performance liquid chromatography (HPLC) to remove small amounts of by-products (e.g., vindesine $N_{b'}$-oxide) together with other details of its preparation on a larger scale, have been reported (Barnett *et al.*, 1978).

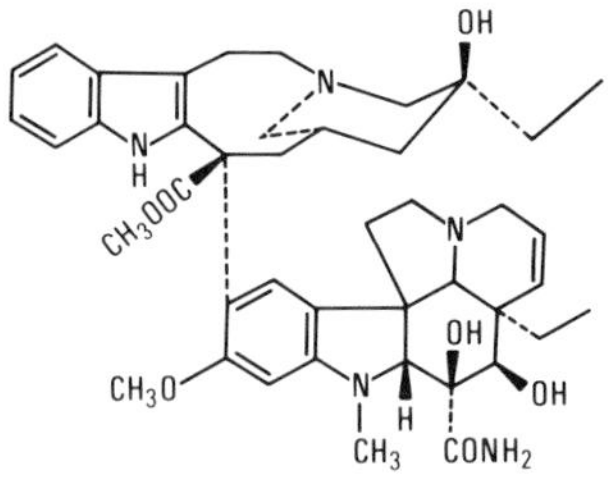

Figure 4. Molecular structure of vindesine (VDS, **3**).

The crystalline hydrazide (**5**), a versatile intermediate, upon nitrosation furnishes desacetyl vinblastine azide (V-CON₃, **6**), from which vindesine is obtained in good yield by reaction with excess ammonia gas in chloroform solution. Preferential aminolysis of VLB using methyl amine gives *N*-methyl VDS (V-CONH-CH₃, **7**) but this reaction with homologous or substituted alkyl ammes fails to give the desired products. *N*-Methyl VDS (**7**) and a number of *N*-substituted vindesines (V-CONH-R, **8–26**) were readily prepared from the azide (**6**) and suitable amines.

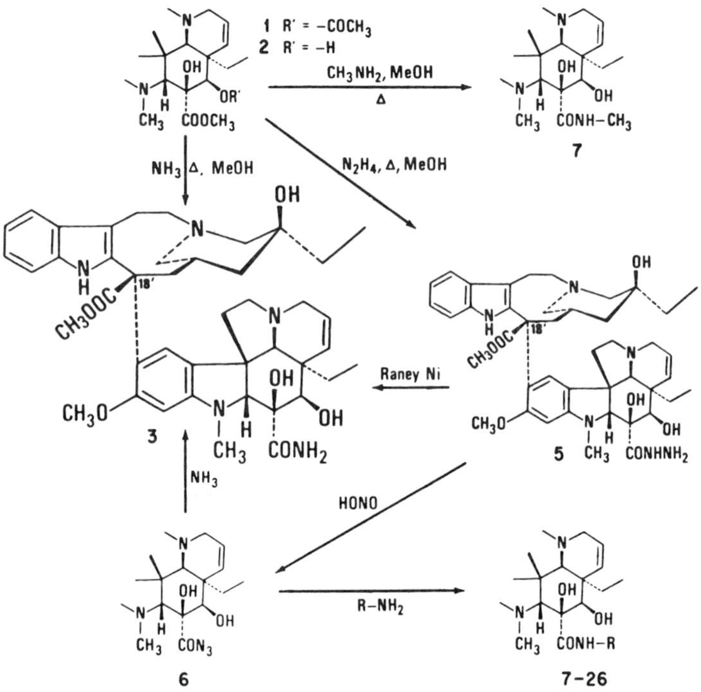

Scheme I

The functional changes associated with the conversion of vinblastine to vindesine—ester to amide and acetate to alcohol—generate additional hydrogen-bonding sites in the region of modification (see Fig. 5). This change in regional functionality is illustrated in molecular models (see Fig. 6).

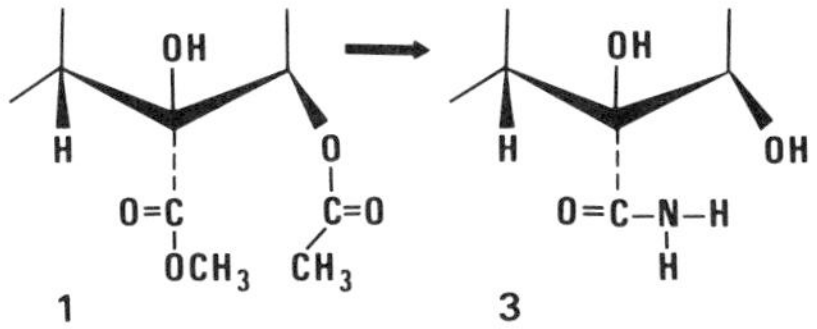

Figure 5. Chemical functionality changes in the conversion of vinblastine (1) to vindesine (3).

The behavior of VDS in thin layer chromatography (TLC) indicated greater polarity for VDS relative to VLB and desacetyl VLB (4), an observation confirmed by comparison of the apparent partition coefficient of VDS (P* = 2.6) determined between n-octanol and water at pH 7.4 (Barnett et al., 1978). For VCR, the partition coefficient P* is 2.15; for VLB, P* is 2.9.

At the outset, the increase in polarity for VDS, relative to that of VLB, was held to be desirable. It was assumed that low lipophilicity would tend to reduce uptake by peripheral nerve cells, thereby minimizing the potential for neurotoxicity. This assumption is based on reports indicating enhanced transport of lipophilic agents across cell membranes of the blood–brain barrier and of peripheral nerve. It generally pertains to passive transport of molecules not ionized at physiological pH (Oldendorf, 1974, 1978; Rall and Zubrod, 1962; Rapoport, 1976; Ritchie and Greengard, 1966). A further discussion of polarity-neurotoxicity aspects is deferred to Section III,F.

Turning to the biological assay of modification products available in adequate amounts, VDS (V-CONH$_2$, 3), N-methyl VDS (7), and desacetyl VLB (4) were converted to water soluble sulfate salts for i.p. administration. The hydrazide (5) was tested as free base because of instability in acid medium.

Following the demise of the sensitive P-1534 leukemia assay, the search for a suitable test system led to selection of the Ridgway osteogenic sarcoma (ROS; Sugiura and Stock, 1952) as the primary assay system. Subsequent evaluation made use of the Gardner lymphosarcoma (GLS; Gardner et al., 1944), the P-388 leukemia, and B-16 melanoma (Geran et al., 1972). A newer P-1534(J) leukemia strain implanted s.c. (Miller et al., 1977) and the P-388/VCR leukemia strain (Wolpert-DeFilippes et al., 1975) are systems used more recently.

The typical response of the ROS to six dimeric alkaloids is summarized in Table I. The inactivity of VLB (1), mentioned previously, is contrasted with the 75–100% inhibition of tumor growth shown by vindesine (3) and N-methyl vindesine (7) at the dose regimen of 0.3–0.5 mg/kg × 8. VCR (2), twice as

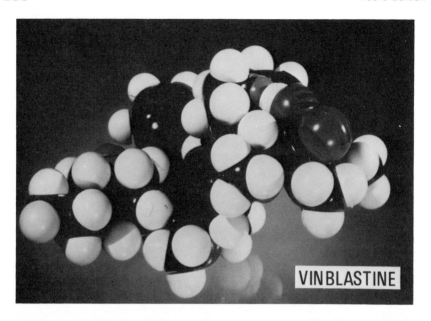

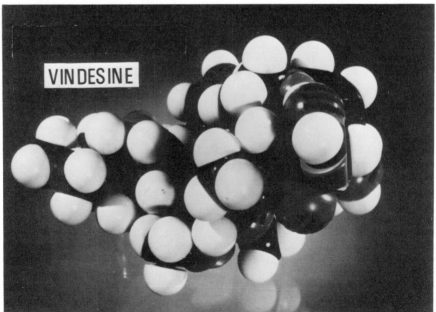

Figure 6. Three-dimensional molecular models of VLB (**1**, top) and VDS (**3**, bottom), showing in middle right section the functionality changes depicted in Fig. 5.

TABLE I

Inhibition of Growth of the Ridgway Osteogenic Sarcoma in AKR Mice by VDS, N-Methyl VDS, and Other Dimeric Alkaloids[a]

Alkaloid agent sulfates salts	Dose in mg/kg i.p. for 8 days							
	0.1	0.15	0.2	0.25	0.3	0.4	0.5	1.0
(1) VLB	−	−	−	−	−	−		
(2) VCR	++	+++	+++	+++	+++			
(3) VDS	±	++	++	++	+++	Tx[b]		
(7) N-Methyl VDS	±		++	++	+++	+++	+++	Tx
(4) Desacetyl VLB		±	+		+++	+++		
(5) Desacetyl VLB hydrazide[c]			++		+++	+++		

[a] +++ = 75–100% inhibition of tumor growth compared to that of controls; ++ = 50–75% inhibition; + = 25–50%; ± = less than 25% inhibition; − = No measurable inhibition.

[b] Tx = toxicity; ≥ 50% of the mice died before the ninth day.

[c] Free base.

active as vindesine, shows this inhibition in the 0.15–0.3 mg/kg × 8 range. The activity of the monohydrazide (5) equals that of VDS, but desacetyl VLB (4) is slightly less active.

The resulting ranking in the ROS model, namely VCR > (VDS = N-methyl VDS) > desacetyl-VLB >> VLB, is seen also in the ranking of these alkaloids assayed against the Gardner lymphosarcoma (Barnett *et al.*, 1978; Conrad *et al.*, 1979; Sweeney *et al.*, 1978); see Table II, Section III,C.

That vindesine, N-methyl VDS (7), and the hydrazide (5) effectively inhibit the ROS–GLS systems, which are insensitive to vinblastine, constitutes a *qualitative* difference in antitumor activity between these agents and the parent alkaloid. A qualitative difference is viewed as strong support for a clinical candidate.

Of these three alkaloids, the monohydrazide (5) is not an attractive candidate agent, because it is instable in acid medium and lacks water solubility as the free base. N-methyl VDS (7), though equally active as vindesine, was considered less attractive because of observations made in a chronic IV administration experiment. Mice receiving vindesine for eight weeks continued to gain weight, while those receiving N-methyl VDS gained less and began to lose weight after six weeks (Conrad *et al.*, 1979).

C. Further Vindesine Congeners

The experimental antitumor activity of vindesine, then emerging as an attractive candidate for clinical evaluation, was further compared to that of a larger number of VDS congeners (V-CONH-R, **8–26**) prepared from the azide (V-CON$_3$, **6**) and the appropriate amine (see Scheme I). Limitations of practical nature preclude the comparative assay of several dozen available congeners (Conrad *et al.*, 1979) against several—about six—tumor models within a reasonable time period.

Instead, these compounds were assayed against a single model—the Gardner lymphosarcoma—chosen for its reliability and reproducibility of results. A small number of these congeners with superior activity against the GLS were subsequently assayed against the other tumor models. Details of these comparative assays and the experimental antitumor activity of VDS—summarized below—have been reported previously (Barnett *et al.*, 1978; Conrad *et al.*, 1979; Sweeney *et al.*, 1978).

The GLS model thus served as primary assay for N-alkyl- and N-aralkyl-substituted vindesines (**7–14**, Table II), VDS congeners with N-alkyl groups bearing polar functions (**15–21**, Table III), and functions of moderate chemical reactivity (**22–26**, Table IV).

While the activity of N-methyl VDS (7) resembles that of VDS, a gradual loss of activity and increase of toxicity accompanies the introduction of larger alkyl (**8, 9, 12**), benzyl (**13**), and p-hydroxy-phenethyl (**14**) substituents. N-Allyl VDS

TABLE II

Inhibition of Growth of the Gardner Lymphosarcoma in C$_3$H Mice by VDS, N-Alkyl VDS, N-Aralkyl VDS, and Parent Dimeric Alkaloids[a]

	Dose in mg/kg i.p. for 9 days							
	0.05	0.1	0.15	0.2	0.25	0.3	0.4	0.5
(3) V-CONH$_2$ (VDS)[b]	−	++	+	+++	+++	+++	+++	Tx
(7) -CONH-CH$_3$	−	−		+++		+++		
(8) -CONH-CH$_2$CH$_3$		−		+++		+++	Tx	
(9) -CONH-CH(CH$_3$)$_2$		−		Tx			Tx	
(10) -CONH-CH$_2$CH=CH$_2$				+++			Tx	
(11) -CONH-CH$_2$C≡CH		++		+++			Tx	
(12) -CONH-(CH$_2$)$_5$CH$_3$				+				
(13) —CONH—CH$_2$—C$_6$H$_5$		+					−	
(14) —CONH—CH$_2$CH$_2$—C$_6$H$_4$OH		−		−			±	
(1) VLB[b]		±			+	++	−	
(2) VCR[b]	±	++	+++	+++	Tx	Tx	±	
(4) Desacetyl VLB[b]	−	−	−	+++	+	+++	+++	+++

[a] See Table I for activity and toxicity notation. Alkaloids (8–14), prepared from the azide (V-CON$_3$, 6) and the appropriate amine, were converted to the sulfate salts for i.p. administration.

[b] Treatment lasted 8 days.

TABLE III

Inhibition of Growth of Gardner Lymphosarcoma in C_3H Mice by VDS, N-CH_2CH_2 X Vindesines with Polar Groups X[a]

Alkaloid agent V-CONH-R	Dose in mg/kg i.p. for 9 days					
	0.05	0.1	0.2	0.3	0.4	0.5
(3) V-CONH₂ (VDS)	−	++	+++	+++	+++	Tx
(15) -CONH-CH₂CH₂NH₂		−	−		−	
(16) -CONH-CH₂CH₂NHCOCH₃[b]		+++	+++		+++	
(17) -CONH-CH₂CH₂OH	+++	+++	+++		+++	
(18) -CONH-CH₂CH₂OCO(CH₂)₂CH₃[b]	−	++	+++		Tx	
(19) -CONH-CH₂CH₂SH[c]			+	++	+++	
(20) -CONH-CH₂CH₂-SCH₃		−	−		Tx	
(21) -CONH-CH₂CH₂CN	−	−	+++		+++	

[a] See Table I for activity and toxicity notation. Alkaloids (15), (17), (19), (20), and (21), prepared from the azide (6) and the appropriate amine.
[b] Compounds (16) and (18) prepared by acylation of (15) and (17), respectively.
[c] See text for chemical details.

(10) and *N*-propargyl-VDS (11) appear superior to *N*-isopropyl VDS (9). Clearly, increased lipophilicity does not improve antitumor activity against the GLS relative to that of VDS. Whether these lipophilic agents (e.g., compounds 7–9, and 12) have greater potential for neurotoxicity than VDS has not been established.

The introduction of a polar function in the N-ethyl group of compound (8) affects anti-GLS activity in several ways. *N*-β-amino-ethyl VDS (15), a compound presumably extensively protonated at physiological pH, has negligible activity.

The activity of *N*-β-hydroxyethyl VDS (V-CONHCH$_2$CH$_2$OH, 17) equals or slightly surpasses that of VDS, and even VCR, in terms of its therapeutic dose range. While acylation of *N*-β-hydroxyethyl VDS to give the butyrate ester (18) results in slightly diminished GLS activity, conversion of *N*-β-aminoethyl VDS (15) to *N*-β-acetamidoethyl VDS (16) restores activity approaching that of VDS. Moderate GLS activity for *N*-β-mercaptoethyl VDS (19) has been observed in a single experiment; the susceptibility of (19) to air oxidation, leading to disulfide formation (Section III,J), has limited its further testing.

The profile of anti-GLS activity noted for the compounds in Table III (15–21) appears compatible with the need for hydrogen bonding capability at the β-ethyl site. In view of the superior activity of *N*-β-hydroxyethyl (17) against the GLS—also seen against the ROS—is 17 a more attractive candidate for clinical trial than VDS? The answer to this question is discussed after presentation of the test results in the B-16 melanoma.

Among the VDS congeners bearing a weakly reactive chemical function (Table IV), only the *N*-acetaldehyde VDS congener (23) and its dimethyl acetal (24) have a fair therapeutic dose range of GLS activity, suggesting that—at least against the GLS—no therapeutic improvement results from such reactive and potentially alkylating functionalities. A difficult choice is thereby avoided.

Although chemical reactivity present in a drug molecule can generate or contribute to antitumor activity, such reactivity does carry an unknown burden of potential toxicity and/or carcinogenicity, perhaps to emerge only at the clinical stage. When no alternate agent is available, the acceptance of this burden may be unavoidable. When alternate, equally effective candidate agents—here VDS or *N*-β-hydroxyethyl VDS—are available, which do not bear chemically reactive functions and which interact—apparently effectively—in a noncovalent manner with the target molecule, the deliberate introduction of a potential alkylating functionality into a parent Vinca alkaloid appears medicinally undesirable.

Further comparative assay of VDS (3) and *N*-β-hydroxyethyl VDS (17) demonstrated activity of the order of VLB and VCR in prolonging survival time of mice bearing the P-388/S and P-1534(J) leukemias implanted i.p. and s.c., respectively. Although individual differences in activity do exist, no useful differentiation in the ranking of VDS, *N*-β-hydroxyethyl VDS (17), or other al-

TABLE IV

Inhibition of Growth of Gardner Lymphosarcoma in C₃H Mice by VDS and N-Substituted VDS Bearing Alkyl Groups with Functions of Moderate Chemical Reactivity[a]

	Dose in mg/kg i.p. for 9 days				
(3) V-CONH₂ (VDS)	±	++	+++	+++	+++
(22) -CONH-CH₂CH₂OCO-CH=CH₂	++	+++	Tx		Tx
(23) -CONH-CH₂CHO		+	+++		+++
(24) -CONH-CH₂CH(OCH₃)₂[b]	±	+++	+++		Tx
(25) -CONH-CH₂CH₂C(C=NH)OCH₃[c]	−	−	±		±
(26) -CONH-CH-COS-CH₂CH₂[d]	−	−	Tx		Tx

[a] Activity and toxicity notation, see Table I, footnote a.
[b] Free base.
[c] Hydrochloride salt.
[d] Prepared from azide (6) and homocysteine thiolactone.

kaloids resulted from this comparison. (Barnett et al., 1978; Conrad et al., 1979; Montgomery, 1974; Johnson, 1975; Sweeney et al., 1978.

Differentiation of activity of vindesine and other agents did, however, result from assays against the B-16 melanoma. Activity of VDS and congeners against the B-16 melanoma i.p. is reported (Table V) from observations made in Experiment II at the National Cancer Institute (Johnson, 1975) and Experiment I at the Lilly Research Laboratories. Depending on conditions of the B-16 melanoma assay, variations are noted between separate assays (see Sweeney et al., 1978), but these do not—in our experience—change the ranking order observed for the agents concerned.

In terms of increased life span and of mice surviving the 60-day and 56-day duration of Experiment I and II, respectively, vindesine has superior activity against the B-16 melanoma relative to that of N-β-hydroxyethyl VDS (17). This constitutes a reversal of the ranking order for these two compounds against the ROS-GLS system. Similarly, N-β-acetamidoethyl VDS (16), while active at the level of VDS against the ROS–GLS models, is less active against the B-16 melanoma (single test).

The contrast in ranking order for vindesine and N-β-hydroxyethyl VDS, observed in comparing the ROS–GLS versus the B-16 melanoma model, poses a question for which an answer is found in the pertinent chemotherapy literature. It is recognized that the choice to be made at this point is between the ROS–GLS and B-16 tumor models, their relative significance and relative predictive merit (Venditti, 1975), rather than between the two agents here under consideration.

From an extensive analysis of the performance by several oncolytic agents, Venditti (1975) presents evidence for a strong positive correlation between clinical efficacy and activity against the B-16 melanoma and ascribes high predictive

TABLE V

Increase in Lifespan of Mice Bearing the B-16 Melanoma by VDS, VDS Congeners, and Parent Alkaloids[a]

Alkaloid agent V-CONH-R	Experiment I[b] Dose in mg/kg[d]		
	0.6	0.9	1.2
(3) V-CONH$_2$ (VDS)	117(2)	114(2)	155(4)
(17) -CONH-CH$_2$CH$_2$OH	51(1)	97(0)	93(0)
(16) -CONH-CH$_2$CH$_2$NHCOCH$_3$	0(0)	0(0)	4(0)
(19) -CONH-CH$_2$CH$_2$SH[e]	170(4)	141(1)	Tx
(20) -CONH-CH$_2$CH$_2$SCH$_3$	163(6)	58(2)	Tx
(14) —CONH—CH$_2$CH$_2$—⬡—OH	125(2)	152(4)	190(8)

Alkaloid agent	Experiment II[c] Dose in mg/kg[d]				
	0.4	0.65	1.1	1.8	3.0
VDS	85(0)	75(0)	110(1)	142(2)	138(1)
(17)	28(0)	60(0)	62(0)	68(1)	75(0)
VLB	60(0)	68(0)	78(1)	92(0)	102(1)
VCR	20(0)	30(0)	52(0)	62(0)	75(0)

[a] Data recorded as percent increase of lifespan (ILS) over that of controls (survivors at end of experiment/10). Survivors not included in ILS calculation.

[b] Experiment I (Lilly Research Laboratories) using C57BL/6 mice, 10^6 cell inoculum implanted i.p. Average day of death control mice 19.1 days (16–22 days range). Duration Exp. I 60 days (Conrad et al., 1979).

[c] Experiment II (National Cancer Institute; Johnson, 1975) using B6D2F$_1$ male mice; 0.25 ml of a 20% w/v brei of B-16 melanoma implanted i.p. Median day of death controls 20.0 (range 16–36 days). Duration Exp. II 56 days (see Barnett et al., 1978; Sweeney et al., 1978).

[d] Dose i.p. on days 1, 5, and 9.

[e] See text for chemical details.

merit to the B-16 model. Vindesine, already attractive because of its activity against the VLB-insensitive ROS–GLS systems, was favored for clinical evaluation because of its superior activity—relative to that of the hydroxyethyl congener (**17**)—against the B-16 melanoma.

In summary, vindesine was selected because it is deemed to have optimum *collective activity,* defined as the sum of antitumor activity against the tumor models used with due consideration of their respective predictive merit. This selection, based on antitumor activity, is conditional on subsequent toxicological examination of vindesine and assessment of its potential for neurotoxicity relative to that of VCR (see Section III,E).

Turning to compound (**14**), N-β-hydroxyphenethyl VDS, B-16 melanoma activity shown by this congener (Table V) surpasses that of VDS. This vindesine congener (**14**), although ranking below VDS in collective activity because of negligible GLS activity, is of special interest as a potential substrate for tyrosinase and because of the possibility of metabolic activation by this enzyme (see Vogel *et al.,* 1977, and references cited therein). Metabolic activation of γ-L-glutaminyl-4-hydroxybenzene, the "mushroom factor" of *Agaricus bisporus,* has been reported to generate specific inhibitors of melanogenic cells (see Vogel *et al.,* 1977, and references cited therein), an observation indicating a wide structural tolerance for substrate enzyme interaction.

No reduction of tumor size relative to controls was seen when compound (**14**) was assayed against melanoma inoculae (s.c.) of human origin in athymic "nude" mice using the optimum dose regimen based on the B-16 melanoma results (Vogel, 1978).

Vindesine, however, lacks activity against the s.c. form of the B-16 melanoma but has good activity against the P-1534(J) leukemia implanted s.c. Likewise, vindesine, though active against the B-16 melanoma (i.p.), was without activity against the melanoma xenograft (Vogel, 1978).

The role of N-β-p-hydroxyphenethyl VDS (**14**) as a potential agent for melanoma control thus stands in need of further clarification.

D. Auxiliary Assay Methods

Auxiliary *in vitro* systems, employed for assay of small, initial research samples of *Catharanthus* alkaloids, VDS, its congeners, and related products, usually in the form of their water-soluble sulfate salts, have greatly facilitated the current investigation of structure-activity relationships of these agents.

The principal *in vitro* assay, the mitotic arrest of Chinese hamster ovary cells in tissue culture, often furnishes a first indication of biological potency (Boder, 1975; Sweeney *et al.,* 1978). The second *in vitro* assay, used to provide additional information on analogs of special biological or structural consequence, is the avidity of the alkaloids to bind to tubulin of varied origin (e.g., rat, pig brain, or other) (Owellen *et al.*, 1976; Wilson *et al.*, 1978). The radioimmunoassay, developed (Root *et al.*, 1975) for VLB, VCR, and VDS in clinical serum samples (Dyke and Nelson, 1977), constitutes a third, if less direct, method which reflects potency and is used for modification products of special interest.

In these auxiliary systems, it was noted that the potency profile of the various alkaloid agents evidences a broad correspondence with the profile of *in vivo* antitumor activity (Tables VI and VII, Fig. 7). That the Vinca alkaloids cause their cytolytic and—presumably—their *in vivo* antitumor effects through interaction with microtubular protein is a widely held notion (Dahl *et al.*, 1976; DeConti and Creasey, 1975; Donoso *et al.*, 1977; Dustin, 1978; Himes *et al.*, 1976; Owellen *et al.*, 1976; Wilson, 1975), which is wholly compatible with the correspondence, referred to above, between antitumor activity and mitotic arrest, on the one hand, and tubulin binding on the other.

For the alkaloids listed in Table VI, the relative potency to cause mitotic arrest of Chinese hamster ovary (CHO) cells in culture does correspond to that of their antitumor activity—e.g., the P-388 and P-1534(J) leukemia model—in a qualitative (semiquantitative) way.

The alkaloids having *in vivo* activity of the order of VLB (**2–4, 7, 17**) produce 15–50% cells arrested in mitosis, while those alkaloids with lesser *in vivo* activity—e.g., leurosidine, leurosine, vindesine $N_{b'}$-oxide—fail to do so.

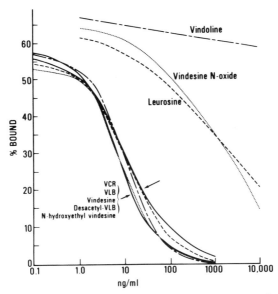

Figure 7. RIA calibration curves for Vinca alkaloids showing effect of increasing concentrations of alkaloid on percentage of antibody-bound ³H-VLB.

TABLE VI

Mitotic Accumulation of Chinese Hamster Ovary Cells in Culture by VDS, VDS Congeners, Parent Alkaloids, Related Products[a]

Alkaloid agent V-CONH-R	Mitotic arrest[b] (dose in μg/ml)			
	2.0	0.2	0.02	0.002
(3) V-CONH$_2$ (VDS)		+ +	+ +	+
(7) V-CONH-CH$_3$			+ +	−
(17) V-CONH-CH$_2$CH$_2$OH		+ +	+ +	−
(20) V-CONH-CH$_2$CH$_2$SCH$_3$		+ +	+ +	−
VDS-$N_{b'}$-oxide[c]	+ +	±		
(1) VLB		+ + +	+ +	−
(2) VCR			+ + +	+
(4) Desacetyl VLB	+	+ +	+ + +	±
Leurosidine[d]	+ +	−		
Leurosine[e]	+ +	−		
Vindoline[f]	−			
Colchicine		+ +	−	

[a] Test results kindly provided by George B. Boder, The Lilly Research Laboratories (see Conrad *et al.*, 1979; Owellen *et al.*, 1976; Sweeney *et al.*, 1978).

[b] + + + = increase of 40–50% cells in mitotic arrest over control values; + + = 15–40%; + = 10–15%; ± = 7–10%; − = 3–7%.

[c] Vindesine $N_{b'}$-oxide inhibits the ROS–GLS only to the extent of 25–50% in the dose range of 0.5–0.8 mg/kg × 9. Used as the free base (Barnett *et al.*, 1978).

[d] Fig. 3

[e] Fig. 2.

[f] "Lower" moiety.

That the observed correspondence lacks full quantitative character is illustrated by two instances. First, the high potency of VLB in arresting CHO cells is a "false" positive in predicting its behavior (negative) in the ROS–GLS model, though a "true" positive for other systems and true in its "prediction" of clinical efficacy. Second, the difference in VDS-vs.-(17) ranking orders (Section III,C) in the ROS–GLS and the B-16 melanoma models, respectively, is not reflected in the mitotic assays for these two agents.

Despite these limitations and lack of quantitative correspondence, the mitotic assay has *materially contributed* to the delineation of the SAR of VLB modification products.

The affinity—relative to that of VLB—with which the several alkaloids bind to pig brain tubulin preparations (Table VII) was determined by competitive exchange with tubulin bound ^{3}H—VLB (Bromer and Kirk, 1975). Similar to the mitotic assay, the profile of binding affinity, observed under the conditions of the experiments, corresponds to *collective* experimental antitumor activity in a qualitative (semiquantitative) way.

It is of interest to note that the relative affinity value is the same for desacetyl VLB (**4**) and VDS (**3**), indicating that differences seen between these two agents at the experimental level reflect differences in distribution, chemical or metabolic alteration, or cell entry, rather than differences in binding affinity to a tubulin receptor site. It is also of interest to compare the affinity values of the parent alkaloids, VDS and its congeners (Table VII), all of which have approximately the same basicity ($N_{b'}$-pK'_a = 7.5 in 66% DMF), with the values of the less active agents, leurosidine (pK'_a = 8.8) and vindesine $N_{b'}$-oxide (pK'_a = 6.6). Whether the apparently maximal binding for agents in the pK'_a = 7.5 range is due to the basicity or to steric factors is a question at present without answer.

The question has been raised (Himes *et al.*, 1976), do the "minor" differences seen in relative affinity values—e.g., 3.9 for VCR versus 1 for VLB—

TABLE VII

Interaction of VDS, VDS Congeners, Parent Alkaloids with Tubulin[a]

Alkaloid agent V-CONH-R	Relative binding affinity[b]
(**3**) V-CONH$_2$ (VDS)	1.8
(**10**) -CONH-CH$_2$CH=CH$_2$	1.1
(**16**) -CONH-CH$_2$CH$_2$NHCOCH$_3$	1.0
(**17**) -CONH-CH$_2$CH$_2$OH	1.4
(**20**) -CONH-CH$_2$CH$_2$SCH$_3$	1.3
(**14**) —CONH—CH$_2$CH$_2$—⟨ ⟩—OH	1.0
(**1**) VLB	1.0
(**2**) VCR	3.9
(**4**) Desacetyl VLB	1.8
VDS $N_{b'}$-oxide[c]	0.07
Leurosidine[d]	0.2
Leurosine[e]	0.5
Vindoline[f]	0.001

[a] Test results kindly provided by W. W. Bromer and J. W. Kirk, The Lilly Research Laboratories. Tubulin source is pig brain.

[b] Relative to value for VLB set arbitrarily at 1. Binding constant of VLB with the pig brain preparations used is of the order of 10^6 liter/mole. Relative affinity data obtained by competitive exchange with tubulin bound ^{3}H-VLB (^{3}H-VLB from Searle/Amersham Corporation, Chicago, Illinois). See related studies by Owellen *et al.* (1976) and Wilson *et al.* (1978).

[c] Vindesine $N_{b'}$-oxide inhibits the ROS–GLS only to the extent of 25–50% in the dose range of 0.5–0.8 mg/kg × 9. Used as the free base (Barnett *et al.*, 1978).

[d] Fig. 3.

[e] Fig. 2

[f] "Lower" moiety.

have actual significance in predicting the differences seen at the experimental and especially at the clinical level? Considerable variation in binding constants of VLB to tubulin preparations of different origin has been reported from different laboratories employing different methods (Hains *et al.*, 1978). The relative affinity values reported here (Table VII) showing differential effects for various alkaloids, obtained with pig-brain tubulin preparations under similar conditions, are considered to be internally consistent and useful for purpose of comparison.

Himes and associates (1976) have examined the effect of VLB, VCR, and VDS on the assembly of bovine-brain tubulin into microtubules *in vitro* and, noting only small (50%) differences, report that the alkaloids inhibit assembly with near-equal effectiveness. On the basis of their observations and those of others on relative tubulin-binding affinity (Owellen *et al.*, 1972), the authors (Himes *et al.*, 1976) concluded that the different oncological effects and clinical toxicities (Donoso *et al.*, 1977) *do not follow* directly from differential effects on tubulin or microtubules, but instead from other biological processes.

The data presented above on a larger group of alkaloids (Table VII) as well as other similar reports (Owellen *et al.*, 1976; Wilson *et al.*, 1978) do show that measurable differences in relative tubulin-binding affinity and tubulin assembly can be experimentally demonstrated. In our view, the differences manifested at the *in vitro* tubulin level are in part, if not largely, responsible for activity differences at the experimental level and, in turn, for the enhanced differences seen at the clinical level in efficacy and especially in the nature of dose-limiting toxicity.

A further discussion of the possible relationship between tubulin binding and neurotoxicity is deferred to the next section (III,E).

Turning to the third auxiliary method, the sensitivity profile (see Fig. 7) in the ^{3}H-VLB-based radioimmunoassay (RIA) for Vinca alkaloids in clinical use again corresponds to that of experimental antitumor activity in a qualitative way. An inspection of the calibration curves for several alkaloids (Fig. 7) shows the RIA to be equally sensitive to VLB, VCR, VDS, desacetyl VLB (**4**), and *N*-β-hydroxyethyl VDS (**17**), alkaloid agents with good collective antitumor activity (Root, 1976). The RIA is less sensitive to the considerably less active species—e.g. leurosine, vindesine $N_{b'}$-oxide (Barnett *et al.*, 1978)—and essentially unresponsive to vindoline, a monomeric alkaloid lacking antitumor activity.

Presence of this sensitivity–activity correlation in the RIA, developed primarily for analytical purposes, does point to a biological significance for this assay, a significance perhaps not entirely unexpected *a posteriori* from the chemical design for the antigen employed.

The RIA, developed to determine serum levels of dimeric alkaloids in the nanogram range in patients receiving VLB, VCR, or VDS, utilizes the antigen prepared by coupling *chemically activated* desacetyl VLB acid azide (V-CON$_3$,

6) to bovine serum albumin (BSA) preferentially through the ε-amino group of lysine units present (Bromer and Kirk, 1975; Conrad *et al.*, 1979). Ultraviolet spectral analysis indicates the presence of 25–35 units of vindesine per BSA molecule, a number considered adequate for effective antibody formation. Because VDS units in the antigen complex are coupled to BSA through the vindoline carboxamide function, thereby leaving the "upper" moiety common to VLB, VCR, and VDS unaltered and suitably exposed, antibodies to this antigen produced by the rabbit may be expected to respond equally to the three alkaloids (Fig. 8).

The RIA, employing displacement of ^{3}H-VLB from its complex with antibody by the alkaloid in serum samples, has been found sensitive to VLB, VCR, and VDS in serum concentrations down to 0.1 ng/ml (Root *et al.*, 1975). The RIA has been used successfully in the phase I clinical trial of vindesine to determine its pharmacokinetics in comparison with VLB and VCR (Dyke and Nelson, 1977; Owellen *et al.*, 1977b).

Teale and associates (1977) have reported an RIA with differing sensitivities for VLB and VCR in the 2–4 ng/ml serum range employing an antigen prepared by a Mannich-type condensation reaction between VLB, formaldehyde, and BSA. The VLB–BSA conjugate thus prepared contained 8 moles vinblastine per mole BSA.

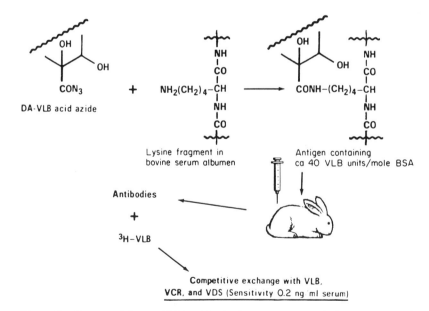

Figure 8.   Reaction of desacetyl VLB azide (**6**) with ε-amino group of BSA lysine unit to form VDS–BSA antigen for the production of the antibodies used in alkaloid RIA.

In summary, the three auxiliary assay methods used positively respond to the alkaloids which do show activity of the order of VLB in one or more of the tumor systems used. Among the group of parent alkaloids, VDS and its congeners, correspondence between *in vitro* response and *in vivo* activity has met with few exceptions, the most conspicuous one of which is the "false" positive of VLB in the face of ROS–GLS inactivity of this agent. When the notion of collective activity, used throughout this discussion, is considered, the VLB exception, too, disappears, demonstrating the general usefulness of these assays in SAR studies for this group of Vinca alkaloids.

E. Toxicological Aspects; Neurotoxicity Potential

Selection of vindesine for clinical evaluation on the basis of its optimum collective antitumor activity was followed by a preclinical toxicological investigation in mice, rats, and dogs (Todd *et al.*, 1976). Comparative acute toxicity studies of vindesine given i.v. to mice and rats showed that the LD_{50} values for vindesine are between those of VLB and VCR in both species (Table VIII). Preliminary studies in dogs and rats demonstrated that large total doses were better tolerated when administered once or twice weekly rather than daily.

Groups of rats and dogs were given a range of i.v. doses of VDS for 90 days. Compared with VCR, dogs given larger doses of VDS and for a longer time had fewer signs of toxicity. The primary toxicity of vindesine in the animals studied was seen in the rapidly proliferating cells that generate the blood cells, intestinal mucosa, sperm cells, etc. (Todd *et al.*, 1976).

Because the experimental antitumor spectrum of VDS resembles that of VCR rather than that of the parent VLB, special attention was given to the experimen-

TABLE VIII

Acute LD$_{50}$ Values of VDS, VDS Congener, Parent Alkaloids[a]

Alkaloid agent sulfate salts	$LD_{50} \pm$ S.E. (mg/kg i.v.)	
	Mouse	Rat
Vinblastine	10.8 ± 0.8	2.9 ± 1.5
Vindesine	6.3 ± 0.5	2.0 ± 0.2
Vincristine	2.1 ± 0.1	1.0 ± 0.1
Desacetyl VLB (**4**)	5.8 ± 0.5	—
N-β-Hydroxyethyl VDS (**17**)	4.4 ± 0.3	—
Vindesine $N_{b'}$-oxide[b]	[c]	

[a] Results kindly provided by Glen C. Todd, The Lilly Research Laboratories (Todd *et al.*, 1976).
[b] Used as the free base (Barnett *et al.*, 1978).
[c] $LD_0 > 10.0$.

tal estimation of vindesine's potential to produce neurotoxicity relative to that of VCR. In order to be useful in predicting clinical neurotoxicity, it is necessary that the experimental model used express the potential for neurotoxicity in a manner that corresponds—at least qualitatively—to the clinical neurotoxicity seen with VCR, VLB, and—eventually—vindesine.

For the purpose of the present discussion, vincristine neurotoxicity seen clinically is described as progressive and frequently dose-limiting, that of VLB as minimal, and that of vindesine as intermediate, when using acceptable, therapeutically indicated dose regimens (Dyke and Nelson, 1979).

Inhibition of axoplasmic transport in the cat sciatic nerve by VLB and VCR, measured *in vitro*, revealed a potency ratio corresponding in a qualitative way with that of the clinical neurotoxicity of these agents. In this model using initially the intact nerve (Ochs and Worth, 1974) and more recently the desheathed sciatic nerve preparation (Chan *et al.*, 1978, 1980), vindesine reproducibly was found to be less potent—by a factor of 2–3—than VCR in inhibiting axoplasmic transport. Reduced potency in this system is held to be indicative of a reduced neurotoxicity potential for VDS, an indication requiring additional reinforcement from *in vivo* models.

The use of axoplasmic transport as a model predictive for clinical neuropathy tacitly implies two assumptions, namely that inhibition of transport by the alkaloids is linked on the one hand to neurotoxicity, and to an interference with microtubular function on the other. Published reports and references listed therein, dealing with the relation of neurotoxicity, axoplasmic transport, and function of neurotubules support both assumptions as reasonable working hypotheses (Bensch and Malawista, 1969; Bradley, 1970; Donoso *et al.*, 1977; Dustin, 1978; Paulson and McClure, 1975; Weiss *et al.*, 1974).

To verify the reduced potential for neurotoxicity suggested for VDS by the axoplasmic transport model, meaningful *in vivo* assays, utilizing the chicken and monkey as test animals, have been developed by Todd and associates (1979). Both models support the notion of a reduced potential for neurotoxicity of VDS relative to that of VCR.

Briefly, chickens receiving biweekly i.v. doses of VDS ranging from 0.8–2.4 mg/kg given over a 4-week period showed signs of general toxicity (hypoactivity, anorexia, lethargy, reduced body weight) without signs of neurotoxicity, while those receiving VCR in doses ranging from 0.4–0.8 mg/kg developed signs of neuromuscular toxicity consisting of ataxia, wing-drop, leg-weakness, and inability to stand, walk, or to hold themselves erect. VLB given in doses of 0.4–0.8 mg/kg produced marked pallor and death without physical signs of neurotoxicity. The mortality incidence in these regimens was VDS 3/8, VLB 7/8, VCR 7/8, control 0/8 (Todd, 1979).

The monkey appears closely to resemble man in the response to vincristine because the burden of toxicity is on the peripheral nerves (Bradley, 1970;

Johnson *et al.*, 1963) in both. Young adult rhesus monkeys (two per group) were given once weekly intravenous doses of VDS ranging from 0.1–0.3 mg/kg and of VCR from 0.1–0.25 mg/kg over a period of 9 months, regimens causing a 50% or greater reduction in leukocyte count. Route of administration and dose regimen selected were designed to approximate closely the clinical conditions under which these agents are used.

Both monkeys receiving VCR from 5 to 7 months developed signs of neurotoxicity, progressing from marked incoordination to contraction of the flexor muscle and formation of a permanent fist, clinical symptoms closely resembling those encountered in patients receiving the drug. Neither monkey receiving VDS for nine months showed signs of neurotoxicity (Todd, 1978).

The three models, *in vitro* and *in vivo*, uniformly predicting reduced neurotoxicity for VDS relative to that of VCR, thus adequately serve in the estimation of the potential of single, selected agents. Because of the expertise required and the duration of the *in vivo* experiments, these models are not well suited for routine estimation of many products.

For the rapid estimation of alkaloid samples for potential neurotoxicity, a tissue culture model utilizing newborn rat midbrain cells was developed recently by King and Boder (1979) at the Lilly Research Laboratories. In this assay, after 24 hours of exposure to alkaloid concentrations between 0.004 and 0.1 μg/ml, there is loss of cell processes and swelling of the cell body. These cell features, expressed in percent of normal cells remaining after exposure, are well suited for quantitative assay purposes.

Results obtained in this tissue culture assay with VLB, VCR, and VDS, correspond with the respective clinical neurotoxicity in a *near-quantitative way* (King and Boder, 1978). Vindesine potency falls between that of vincristine (higher) and vinblastine (lower).

Two further points pertaining to neurotoxicity models deserve mention. First, it has been pointed out (King and Boder, 1979) that differences in neurotoxicity of VCR, VLB, and VDS may be attributed either to differences in their action at the cellular level, including cell entry, effect on microtubular function, and others, or to differences in their pharmacokinetics. Determination of the pharmacokinetics for the three drugs in patients (Dyke and Nelson, 1977) indicate that vincristine has the lowest and vinblastine the highest body clearance, with vindesine intermediate. Thus it follows that both the pharmacokinetics and cellular effects of VCR may contribute to its greater neurotoxicity, though the ratio of the respective contributions cannot be ascertained at present.

Second, the four assays satisfactorily measure neurotoxicity potential, yet little is gained in understanding the relation between alkaloid neurotoxicity and its chemical structure and properties. Although polarity of the molecule is recognized as an important factor, its precise role is in question. Aside from the need for a basic understanding, an answer to the polarity question is needed primarily *to aid in the design* of products with minimal neurotoxicity.

F. Role of Polarity

The role of polarity was examined further because a better understanding of the relation of polarity with cell entry, with neurotoxicity, as well as with antitumor activity (see Section III,H) is thought to be essential in the rational design of VLB modification products.

Neither in the axoplasmic transport model nor in the tissue culture assay for neurotoxicity, *in vitro* systems which are much less complex than the *in vivo* models used, is it clear whether a difference in alkaloid uptake by the cell or a difference in binding affinity for subcellular elements (tubulin) is primarily responsible for the different effects in transport inhibition and neuronal cell changes seen with VLB, VCR, and VDS.

Two recent publications furnish welcome information on comparative *in vitro* alkaloid uptake, namely, uptake of ^{3}H-VLB, -VCR, and -VDS by cat sciatic nerve (Iqbal and Ochs, 1978, 1980) and of ^{3}H-VLB and -VCR by platelets, rat lymphoma, and cultured L-5178Y cells (Gout *et al.*, 1978). The results obtained in these studies at the Indiana University School of Medicine and at the Cancer Research Center, University of British Columbia, respectively, are of special interest and caution against an oversimplification of the relation between polarity and cell entry, and consequently the relation between polarity and neurotoxicity.

Accepting the notion that of the Vinca alkaloid pair, it is VLB, the more lipophilic member, that would be expected to penetrate more readily across nerve cell membranes (Oldendorf, 1974, 1978; Rall and Zubrod, 1962; Rapoport, 1976; Ritchie and Greengard, 1966), Donoso and associates (1977) point to the seeming contradiction between this widely held notion and the observed greater neurotoxicity of the more polar member, VCR (Gerzon *et al.*, 1979).

A tabulation of the relative uptake of tritiated alkaloids (Table IX), their apparent partition coefficients P* (Barnett *et al.*, 1978), and relative tubulin binding affinity values, permits a look at this seeming contradiction under simpler circumstances. The 3–4-fold higher uptake listed for the more polar VCR (P* = 2.15) compared to that for VLB (P* = 2.9), determined in the presence of 3mM calcium, *contrasts* with the higher uptake anticipated for VLB on the basis of the earlier notion of an enhanced uptake associated with lipophilic character (Oldendorf, 1974, 1978).

No immediate explanation for the enhanced uptake of VCR is presently in hand. In a study of uptake and binding of VCR by murine leukemia cells (Bleyer *et al.*, 1975), evidence was obtained for a carrier-mediated transport mechanism translocating the drug into the cells. A comparison with VLB was not included in

TABLE IX

Comparative Uptake of ^{3}H-VLB, ^{3}H-VCR, and ^{3}H-VDS by the Cat Sciatic Nerve in Vitro, in Relation to Apparent Partition Coefficient (P*) and Tubulin-Binding Affinity

Alkaloid	Relative uptake[a]	P*[b]	Relative tubulin binding affinity[c]
^{3}H-Vinblastine	1	2.9	1
^{3}H-Vincristine	3.9	2.15	3.9
^{3}H-Vindesine	~1	2.6	1.8

[a] Uptake measured relative to VLB = 1. Two hour incubation in ^{3}H-alkaloid, 3 mM Ca^{2+} solution (Iqbal and Ochs, 1978, 1980). The cooperation of Drs. Ochs and Iqbal, Department of Physiology, Indiana University School of Medicine, is gratefully acknowledged.

[b] Apparent partition coefficient P* determined between n-octanol and aqueous buffer solution at pH 7.2.

[c] Values relative to VLB = 1, from Table VII.

this investigation. Whether an active transport of the alkaloid or a passive transport is involved in the nerve cell preparation used, remains to be ascertained. In either case, the (higher) tubulin-binding affinity of VCR (cf. Table VII) cannot be without effect on alkaloid uptake.

The higher uptake of VCR by nerve cells conceivably contributes to the higher potency of the alkaloid in causing a block of axoplasmic transport, to its effect on neuronal cells in culture, and to the clinical neuropathy.

In an elegant study of ^{3}H-VLB and ^{3}H-VCR uptake by rat lymphoma and L-5178Y cells, Gout and associates (1978) found that VCR is preferentially retained by the cells. When placed in an alkaloid-free medium, there is rapid efflux of VLB, VCR being retained tenaciously. The resulting high intracellular levels of VCR, according to these authors (Gout *et al.*, 1978), may be largely responsible for the greater antitumor activity and greater toxicity seen with this agent.

These uptake studies are of considerable interest. While they point to further research possibilities, no simple guideline for using polarity values as an aid in drug design has emerged. Within the narrow range of P* values (2.15–2.9), an increase of polarity appears to be associated with an increase of neurotoxicity. Whether an *inverse relationship between lipophilicity and neurotoxicity* exists outside this fairly narrow range of P* values or even outside this small group of alkaloids is not known at this time. A partial answer to this query comes from activity data of a small group of vincristine amide congeners (Section III,H), generally more polar than VCR. Briefly, the antitumor activity of the vincristine amide derivatives (*N*-formyl-*N*-desmethyl vindesines) is decreased with respect to their vindesine counterparts and to VCR.

In concluding the discussion of the role of polarity in activity and toxicity, one helpful indication emerges. In the absence of a more complete understanding of the role of polarity, it appears prudent in the design of VLB modifications *to favor products with P* values between 2.15 and 2.9.* Vindesine (P* = 2.6), desacetyl VLB, and N-β-hydroxyethyl VDS (**17**) products with a similar polarity in the middle of this range are among the more active alkaloids reported. Though restricted to Vinca alkaloid modification, this polarity range is not unlike that mentioned for agents with different pharmacological action (Filer *et al.*, 1977; Peng *et al.*, 1975).

G. Preclinical Pharmacology; Preliminary Clinical Reports of Vindesine

As part of the preclinical pharmacological studies, the effects of vindesine administered i.v. on blood pressure, respiration, and heart rate have been examined in the cat (Nickander, 1976).

The disposition and tissue distribution of ^{3}H-vindesine given i.v. was studied in the rat by McMahon and associates (Culp *et al.*, 1977). No evidence of significant metabolism of VDS in the animal was noted. Trace amounts of polar products, observed to remain near the origin of TLC plates of bile samples, sometimes interpreted as polar metabolites (Owellen *et al.*, 1977c), resemble such polar by-products frequently seen in chemical experimentation with vinca alkaloids. One of these polar products, tentatively identified as desacetyl VLB acid (Hargrove, 1964) sodium salt, has failed to show demonstrable antitumor activity (Cullinan and Gerzon, 1974).

Uptake of ^{3}H-VDS in the sciatic nerve of the rat was relatively low compared to some other tissues (spleen, lung, liver, etc.) but 5-10 times higher than in the brain. There were no significant differences in alkaloid levels in the sciatic nerve of rats given ^{3}H-VDS, ^{3}H-VLB, and ^{3}H-VCR, respectively (Culp and McMahon, 1978).

Urine levels of labeled alkaloid in rats given VDS p.o. amounted to only 1-3% of those resulting from i.v. administration, an observation indicating minimal absorption of vindesine from the gastrointestinal tract.

Preclinical studies of vindesine were concluded in 1974, and clinical evaluation was initiated in 1975 at the Lilly Laboratories for Clinical Research and continued at cancer centers in the United States and abroad.

In support of phase I studies, alkaloid levels in serum from patients receiving VDS for treatment were measured using the RIA developed for this purpose. The pharmacokinetic data obtained were compared with those for VLB and VCR, showing that VDS has a body clearance intermediate between that of VLB (higher) and VCR (lower) (Dyke and Nelson, 1977).

Phase I trial established that adult patients tolerate a dose of 3 mg per square meter of body surface area given once weekly by rapid injection. Phase I-II trial has produced a complete remission (16 months) in a patient with metastatic (skin and lymph nodes) malignant melanoma. Clinically, vindesine appears to be less neurotoxic than VCR, and, generally, its administration has not had to be discontinued because of neurotoxicity (Dyke and Nelson, 1977).

Mathé and associates (1978) report that a high proportion of remissions was obtained in patients with acute lymphoid leukemia. The most remarkable characteristic of vindesine was stated to be the absence of cross-resistance with vincristine as documented in acute lymphoid leukemia (Mathé et al., 1978).

From a phase II study of VDS in the treatment of breast carcinoma, malignant melanoma, and other tumors, it was concluded that vindesine is a clinically active agent with a spectrum of toxicity between that of VCR and VLB (Smith et al., 1978).

Activity of VDS in the treatment of sqamous-cell carcinoma of the esophagus (Kelsen et al., 1979) and of VDS in combination with cis-dichloro-diammine-platinum in the treatment of non-small-cell lung cancer (Casper et al., 1979) has been reported recently.

In terms of feedback information, the absence of resistance reported for VDS in patients resistant to VCR represents a clinical benefit not predicted by the collective activity seen with VDS in the experimental tumor models used.

H. Vincristine Amides

In an extension of the VDS congener series, a limited number of desacetyl VCR amides (N_a-formyl-N_a-desmethyl vindesines) were prepared for biological evaluation. These amides, though accessible from vincristine by routes similar to those used for the preparation of VDS congeners from VLB, preferably are prepared by the elegant, low-temperature CrO_3 oxidation (Richter, 1973) of the corresponding N-substituted vindesine in acetic-acid-actone solution (Barnett, 1976).

The experimental antitumor activity of the desacetyl VCR amides as exemplified by compounds (**27-29**) against the GLS (Table X) was found to be uniformly lower than that of the corresponding VDS congener and lower also than VCR. Based on TLC behavior (silica plates, system ether—CH_3OH—40% CH_3NH_2 solution, 20-4-1), desacetyl VCR amide (27) is a more polar compound than either VDS or VCR, a relationship compatible with the relevant structural changes.

Despite the limited nature of data on desacetyl VCR amides, it appears nevertheless that compounds more polar than VCR are less active than those in the P* range 2.15-2.9. Taken together with decreased activity noted (Table II) for the N-higher-alkyl vindesines, this observation reinforces the stated preference for the VLB-VCR polarity range as an area for modification studies.

TABLE X

Inhibition of Growth of the Gardner Lymphosarcoma by Desacetyl Vincristine Amides[a] and Other Dimeric Alkaloids[b]

	Dose in mg/kg i.p. for 8–10 days							
	0.1	0.2	0.25	0.3	0.4	0.5	1.0	2.0
(27) -CONH$_2$	±	±				±	++	+++
(28) -CONH-CH$_3$			+	++	+++	+++	+++	+++
(29) -CONH-CH(CH$_3$)$_2$						±	+	+++
(3) VDS	+	+++	+++	+++	+++	Tx	Tx	
(7) N-CH$_3$ VDS	−	+++		+++				
(2) VCR	++	+++		Tx				
Desacetyl VCR[c]	+	++		+++	+++	+++	Tx	

[a] Prepared for C. J. Barnett, The Lilly Research Laboratories.
[b] See Table I for activity and toxicity notation.
[c] Prepared by formylation of desacetyl N-desmethyl vinblastine (Gorman, 1967).

J. Bis-(N-ethylidene-vindesine)-disulfide

Introduction of a sulfhydryl function in the vindesine molecule was deemed desirable for several reasons. In a study of the role of tubulin sulfhydryl functions, Mellon and Rehbun (1976) conclude that polymerizability and free sulfhydryl content of tubulin are closely correlated. From the same laboratory it was suggested that inhibition of mitosis in sea urchin eggs by methylxanthines is brought about by disturbance of *in vivo* thiol-disulfide levels in the cell, including a primary effect on glutathione levels (Nath and Rehbun, 1976). Also, from *in vitro* polymerization studies of porcine brain tubulin, it has been concluded that two -SH functions in the tubulin molecule may function as binding sites for microtubule assembly (Kuriyama and Sakai, 1974). Finally, a correlation between growth rate and levels of sulfhydryl and disulfide functions in tumor cells has been demonstrated for a series of rat hepatomas of differing growth rates (Stratman *et al.*, 1975).

Experimental antitumor activity has been demonstrated by sulfhydryl agents as well as by agents reacting with such functions in target molecules. Thus, tumor rejection in experimental animals treated with radioprotective thiols has been observed (Apffel *et al.*, 1975), a rejection said to be caused by disturbance of the interchange between thiols and disulfide functions required for the free proliferation of murine tumor cells (Apffel and Walker, 1973). On the other hand, a large number of potential antitumor agents have been shown to act by S-alkylation of sulfhydryl functions of target enzymes or coenzymes (Fujita and Nagao, 1977).

It seems, therefore, not unreasonable to assume that a VDS-linked sulfhydryl or disulfide function—while the alkaloid resides at the tubulin binding site—could profitably interact with these functions in the tubulin molecule. Such intermolecular interaction involving sulfhydryl and disulfide functions only, basically differs from the *S-alkylation* reaction discussed before.

To explore the consequence of introducing an —SH or —S—S— function on biological activity, *N*-β-mercaptoethyl VDS (**19**), mentioned previously (Table III), and the corresponding disulfide (V-CONH-CH$_2$CH$_2$S-)$_2$ (**30**, Fig. 9), both were prepared pure from the azide (**6**) by the route shown in Scheme II.

$$\underset{\text{V-CON}_3}{+} \text{NH}_2\text{CH}_2\text{CH}_2\text{S-C(C}_6\text{H}_5)_3 \rightarrow \text{V-CONH-CH}_2\text{CH}_2\text{S-C(C}_6\text{H}_5)_3 \xrightarrow[\text{2. H}_2\text{S}]{\text{1. Hg(OAc)}_2} \text{V-CONH-CH}_2\text{CH}_2\text{-SH}$$

$$(\mathbf{19})$$

$$\underset{+}{} (\text{NH}_2\text{CH}_2\text{CH}_2\text{S-})_2 \rightarrow \text{V-CONH-CH}_2\text{CH}_2\text{S-S-CH}_2\text{CH}_2\text{NHCO-V} \underset{(\text{O}_2)}{\overset{(\text{H}_2)}{\rightleftarrows}} \text{V-CONH-CH}_2\text{CH}_2\text{-SH}$$

$$(\mathbf{30})$$

Scheme II

N-β-Mercaptoethyl VDS (**19**), readily obtained in the pure state by mercuric acetate–hydrogen sulfide de-tritylation of the S-trityl intermediate, is readily oxidized as the free base to produce a mixture of (**19**) and the disulfide (**30**). Acquisition of meaningful assay results for the -SH compound (**19**) is thereby precluded. The disulfide, on the other hand, is readily prepared in pure form from the reaction of the azide (2 moles) and cystamine, followed by conversion of the free base to the disulfate salt (Conrad *et al.*, 1978, 1979).

In biological evaluation, the disulfide (Fig. 9) resembles VDS in its activity against the GLS, the P-1534J, and P-388 tumor models and in its acute i.v. toxicity in mice ($LD_{50} = 6.9 \pm 0.6$). Against the B-16 melanoma, the disulfide (MW = 1624) on a milligram for milligram basis resembles or slightly surpasses VDS (MW 753) in its activity. Its tubulin-binding affinity resembles that of VLB, while in its inhibition of CHO cell mitosis and in the sensitivity of the RIA response, the disulfide ranks below VDS.

Noteworthy is the activity shown by the di-sulfide (**30**) against the P-388/VCR murine leukemia (Table XI), a strain reportedly resistant to maytansine as well as to VCR (Wolpert-DeFilippes *et al.*, 1975).

First observed at the National Cancer Institute (Experiment I, Table XI), the lack of complete resistance of the P-388/VCR leukemia strain towards the disulfide (**30**) was subsequently repeated at the Lilly Research Laboratories using the N.C.I. strain (Experiment II). To confirm this observation, the disulfide (**30**) was assayed at the Southern Research Institute, Birmingham, Alabama, against the leukemia strain designated P-388/VCR/I/63 (Experiment III).

In addition to the increase in life span, the therapeutic effect in this experiment was reported in terms of tumor-cell kill, thought to be a more significant criterion than the increase in lifespan (Schabel *et al.*, 1977). At optimum dosage, the disulfide produced a 110–145% increase of lifespan compared on a milligram basis to a 100% increase with VDS. Significantly, the data show that the disulfide treatment reduces the vincristine-resistant P-388 leukemia cell population by 2–3 logs, compared to 1.0 log for VDS.

That VCR-resistance of the P-388/VCR leukemia cell is due at least in part to impaired accumulation and binding within the cell has been suggested by studies of alkaloid uptake (Bleyer *et al.*, 1975). Entrance of the larger, bridged bis-vindesine disulfide molecule (**30**) into these cells which exclude vincristine, however, does not appear a likely event.

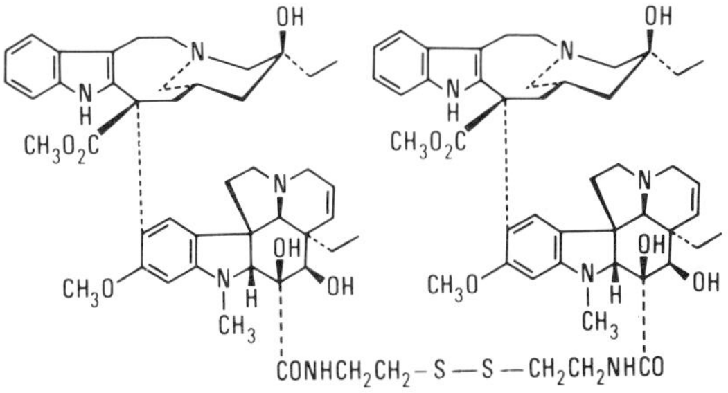

Figure 9.   Molecular structure of bridged bis-vindesine disulfide (**30**), (V-CONH-CH$_2$CH$_2$S-)$_2$.

TABLE XI

Response of P-388/VCR (Experiments I and II) and P-388/VCR/I/63 (Experiment III) Murine Leukemia Strain to the Disulfide ($V\text{-}CONH\text{-}CH_2CH_2S\text{-}I)_2$ and Other Alkaloids[a]

Alkaloid[e]	Experiment I[b] Dose[f]	ILS[g]	Experiment II[c] Dose[f]	ILS[g]	Experiment III[d] Dose[f]	ILS[g]	Approx. cell kill/dose[h]	Approx. no. cells alive[i]
Vincristine	1.7	Tx[j]	1.2	0	1.5	50	1.0	2.7×10^8
	1.0	0	0.9	0				
	0.6	0	0.6	5				
	0.4	0						
Vindesine	3.0	Tx	1.2	3	2.0	100	2.0	2.5×10^5
	1.8		0.9	5				
	1.1		0.6	0				
V-CONH-CH₂CH₂-S	0.65				1.6[k]	145	2.9	4.5×10^2
	3.0	Tx			1.0	115	2.3	3.0×10^4
V-CONH-CH₂CH₂-S	1.8	33	1.2	22				
	1.1	46	0.9	54				
			0.6	35				
Vinblastine	0.65	50			0.7	85	1.7	2.0×10^6
	3.0	TX						

1.8	17	1.2	10
1.1	12[l]	0.9	9
0.65	17	0.6	9

[a] Inoculum of 10^6 cells implanted i.p. on day zero. ILS reported as percent increase in lifespan of treated mice over that of controls. No survivors encountered except for one survivor by VLB at 1.1 mg/kg.

[b] Experiment I done at the National Cancer Institute; courtesy of Dr. Randall K. Johnson (1977). Cell strain P-388/VCR, NCI. Groups of eight male $CD2F_1$ mice; experiment terminated on day 49.

[c] Experiment II done at The Lilly Research Laboratories using the P388/VCR strain received from the NCI. Groups of 10 $B6D2F_1$, mice; experiment terminated at day 45 (Conrad et al., 1979).

[d] Experiment III done at the Southern Research Institute, Birmingham, Alabama; courtesy of W. Russell Laster and John A. Montgomery (1977). Cell strain P-388/VCR/I/63. Groups of 10 male Dublin-CDF_1 mice; experiment terminated at 45 days; cell doubling time determined as 0.49 days.

[e] Disulfide agent in the forms of the disulfate salt, others monosulfates.

[f] Dose in mg/kg given i.p. on days 1, 5, and 9.

[g] Experiment I: ILS expressed in percent increase of the median survival time of treated mice as compared to the median day of death of controls (12.0 days). Experiment II: ILS expressed in percent increase of the average day of death of treated mice as compared to the average day of death of treated mice as compared to the average day of death of controls (10.1 days). Experiment III: ILS as in Experiment I. Controls 10.0 days.

[h] Cell kill expressed in logs.

[i] Number of cells alive at the end of treatment (day 10).

[j] Toxicity.

[k] LD_{20} dose.

[l] One survivor at day 49.

305

In the investigation of a group of bridged bis-acridines, a highly significant correlation was found to exist between their *in vivo* antitumor activity and effects on phenomena associated with plasma membrane-related intercellular interactions (Fico *et al.*, 1977). This observation prompted an attempt by these workers to develop antitumor compounds which primarily affect membrane-related interactions. Among the group of bridged bis-acridines, those with C_6–C_8 connecting chains appeared to have optimum activity (Fico and Canellakis, 1977). It is conceivable that the action of the disulfide (**30**), linking two vindesine moieties by means of the —CH_2CH_2S-S-CH_2CH_2— bridge, also involves membrane-related effects similar to those suggested for the bis-acridines, but no evidence to support this notion exists at present.

In order to gain further information about the structural specificity of the connecting chain, an additional limited number of bisvindesines linked by diverse bridges (Fig. 10), containing sulfur, selenium, oxygen, nitrogen, and carbon atoms were prepared from the azide (**6**) and the appropriate diamine (Conrad and Gerzon, 1978). Preliminary testing of these analogues indicates activity against the B-16 melanoma of the order of VDS for the propylidene congener (Fig. 10, penultimate compound listed) and activity against the P-1534J leukemia ranking below that of VDS for the diselenide (Fig. 10).

In summary, the disulfide (V-CONH-CH_2CH_2S-$)_2$, because of its novel molecular structure and its activity against the P-388/VCR leukemia strain, represents a unique alkaloid agent well worthy of further evaluation. Lower potency (relative to that of VLB and VDS) of this disulfide (**30**) in the axoplasmic transport inhibition assay (Chan *et al.*, 1978, 1980; Ochs, 1978) and also in the neural cell culture assay (Boder, 1979; King and Boder, 1978, 1979) is positive

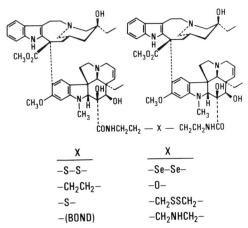

X	X
—S–S–	—Se–Se–
—CH_2CH_2–	—O–
—S–	—CH_2SSCH_2–
—(BOND)	—CH_2NHCH_2–

Figure 10. Bis-vindesines linked by carbon-, oxygen-, sulfur-, selenium-, and/or nitrogen-containing bridges (**30**, -S-S-; **31**, -Se-Se-).

information which, however, stands in contrast to the report on toxicity of the disulfide (**30**) in the chicken model approaching that of VCR (Todd, 1979; Todd *et al.*, 1979).

To assess the prospective therapeutic merit of the disulfide (V-CONH-CH_2CH_2S-)$_2$, determination of its pharmacokinetic behavior in the monkey by means of the RIA method appears highly desirable.

K. Vinblastine Oxazolidinediones

N-Methyl VLB oxazolidinedione (**31**; Fig. 11, R = CH_3), a representative member of a group of such VLB-heterocyclic derivatives (Fig. 11, R = n-C_3H_7, $ClCH_2CH_2$, etc.), has been prepared in a one-step reaction from VLB and methyl isocyanate (Boder *et al.*, 1977a) and has been evaluated against several tumor models by the i.p. and p.o. routes of administration. In tests done at the NCI, Drug Evaluation Branch, N-methyl VLB oxazolidinedione (**31**) given i.p. and p.o. in a single dose was shown to have activity against the P-388/S leukemia model of the order of VLB (Abbott, 1976). The alkaloid (**31**) given i.p. has been reported active also against the murine melanoma, the GLS, and the P-1534J leukemia models (Miller *et al.*, 1978).

The RIA was shown to be somewhat less sensitive to the oxazolidine derivative (**31**) than to VLB, VCR, and VDS, but useful calibration curves have been obtained (Root, 1976). A comparison of RIA-determined serum levels on N-methyl VLB oxazolidinedione (**31**) and VLB (**1**) has been done in monkeys following i.p. and oral dosing. From this determination the half-life of (**31**) was found to be 16.0 hours, and that of VLB 9.6 hours (Miller *et al.*, 1978).

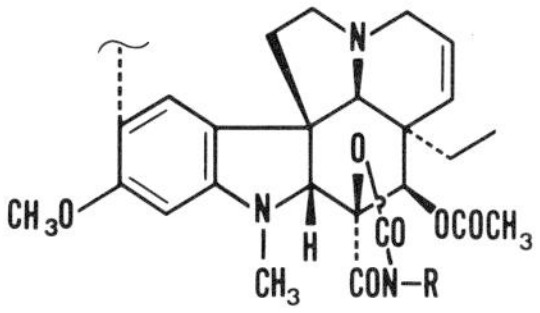

Figure 11.   Partial structure of N-methyl vinblastine oxazolidinedione (**31**; R = CH_3).

IV. CHEMICAL MODIFICATION: VELBANAMINE ("UPPER") MOIETY

Among the modification products derived from VLB by chemical alteration of the carbomethoxy-velbanamine portion, 3′,4′-anhydro vinblastine (**32**; Fig. 12, R = vindoline), obtained first by C-4′ dehydration of VLB (Miller *et al.*, 1977), holds special interest for several reasons. Dehydration is effected by

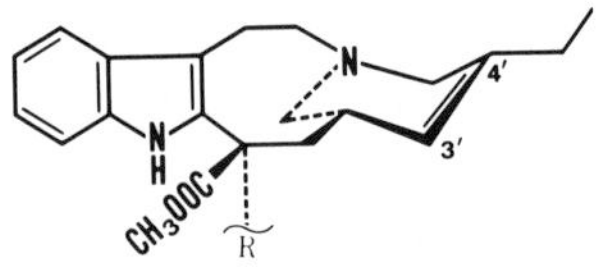

Figure 12.   Partial structure of 3′,4′-anhydro vinblastine (**32**); R = vindoline.

reaction with concentrated sulfuric acid (!), yielding the intermediate desacetyl 3′,4′-anhydro VLB (**33;** Fig. 12, R = desacetyl vindoline), which upon reacetylation furnishes anhydro vinblastine (**32**), a compound relatively unstable, especially in solution, but more stable as the hydrochloride salt.

Anhydro VLB (**32**), not previously reported to occur in *Catharanthus roseus* extracts, has recently been shown to be a "natural" product (Scott *et al.,* 1978), presumably arising in the plant by the reaction of catharanthine and vindoline, but lacking in stability to survive extraction procedures. Moreover, anhydro VLB has been obtained in France and in Canada by a partial synthesis utilizing vindoline and catharanthine, the monomeric Vinca alkaloids freely available from *C. roseus* extracts (Potier *et al.,* 1975; Kutney *et al.,* 1975).

An adequate amount of desacetyl anhydro VLB (**33**) was available for biological assay. Slight inhibition of mitosis of CHO cells and of the growth of the GLS was seen at doses 5–10 times higher than those of VLB needed for optimum effect in these systems (Miller *et al.,* 1977).

Vincadioline (**34**), a natural *Catharanthus roseus* alkaloid, has been characterized as *trans*(?)-3′-hydroxy VLB by nmr spectral analysis and by its behavior in acetylation studies. Vincadioline (**34**) demonstrates metaphase arrest activity in the range of 2 × 10^{-2} to 2 × 10^{-4} mcg/ml (Jones and Cullinan, 1975). Further chemical and biological data for this alkaloid (**34**) will be communicated in a future publication (Thompson, 1980).

(C-4′)-Deoxy vinblastine (**35;** Fig. 13), a natural alkaloid (Neuss *et al.,* 1968), and (C-4′)-deoxy leurosidine (**36;** Fig. 14), prepared by Raney nickel treatment of leurosine (Fig. 2) (Neuss *et al.,* 1968), have been assayed as mitotic inhibitors of CHO cells in culture, against the B-16 melanoma and other tumor models. While deoxy leurosidine (**36**) was appreciably less active than VLB as a mitotic inhibitor, against the B-16 melanoma surprisingly, alkaloid (**36**) has activity of the order of VLB (Boder *et al.,* 1977b). Further research in this area will be reported (Thompson, 1980).

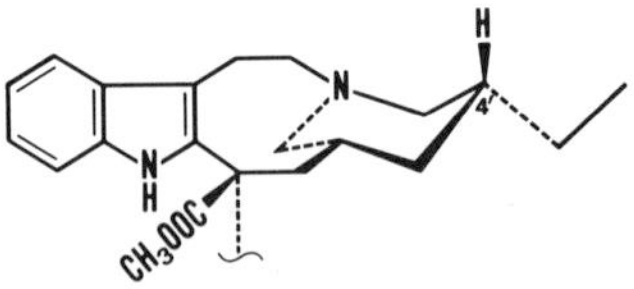

Figure 13.   Partial structure of 4′-deoxy vinblastine (**35**).

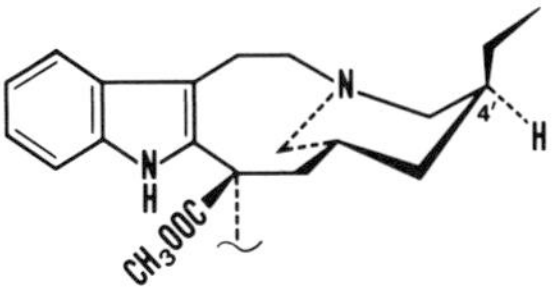

Figure 14. Partial structure of 4′-deoxy leurosidine (**36**).

N-Formyl-N-desmethyl leurosine (**37**, Fig. 15), a leurosine modification product prepared by low-temperature CrO_3 oxidation (Richter, 1973), has been evaluated for acute toxicity and experimental antitumor activity at the National Institute of Oncology, Budapest, Hungary (Somfai-Relle et al., 1975). Chemically, formyl leurosine (**37**) combines features of vincristine (**2**) and leurosine (Fig. 2).

From the reported acute toxicity values (LD_{50} i.p. in the rat is 12.6 mg/kg, and 20 mg/kg in the mouse) and the optimal dose range (0.5–5.0 mg/kg) formyl leurosine (**37**) appears to be about 10 times less potent than VLB. Preliminary reports describing the clinical evaluation of formyl leurosine (**37**) have been presented indicating formyl leurosine to be an active oncolytic agent with negligible neurotoxicity (Eckhardt et al., 1975; Farkas and Eckhardt, 1975). The continued clinical evaluation of formyl leurosine, in progress under the direction of Sandor Eckhardt, M.D., Director, National Institute of Oncology, Budapest (Eckhardt, 1978), is considered to be an auspicious venture. The outcome of this important undertaking may well provide a helpful answer to the question raised earlier: can alkaloids of lesser experimental potency (relative to VLB and VCR) be considered good candidates for clinical evaluation?

Hopefully, clinical efficacy of formyl leurosine in the control of cancer would be a first example of a positive answer to this query.

V. CONCLUSIONS AND SUMMARY; RESEARCH OPPORTUNITIES

The review of vinblastine modifications presented in this chapter focused on two main topics: first, structure–activity and structure–toxicity relationships (SAR) of natural and modified dimeric *Catharanthus* alkaloids, and second, the experimental assay models in their performance to predict—or at least anticipate—clinical activity and clinical toxicity of new products.

While structure modifications in the vindoline (''lower'') moiety have yielded useful information on the SAR of dimeric alkaloids, such modification of the ''upper'' moiety has thus far yielded only limited useful information, due in part to less accessible chemical functionality.

The choice of the GLS as primary assay has been a fortunate one because of

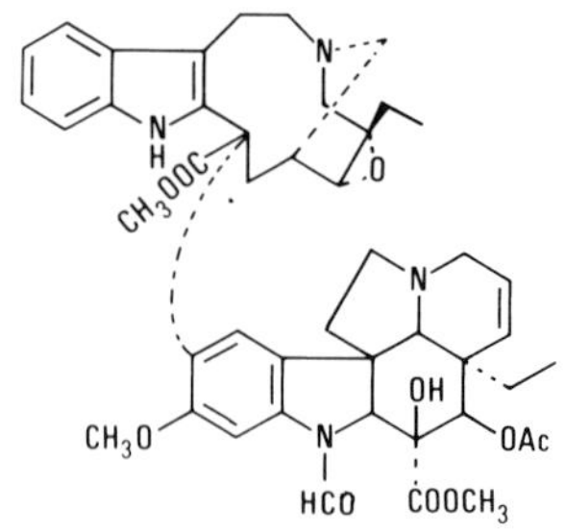

Figure 15. Molecular structure of *N*-formyl-*N*-desmethyl leurosine (**37**).

the qualitative difference in its response to vindesine, most vindesine congeners, and desacetyl VLB (**4**), on the one hand, and the parent alkaloid, VLB, on the other. A qualitative difference in response, especially when shown by a tumor model of adequate predictive merit such as the ROS, B-16 melanoma, or Lewis lung carcinoma, (Geran *et al.*, 1972) should constitute strong recommendation for clinical evaluation of a Vinca alkaloid modification product. In restricting discussion to Vinca alkaloids, it is evident that encouraging clinical results have been achieved with modification products—one thus far—showing a qualitative difference in antitumor activity *while retaining activity and potency* of the parent alkaloid. Future research may show whether promising results can derive also from products showing experimental advantages (e.g., P-1534 survivors with leurosidine) at much reduced potency (equivalent to higher dose) relative to that of the established clinically useful alkaloids.

With a degree of caution, reference is made to the expression "clinical enhancement factor," coined to describe differences in magnitude of related observations at the experimental and at the clinical level. A pertinent example of this enhancement phenomenon is seen by comparing the VLB:VCR potency ratio at the experimental level to that at the clinical level. Expressed in terms of dose required for optimum antitumor effect, the dose of VLB is twice that of VCR at the experimental level, compared to five or more times at the clinical level (Calabresi and Parks, 1975). While the present discussion does not allow analysis of additional examples, two implications of the "clinical enhancement factor" are mentioned briefly. The realization that a small experimental advantage may translate to a substantial clinical advantage gives welcome encouragement to a continued modification effort. On the other hand, realization that, possibly, a "small" experimental disadvantage could likewise be enhanced in patient treatment is a sobering thought indeed.

The term collective antitumor activity, used in this discussion, expresses in useful short form an observation made by Venditti (1975) in a discussion of the relevance of transplanted tumor models. After establishing criteria for individual models, the author states (p. 276): "Priorities for development to clinical trial among agents that meet one of these criteria are established on the basis of the qualitative and quantitative aspects of responses *among all the animal models.*"

Turning to the models used for estimation of potential for neurotoxicity, the axoplasmic transport assay and the chicken and monkey model assisted effectively in the selection of vindesine for clinical evaluation. The neural tissue culture cell assay has done so *ex post facto.* Whether these models will serve equally well in anticipating clinical behavior of other Vinca modification products or antitumor agents of different type remains to be demonstrated.

Vindesine clinical evaluation, hopefully, will continue to provide clinical feedback information for guidance in structure design and in the rational selection of meaningful new tumor models. The value of feedback information from clinical observations is well exemplified by the finding, many years after its discovery, of the beneficial role played by VLB in combination with bleomycin and other agents against testicular tumors (Einhorn and Donahue, 1977). An increased awareness of the need for the chemist and biologist to receive and for the clinical investigator to provide such feedback information is felt to be critical in furthering cooperative chemotherapeutic efforts.

In the course of the above VLB modification studies, certain research areas inviting attention have become apparent. These deal with the chemistry, biochemistry, mode of action, and biological aspects of the dimeric alkaloids.

Chemical exploration of structural alterations, emphasizing those in the "upper" moiety of Vinca alkaloids, is in progress and will be the subject of future publications (Thompson, 1980). "Upper" moiety alterations may also serve the need of complementing the presently used RIA. It is recognized that the RIA, useful in detecting nanogram quantities of VLB, VCR, and VDS in serum samples, does not now distinguish between the alkaloids themselves or their potential, active, or inactive metabolites, if any, which have retained the common "upper" moiety. An RIA based on an antigen uniquely linked to BSA through "upper" moiety functionality could be expected to distinguish between VLB, VCR, VDS, and other agents modified in the vindoline moiety with high sensitivity.

Biochemical areas of investigation include two related topics of tubulin research: first, photogeneration-labeling with Vinca alkaloid derivatives, equipped with suitable photosensitive functionality, of the tubulin receptor-site, and second, affinity–chromatographic approaches to the isolation of tubulin from normal (e.g., nerve) cells and from Vinca alkaloid sensitive and resistant tumor cells. Plans for both projects have been formulated, based on chemistry developed in the course of the VLB modification program (Gerzon and Wilson, 1979). These techniques may permit detection of a structural basis and confirmation for the intriguing suggestion attributing changes in tubulin function in fungal cells as the cause of resistance acquired against the action of a tubulin-binding fungicide (Davidse and Flach, 1977). It is further hoped that these two projects may shed light on differential tubulin structure including that of the alkaloid-binding site. Such information is needed to fathom the *biochemical basis* for antitumor specificity displayed by these alkaloids.

Mode-of-action studies, particularly an extension of alkaloid uptake, retention, and efflux studies (cf. Section III,E) for various cell types, including those of peripheral nerve (Ochs, 1979), may provide a clue to the *cellular basis* for antitumor specificity.

The minimal, if not absent, carcinogenic risk associated with Vinca alkaloid therapy has been noted (Section II,B). The significance of mutagenic activity recently reported for vincristine in an L-5178Y mouse lymphoma cell assay needs to be determined in the face of negative activity of this agent in the Ames *Salmonella*/microsome assay (Matheson *et al.*, 1978) and its lack of carcinogenic activity in the *in vivo* assay reported by Weisburger (1977). Lack of carcinogenic activity of the Vinca alkaloids (Schmähl and Osswald, 1970), an Aesculapian virtue, further urges full exploitation of the potential benefits possibly to emerge from modification of this class of oncolytic agents. Two general approaches are indicated.

First, more extensive structural alterations than those used in vindesine and its congeners can be envisaged to generate far-reaching changes in biological activity. Having shown the feasibility of ''minor'' alterations which lead in vindesine to an expansion of experimental antitumor spectrum and reduced neurotoxicity potential, other agents with ''major'' structural alteration which retain alkaloid activity appear worthy of investigation. The disulfide $(V\text{-}CONH\text{-}CH_2CH_2S\text{-})_2$ is one example fitting this specification; others can be envisaged. A study of the disulfide (**30**) and other bridged bis-vindesines would profit from an analysis of possible membrane-related effects involved in their action (Fico and Canellakis, 1977).

A second approach is exemplified by the potential tyrosinase substrate, *N*-β-*p*-hydroxyphenethyl VDS (**14**), an agent with therapeutic potential in melanoma control. Additional examples of such tumor-type specific modification products are sought which would exploit biochemical and enzymatic differences of tumor cells, an approach advocated as enzyme-pattern directed chemotherapy (Jackson and Weber, 1976).

Finally, a possible topic of future Vinca alkaloid research invites brief mention. This concerns the possible existence in mammalian cells of oligopeptides relating in structure and physiological function to the Vinca alkaloids in the manner enkephalins relate to the morphine group of alkaloids.

The role of such VLB–VCR-like peptides, if found to exist, in the control of microtubule assembly and function, and their presumed effect on cell division processes, would present areas of considerable research interest. Also, the existence of such VLB–VCR-like peptides and of cellular provisions for their transport would reflect a physiological basis for the carrier-mediated transport mechanism proposed for VCR (see Section III,F; see also Bleyer *et al.*, 1975).

Recently, Lockwood (1979) has reported the presence in calf-brain cells of colchicine-like peptides which function in microtubule regulation by binding to the colchicine site of tubulin. The possibility that other microtubule-active agents such as the Vinca alkaloids may also have cellular analogues is mentioned in this significant publication (Lockwood, 1979).

This chapter halts here on a note echoing the sound of an early oratorio (John, 0090).

Appendix I: NSC Numbers of Certain Alkaloid Agents in This Chapter

Vinblastine (**1**), 49842; vincristine (**2**), 67574; vindesine (**3**), 245467; N-β-p-hydroxyphenethyl VDS (**14**), 294582; N-β-hydroxyethyl VDS (**17**), 245468; N-β-methylmercaptoethyl VDS (**20**), 283840; (V-CONH-CH$_2$CH$_2$S-)$_2$ (**30**), 277096; N-methyl VLB oxazolidinedione (**31**), 280865; 3′,4′-anhydro VLB (**32**), 258033; 4′-deoxyleurosidine (**36**), 258034.

Appendix II: Abbreviations

The traditional term "Vinca" alkaloids is used in this chapter to denote dimeric *Catharanthus* alkaloids with experimental antitumor activity.

Section III,A: Vinblastine (1), VLB; vincristine (2), VCR; vindesine (3), VDS; VDS without CONH$_2$ group, V-; V-CONH$_2$ is VDS; V-CONH-CH$_2$CH$_2$OH (Table III) is N-β-Hydroxyethyl VDS; Gardner lymphosarcoma, GLS; Ridgway osteogenic sarcoma, ROS; high performance liquid chromatography, HPLC; thin layer chromatography, TLC.

Section III,D: Radioimmunoassay, RIA; bovine serum albumin, BSA.

Section III,F: National Cancer Institute, N.C.I.

Section V: Structure–activity relationships, SAR.

Acknowledgments

The main part of the research reported in this chapter was done at The Lilly Research Laboratories, Eli Lilly and Company, Indianapolis, Indiana.

The author acknowledges important contributions made to the modification program by: Gerald A. Poore, George B. Boder, Jean C. Miller, William W. Bromer, Mary Root, Charles J. Barnett, William E. Jones, Glen C. Todd (The Lilly Research Laboratories), Randall K. Johnson (Arthur D. Little Co., work done at the N.C.I.), John A. Montgomery, W. Russell Laster, Jr. (Southern Research Institute), Sidney Ochs (Indiana University Medical School), their associates, and others mentioned for personal communications in the reference section.

George J. Cullinan and Robert A. Conrad are remembered for tangible contributions toward development of VDS and for good partnership.

Special thanks go to Irving S. Johnson for guidance and counsel; to Richard W. Dyke, M.D., for the deft touch in clinical evaluation of VDS; and to directors of the Lilly Research Laboratories for encouragement and constant support. The assistance generously given by Janie Siccardi in the preparation of the manuscript is remembered with appreciation.

Met by IFF, invincible pharmacognosist Gordon H. Svoboda stands as friend in the battle against common foe.

REFERENCES

Abbott, B. J. (1976). Personal communication. Drug Eval. Branch, Natl. Cancer Inst.

Abraham, D. J. (1975). In "The Catharanthus Alkaloids" (W. I. Taylor and N. R. Farnsworth, eds.), pp. 125–140. Dekker, New York.

Abraham, D. J., and Farnsworth, N. R. (1969). J. Pharm. Sci. **58**, 694.

Apffel, C. A., and Walker, J. E. (1973). J. Natl. Cancer Inst. **51**, 575.

Apffel, C. A., Walker, J. E., and Issarescu, S. (1975). Cancer Res. **35**, 429.

Barnett, C. J. (1976). Unpublished experiments. Lilly Res. Lab.

Barnett, C. J., Cullinan, G. J., Gerzon, K., Hoying, R. C., Jones, W. E., Newlon, W. M., Poore, G. A., Robison, R. L., Sweeney, M. J., Todd, G. C., Dyke, R. W., and Nelson, R. L. (1978). J. Med. Chem. **21**, 88.

Bensch, K. G., and Malawista, S. E. (1969). J. Cell Biol. **40**, 95.

Bleyer, W. A., Frisby, S. A., and Oliverio, V. T. (1975). Biochem. Pharmacol. **24**, 633.

Boder, G. B. (1975). Unpublished experiments. Lilly Res. Lab.

Boder, G. B. (1979). Unpublished experiments.

Boder, G. B., Miller, J. C., and Poore, G. A. (1977a). Am. Chem. Soc., Natl. Meet., 174th, Chicago, Ill. Abstr. MEDI 47.

Boder, G. B., Paschal, G. C., Poore, G. A., and Thompson, G. L. (1977b). Am. Chem. Soc., Natl. Meet., 174th, Chicago, Ill. Abstr. MEDI 48.

Bradley, W. G. (1970). J. Neurol. Sci. **10**, 133.

Bromer, W. W., and Kirk, J. W. (1975). Unpublished experiments. Lilly Res. Lab.

Calabresi, P., and Parks, R. E., Jr. (1975). In "The Pharmacological Basis of Therapeutics" (L. S. Goodman and A. Gilman, eds.), 5th ed., pp. 1284–1287. Macmillan, New York.

Casper, E. S., Gralla, R. J., and Golbey, R. B. (1979). Proc. Am. Soc. Clin. Oncol. **20**, 337.

Chan, S., Worth, R., and Ochs, S. (1978). Proc. Am. Soc. Neurochem. Meet., Washington, D.C. Abstr. 318.

Chan, S., Worth, R., and Ochs, S. (1980). J. Neurobiol. **11**, 251.

Conrad, R. A., and Gerzon, K. (1978). Unpublished experiments.

Conrad, R. A., Gerzon, K., and Poore, G. A. (1978). Am. Chem. Soc., 10th Cent.-12th Great Lakes Reg. Meet., Indianapolis, Indiana Abstr. MEDI 18.

Conrad, R. A., Cullinan, G. J., Gerzon, K., and Poore, G. A. (1979). J. Med. Chem. **22**, 391.

Creasey, W. A. (1977). Personal communication.

Creasey, W. A., Scott, A. I., Wei, C. C., Kutcher, J. J., Schwartz, A., and Marsh, J. C. (1975). Cancer Res. **35**, 1116.

Cullinan, G. J., and Gerzon, K. (1974). Unpublished experiments.

Culp, H. W., and McMahon, R. E. (1978). Unpublished experiments. Lilly Res. Lab.

Culp, H. W., Daniels, W. D., and McMahon, R. E. (1977). Cancer Res. **37**, 3053.

Cutts, J. H., Beer, C. T., and Noble, R. L. (1960). Cancer Res. **20**, 1023.

Dahl, W. N., Oftebro, R., Pettersen, E. O., and Brustad, T. (1976). *Cancer Res.* **36,** 3101.

Davidse, L. C., and Flach, W. (1977). *J. Cell Biol.* **72,** 174.

DeConti, R. C., and Creasey, W. A. (1975). In "The Catharanthus Alkaloids" (W. I. Taylor and N. R. Farnsworth, eds.), pp. 237–278. Dekker, New York.

DeVita, V. T., Jr., Serpick, A. A., and Carbone, P. P. (1970). *Ann. Intern. Med.* **73,** 881.

Donoso, J. A., Green, L. S., and Heller-Bettinger, I. E. (1977). *Cancer Res.* **37,** 1401.

Dustin, P. (1978). "Microtubules." Springer-Verlag, Berlin and New York.

Dyke, R. W., and Nelson, R. L. (1977). *Cancer Treat. Rev.* **4**(2), 135.

Dyke, R. W., and Nelson, R. L. (1979). Personal communication. Lilly Lab. Clin. Res.

Eckhardt, S. (1978). Personal communication.

Eckhardt, S., Hindy, I., and Farkas, E. (1975). *Int. Congr. Chemother., 9th, London* Abstr. C-113.

Einhorn, L. H., and Donahue, J. (1977). *Ann. Intern. Med.* **87,** 293.

Farkas, E., and Eckhardt, S. (1975). *Int. Congr. Chemother., 9th, London* Abstr. C-114.

Fico, R. M., and Canellakis, E. S. (1977). *Biochem. Pharmacol.* **26,** 269, 275.

Fico, R. M., Chen, T. K., and Canellakis, E. S. (1977). *Science* **198,** 53.

Filer, C. N., Granchelli, F. E., Soloway, A. H., and Neumeyer, J. L. (1977). *J. Med. Chem.* **20,** 1504.

Freireich, E. J., Henderson, E. S., Karon, M. R., and Frei, E., III (1968). "The Proliferation and Spread of Neoplastic Cells." Williams & Wilkins, Baltimore, Maryland.

Fujita, E., and Nagao, Y. (1977). *Bioorg. Chem.* **6,** 287.

Gailani, S. D., Armstrong, J. G., Carbone, P. P., Tan, C., and Holland, J. F. (1966). *Cancer Chemother. Rep.* **50**(1/2), 95–103.

Gardner, W. V., Dougherty, T. S., and Williams, W. L. (1944). *Cancer Res.* **4,** 73.

Geran, R. I., Greenberg, N. H., MacDonald, M. M., Schumacher, A. M., and Abbott, B. J. (1972). *Cancer Chemother. Rep., Part 3* **3**(2), 9.

Gerzon, K., and Wilson, L. (1979). Unpublished experiments.

Gerzon, K., Cochran, J. E., Jr., White, L. A., Monahan, R., Krumkalns, E. V., Scroggs, R. E., and Mills, J. (1959). *J. Med. Chem.* **1,** 233.

Gerzon, K., Ochs, S., and Todd, G. C. (1979). *Proc. Am. Assoc. Cancer Res.* **20,** 46 (Abstr. 186).

Gorman, M. (1967). U.S. Patent 3,354,163.

Gout, P. W., Wycik, L. L., and Beer, C. T. (1978). *Eur. J. Cancer* **14,** 1167.

Gröbe, H., and Palm, D. (1972). *Monatsschr. Kinderheilkd.* **120,** 23.

Hains, F. O., Dickerson, R. M., Wilson, L., and Owellen, R. J. (1978). *Biochem. Pharmacol.* **27,** 71.

Hargrove, W. W. (1964). *Lloydia* **27**(4), 340.

Henderson, E. S., and Samaha, R. J. (1969). *Cancer Res.* **29,** 2272.

Himes, R. H., Kersey, R. N., Heller-Bettinger, I. E., and Samson, F. E. (1976). *Cancer Res.* **36,** 3798.

Hodes, M. E., Rohn, R. J., and Bond, W. H. (1960). *Cancer Res.* **20,** 1041.

Hodes, M. E., Rohn, R. J., Bond, W. H., and Yardley, J. (1963). *Cancer Chemother. Rep.* **28,** 53.

Holland, J. F. (1969). *Cancer Res.* **29,** 2270.

Iqbal, Z., and Ochs, S. (1978). *Proc. Am. Soc. Neurochem. Meet., Washington, D.C.* Abstr. 319.

Iqbal, Z., and Ochs, S. (1980). *J. Neurochem.* **34,** 59.

Jackson, R. C., and Weber, G. (1976). *Biochem. Pharmacol.* **25,** 2613.

John, St. (0090). Chap. 21, vs. 25.

Johnson, I. S. (1968). *Cancer Chemother. Rep.* **52**(4), 455.

Johnson, I. S., Wright, H. F., and Svoboda, G. H. (1959). *J. Lab. Clin. Med.* **54,** 830.

Johnson, I. S., Armstrong, J. G., Gorman, M., and Burnett, J. P. (1963). *Cancer Res.* **23,** 1390.

Johnson, R. K. (1975). Personal communication. Natl. Cancer Inst.

Johnson, R. K. (1977). Personal communication.

Jones, W. E., and Cullinan, G. J. (1975). U.S. Patent 3,887,565.

Kelsen, D. P., Bains, M., Golbey, R. B., and Woodcock, T. (1979). *Proc. Am. Soc. Clin. Oncol.* **20,** 338 (Abstr. C-193).

King, K. L., and Boder, G. B. (1978). *J. Cell Biol.* **79,** 97A.

King, K. L., and Boder, G. B. (1979). *Cancer Chemother. Pharmacol.* **2,** 239.

Kuriyama, R., and Sakai, H. (1974). *J. Biochem. (Tokyo)* **76,** 651.

Kutney, J. P., Ratcliffe, A. H., Treasurywala, A. M., and Wunderly, S. (1975). *Heterocycles* **3,** 639.

Laster, W. R., and Montgomery, J. A. (1977). Personal communication.

Livingston, R. B., and Carter, S. K. (1970). "Single Agents in Cancer Chemotherapy." IFI/Plenum, New York.

Lockwood, A. H. (1979). *Proc. Natl. Acad. Sci. U.S.A.* **76,** 1184.

Mathé, G., Schneider, M., Band, P., Amiel, J. L., Schwarzenberg, L., Cattan, A., and Schlumberger, J. R. (1965). *Cancer Chemother. Rep.* **49,** 47–49.

Mathé, G., Schneider, M., Schwarzenberg, L., Amiel, J. F., Cattan, A., Schlumberger, J. R., and Gracia, A. (1966). *Excerpta Med. Int. Congr. Ser.* **106,** 97.

Mathé, G., Misset, J. L., DeVassal, F., Gouveia, J., Hayat, M., Machover, D., Belpomme, D., Pico, J. L., Schwarzenberg, L., Ribaud, P., Musset, M., Jasmin, C., and DeLuca, L. (1978). *Cancer Treat. Rep.* **62**(5), 805.

Matheson, D., Brusick, D., and Carrano, R. (1978). *Drug Chem. Toxicol.* **1**(3), 277.

Mellon, M. H., and Rehbun, L. I. (1976). *J. Cell Biol.* **70,** 226.

Miller, J. C., Gutowski, G. E., Poore, G. A., and Boder, G. B. (1977). *J. Med. Chem.* **20,** 409.

Miller, J. C., Nelson, R. L., Poore, G. A., Root, M. A., and Todd, G. C. (1978). *Am. Chem. Soc., 10th Cent.-12th Great Lakes Reg. Meet., Indianapolis, Indiana* Abstr. MEDI 17.

Moncrief, J. W., and Lipscomb, W. N. (1965). *J. Am. Chem. Soc.* **87,** 4963.

Montgomery, J. A. (1974). Personal communication. South. Res. Inst.

Moxley, J. H., III, DeVita, V. T., Brace, K., and Frei, E., III (1967). *Cancer Res.* **27,** 1258.

Nath, J., and Rehbun, L. I. (1976). *J. Cell Biol.* **68,** 440.

Neuss, N., Gorman, M., Hargrove, W. W., Cone, N. J., Biemann, K., Büchi, G., and Manning, R. (1964). *J. Am. Chem. Soc.* **86,** 1440.

Neuss, N., Gorman, M., Cone, N. J., and Huckstep, L. L. (1968). *Tetrahedron Lett.* p. 783.

Nickander, R. (1976). Unpublished experiments. Lilly Res. Lab.

Ochs, S. (1978). Unpublished experiments. Indiana Univ. Sch. Med.

Ochs, S. (1979). Unpublished experiments.

Ochs, S., and Worth, R. (1974). *Proc. Am. Assoc. Cancer Res.* **16,** 70 (Abstr. 278).

Oldendorf, W. H. (1974). *Annu. Rev. Pharmacol.* **14,** 239.

Oldendorf, W. H. (1978). *In* "Transport Phenomena in the Nervous System" (G. Levi, L. Battistin, and A. Lajtha, eds.), pp. 103–109. Plenum, New York.

Owellen, R. J. (1975). *Fed. Proc., Fed. Am. Soc. Exp. Biol.* **34,** 808.

Owellen, R. J., Owens, A. H., Jr., and Donigian, D. W. (1972). *Cancer Res.* **32,** 685.

Owellen, R. J., Hartke, C. A., Dickerson, R. M., and Hains, F. O. (1976). *Cancer Res.* **36,** 1499.

Owellen, R. J., Donigian, D. W., Hartke, C. A., and Hains, F. O. (1977a). *Biochem. Pharmacol.* **26,** 1213.

Owellen, R. J., Hartke, C. A., and Hains, F. O. (1977b). *Cancer Res.* **37,** 2597.

Owellen, R. J., Root, M., and Hains, F. O. (1977c). *Cancer Res.* **37,** 2603.

Paulson, J. C., and McClure, W. O. (1975). *Ann. N.Y. Acad. Sci.* **235,** 517.

Peng, G. W., Marquez, V. E., and Driscoll, J. S. (1975). *J. Med. Chem.* **18,** 846.

Potier, P., Langlois, N., Langlois, V., and Guéritte, F. (1975). *Chem. Commun.* p. 670.

Rall, D. P., and Zubrod, C. G. (1962). *Annu. Rev. Pharmacol.* **2,** 109.

Rapoport, S. I. (1976). "Blood-Brain Barrier in Physiology and Medicine." Raven, New York.

Richter, G. V. (1973). Belg. Patent BE811,110.

Ritchie, J. M., and Greengard, P. (1966). Annu. Rev. Pharmacol. **6**, 405.

Root, M. A. (1976). Unpublished experiments. Lilly Res. Lab.

Root, M. A., Gerzon, K., and Dyke, R. W. (1975). Fed. Anal. Chem. Spectrosc. Soc., Natl. Meet., 2nd, Indianapolis, Indiana Abstr. 183.

Rosenthal, S., and Kaufman, S. (1974). Ann. Intern. Med. **80**, 733.

Sartorelli, A. C., and Creasey, W. A. (1969). Annu. Rev. Pharmacol. **9**, 51.

Schabel, F. M., Jr., Griswold, wd. p., jr., Laster, W. R., Jr., Corbett, T. H., and Lloyd, H. H. (1977). Pharmacol. Ther. A **1**, 415.

Schmähl, D., and Osswald, H. (1970). Arzneim.-Forsch. **20**, 1461.

Scott, A. I., Guéritte, F., and Lee, S. L. (1978). J. Am. Chem. Soc. **100**, 6253.

Seino, Y., Nagao, M., Yahagi, T., Hoshi, A., Kawachi, T., and Sugimura, T. (1978). Cancer Res. **38**, 2148.

Smith, I. E., Hedley, D. W., Powles, T. J., and McElwain, T. J. (1978). Cancer Treat. Rep. **62**(10), 1427.

Somfai-Relle, S., Nemeth, L., Jovanovich, K., Szasz, K., Gal, F., Bence, J., Toth, K., and Kellner, B. (1975). Int. Congr. Chemother., 9th, London Abstr. C-66.

Stratman, F. W., Hochberg, A. A., Zahlten, R. N., and Morris, H. P. (1975). Cancer Res. **35**, 1476.

Sugiura, K., and Stock, C. C. (1952). Cancer (Philadelphia) **5**, 382.

Svoboda, G. H. (1961). Lloydia **24**, 173.

Svoboda, G. H., Johnson, I. S., Gorman, M., and Neuss, N. (1962). J. Pharm. Sci. **51**, 707.

Sweeney, M. J., Boder, G. B., Cullinan, G. J., Culp, H. W., Daniels, W. D., Dyke, R. W., Gerzon, K., McMahon, R. E., Nelson, R. L., Poore, G. A., and Todd, G. C. (1978). Cancer Res. **38**, 2886.

Teale, J. D., Clough, J. M., and Marks, V. (1977). Br. J. Clin. Pharmacol. **4**, 169.

Thompson, G. L. (1980). In preparation.

Todd, G. C. (1978). Personal communication. Lilly Res. Lab.

Todd, G. C. (1979). Unpublished experiments. Toxicol. Div., Lilly Res. Lab.

Todd, G. C., Gibson, W. R., and Morton, D. M. (1976). J. Toxicol. Environ. Health **1**, 843.

Todd, G. C., Griffing, W. J., Gibson, W. R., and Morton, D. M. (1979). Cancer Treat. Rep. **63**(1), 35.

Venditti, J. (1975). "Pharmacological Basis of Cancer Chemotherapy," p. 245. Williams & Wilkins, Baltimore, Maryland.

Vogel, F. S. (1978). Personal communication.

Vogel, F. S., Kempter, L. A. K., Jeffs, P. W., Cass, M. W., and Graham, O. G. (1977). Cancer Res. **37**, 1133.

Weisburger, E. K. (1977). Cancer (Philadelphia) **40**(4), 1935.

Weiss, H. D., Walker, M. D., and Wiernik, P. H. (1974). N. Engl. J. Med. **291**, 127.

Wenkert, E., Hagman, E. W., Lal, B., Gutowski, G. E., Katner, A. S., Miller, J. C., and Neuss, N. (1975). Helv. Chim. Acta **58**, 1560.

Wilson, L. (1975). Ann. N.Y. Acad. Sci. **253**, 213, 231.

Wilson, L., Morse, A. N. C., and Bryan, J. (1978). J. Mol. Biol. **121**, 255.

Wolpert-DeFilippes, M. K., Adamson, R. H., Cysyk, R. L., and Johns, D. G. (1975). Biochem. Pharmacol. **24**, 751.

Zubrod, C. G. (1974). In "Anti-Neoplastic and Immunosuppressive Agents" (A. C. Sartorelli and D. G. Johns, eds.), Part I, pp. 1–11. Springer-Verlag, Berlin and New York.

CHAPTER 9

Podophyllotoxins

IAN JARDINE

I. INTRODUCTION

Podophyllotoxins are a particularly instructive class of natural products for consideration in the design and synthesis of potential anticancer agents based upon natural product prototypes. They have a long and fascinating history as medicinals and this has relatively recently culminated in the synthesis of analogues which are indeed clinically efficacious for the treatment of cancers and which are currently being established in the armamentarium of antineoplastic drugs. Their story is made more interesting by the fact that the most useful active analogues appear to have a different and new mechanism of action than the parent compounds and, further, that the development of new active analogues may be a real possibility.

Anticancer Agents Based on Natural Product Models
Copyright © 1980 by Academic Press, Inc.
ISBN 0-12-163150-8

Ian Jardine

II. HISTORY

The North American plant *Podophyllum peltatum* Linnaeus, commonly known as the American mandrake or May apple, and the related Indian species *Podophyllum emodi* Wallich have long been known to possess medicinal properties. A fascinating account of this history has been compiled by Kelly and Hartwell (1954) in a review. The dried roots and rhizomes of these plants is known as podophyllum, and when the podophyllum is extracted with alcohol the resin produced is called podophyllin.

Although variously used in the 19th century as medicinals (e.g., cathartics, cholagogues), podophyllum and podophyllin were largely replaced by more effective agents. However, Kaplan (1942) revived interest in these agents when he cured the venereal wart condyloma acuminatum with topical application of podophyllin in oil. This led to studies of the action of podophyllin on tumor tissue and to intensive chemical examination of the constituents of podophyllin. Thus, King and Sullivan (1946, 1947) demonstrated that podophyllin produced its therapeutic effect by acting, like colchicine, as a mitotic poison, and Belkin (1947), and Hartwell and Shear (1947) reported that podophyllin had a destructive effect on cells of experimental cancer in animals.

The chemistry and antineoplastic activity of the many constituents of podophyllin were then actively investigated in a number of laboratories and particularly by J. L. Hartwell in the 1940s at the National Cancer Institute. A superb review of this later history and the chemistry and pharmacology of podophyllum and its constituents to about 1957 was prepared by Hartwell and Schrecker (1958). This review discusses, in depth, the isolation of the constituents of podophyllum, their chemical properties and reactions, their structural elucidation (mainly by chemical methods), their stereochemical assignments, and synthetic approaches to these compounds.

The crystalline substance podophyllotoxin (**1**), illustrated in Fig. 1, was actu-

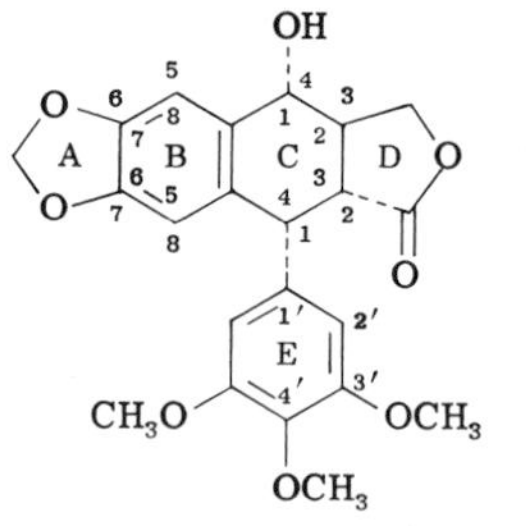

1 Podophyllotoxin

Figure 1. Structure of podophyllotoxin. Note that some authors use the alternative numbering system shown on the exterior of rings B and C. For example, see Fig. 14.

ally first isolated from podophyllin by Podwyssotzki (1880). Its structure was proposed independently by Borsche and Niemann (1932) and by Späth *et al.* (1932) and was revised by Hartwell and Schrecker (1951). A variety of other lignans and lignan glycosides (Fig. 2) were extracted from Podophyllum species and characterized, as discussed by Hartwell and Schrecker (1958). Picropodophyllin glycoside, an isomer of podophyllotoxin glycoside with a *cis*-fused lactone ring (opposite configuration at carbon-3; see Fig. 4) was also isolated. More recently (Pettit, 1977; Pettit and Cragg, 1978; Pettit and Ode, 1979), a series of related lignans, including deoxypodophyllotoxin (**5**), have been isolated from the Japanese evergreen tree *Thujopsis dolabrata* (L. fil.) Sieb. and Zucc. (Cupressaceae) (Akahori *et al.*, 1972). The 3'-desmethyl podophyllotoxin derivative has been isolated by Weiss *et al.* (1975), along with podophyllotoxin and the peltatins (Fig. 2) from *Linum album* Kotschy ex Boiss (Linaceae). Podophyllotoxins have also been isolated from the Polygalaceae plant family (Hoffmann *et al.*, 1978; Hokanson, 1978).

The structurally similar steganacins (Section VI,C, and Fig. 16) have been isolated by Kupchan and co-workers (Kupchan *et al.*, 1973) from the wood and stems of *Steganotaenia araliacea* Hochst.

III. DEVELOPMENT INTO CLINICAL AGENTS

Scattered early attempts of the largely unsuccessful use of podophyllotoxins to treat human neoplasia are summarized by Kelly and Hartwell (1954).

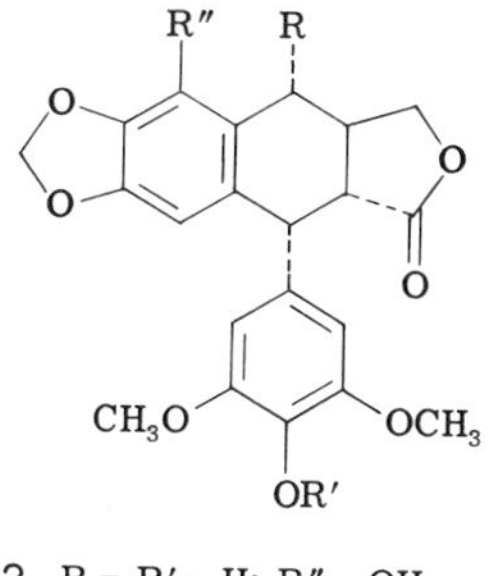

2 R = R' = H; R" = OH. α-Peltatin.

3 R = H; R' = CH$_3$; R" = OH. β-Peltatin.

4 R = OH; R' = R" = H. 4'-Demethylpodophyllotoxin.

5 R = R" = H; R' = CH$_3$. Deoxypodophyllotoxin.

6 R = O-D-glucosyl; R' = CH$_3$; R" = H. Podophyllotoxin glucoside.

7 R = R' = H; R" = O-D-glucosyl. α-Peltatin glucoside.

8 R = H; R' = CH$_3$; R" = O-D-glucosyl. β-Peltatin glucoside.

9 R = O-D-glucosyl; R' = R" = H. 4'-Demethylpodophyllotoxin glucoside.

Figure 2.   Structures of naturally occurring podophyllotoxins.

In a more comprehensive study, α-peltatin was examined by Greenspan *et al.* (1954) as a clinical agent, since it had the greatest activity in several transplantable rodent tumors (Greenspan *et al.*, 1950). After intravenous administration to 45 patients with various metastatic neoplasms, some temporary response was observed but not of significant therapeutic value.

The glycosides of podophyllotoxin, α- and β-peltatin and 4'-demethyl-podophyllotoxin, were found by workers at the Sandoz Laboratories in Basel, Switzerland, to generally inhibit the growth of cell cultures as well as experimentally induced tumors in mice by stopping cell division in early metaphase (Emmenegger *et al.*, 1961). *In vitro,* the glycosides were found to be 10^2–10^4 times less active than their corresponding aglycones, but only 5–20 times less active against Ehrlich's murine ascites tumors. In doses displaying pronounced cytostatic effect, the podophyllum glycosides did not produce the noxious side effects, such as nausea, vomiting, diarrhea, and damage to normal tissue, of their corresponding aglycones. However, these glycosides did not act satisfactorily in clinical trials because of nonspecific side effects.

Complete esterification of the sugar produced derivatives which were, in contrast to the glycosides, only slightly soluble in water and resistant to glycosidases. The antimitotic activity of these derivatives was considerably lower than that of the nonesterified glycosides.

When the glucose residue was only partially substituted by condensation with various aldehydes, however, the specific antimitotic activity of some compounds of this group was found to correspond, *in vitro,* to about that of podophyllotoxin. Significantly, however, they had remarkably low toxic side effects. Further, the extensively tested benzylidene derivative, (**10**), shown in Fig. 3, for example, was found to be well absorbed by the intestinal tract as compared to podophyllotoxin glycoside, and the compound had an increased lifetime in the body.

Alkylation of glycosides of 4'-demethyl podophyllotoxin or α-peltatin produced derivatives which, when tested in Ehrlich's murine ascites tumors, were found to have superior cytostatic properties than the respective glycosides, with unspecific cytotoxicity to normal tissue cells.

Aldehyde condensation products of 4'-demethylpodophyllotoxin β-D-gluco-pyranoside (e.g., **11**, Fig. 3) were found to have a pronounced increase in antimitotic effect in experimental animals when administered in well-tolerated doses, and only a few toxic side effects were noticed.

Neither the 2'-bromo derivative of podophyllotoxin nor its glycoside exhibited any antimitotic activity.

The picro derivatives of podophyllotoxins, obtained by mild base-catalyzed conversion of the *trans*-fused lactone to the more stable *cis*-fused form via the enolate (Fig. 4), as well as some compounds having an open lactone ring, the epimeric glycosido-podophyllinic acids (e.g., **13**, Fig. 5), and hydrazide derivatives (e.g., **14**, Fig. 6) were also tested. The picro compounds displayed a

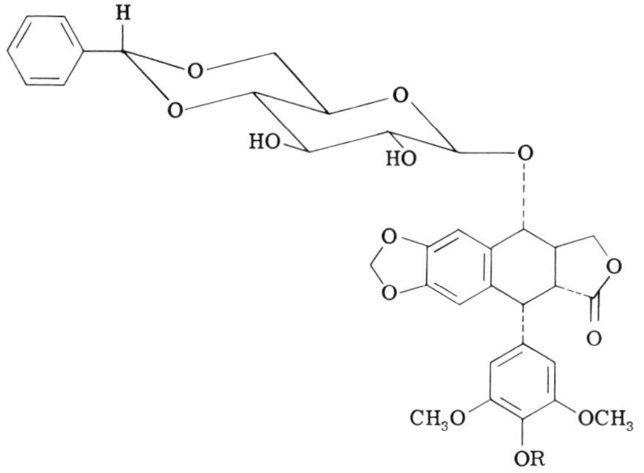

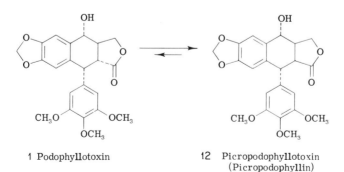

10 R = CH₃. Podophyllotoxin benzylidine-β-D-glucopyranoside.

11 R = H. 4'-Demethylpodophyllotoxin benzylidine-β-D-glucopyranoside.

Figure 3. Structures of semisynthetic podophyllotoxin glucopyranoside derivatives.

specific antimitotic effect but were less active than the corresponding *trans* compounds, and the open lactone derivatives displayed a specific antimitotic effect in cell cultures but only at relatively high concentrations. In animal tests, they were found to be cytostatically ineffective.

Podophyllinic acid ethylhydrazide (**14,** Fig. 6) and podophyllotoxin benzylidine-β-glycopyranoside (**10,** Fig. 3) underwent clinical trials (Chakravort *et al.,* 1967; Lettre and Witte, 1967; Stähelin and Cerletti, 1964; Vaitkevicius and Reed, 1966) but were found to have minimal antitumor activity and were too toxic for general clinical use.

The total synthesis of podophyllotoxin was achieved by Gensler and Gatsonis (1966b) of Boston University, and of podophyllotoxin-β-D-glucopyranoside by

1 Podophyllotoxin

12 Picropodophyllotoxin
    (Picropodophyllin)

Figure 4. The podophyllotoxin–picropodophyllotoxin equilibrium.

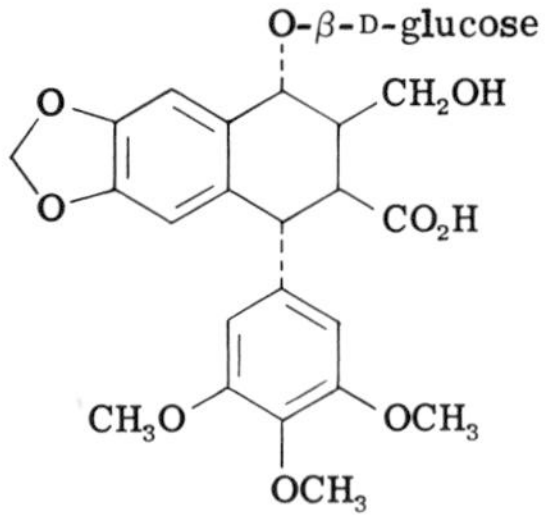

O-β-D-glucose

13 Picropodophyllinic
acid glycoside

Figure 5. Structure of picropodophyllinic acid glycoside.

Kuhn and von Wartburg (1968a) of Sandoz. During the course of extensive studies which led to the latter synthesis, these workers developed a new glycosidation procedure for epipodophyllotoxins (e.g., Fig. 7) which are epimeric to podophyllotoxins at C-1 (Kuhn and von Wartburg, 1968a, 1969) (e.g., **15, 16, 17, 18;** Fig. 7). The resulting epimeric glycoside had no significantly different biological activity compared to the natural podophyllotoxin glycoside.

The aldehyde condensation products of the epipodophyllotoxin compounds were then synthesized (Keller-Juslén *et al.*, 1971). These cyclic acetals of epipodophyllotoxin β-glucopyranoside revealed no dramatic increase in biological activity. However, some cyclic acetals and ketals of 4′-demethylepipodophyllotoxin β-D-glucopyranoside not only exhibited high activity in *in vitro* tests (P-815 mastocytoma cells of mouse in culture) but also gave a significant survival time increase in the mouse lymphocytic leukemia L-1210 test. Two of the most outstanding of these derivatives were selected for clinical trials. These were 4′-demethyl-1-0 - [4,6-0 - (2-thenylidene) - β-D-glucopyranosyl] - epipodophyllotoxin (**19,** Fig. 8) or VM 26 (NSC-122819), and 4′-demethyl-1-0-[4,6-0-(ethylidene-β-D-glucopyranosyl]epipodophyllotoxin (**20,** Fig. 8) or VP 16-213 (NSC-141540) (Stähelin, 1970, 1973).*

VM 26 and VP 16-213 have since been found to be active in Ehrlich ascites tumor, sarcoma 37 and 180, Walker carcinosarcoma, mouse ependymoblastoma, and a variety of murine leukemias as well as L-1210 (Avery *et al.*, 1973; Dombernowsky and Nissen, 1973; Geran *et al.*, 1974; Stähelin, 1970, 1973; Venditti, 1971).

Clinical trials of these agents were undertaken in Europe (Dombernowsky *et al.*, 1972; Jungi and Senn, 1975; Nissen *et al.*, 1972; Sonntag *et al.*, 1974) and the United States (Creaven *et al.*, 1974; Muggia *et al.*, 1971; Rivera *et al.*, 1975a; Sklansky *et al.*, 1974). So far, no significant difference has been found in

---

*VM 26 and VP 16-213 are now also known as teniposide and etiposide, respectively.

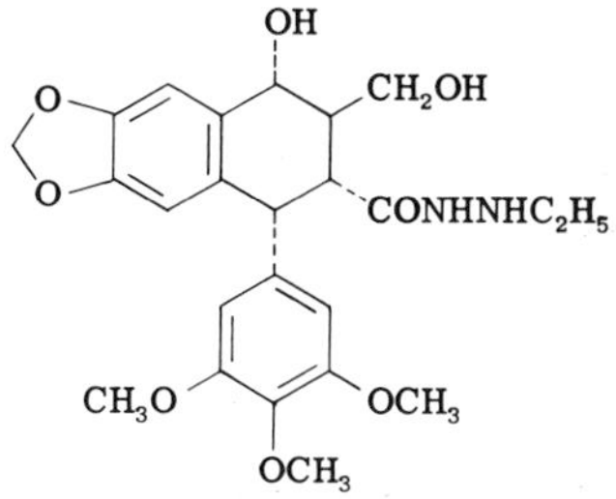

14 Podophyllinic acid
ethylhydrazide

Figure 6.   Structure of podophyllinic acid ethylhydrazide.

the spectrum of clinical antitumor activity of these two compounds (Carter and Slavik, 1976; Goldsmith and Carter, 1973; Rozencweig *et al.*, 1977). VM 26 and possibly VP 16-213 are active in bladder cancer; VP 16-213 is active in acute nonlymphocytic leukemia, in small-cell lung cancer, and possibly in ovarian and thyroid cancer, whereas VM 26 has not been examined much in these tumor types; VM 26 and VP 16-213 are both effective in Hodgkin's disease and non-Hodgkin's lymphomas, especially reticulum-cell sarcoma; VM 26 shows definite antitumor activity in brain tumors; both drugs are ineffective for breast, head, and neck cancers, and soft tissue sarcomas (Carter and Slavik, 1976; Cavalli *et al.*, 1977, 1978; Chiuten *et al.*, 1979; Eagen *et al.*, 1978; Issell and Crooke, 1979; Rozencweig *et al.*, 1977; Sullivan *et al.*, 1979). Also, these compounds are not very myelosuppressive. Further, VM 26, at least in treatment of lymphomas, apparently exhibits no cross-resistance to the standard therapeutic agents

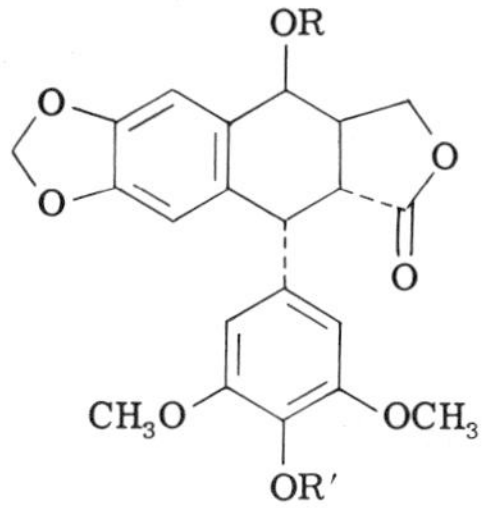

15   R = H, R' = CH$_3$. Epipodophyllotoxin.

16   R = R' = H. 4'-Demethylepipodophyllotoxin.

17   R = β-D-glucopyranoside, R' = CH$_3$.
     Epipodophyllotoxin β-D-glucopyranoside.

18   R = β-D-glucopyranoside, R' = H.
     4'-Demethylepipodophyllotoxin β-D-glucopyranoside.

Figure 7.   Structures of epipodophyllotoxin derivatives.

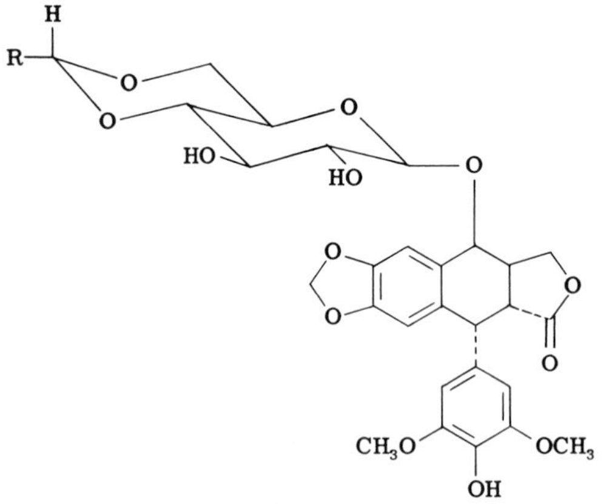

19 R = [thiophene structure]   VM 26 (NSC-122819)

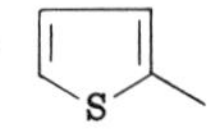

4'-Demethyl-1-0-[4,6-0-(2-thenylidene)-β-D-glucopyranosyl]epipodophyllotoxin.

20 R = CH₃   VP 16-213 (NSC-141540)
4'-Demethyl-1-0-[4,6-0-(ethylidene)-β-D-glucopyranosyl]epipodophyllotoxin.

Figure 8. Structures of VM 26 and VP 16-213.

used for this cancer such as alkylating agents, Vinca alkaloids, and procarbazine and otherwise shows overlapping toxicity with the Vinca alkaloids only in alopecia (Carter and Slavik, 1976; Wilkoff and Dulmadge, 1978).

Currently various cooperative groups are investigating these compounds more extensively (Chiunten *et al.*, 1979, Rozencweig *et al.*, 1977). It is hoped that a useful spectrum of clinical activity will soon be established for these compounds, perhaps especially when they are exploited in combination regimens (Dombernowsky and Nissen, 1976; Osswald, 1978; Pouillart *et al.*, 1978; Rivera *et al.*, 1975b; Roberts and Hilliard, 1978).

IV. STRUCTURE AND CHEMISTRY

A. Podophyllotoxin

The chemistry of podophyllum was reviewed by Hartwell and Schrecker (1958). This excellent review covers isolation procedures, properties, structure determination including stereochemistry and absolute configuration, and synthetic approaches, all mainly for podophyllotoxin. Deoxypodophyllotoxins, dehydropodophyllotoxins, 4'-demethylpodophyllotoxins, the peltatins, sikkimotoxin, lignan glycosides, and even synthetic water-soluble ionic derivatives of podophyllotoxins and peltatins are also covered in depth.

The structure and configuration of podophyllotoxin was established by chemical methods (Hartwell and Schrecker, 1958). A series of chemical interconversions and comparisons of optical rotations revealed a spatial correlation with several related natural lignans of the diarylbutane and aryltetrahydronaphthalene groups (Schrecker and Hartwell, 1955). A key intermediate in these stereochemical deductions, guaiaretic acid dimethyl ether, was correlated with L-3,4-dihydroxyphenylalanine to establish the absolute configuration of podophyllotoxin and related lignans (Schrecker and Hartwell, 1956, 1957) (see Fig. 1).

The first x-ray crystal-structure analysis of one of these lignans, 5' - demethoxy- β- peltatin, was published in 1972 (Bates and Wood, 1972). While confirming the chemically assigned structure of this compound (Bianchi *et al.*, 1969), some doubt was raised regarding the assignment of absolute configuration. A subsequent analysis of the crystal structure of easily obtainable 2' -bromopodophyllotoxin (Kofod and Jorgenson, 1955) by Petcher *et al.* (1973) confirmed the original assignment of absolute configuration of podophyllotoxin as (1*R*, 2*R*, 3*R*, 4*R*). In this latter study, the position of the bromine atom and retention of the configuration at the asymmetric centers were unequivocally determined by chemical degradations and by nmr spectroscopy before the x-ray study.

Hartwell and Schrecker (1958) pointed out that the podophyllum lignans were unique in their antimitotic and tumor-damaging activity, which was closely associated with their unique configuration at C-2, C-3, and C-4, as in podophyllotoxin with its highly strained, *trans*-fused γ-lactone system.

In the presence of mild base catalysis, podophyllotoxin epimerizes smoothly to picropodophyllin (**12**, Fig. 4), which shows little or no cytotoxic activity. This is, in fact, an equilibrium which was examined by Gensler and Gatsonis (1966a) in some detail. They found that podophyllotoxin and picropodophyllin are interconvertible when dissolved in *t*-butyl alcohol with 0.1 *M* piperidine. The equilibrium constants, standard free-energy changes, and standard enthalpy and entropy changes were calculated for this system and compared with qualitative predictions drawn from scale models.

A scale Dreiding model of podophyllotoxin can be built but only with some difficulty. Coplanarity of the methylenedioxybenzene ring with carbon atoms 1 and 4, coupled with the geometric demands of the *trans*-locked lactone ring, produced a strained and inflexible molecule. In contrast, models of picropodophyllin show considerably less rigidity and strain. With models, the latter molecule can, in fact, adopt four easily interconvertible, limiting conformations. Also, space-filling models suggested that in podophyllotoxin the trimethoxy ring cannot rotate entirely freely, whereas in picropodophyllin this group can rotate reasonably freely in most conformations. This rigidity of podophyllotoxin and less strained picropodophyllin predict that the podo-picro equilibrium will show a negative enthalpy and a positive entropy change favoring picropodophyllin, and this was, in fact, quantitatively found (Gensler and Gatsonis, 1966a).

The isomerization of cytotoxic podophyllotoxin to picropodophyllin is thought to occur under physiological conditions (Emmenegger et al., 1961; Kelly et al., 1951; Kocsis et al., 1957) and, since the picro compound is inactive, this would constitute a detoxification by epimerization. However, since the reaction is reversible, Gensler suggested (Gensler and Gatsonis, 1966a) that the low level of activity sometimes observed for picropodophyllin and its analogues might be attributable to the small amount of podophyllotoxin generated by equilibration. However, Gensler (Brewer et al., 1979; Gensler et al., 1977) has modified this suggestion after a conformational study by 60-MHz and 360-MHz ^{1}H nmr on podophyllotoxin and some of its congeners including picropodophyllin. The earlier 60-MHz nmr work of Ayres and co-workers (1972; Ayres and Lim, 1972), which showed that picropodophyllin exists in two conformations, was confirmed. In one of these conformations, the E ring of picropodophyllin is equatorial and perpendicular to the edge of the ABCD ring system, while in the second conformation the E ring flips down into a quasi-axial position, similar to that found for podophyllotoxin. It was, therefore, suggested that it may be this latter conformational form of picropodophyllin that is active (inhibits microtubule assembly), but that this conformation is present in only small amounts of aqueous solutions of picropodophyllin.

This recent study (Brewer et al., 1979) has also found that the 2', 6' proton resonances and the 3', 5' methoxy protons in podophyllotoxin and its congeners are observed as single resonance peaks, suggesting a rapid exchange of the E ring between two conformations, presumed to differ by a 180° rotation about the C-1'-C-4 bond.

Podophyllotoxin is readily epimerized to picropodophyllin in base. Basic hydrolysis of podophyllotoxin, therefore, yields the cis hydroxy acid instead of the trans hydroxy acid (podophyllinic acid). Methanolysis of podophyllotoxin with $ZnCl_2$ as catalyst, however, yields podophyllinic acid methyl ester and neopodophyllotoxin (Kuhn and von Wartburg, 1963; Renz et al., 1965). Neopodophyllotoxin, which contains the lactone group in the 1,3-position can be stereospecifically hydrolyzed by alkali to podophyllinic acid.

Podophyllotoxin has been synthetized by Gensler and co-workers (Gensler et al., 1954; Gensler and Wang, 1954; Gensler and Gatsonis, 1966b), as outlined in Fig. 9 and 10. The last step of the synthesis depends upon the conversion of picropodophyllin to podophyllotoxin. That is, the action of triphenylmethylsodium on the tetrahydropyranyl derivative of picropodophyllin or podophyllotoxin produces the enolate common to both. Irreversible protonation of the enolate with glacial acetic acid followed by removal of the protective group with dilute aqueous acid gives a 45:55 mixture of podophyllotoxin and picropodophyllin which can be chromatographically separated.*

Aiyar and Chang (1975, 1977) have since synthesized the eight possible diastereomeric C-1 alcohols and four possible C-1 ketones of the naturally occurring podophyllotoxin L series.

Schreier (1963, 1964) has synthesized stereoisomeric 6,7-dimethoxy analogues of podophyllotoxins (sikkimotoxins).

Kende *et al.* (1977) have recently described a new and efficient synthesis of (±)-picropodophyllone, which can formally be converted to podophyllotoxin.

B. VM 26 and VP 16-213

Kuhn and von Wartburg (1968a) synthesized podophyllotoxin-β- D-glucopyranoside (**6**, Fig. 11). They reacted podophyllotoxin (**1**) with tetra-*O*-acetyl-α-D-glucopyranosyl bromide in acetonitrile in the presence of Hg(CN)$_2$ to yield tetra-*O*-acetyl-podophyllotoxin-β-D-glucoside (**21**), which was converted into podophyllotoxin- β- D- glucoside by ZnCl$_2$-catalyzed methanolysis. They reported that this *trans* esterification is an advantageous method for the preparation of acid- and base-sensitive glycosides from their corresponding acetyl derivatives.

Kuhn and von Wartburg (1968b) also developed a new glycosidation procedure for the previously unknown glycosides of the epipodophyllotoxin type. Epipodophyllotoxin (**15**, Fig. 11) was shown to react stereoselectively with 2,3,4,6- tetraacetyl- β- D- glucopyranose at low temperature in the presence of BF$_3$-etherate to form the tetraacetate of epipodophyllotoxin-β-D-glucopyranoside (**22**). This acid- and base-sensitive compound was converted to the free glucoside (**17**) by zinc-acetate–catalyzed methanolysis. When podophyllotoxin is glycosidated under the same conditions, inversion of configuration at C-1 of the aglycone moiety occurs leading also to the acetylated epi glucoside. This is presumed to proceed through a common carbonium ion at C-1 generated by BF$_3$ with subsequent substitution by the pyranose group from the less hindered side to give exclusively the epi derivative. The stereochemistry of the glycosidic linkage is determined by the configuration at C-1 of the original glycosidating pyranose moiety. This reaction was shown to be generally useful for the synthesis of hexapyranosides of the epi isomers in the podophyllotoxin series, e.g., 4′- de-methylepipodophyllotoxin-β- D- glucopyranoside and 4′-demethyl-epipodophyllotoxin - β- D- galactopyranoside (Kuhn and von Wartburg, 1959). That is, 4′-demethylepipodophyllotoxin-β-D-glucopyranoside was synthesized (**18**, Fig. 12) by the reaction of the aglycone (**23**) or (**24**) with the corresponding tetra-*O*-

*Podophyllotoxin is available from the Aldrich Chemical Company, Milwaukee, Wisconsin.

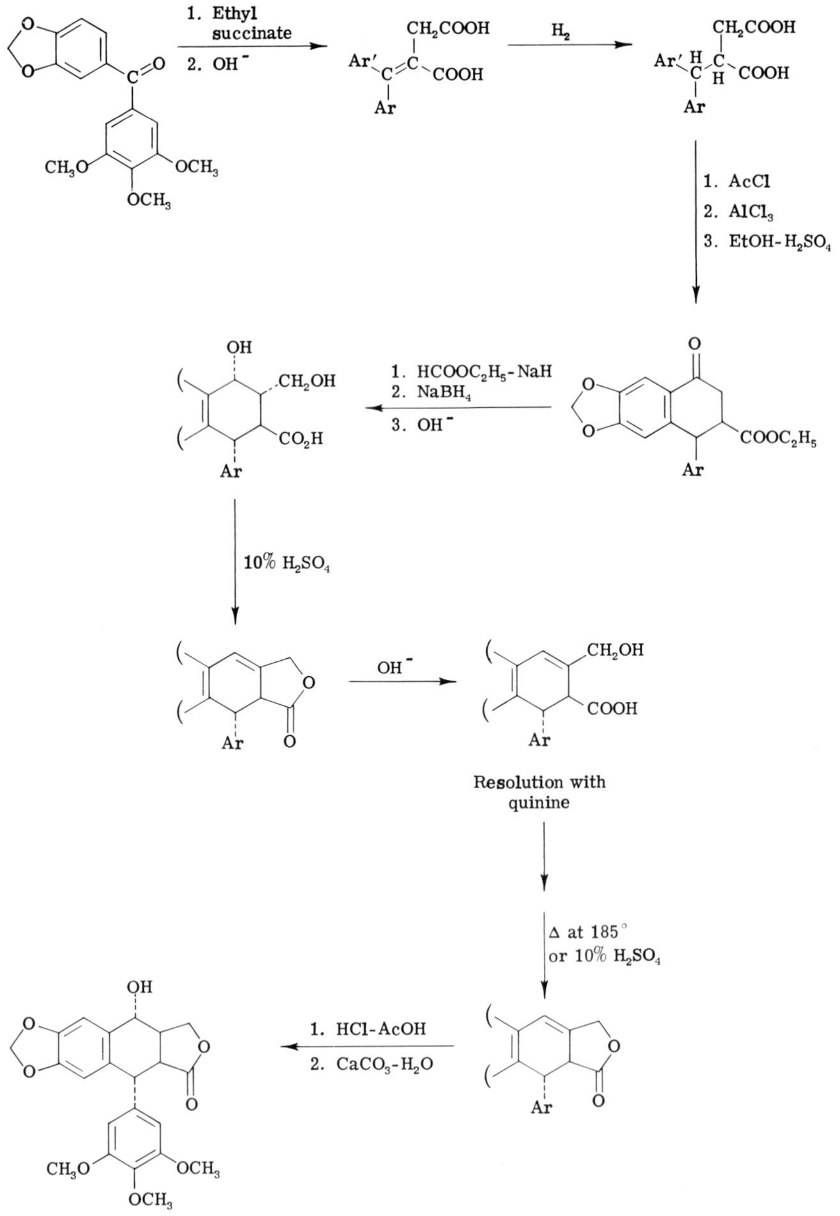

12 Picropodophyllin

Figure 9. Synthesis of picropodophyllin. (From Gensler *et al.*, 1954; Gensler and Wang, 1954.)

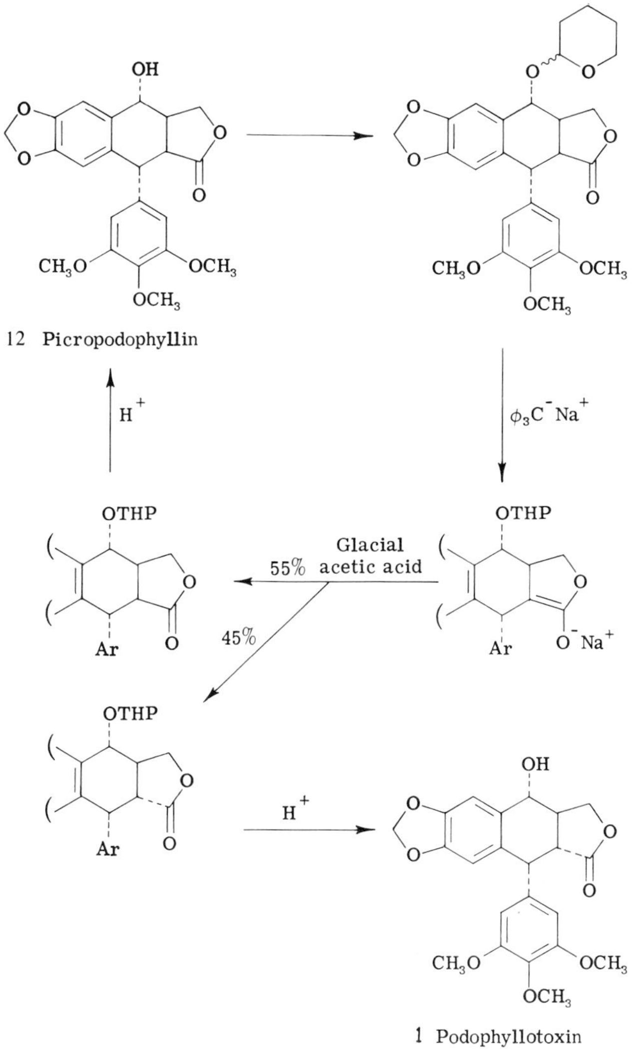

Figure 10. Conversion of picropodophyllin to podophyllotoxin. (From Gensler and Gatsonis, 1966b.)

acetyl-β-D-hexapyranose in the presence of BF_3-etherate, followed by removal of the protecting groups (Fig. 12). The suggested configuration at C-1 of the aglycone moiety in the glycosides was confirmed by nmr. 4'-Demethylated podophyllotoxin can be prepared from podophyllotoxin by a published procedure (Sandoz, 1968).

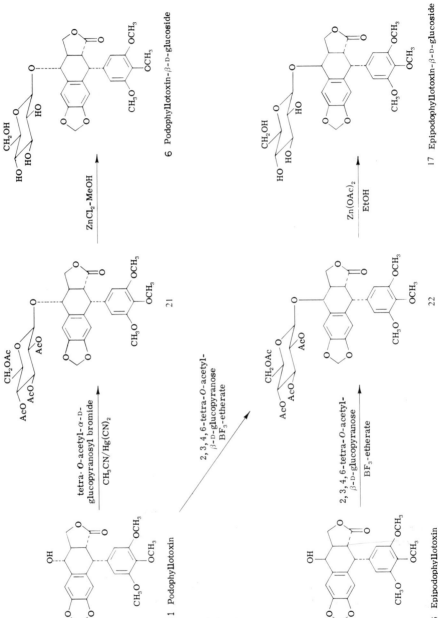

Figure 11. Synthesis of podophyllotoxin-β-D-glucoside and epipodophyllotoxin-β-D-glucoside. (From Kuhn and von Wartburg, 1968a,b.)

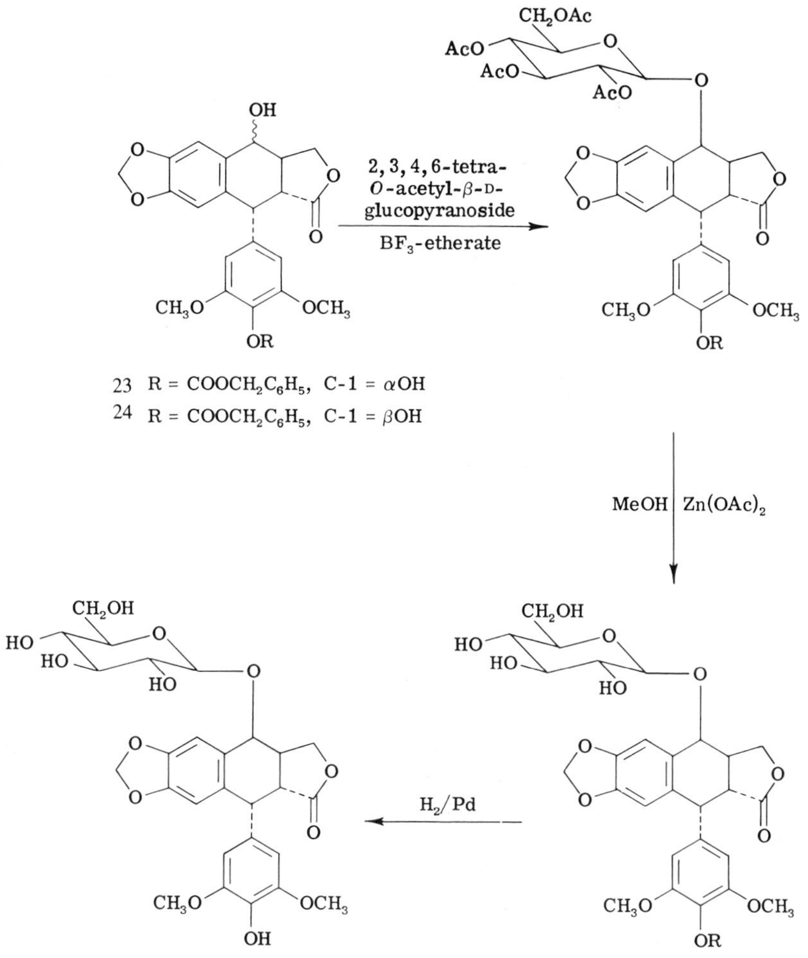

23 R = $COOCH_2C_6H_5$, C-1 = αOH
24 R = $COOCH_2C_6H_5$, C-1 = βOH

18 4′-Demethylepipodophyllotoxin-β-D-glucopyranoside

Figure 12. Synthesis of 4′-demethylepipodophyllotoxin-β-D-glucopyranoside. (From Kuhn and von Wartburg, 1969.)

Conversion of lignan glycosides to cyclic acetals was accomplished (Keller-Juslén *et al.*, 1971) by acid-catalyzed reaction with appropriate carbonyl compounds or their respective acetals or ketals (Fig. 13), the transacetalization reaction being used particularly with simple aliphatic carbonyl compounds. The condensation reaction generally took place on the C-4 and C-6 hydroxyls of the hexapyranose moiety and, with aldehydes, the isomer with an equatorial bond to the aldehyde residue predominates (Fig. 13). A characteristic chemical shift of

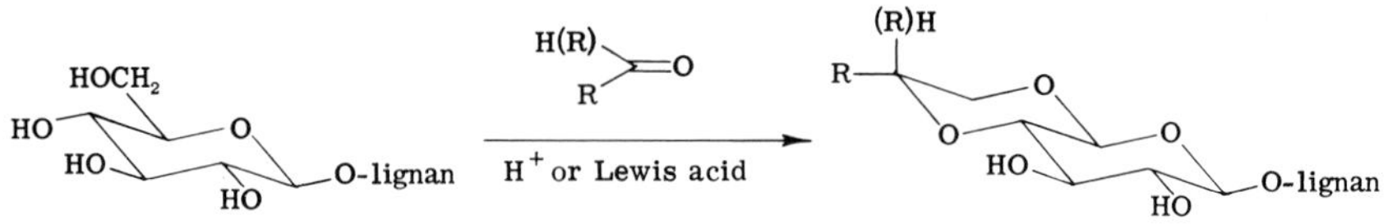

Figure 13. Conversion of lignan glycosides to cyclic acetals. (From Keller-Juslén *et al.*, 1971.)

the axial proton in the nmr spectrum of such sugar acetals can be recognized, especially with acetals of aromatic aldehydes (Bagget *et al.*, 1964).

V. MODE OF ACTION

A. Podophyllotoxin

1. Effect on Microtubules

An essential constituent of the cellular apparatus for cell division (mitosis) (McIntosh *et al.*, 1975) is the microtubules which are hollow, tubelike filaments found in mitotic spindles (Snyder and McIntosh, 1976; Wilson and Bryan, 1974). The subunits of the microtubules are composed of a protein, tubulin, which is capable of reversible polymerization and depolymerization. Tubulin has been isolated in unpolymerized form from a number of cell types, particularly from the brain, where microtubules are constantly formed and dissociated. Tubulin has a molecular weight of 120,000 and consists of two major nonidentical polypeptide chains of the same size, designated α and β. Assembly of microtubules from tubulin can be demonstrated *in vitro* and magnesium and guanosine triphosphate (GTP) bind to tubulin, apparently stabilizing its native conformation and promoting its assembly.

Colchicine (**25,** see Fig. 15 in Section VI,A) is the classic example of a spindle poison which at low concentrations ($\sim 10^{-8}M$) disrupts the assembly and function of microtubules and produces metaphase arrest in dividing cells (Creasey, 1975; Dustin, 1963; Eigsti and Dustin, 1955). This results from the noncovalent binding of colchicine to tubulin, which contains one high-affinity colchicine-binding site (binding constant $\sim 2 \times 10^6 \ M^{-1}$). This binding does not disrupt intact microtubules but rather prevents the polymerization reaction. This colchicine-binding site on tubulin is also a high-affinity binding site for podophyllotoxin, which is thus similarly a metaphase poison and produces similar morphological effects which are practically indistinguishable from those of colchicine (Cornman and Cornman, 1951; Kelly and Hartwell, 1954).

That the mechanism of action of podophyllotoxin was very similar to that of colchicine was first suggested by the finding that podophyllotoxin prevented the binding of colchicine to grasshopper embryo tubulin (Wilson and Friedkin, 1967). Picropodophyllotoxin was much less potent than podophyllotoxin. Similarly, in crude chick-embryo brain extracts, when podophyllotoxin is added prior to or simultaneously with colchicine, the binding of colchicine to tubulin is prevented in a concentration-dependent manner (Wilson, 1970). Tubulin from several different sources has since been used to show that podophyllotoxin competes with colchicine for the colchicine-binding site (Bryan, 1972; Flavin and Slaughter, 1974; Kelleher, 1977; McClure and Paulson, 1977; Wilson and Bryan, 1974; Wilson and Meza, 1973; Wilson et al., 1974; Wilson, 1975; Zweig and Chignell, 1973).

Inhibition of colchicine binding to tubulin by podophyllotoxin is competitive. However, although podophyllotoxin binds at the same site on tubulin as colchicine, the mechanisms of binding are somewhat different. Podophyllotoxin binds rapidly to tubulin, and this binding is rapidly reversible, in contrast to colchicine (Wilson, 1975). Also, podophyllotoxin binds to tubulin in a less temperature-dependent fashion than colchicine, and the relative affinities of these two compounds vary for tubulins from different sources. Further, colchicine binding to tubulin results in positive enthalpy and entropy changes and a relatively large favorable free-energy change which Bryan (1972) suggests indicates that the binding site is located in a hydrophobic or nonpolar pocket. However, it has been demonstrated (Bhattacharyya and Wolff, 1974; Cortese et al., 1976) that tropolone inhibits the binding of colchicine but not podophyllotoxin to tubulin. It has been suggested, therefore, that these drugs occupy overlapping rather than identical sites on tubulin, and that hydrophobic forces play a more significant role in colchicine binding.

β-Peltatin and steganacin (Section VI,C) also have high affinities for the colchicine binding site on tubulin (Kelleher, 1977; Schiff et al., 1978; Wang et al., 1977).

The activity of these compounds can also be conveniently compared by their effects on microtubule assembly (Loike et al., 1978; Wilson and Bryan, 1974) and by their effects on cell division using sea urchin eggs or mammalian cell cultures (Cornman and Cornman, 1951; Schiff et al., 1978; Schindler, 1965; Stähelin, 1970; Wang et al., 1977; Wilson and Bryan, 1974).

In addition to their effect on mitosis, colchicine and podophyllotoxin inhibit other microtubule-dependent processes, such as fast axoplasmic transport (Paulson and McClure, 1975), long saltatory intracellular movements (Freed and Lebowitz, 1970), and cilia regeneration (Makrides et al., 1970).

It should be noted that several studies have reported that the concentration of colchicine active *in vivo* is significantly below that required for activity *in vitro* (Olmsted and Borisy, 1973; Wilson, 1975). Also, colchicine and podophyllotoxin, at about only $10^{-8}M$, produce 50% inhibition of mouse tumor cell growth in culture (Stähelin, 1970), which is about 1% of reported dissociation constants (e.g., Kelleher, 1977). These data suggest that tubulin-binding drugs disrupt motosis by preferential binding to only a small fraction of the tubulin molecules which are critical for polymerization (Kelleher, 1977).

2. Other Biological Effects of Podophyllotoxin

As with colchicine, podophyllotoxin acts on cells in ways which are probably unrelated to any effect on microtubules. That is, podophyllotoxins competitively inhibit nucleoside transport in mammalian cells (e.g., HeLa cells) by inhibiting the facilitated diffusional component of nucleoside transport (Loike and Horwitz, 1976a; Mizel and Wilson, 1972). This effect of podophyllotoxin is greater than that of colchicine but occurs at higher concentration than that required to arrest cell division in mitosis. Picropodophyllotoxin and 4'-demethylepipodophyllotoxin also inhibit nucleoside transport in vitro.

Interestingly, 4'-demethyl analogues of podophyllotoxin, including 4'-demethylpodophyllotoxin, 4'-demethylepipodophyllotoxin, 4'-demethyldeoxypodophyllotoxin, α-peltatin, as well as VM 26 and VP 16-213, induce the intracellular degradation of DNA in HeLa cells (Loike and Horwitz, 1976b).

B. VM 26 and VP 16-213

In contrast to the arrest of cells in metaphase produced by podophyllotoxin and its earlier derivatives (Stähelin, 1972; Stähelin and Poschmann, 1978), VM 26 and VP 16-213 at low concentrations (e.g., 0.005–0.01 μg/ml for VM 26) apparently prevent cells from entering mitosis and arrest cells in the late S or G_2 phase of the cell cycle (Grieder et al., 1974; Huang et al., 1973; Krishan et al., 1975; Loike and Horwitz, 1976a; Misra and Roberts, 1975; Stähelin, 1970, 1972, 1973). In in vivo experiments, Krishan et al. (1975) have also shown that cells accumulate in G_2 after treatment with these epipodophyllotoxin derivatives. The precise biochemical basis of this main effect of VM 26 and VP 16-213, i.e., inhibition of entry of cells into mitosis, is unknown.

In a study on the early effects of VM 26 and VP 16-213 and other cytostatic agents, incuding the podophyllotoxin derivative podophyllotoxin-β-D-benzylidene glucoside, mechlorethamine, 1-β-D-arabinofuranosylcytosine, 6-mercaptopurine, methotrexate, colchicine, vincristine, bleomycin, 1,2-bis(3,5-dioxopiperazin-1-yl) propane (ICRF 159), and x-rays, on mastocytoma cell cultures, it was concluded that only x-ray-induced delay in cell division results in accumulation of cells in the G_2 phase of the cell cycle similar to the G_2 arrest of cells induced by VM 26 and VP 16-213 (Grieder et al., 1977). However, whereas irradiated cells soon resumed proliferation, the drug-induced G_2 arrest was poorly reversible. It was concluded, therefore, that VM 26 and VP 16-213 possess a unique mechanism of action.

At higher drug concentrations, which are probably irrelevant *in vivo*, VM 26 and VP 16-213 arrest cells in metaphase (Stähelin, 1973; Stähelin and Poschmann, 1978). Also at higher drug concentrations in cell cultures, VM 26 and VP 16-213 inhibit the incorporation of tritiated thymidine (Grieder *et al.*, 1974; Loike and Horwitz, 1976a; Muggia *et al.*, 1971; Stähelin, 1970, 1973). VM 26 has also been shown to impair mitochondrial electron transport in the respiratory chain at the NADH dehydrogenase level (Gosalvez *et al.*, 1972, 1974, 1975; Gotzos *et al.*, 1979), and VP 16-213 has been shown to induce single-stranded breaks in DNA (Horwitz and Loike, 1977; Huang *et al.*, 1973; Loike and Horwitz, 1976b). However, these compounds do not bind to microtubules (Loike and Horwitz, 1976a).

VP 16-213 is generally less active *in vitro* and *in vivo* than is VM 26, but it does show a more pronounced therapeutic effect in leukemic mice (Huang *et al.*, 1973; Stähelin, 1972, 1973). A complete comparative analysis of the experimental and clinical features of these compounds by Rozencweig *et al.* (1977) concluded, however, that there was no significant difference between these drugs, at least as far as clinical application is concerned. In humans the maximum tolerated dose of VP 16-213 is, however, three to four times that of VM 26 (Creaven *et al.*, 1974; Muggia *et al.*, 1971). This difference may be attributable to the larger renal clearance of VP 16-213 versus VM 26 and to the smaller degree of serum protein binding of the former (Allen and Creaven, 1975; Creaven and Allen, 1975a,b; Pelsor *et al.*, 1978). Both drugs bind extensively to serum protein, but VM 26 has a nearly 10-fold higher affinity for human serum albumin than does VP 16-213. At the cellular level, the apparent increased antineoplastic activity of VM 26 over VP 16-213 may be accounted for by an increased cellular accumulation and retention of VM 26 over VP 16-213 or perhaps to a greater intrinsic specificity of VM 26 for receptor biomolecules (Allen, 1978a).

Metabolism of these epipodophyllotoxin drugs can be extensive (Allen and Creaven, 1975; Creaven and Allen, 1975a,b), and a major metabolite of VP 16-213, the lactone-ring opened hydroxy acid, has been identified (Allen *et al.*, 1976). A multicomponent pharmacokinetic model of VP 16-213 has been developed for humans to account for drug plus metabolite(s) sequestered in the body (Allen, 1978b; Allen and Creaven, 1975; Pelsor *et al.*, 1978).

The kinetics of the cytotoxicity of VM 26 and VP 16-213 on L-1210 leukemia and hematopoietic stem cells have been investigated by Vietti *et al.* (1978).

A convenient reverse-phase high-performance liquid chromatographic method has been developed for the analysis of these drugs and their metabolites at physiological concentrations (Strife *et al.*, 1980). Ultraviolet detection or, for greater sensitivity, fluorescence detection (Udenfriend *et al.*, 1957) are employed in these essays.

It should be noted that delayed toxic deaths may occur in mice and rats without noticeable acute toxicity after intraperitoneal administration of VM 26 or VP

16-213 (Avery *et al.*, 1973). Hacker and Roberts (1975, 1977) have suggested that this toxicity is caused by drug-induced compromise of hepatic intermediary metabolism. This toxicity, however, does not apparently appear in humans or primates, has been shown to be due to a local effect after intraperitoneal injection, and does not occur after intravenous or oral administration (Hacker and Roberts, 1977; Stähelin, 1976).

VI. STRUCTURE–ACTIVITY RELATIONSHIPS

A. Podophyllotoxin

The high activities of podophyllotoxin, deoxypodophyllotoxin, 4'-demethylpodophyllotoxin, α-peltatin, and β-peltatin were demonstrated in mice bearing sarcoma 37 (Hartwell and Schrecker, 1958). The relative inactivities of many other analogues such as epipodophyllotoxin, picropodophyllin, and podophyllotoxin β-D-glycoside were also demonstrated in this system. These observations were also confirmed by examination of the antimitotic effect and toxicity to rats of many of these compounds (Maturova *et al.*, 1959; Seidlova-Masinova *et al.*, 1957).

More recently, Kelleher (1977) has examined the ability of podophyllotoxin analogues to inhibit colchicine binding to mouse brain tubulin. Mouse brain tubulin was chosen because podophyllotoxin analogues had been extensively analyzed in mouse tumors *in vitro* (Stähelin, 1970) and *in vivo* (Leiter *et al.*, 1950). Podophyllotoxin was found to bind to tubulin more rapidly and in a less temperature-dependent fashion than colchicine, as had been found in other studies (Wilson, 1975). Table I, column A, lists the tubulin-binding affinities found in the Kelleher study for podophyllotoxin analogues with substitutions in rings B, C, and E, whereas Table II, column A, compares the binding affinities of podophyllotoxin analogues with substitutions on the lactone ring. All active analogues in this study were competitive inhibitors. A number of conclusions can be drawn from these results. That is, β-peltatin has a significantly greater affinity for mouse brain tubulin than either podophyllotoxin or colchicine ($K_i = 1.1 \mu M$ in this assay). The isomer epipodophyllotoxin is less active than podophyllotoxin but deoxypodophyllotoxin is of comparable activity to the parent compound, suggesting geometric constraints on the C-1 position at the receptor site. In fact, the above observations taken together suggest that the increased tubulin binding activity of β-peltatin is related more to the presence of a hydroxyl group on ring B than to the lack of one on ring C. This is further supported by the decreased activity, to approximately that of podophyllotoxin, of β-peltatin-B-methyl ether relative to β-peltatin.

Conversion of the C-4' methoxyl group to a hydroxyl as in 4'-demethyl-

	R^1	R^2	R^3	R^4	A^a, K_1 (μM)	B^b, ID_{50} (μM)
Podophyllotoxin	OCH₃	OH	H	H	0.51	0.6
Epipodophyllotoxin	OCH₃	H	OH	H	1.2	5
Deoxypodophyllotoxin	OCH₃	H	H	H	0.54	0.5
β-Peltatin	OCH₃	H	H	OH	0.12	0.7
β-Peltatin-B-methyl ether	OCH₃	H	H	OCH₃	0.57	—
4'-Demethylpodophyllotoxin	OH	OH	H	H	0.65	0.5
4'-Demethylepipodophyllotoxin	OH	H	OH	H	—	2
4'-Demethyldeoxypodophyllotoxin	OH	H	H	H	—	0.2
α-Peltatin	OH	H	H	OH	—	0.5
VM 26	OH	H	Thenylidene glucoside	H	—	c
VP 16-213	OH	H	Ethylidene glucoside	H	—	c

[a] Inhibition constants (K_1) were determined for the inhibition of [³H]-colchicine binding to mouse brain tubulin. From Kelleher (1977).

[b] Chicken brain tubulin (1.0 mg/ml) was incubated at 37° in a standard reaction mixture with various concentrations of drug to determine the ID_{50} or 50% inhibitory dose (the concentration of drug necessary to inhibit microtubule assembly by 50% during a 60-min polymerization reaction). [From Loike et al. (1978). Reprinted with permission; copyright by Cancer Research Journal.]

[c] No inhibition of microtubule assembly was seen at 100 μM.

TABLE II

Tubulin-Binding Affinities and Inhibition of in Vitro Microtubule Assembly by Podophyllotoxin Derivatives Modified in Ring D[c]

	R^2	R^5	R^6	A^{a}, K_i (μM)	B^{b}, ID_{50} (μM)
Podophyllotoxin	OH	O	C=O	0.51	0.6
Deoxypodophyllotoxin	H	O	C=O	1.2	0.5
Podophyllotoxin cyclic ether	OH	O	H₂	5.2	1
Deoxypodophyllotoxin cyclic ether	H	O	H₂	—	0.8
Deoxypodophyllotoxin-cyclopentane	H	H₂	H₂	—	5
Deoxypodophyllotoxin-cyclopentanone	H	C=O	H₂	—	5
Podophyllotoxin cyclic sulfide	OH	S	H₂	—	10
Deoxypodophyllotoxin cyclic sulfide	H	S	H₂	—	10
Podophyllotoxin cyclic sulfone	OH	SO₂	H₂	—	d
Deoxypodophyllotoxin cyclic sulfone	H	SO₂	H₂	—	d
Podophyllic acid 2-ethylhydrazide	OH	(see Fig. 6)		4.5	—
Picropodophyllin	OH	(see Fig. 4)		10	30
Picropodophyllic acid	OH	(see Fig. 5)		Inactive	—

[a] Inhibition constants (K_i) were determined for the inhibition of [³H]-colchicine binding to mouse brain tubulin. From Kelleher (1977).

[b] Chicken brain tubulin (1.0 mg/ml) was incubated at 37°C in a standard reaction mixture with various concentrations of drug to determine the ID_{50} or 50% inhibitory dose (the concentration of drug necessary to inhibit microtubule assembly by 50% during a 60-min polymerization reaction). [From Loike et al. (1978). Reprinted with permission; copyright by Cancer Research Journal.]

[c] The numbering system for ring D is taken from Loike et al. (1978).

[d] No inhibition of microtubule assembly was seen at 100 μM.

podophyllotoxin results in only a slight decrease in activity. All podophyllotoxin analogues with alterations to the lactone ring have reduced affinity for tubulin. Hydrophilic substitutions, as with glucopyranosides, decrease tubulin binding considerably (results not given), although the bulk of these groups may be significant.

It would appear that the lactone ring area may be involved in the interaction with tubulin. In fact, Gensler *et al.* (1977) synthesized several delactonized derivatives of podophyllotoxin to block deactivating epimerization, by changing the lactone carbonyl to methylene. Several biological assays, namely three cell culture cytotoxicity tests, binding to tubulin, and inhibition of microtubule assembly, showed that, surprisingly, most of these nonenolizable derivatives retain activity although this is generally reduced with respect to podophyllotoxin.

This latter study has been extended (Loike *et al.*, 1978) to examine the inhibition of microtubule assembly *in vitro* by a number of podophyllotoxin analogues, including nonenolizable derivatives of podophyllotoxin and deoxypodophyllotoxin, epipodophyllotoxin, picropodophyllotoxin, several 4'-demethyl compounds, as well as VM 26 and VP 16-213. Tables I and II, column B, list the results obtained for inhibition of chicken brain microtubule assembly *in vitro* by podophyllotoxin and these analogues.

Podophyllotoxin, deoxypodophyllotoxin, and β-peltatin have similar potencies in this assay, and when the 4'-methyl group is removed from these compounds the resulting analogues, 4'-demethylpodophyllotoxin, 4'-demethyldeoxypodophyllotoxin, and α-peltatin, are more active. 4'-Demethyldeoxypodophyllotoxin is, in fact, the most active compound in this study. Similarly, although epipodophyllotoxin is an order of magnitude less active than podophyllotoxin, suggesting unfavorable interactions between tubulin and the hydroxy group of epipodophyllotoxin, 4'-demethylepipodophyllotoxin is only about 3-4 times less active than podophyllotoxin. Thus, in general, the 4'-hydroxy analogues are at least as active as their respective 4'-methoxyl analogues. As mentioned previously, 4'-demethyl derivatives of podophyllotoxin, epipodophyllotoxin, and deoxypodophyllotoxin also induce the intracellular degradation of DNA in HeLa (Loike and Horwitz, 1976b).

Whereas substitutions in ring B of deoxypodophyllotoxins to give the peltatins do not alter the activity to a great extent, substitutions in ring D of podophyllotoxins decrease the activity. Equivalent substitutions also have parallel effects in decreasing the activity of podophyllotoxins and deoxypodophyllotoxins (Table II). However, the ring D cyclic ether derivatives of both of these compounds still retain considerable activity.

Although the lactone group of ring D is not required for activity, substitutions at position 12 in ring D of compounds with a methylene group at position 13 show considerable losses of activity which become more severe as the substitutions become more bulky. This suggests strict steric requirements for the interaction of position 12 with tubulin.

These authors (Brewer *et al.*, 1979) have also completed a 350-MHz nmr examination of the conformations of podophyllotoxin and these congeners (Tables I and II, column B) and have found that the conformations of these derivatives are identical with the exception of picropodophyllin (see Section IV,A). The 360-MHz proton nmr spectra of podophyllotoxin, picropodophyllotoxin, epipodophyllotoxin, and VP 16-213 are reproduced in Fig. 14. Note that an alternative numbering system is used in this work.

From the above structure–activity relationship studies it can be concluded that the relationship between the structural modifications and the resulting antimitotic activity supports the suggestion that the C and D rings of these compounds are involved in their interaction with tubulin. Specifically (Brewer *et al.*, 1979), the activity of these compounds is sensitive to the configuration, size, and/or hydrophilic character of substituents at the C-1 position in the C ring and to the steric features of substituents at the 12 position of the D ring.

Colchicine has been studied extensively with respect to structure–activity relationships (Dustin, 1963; Eigsti and Dustin, 1955; Kelleher, 1977; Margulis, 1974, 1975; Schindler, 1965; Zweig and Chignell, 1973) and, since colchicine and podophyllotoxin both bind to the same site on tubulin, comparative structural studies of these compounds may be considered relevent regarding an examination of podophyllotoxin.

For example, from x-ray structural data, Margulis (1974, 1975) has suggested that the major points of structural overlap of colcemid (*N*-desacetyl-*N*-methyl-colchicine) **(26)** and podophyllotoxin are those labeled *a*, *b*, and *c* in Fig. 15. The points labeled *a* are similar in that they are groups capable of providing the hydrogen atom in a hydrogen bond. However, loss of these groups as in deacetylaminocolchicine (Schindler, 1965) and deoxypodophyllotoxin (Loike *et al.*, 1978) does not diminish the activity of these compounds. Kelleher (1977) suggests that the lower oxygen on the lactone ring of podophyllotoxin, instead of the methylenedioxy-oxygen, is structurally analogous to the tropolone methoxy group on colchicine (or colcemid). The fact that colchicine analogues without OCH_3 or SCH_3 on the tropolone ring (Kelleher, 1977; Zweig and Chignell, 1973) are inactive as inhibitors of colchicine binding are suggested as supportive of this view (Kelleher, 1977). However, since the carbonyl oxygen of the lactone in podophyllotoxin is not required for activity, this suggestion may be doubtful. The question, therefore, of the molecular relationship of colchicine and podophyllotoxin with respect to their tubulin-binding activity would seem to be not completely resolved at this point, although the suggestion of Margulis is the most convincing.

Some other aspects of structure–activity relationships of podophyllotoxins—for example, the relative inactivity of picropodophyllotoxin—have been mentioned in previous sections of this chapter.

Other structural analogues of podophyllotoxin, some of which only vaguely

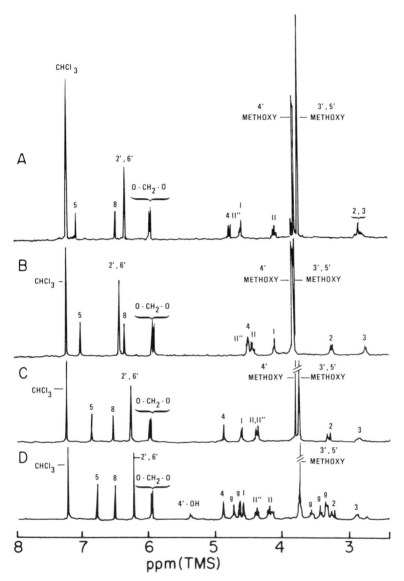

Figure 14. 360-MHz proton nuclear magnetic resonance spectra of: A, podophyllotoxin; B, picropodophyllotoxin; C, epipodophyllotoxin; and D, VP 16-213. Note that the alternative numbering system referred to in Fig. 1 is used here. (Reprinted with permission from Brewer *et al.*, 1979, copyright by the American Chemical Society.)

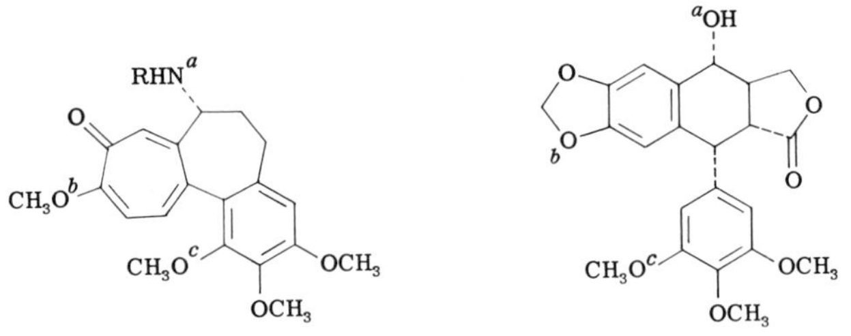

25 R = CH₃CO Colchicine   1 Podophyllotoxin
26 R = CH₃   Colcemid

Figure 15. Suggested points of structural overlap of colchicine and colcemid with podophyllotoxin. (From Margulis, 1974.)

resemble the parent compound, have been tested and have been generally found to be without biological activity (Kelleher, 1977; Loike *et al.*, 1978; Smissman *et al.*, 1976).

B. VM 26 and VP 16-213

The inactivity of VM 26 and VP 16-213 in the inhibition of microtubule assembly has been noted above (Table I). One of the preferred rotamer conformations for the ethylidine glucoside moiety of VP 16-213 is over a C-11 proton of ring D of this compound (Brewer *et al.*, 1979; Loike *et al.*, 1978). This suggests that the bulky glucoside moiety could sterically interfere with the ability of ring D of VP 16-213 and other similar analogues to bind to tubulin.

When Keller-Juslén *et al.* (1971) synthetized VM 26 and VP 16-213, they made a large number of different aldehyde and ketone condensation products. The two compounds selected, VM 26 and VP 16-213, had the highest activities against P-815 mastocytoma cells of the mouse *in vitro* (VM 26, $ED_{50} = 0.0048$ mg/liter) and against mouse leukemia L-1210 (VP 16-213, survival time increase $= 167\%$), respectively. These compounds also possessed high activity in the other test. That is, VM 26 produced 121% survival time increase in the L-1210 test, and VP 16-213 had an ED_{50} of 0.031 mg/liter in the P-815 test. Many other analogues possessed considerable activity in both these screens. A sample of these is listed in Table III.

It might be noted that comparatively small changes in R^1 can produce fairly dramatic differences in activities in these two tests. For example, in going from CH_3 to C_2H_5 the ED_{50} for P-815 changes from 0.031 to 0.0085 mg/liter, a substantial increase in activity in this assay. However, the L-1210 survival drops from 167% to 97%. Similarly, the ED_{50}s for R^1 as $(CH_3)_2CH$, $C_2H_5CHCH_3$, $(CH_3)_2CHCH_2$, $n\text{-}C_4H_9$, and C_5H_9 are comparable for these highly active deriva-

TABLE III

Activity of Condensation Products of 4'-demethylepipodophyllotoxin β-D-glucopyrano-side with Aldehydes and Ketones

R^1	R^2	P-815[a] Mastocytoma cells of the mouse, *in vitro* ED_{50}, mg/liter	Mouse[a] leukemia L-1210, survival time increase (%)
CH_3	H	0.031	167
C_2H_5	H	0.0085	97
$CH_3CH = CH$	H	0.016	121
$(CH_3)_2CH$	H	0.0055	121
$C_2H_5CHCH_3$	H	0.0055	84
$(CH_3)_2CHCH_2$	H	0.0048	36
$n\text{-}C_4H_9$	H	0.0062	85
C_5H_9	H	0.0047	39
(furyl group)	H	0.018	136
(thienyl group)	H	0.0048	121
C_6H_5	H	0.0068	97
$p\text{-}H_3CC_6H_4$	H	0.0086	64
(o-nitrostyryl group) —CH=CH	H	0.0093	29
1-Naphthyl	H	0.013	95
CH_3	CH_3	0.015	106
C_2H_5	CH_3	0.0060	69

[a] From Keller-Juslén *et al.* (1971). Reprinted with permission; copyright by the American Chemical Society.

tives in this P-815 assay, but the L-1210 activities vary considerably. Also noteworthy are substantial activity changes produced by varying R^1 as aromatic moieties and by employing closely related ketones for R^1 and $R^{2.}$

The biological activity of these compounds was also dependent on the nature of the sugar moiety (Keller-Juslén et al., 1971). For example, cyclic acetals of 4'-demethylepipodophyllotoxin β-D-galactopyranoside did not give comparable biological results. Although isomeric to the comparable glucopyranosides, these analogues are geometrically significantly different at the aldehyde condensation site.

These data suggest that the whole region of these molecules around the sugar moiety is somewhat critical with respect to their activities.

C. Steganacin

Stegancin (Fig. 16) (Kupchan et al., 1973), a structurally similar compound to podophyllotoxin, has been reported (Wang et al., 1977) to inhibit cleavage of sea urchin eggs. It also inhibits, in vitro, calf and rabbit tubulin polymerization and causes a slow depolymerization of existing microtubules. Further, steganacin blocks HeLa cell replication in mitosis, inhibits chicken brain microtubule assembly in vitro, and competitively inhibits colchicine binding to purified tubulin (Schiff et al., 1978). In this latter work, a number of analogues of steganacin were also examined for inhibiting and binding activity. The natural product was the most potent compound tested.

The total synthesis of steganacin and a series of derivatives has been described (Damon et al., 1976; Hughes and Raphael, 1976; Kende and Liebeskind, 1976; Kende et al., 1976; Krow et al., 1978; Ziegler and Schwartz, 1978).

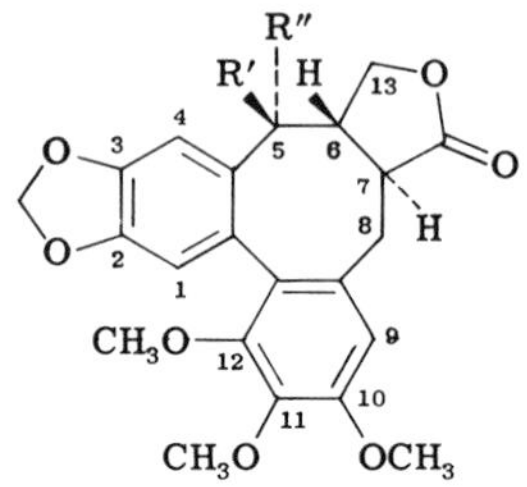

27 $R' = OAc$; $R'' = H$. Steganacin.

28 $R' = O_2C$ ⟨⟩ ; $R'' = H$. Steganangin.

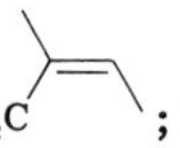

29 $R' = OH$; $R'' = H$. Steganol.

30 $R' + R'' = \overset{O}{\|}$. Steganone.

Figure 16. Structures of steganacin and related compounds. (From Kupchan et al., 1973.)

VII. SUMMARY

The podophyllotoxins have been developed considerably since their discovery, isolation, structure elucidation, and synthesis. They have emerged through a long and at many times frustrating history to become not just experimental tools, which is a major application of podophyllotoxin and many of its derivatives, but also clinically effective agents in the forms of VM 26 and VP 16-213. These latter agents are currently being established as useful anticancer drugs.

There is, however, much still to be accomplished in this area. The fact that the precise mechanism of action of the clinically effective agents VM 26 and VP 16-213 is still a mystery suggests that when this problem is solved it should be possible to take advantage of that information to further the design of even more efficacious clinical analogues based on sound molecular considerations. Further, since other analogues of the aglycone moiety of VM 26 and VP 16-213 are available, such as other podophyllotoxin, colchicine, and steganacin analogues, there would appear to be many compounds which could be synthesized and tested as analogues of the effective agents VM 26 and VP 16-213.

Acknowledgments

The author is grateful to Eli Lilly and Company for a Research Fellowship, to the American Cancer Society through the Purdue Cancer Center for support, and to Robert J. Strife for a critical reading of the manuscript.

REFERENCES

Aiyar, V. N., and Chang, F. C. (1975). *J. Org. Chem.* **40,** 2384.
Aiyar, V. N., and Chang, F. C. (1977). *J. Org. Chem.* **42,** 246.
Akahori, A., Yasuda, F., Ando, M., Hori, K., and Okanishi, T. (1972). *Chem. Pharm. Bull.* **20,** 150.
Allen, L. M. (1978a). *Cancer Res.* **38,** 2549.
Allen, L. M. (1978b). *Drug Metab. Rev.* **8,** 119.
Allen, L. M., and Creaven, P. J. (1975). *Eur. J. Cancer* **11,** 697.
Allen, L. M., Marcks, C., and Creaven, P. J. (1976). *Proc. Am. Assoc. Cancer Res.* **17,** 6.
Avery, T. L., Roberts, D., and Price, R. A. (1973). *Cancer Chemother. Rep.* **57,** 165.
Ayres, D. C., and Lim, C. K. (1972). *J.C.S. Perkin I,* p. 135.
Ayres, D. C., Harris, J. A., Jenkins, P. N., and Phillips, L. (1972). *J.C.S. Perkin I,* p. 1343.
Bagget, N., Duxbury, J. M., Foster, A. B., and Webber, J. M. (1964). *Chem. Ind. (London)* p. 1832.
Bates, R. B., and Wood, J. B. (1972). *J. Org. Chem.* **37,** 562.
Belkin, M. (1947). *Fed. Proc., Fed. Am. Soc. Exp. Biol.* **6,** 308.
Bhattacharyya, B., and Wolff, J. (1974). *Proc. Natl. Acad. Sci. U.S.A.* **71,** 2627.
Bianchi, E., Seth, K., and Cole, J. R. (1969). *Tetrahedron Lett.,* 2759.

Borsche, W., and Niemann, J. (1932). *Justus Liebigs Ann. Chem.* **494**, 126.

Brewer, C. F., Loike, J. D., Horwitz, S. B., Sternlicht, H., and Gensler, W. J. (1979). *J. Med. Chem.* **22**, 215.

Bryan, J. (1972). *Biochemistry* **11**, 2611.

Carter, S. K., and Slavik, M. (1976). *Cancer Treat. Rev.* **3**, 43.

Cavalli, F., Tschopp, L., Gerber, A., Sonntag, R. W., Ryssel, H. J., and Brunner, K. W. (1977). *Schweiz. Med. Wochenschr.* **107**, 1960.

Cavalli, F., Sonntag, R. W., Jungi, F., Senn, H. J., and Brunner, K. W. (1978). *Cancer Treat. Rep.* **62**, 473.

Chakravort, R., Serkar, S., Sen, S., and Mukeji, B. (1967). *Br. J. Cancer* **21**, 33.

Chiuten, D. F., Bennett, J. M., Creech, R. H., Glick, J., Falkson, G., Brodovsky, H. S., Begg, C. B., Muggia, F. M., and Carbone, P. P. (1979). *Cancer Treat. Rep.* **63**, 7.

Cornman, I., and Cornman, M. E. (1951). *Ann. N.Y. Acad. Sci.* **51**, 1443.

Cortese, F., Bhattacharyya, B., and Wolff, J. (1976). *Fed. Proc., Fed. Am. Soc. Exp. Biol.* **35**, 1483.

Creasey, W. A. (1975). *In* "Antineoplastic and Immunosuppressive Agents" (A. C. Sartorelli and D. G. Johns, eds.), Part II, pp. 670–694. Springer-Verlag, Berlin and New York.

Creaven, P. J., and Allen, L. M. (1975a). *Clin. Pharmacol. Ther.* **18**, 221.

Creaven, P. J., and Allen, L. M. (1975b). *Clin. Pharmacol. Ther.* **18**, 227.

Creaven, P. J., Newman, S., Selawry, O. S., Cohen, M. H., and Primack, A. (1974). *Cancer Chemother. Rep.* **58**, 901.

Damon, R. E., Schlessinger, R. H., and Blount, J. F. (1976). *J. Org. Chem.* **41**, 3772.

Dombernowsky, P., and Nissen, N. I. (1973). *Acta Pathol. Microbiol. Scand.* **81**, 715.

Dombernowsky, P., and Nissen, N. I. (1976). *Eur. J. Cancer* **12**, 181

Dombernowsky, P., Nissen, N. I., and Larsen, V. (1972). *Cancer Chemother. Rep.* **56**, 71.

Dustin, P., Jr. (1963). *Pharmacol. Rev.* **15**, 449.

Eagan, R. T., Ingle, J. N., Creagor, E. T., Frytak, S., Kvols, L. K., Rubin, J., and McMahon, R. T. (1978). *Cancer Treat. Rep.* **62**, 843.

Eigsti, O. J., and Dustin, P., Jr. (1955). "Colchicine in Agriculture, Medicine, Biology and Chemistry." Iowa Coll. Press, Ames.

Emmenegger, H., Stähelin, H., Rutschmann, J., Renz, J., and von Wartburg, A. (1961). *Arzneim.-Forsch.* **11**, 327, 459.

Flavin, M., and Slaughter, C. (1974). *J. Bacteriol.* **118**, 59.

Freed, J. L., and Lebowitz, M. M. (1970). *J. Cell Biol.* **45**, 334.

Gensler, W. J., and Gatsonis, C. D. (1966a). *J. Org. Chem.* **31**, 3224.

Gensler, W. J., and Gatsonis, C. D. (1966b). *J. Org. Chem.* **31**, 4004.

Gensler, W. J., and Wang, S. Y. (1954). *J. Am. Chem. Soc.* **76**, 5890.

Gensler, W. J., Samour, C. M., and Wang, S. Y. (1954). *J. Am. Chem. Soc.* **76**, 315.

Gensler, W. J., Murthy, C. A., and Trannell, M. H. (1977). *J. Med. Chem.* **20**, 635.

Geran, R. I., Congleton, G. F., Dudeck, L. E., Abbott, B. J., and Gargus, J. M. (1974). *Cancer Chemother. Rep., Part 2* **4**, 53.

Goldsmith, M. A., and Carter S. K. (1973). *Eur. J. Cancer* **9**, 477.

Gosalvez, M., Perez-Garcia, J., and Lopez, M. (1972). *Eur. J. Cancer* **8**, 471.

Gosalvez, M., Blanco, M., Hunter, J., Miko, M., and Chance, B. (1974). *Eur. J. Cancer* **10**, 567.

Gosalvez, M., Garcia-Canero, R., and Reinhold, H. (1975). *Eur. J. Cancer* **11**, 709.

Gotzos, V., Cappelli-Gotzos, B., and Despond, J.-M. (1979). *Histochem. J.* **11**, 691.

Greenspan, E. M., Leiter, J., and Shear, M. J. (1950). *J. Natl. Cancer Inst.* **10**, 1295.

Greenspan, E. M., Colsky, J., Schoenbach, E. B., and Shear, M. J. (1954). *J. Natl. Cancer Inst.* **14**, 1257.

Grieder, A., Maurer, R., and Stähelin, H. (1974). *Cancer Res.* **34**, 1788.

Grieder, A., Maurer, R., and Stähelin, H. (1977). *Cancer Res.* **37,** 2998.

Hacker, M., and Roberts, D. (1975). *Cancer Res.* **35,** 1756.

Hacker, M., and Roberts, D. (1977). *Cancer Res.* **37,** 3287.

Hartwell, J. L., and Schrecker, A. W. (1951). *J. Am. Chem. Soc.* **73,** 2909.

Hartwell, J. L., and Schrecker, A. W. (1958). *Fortschr. Chem. Org. Naturst.* **15,** 83.

Hartwell, J. L., and Shear, M. J. (1947). *Cancer Res.* **7,** 716.

Hoffmann, J. J., Weidhopf, R. M., and Cole, J. R. (1978). *J. Pharm. Sci.* **66,** 586.

Hokanson, G. C. (1978). *Lloydia* **41,** 497.

Horwitz, S. B., and Loike, J. D. (1977). *Lloydia* **40,** 82.

Huang, C. C., Hou, Y., and Wang, J. J. (1973). *Cancer Res.* **33,** 3121.

Hughes, L. R., and Raphael, R. A. (1976). *Tetrahedron Lett.,* 1543.

Issell, B. F., and Crooke, S. T. (1979). *Cancer Treatment Reviews* **6,** 107.

Jungi, W. F., and Senn, H. J. (1975). *Cancer Chemother. Rep.* **59,** 737.

Kaplan, I. W. (1942). *New Orleans Med. Surg. J.* **94,** 388.

Kelleher, J. K. (1977). *Mol. Pharmacol.* **13,** 232.

Keller-Juslén, C., Kuhn, M., von Wartburg, A., and Stähelin, H. (1971). *J. Med. Chem.* **14,** 936.

Kelly, M. G., and Hartwell, J. L. (1954). *J. Natl. Cancer Inst.* **14,** 967.

Kelly, M. G., Leiter, J., Bourke, A. R., and Smith, P. K. (1951). *Cancer Res.* **11,** 263.

Kende, A. S., and Liebeskind, L. S. (1976). *J. Am. Chem. Soc.* **98,** 267.

Kende, A. S., Liebeskind, L. S., Kubiak, C., and Eisenberg, R. (1976). *J. Am. Chem. Soc.* **98,** 6389.

Kende, A. S., Liebeskind, L. S., Mills, J. E., Rutledge, P. S., and Curran, D. P. (1977). *J. Am. Chem. Soc.* **99,** 7083.

King, L., and Sullivan, M. (1946). *Science* **104,** 244.

King, L., and Sullivan, M. (1947). *AMA Arch. Pathol.* **43,** 374.

Kocsis, J. J., Walaszek, E. J., and Geiling, E. M. K. (1957). *Arch. Intern. Pharmacodyn. Ther.* **111,** 134.

Kofod, H., and Jorgensen, C. (1955). *Acta Chem. Scand.* **9,** 1327.

Krishan, A., Paika, K., and Frei, E. (1975). *J. Cell Biol.* **66,** 521.

Krow, G. R., Damodoran, K. M., Michener, E., Wolf, R., and Guare, J. (1978). *J. Org. Chem.* **43,** 3950.

Kuhn, M., and von Wartburg, A. (1963). *Experientia* **19,** 391.

Kuhn, M., and von Wartburg, A. (1968a). *Helv. Chim. Acta* **51,** 163.

Kuhn, M., and von Wartburg, A. (1968b). *Helv. Chim. Acta* **51,** 1631.

Kuhn, M., and von Wartburg, A. (1969). *Helv. Chim. Acta* **52,** 948.

Kupchan, S. M., Britton, R. W., Ziegler, M. F., Gilmore, C. J., Restivo, R. J., and Bryan, R. F. (1973). *J. Am. Chem. Soc.* **95,** 1335.

Leiter, J., Downing, V., Hartwell, J. L., and Shear, M. J. (1950). *J. Natl. Cancer Inst.* **10,** 1273.

Lettre, H., and Witte, S. (1967). "Experimental and Clinical Experiences with Podophyllin Derivatives in Tumor Therapy." Schattauer, Stuttgart.

Loike, J. D., and Horwitz, S. B. (1976a). *Biochemistry* **15,** 5435.

Loike, J. D., and Horwitz, S. B. (1976b). *Biochemistry* **15,** 5443.

Loike, J. D., Brewer, C. F., Sternlicht, H., Gensler, W. J., and Horwitz, S. B. (1978). *Cancer Res.* **38,** 2688.

McClure, W. O., and Paulson, J. C. (1977). *Mol. Pharmacol.* **13,** 560.

McIntosh, J. R., Cande, Z., Snyder, J., and Vanderslice, K. (1975). *Ann. N.Y. Acad. Sci.* **253,** 407.

Makrides, E. B., Banerjee, S., Handle, L., and Margulis, L. (1970). *J. Protozool.* **17,** 548.

Margulis, T. N. (1974). *J. Am. Chem. Soc.* **96,** 899.

Margulis, T. N. (1975). *In* "Microtubules and Microtubule Inhibitors" (M. Borgers and M. de Brabander, eds.), pp. 67–78. North-Holland Publ., Amsterdam.

Maturova, M., Malinsky, J., and Santavy, F. (1959). *J. Natl. Cancer Inst.* **22**, 297.

Misra, N. C., and Roberts, D. W. (1975). *Cancer Res.* **35**, 99.

Mizel, S. B., and Wilson, L. (1972). *Biochemistry* **11**, 2573.

Muggia, F., Selawry, O., and Hansen, H. (1971). *Cancer Chemother. Rep.* **55**, 575.

Nissen, I., Larsen, V., Pedersen, H., and Thomsen, K. (1972). *Cancer Chemother. Rep.* **56**, 769.

Olmsted, J. B., and Borisy, G. G. (1973). *Biochemistry* **12**, 4282.

Osswald, H. (1978). *Arzneim.-Forsch.* **28**, 387.

Paulson, J. C., and McClure, W. O. (1975). *J. Cell Biol.* **67**, 461.

Pelsor, F. R., Allen, L. M., and Creaven, P. J. (1978). *J. Pharm. Sci.* **67**, 1106.

Petcher, T. J., Weber, H. P., Kuhn, M., and von Wartburg, A. (1973). *J.C.S. Perkin II*, p. 288.

Pettit, G. R. (1977). "Biosynthetic Products for Cancer Chemotherapy," Vol. 1. Plenum, New York.

Pettit, G. R., and Cragg, G. M. (1978). "Biosynthetic Products for Cancer Chemotherapy," Vol. 2. Plenum, New York.

Pettit. G. R., and Ode, R. H. (1979). "Biosynthetic Products for Cancer Chemotherapy," Vol. 3. Plenum, New York.

Podwyssotzki, V. (1880). *Arch. Exp. Pathol. Pharmakol.* **13**, 29.

Pouillart, P., Palangie, T., Jouve, M., Langlois, A., Garcia-Giralt, E., Regensbert, C., Blic, V., Huguenin, P., Morin, P., Gautier, H., Baron, A., and Dat-Xuang (1978). *Nouv. Presse Med.* **1**, 2235.

Renz, J., Kuhn, M., and von Wartburg, A. (1965). *Justus Liebigs Ann. Chem.* **681**, 207.

Rivera, G., Avery, T., and Pratt, C. (1975a). *Cancer Chemother. Rep.* **59**, 743.

Rivera, G., Avery, T., and Roberts, D. (1975b). *Eur. J. Cancer* **11**, 639.

Roberts, D., and Hilliard, S. L. (1978). *Cancer Res.* **38**, 2317

Rozencweig, M., Von Hoff, D. D., Henney, J. E., and Muggia, F. M. (1977). *Cancer (Philadelphia)* **40**, 334.

Sandoz Ltd. (1968). *C.A.* **68**, 2894x.

Schiff, p. B., Kende, A. S., and Horwitz, S. B. (1978). *Biochem. Biophys. Res. Commun.* **85**, 737.

Schindler, R. (1965). *J. Pharmacol. Exp. Ther.* **149**, 409.

Schrecker, A. W., and Hartwell, J. L. (1955). *J. Am. Chem. Soc.* **77**, 432, 6725.

Schrecker, A. W., and Hartwell, J. L. (1956). *J. Org. Chem.* **21**, 381.

Schrecker, A. W., and Hartwell, J. L. (1957). *J. Am. Chem. Soc.* **79**, 3827.

Schreier, E. (1963). *Helv. Chim. Acta* **46**, 75.

Schreier, E. (1964). *Helv. Chim. Acta* **47**, 1529.

Seidlova-Masinova, V., Malinsky, J., and Santavy, F. (1957). *J. Natl. Cancer Inst.* **18**, 359.

Sklansky, B. D., Mann-Kaplin, R. S., Reynolds, A. F., Rosenblum, M. L., and Walker, M. D. (1974). *Cancer (Philadelphia)* **33**, 460.

Smissman, E. E., Murray, R. J., McChesney, J. D., Houston, L. L., and Pazdernik, T. L. (1976). *J. Med. Chem.* **19**, 148.

Snyder, J. A., and McIntosh, J. R. (1976). *Annu. Rev. Biochem.* **45**, 669.

Sonntag, R. W., Senn, H. J., Nage, G., Giger, K., and Alberto, P. (1974). *Eur. J. Cancer* **10**, 93.

Späth, E., Wessely, F., and Kornfield, L. (1932). *Ber. Dtsch. Chem. Ges.* **65**, 1536.

Stähelin, H. (1970). *Eur. J. Cancer* **6**, 303.

Stähelin, H. (1972). *Planta Med.* **22**, 336.

Stähelin, H. (1973). *Eur. J. Cancer* **9**, 215.

Stähelin, H. (1976). *Eur. J. Cancer* **12**, 925.

Stähelin, H., and Cerletti, A. (1964). *Schweiz. Med. Wochenschr.* **94**, 1490.

Stähelin, H., and Poschmann, G. (1978). *Oncology* **35**, 217.

Strife, R. J., Jardine, I., and Colvin, M. (1980). *J. Chromatogr.* **182**, 211.

Sullivan, M. P., van Eyo, J., Herson, J., Starling, K. A., Ragab, A., and Sexhauer, C. (1979). *Cancer Treat. Rep.* **63,** 155.

Udenfriend, S., Duggan, D. E., Vasta, B. M., and Brodie, B. B. (1957). *J. Pharmacol. Exp. Ther.* **120,** 26.

Vaitkevicius, V., and Reed, M. (1966). *Cancer Chemother. Rep. 50,* 565.

Venditti, J. M. (1971). *Cancer Chemother. Rep., Part 3* **2,** 35.

Vietti, T. J., Valeriote, F. A., Kalish, R., and Coulter, D. (1978). *Cancer Treat. Rep.* **62,** 1313.

Wang, R. W., Rebhun, L. I., and Kupchan, S. M. (1977). *Cancer Res.* **37,** 3071.

Weiss, S. G., Tin-Wa, M., Perdue, R. E., Jr., and Farnsworth, N. R. (1975). *J. Pharm. Sci.* **64,** 95.

Wilkoff, L. J., and Dulmadge, E. A. (1978). *Proc. Am. Assoc. Cancer Res.* **19,** 37.

Wilson, L. (1970). *Biochemistry* **9,** 4999.

Wilson, L. (1975). *Ann. N.Y. Acad. Sci.* **253,** 213.

Wilson, L., and Bryan, J. (1974). *Adv. Cell Mol. Biol.* **3,** 21.

Wilson, L., and Friedkin, M. (1967). *Biochemistry* **6,** 3126.

Wilson, L., and Meza, I. (1973). *J. Cell Biol.* **58,** 709.

Wilson, L., Bamburg, J. R., Mizel, S. B., Grisham, L. M., and Creswell, K. M. (1974). *Fed. Proc., Fed. Am. Soc. Exp. Biol.* **33,** 158.

Ziegler, F. F., and Schwartz, J. A. (1978). *J. Org. Chem.* **43,** 985.

Zweig, M. H., and Chignell, C. F. (1973). *Biochem. Pharmacol.* **22,** 2141.

CHAPTER 10

Maytansinoids

Yasuo Komoda and Toyokazu Kishi

I. INTRODUCTION

Maytansine was first isolated by Kupchan *et al.* (1972a) as an anti-leukemic and cytotoxic substance in a very low yield from *Maytenus serrata*. It was the first member discovered of an entirely new group of natural products called maytansinoids, which is used as a generic term for all derivatives structurally related to maytansine. Maytanside is used as the term for those maytansinoids which contain the macrocyclic ring system but lack the ester moiety at C-3. The name maytansine was derived from *Mayte*nus and its *ansa* structure. Maytansine has a structure resembling that of ansamycin antibiotics. It is the first molecule of this type shown to contain aryl chloride, epoxide, and hydroxycycliccarbamate functions, and is the first of a class of ansa compounds which show significant antitumor activity and cytotoxicity.

Maytansine was found to be one of the most promising antitumor agents which were found in a program developed by the National Cancer Institute (NCI), to investigate the antitumor principles in plants, and it is now undergoing in phase II clinical trials. Interest in maytansine as a tumor inhibitor was stimulated by its

Anticancer Agents Based on Natural Product Models

high (at the level of μg/kg) and wide (from 50- to 100-fold dosage) range of activity against mouse P-388 lymphocytic leukemia.

The extreme scarcity of maytansine from plant sources and its complex structure prompted several groups of organic chemists to try to synthesize it. However, a research group of Takeda Chemical Industries, Japan, found that a microorganism of the *Nocardia* species produced maytansinoids, i.e., ansamitocins (Higashide *et al.*, 1977), which may eliminate the problem of scarcity.

Some of the chemistry and biological activities of maytansinoids have been reviewed previously (Komoda, 1974). This chapter deals with the isolation and structures (Section II), chemical synthesis (Section III), and biological activities (Section IV) of maytansinoids.

II. ISOLATION AND STRUCTURES OF MAYTANSINOIDS

Eight biologically active maytansinoids with an ester at C-3, which are generically called maytanside esters, have been isolated from plants. Five of them, maytansine, maytanprine, maytanbutine, maytanvaline, and maytanacine, have no functional group at C-15. The others, colubrinol, colubrinol acetate, and maytanbutacine, are substituted with a hydroxy or acetoxy group at C-15. Four biologically almost inactive maytansinoids with no ester at C-3, maytansinol, maysine, normaysine, and maysenine, which are generically called maytansides, have been also isolated from plants. In addition to these, five biologically active maytanside esters, maytanacine, maytansinol 3-propionate, and ansamitocins P-3, P-3′, and P-4, have been obtained from the fermentation broth of a microorganism of the *Nocardia* species.

A. Maytansine

Maytansine was first isolated from *Maytenus serrata* (Hochst. ex A. Rich.) R. Wilczek (Celastraceae), formerly known as *M. ovatus* Loes., in a yield of only 0.00002% of the dried plant material (Kupchan *et al.*, 1972a). The reasons for the successful isolation of such an extremely low content are the considerable chemical stability of maytansine and its great biological activity. Even the initial ethanol extract showed significant inhibitory activity in *in vivo* and *in vitro* bioassay systems.

1. Isolation of Maytansine

In the winter of 1961–1962, samples of *M. serrata* were collected in Ethiopia, and its ethanolic extract showed significant inhibitory activity *in vitro* against a culture of KB cells derived from human carcinoma of the nasopharynx

(KB) and *in vivo* against five standard animal tumor systems: mouse L-1210 and P-388 leukemias, mouse sarcoma 180 and Lewis lung carcinoma solid tumors, and rat Walker 256 intramuscular carcinosarcoma.

Fractionation of the ethanolic extract, guided by bioassay against KB and P-388 revealed that the inhibitory activity was concentrated successively in the ethyl acetate layer of an ethyl-acetate–water partition and in the aqueous methanol layer of an aqueous-methanol–petroleum-ether partition. Column chromatography of the aqueous methanol solubles on neutral silica gel was followed by treatment of the active eluate with acetic-anhydride–pyridine which facilitated the subsequent separation without affecting the active principles, and the resulting residue was subjected to column chromatography, first on neutral silica gel and then on alumina. The fraction eluted from the alumina column was then subjected to preparative thin layer chromatography (PTLC) successively on alumina, then silica gel to give two fractions, both of which showed high biological activity.

Further PTLC on neutral silica gel of one of the active fractions afforded a highly enriched concentrate (Fraction A, 1 mg/kg of dried plant) as a solid residue which was homogeneous according to both silica gel and alumina TLC yet resisted all attempts at crystallization. X-ray crystallography was used to elucidate the structure of the active compound, after it had been further purified by preparing heavy atom derivatives.

Elemental analysis of Fraction A indicated the presence of a significant number of nitrogen atoms. Partitioning between 2 *N* hydrochloric acid and diethyl ether, with the active principle remaining in both layers, indicated that none of the nitrogen atoms was strongly basic and/or the active principle was a slightly water-soluble compound. Attempts to prepare a quarternary salt derivative from Fraction A revealed that a common crystalline product, apparently a methyl derivative, was formed in methanolic solution. Similar experiments in other alcohol solutions also afforded common crystalline alkyl derivatives. Thus, when Fraction A was treated with 3-bromopropanol and *p*-toluenesulfonic acid in dichloromethane at room temperature, the crystalline 3-bromopropyl derivative (**2**) was obtained. Treatment of (**2**) with 2 *N* hydrochloric acid in aqueous methanol afforded a crystalline hydrolysis product which was used to seed a solution of Fraction A in dichloromethane and hexane and yield crystalline maytansine (**1**, 0.2 mg/kg of dried plant, 0.00002%): $C_{34}H_{46}ClN_3O_{10}$; m.p. 171–172°C; $[\alpha]_D^{26} -145°$ (c = 0.055, CHCl$_3$). The maytansine obtained showed the strongest inhibitory activity in KB and P-388 among all the fractions obtained in the course of the fractionation.

2. Structure of Maytansine

Due to the extremely small quantity of maytansine obtained and the reversible interrelation of maytansine and 3-bromopropyl derivative, the latter compound was an attractive target for x-ray crystallographic analysis (Bryan *et al.*, 1973). Crystals of 3-bromopropyl ether belong to the orthorhombic system with the space group $p2_12_12_1$ with a = 24.239(4), b = 16.044(4), c = 10.415(2) Å. The unit cell contains four formula units of $C_{37}H_{51}BrClN_3O_{10}$. The structure was solved by the heavy atom method. The structure found for the molecule in the crystal led to structural assignment (1) for maytansine. The absolute configurations of (2) were found to be 3*S*, 4*S*, 5*S*, 6*R*, 7*S*, 9*S*, 10*R*, and 2'*S*.

The disposition of substituents about the various bond axes shows almost perfect minimization of the intramolecular repulsions, so that no strong intermolecular forces seem to be involved in dictating the observed conformation of the molecule. The absence of serious strain and the chemically enforced rigidity of many parts of the molecule suggest that this structure may well be maintained in solution.

The nucleus of the molecule is a roughly rectangular 19-membered ring: that is, the two longer sides of the 19-membered ring are roughly parallel and separated by about 5.4 Å, so that there is a hole in the center of the ring. The two faces of the ring have different characters; the lower face, opposite the ester residue, is predominantly hydrophobic, while the upper face is more hydrophilic. Furthermore, the ester residue is oriented in a manner which would sterically hinder the approach of reactants to the hydrophilic face.

The ester function in the antileukemic maytansinoids may play a key role in the formation of highly selective molecular complexes with growth-regulatory biological macromolecules. Such a molecular complex formation may be crucial for the subsequent selective alkylation of specific nucleophiles by, e.g., the carbinolamide and epoxide functions. In this connection, it is noteworthy that maytansine alkyl ethers in which the reactive carbinolamide is no longer available as a potential alkylating function show no antileukemic activity.

B. Maytansinoids of Plant Origin

As soon as the structure of maytansine was determined, its potent antileukemic activity prompted investigation of the possibility of clinical use. However, extremely low yield of maytansine from *M. serrata* made clinical trials practically impossible. A search of other species in Celastraceae as potential sources of maytansine revealed that *Maytenus buchananii* (Loes.) R. Wilczek, collected in Kenya, gave more than a seven times higher yield of maytansine. *Putterlickia verrucosa* Szyszyl, collected in South Africa, proved to be the richest source of maytansine, yielding more than eight times the amount from *M. buchananii*. However, *M. buchananii* was selected for large-scale extraction because collecting a large amount of *P. verrucosa* is very difficult.

In the course of the isolation and subsequent search for a better plant source of maytansine, various compounds related to maytansine were also found in *M.*

serrata, M. buchananii, M. arbutifolia, and *P. verrucosa.* Also, Wani *et al.* (1973) investigated the antitumor agents contained in *Colubrina texensis* Gray (Rhamnaceae), which does not belong to Celastraceae, and succeeded in isolating colubrinol and colubrinol acetate along with maytanbutine, which was first isolated from *M. serrata.*

Maytansinoids (**1**) and (**3**)–(**9**), maytanside esters, showed excellent antileukemic activity against P-388 in mouse and potent cytotoxicity against KB cells. In contrast, maytansinoids (**11**)–(**14**), maytansides, showed greatly diminished activity against both P-388 and KB relative to maytanside esters. Since (**11**)–(**14**) all lack the C-3 ester moiety, this appears to be necessary for antileukemic activity (Kupchan *et al.*, 1978).

1. Maytanside Esters

a. C-15-Unsubstituted Maytanside Esters (Maytanprine **3**, Maytanbutine **4**, Maytanvaline **5**, Maytanacine **6**)

Maytanprine (**3**) and maytanbutine (**4**) were first isolated from *M. serrata* (Kupchan *et al.*, 1972b). They have also been isolated along with a new maytanside ester, maytanvaline (**5**), from *M. buchananii* in better yield as in the case of maytansine (Kupchan *et al.*, 1974). On the other hand, an ethanolic extract of *P. verrucosa* was fractionated without treatment with acetic anhydride and pyridine to give another new maytanside ester, maytanacine (**6**), along with maytansine, maytanprine, and maytanbutine (Kupchan *et al.*, 1975).

The mass spectral characteristics indicated that (**3**)–(**5**) have macrocyclic structures similar to maytansine except for differences in the *N*-acyl group of the ester sidechains. The nmr spectra of (**3**)–(**5**) differed from that of (**1**) solely in the signals attributed to the terminal *N*-acyl group, as expected from the mass spectral fragmentation patterns. The nmr signals for the *N*-acyl group of (**3**) and (**4**) indicated a —CH₂CH₃ group with nonequivalent methylene protons and a —CH(CH₃)₂ moiety with two nonequivalent methyl groups, respectively. Hydrolysis of (**5**) with sodium carbonate in aqueous methanol yielded maysine (**11**) and *N*-isovaleryl-*N*-methyl-L-alanine, identified as its methyl ester by comparison with an authentic sample.

The mass spectrum of (**6**) indicated that it was a maytanside ester similar to maytansine except for differences in the R¹ group of the ester sidechain. The nmr spectrum of (**6**) contained an acetate methyl signal and lacked the C-2′H, C-2′CH₃ and N-CH₃ signals of the maytansine ester sidechain. The structure of (**6**) was further confirmed by reductively cleaving it to give maytansinol (**14**), and then treating (**14**) with acetic-anhydride–pyridine to yield (**6**).

b. C-15-Substituted Maytanside Esters (Colubrinol **7**, Colubrinol Acetate **8**, Maytanbutacine **9**)

Fractionation of an ethanolic extract

of *C. texensis*, guided by bioassay in KB and P-388, gave colubrinol (**7**) and colubrinol acetate (**8**) along with maytanbutine (**4**) (Wani *et al.*, 1973).

The ir and uv spectra of (**7**) and (**8**) were very similar to those of (**4**). Their molecular formulas suggested that (**7**) and (**8**) were closely related to (**4**) but contained an additional hydroxy or acetoxy group, respectively. In agreement with this, the ester carbonyl absorption in the ir spectrum of (**8**) was more intense than that for (**7**) or (**4**). Mild base-catalyzed methanolysis of (**8**) gave (**7**), and acetylation of the latter gave (**8**), indicating that the hydroxy function in (**7**) was at the same position as the acetoxy group in (**8**).

The nmr and mass spectra of (**4**), (**7**), and (**8**) revealed the presence of the same ester sidechain in all. Furthermore, the doublets assigned to the protons of C-15 in the nmr spectrum of (**4**) were absent in those of (**7**) and (**8**). The singlet assigned to the C-15 proton in (**7**) had shifted downfield in (**8**). Manganese dioxide oxidation of (**7**) gave the conjugated ketone. Therefore, the additional hydroxy or acetoxy function must be located in C-15.

From an ethanolic extract of *M. arbutifolia*, which is a species very closely related to *M. serrata*, a new maytansinoid, maytanbutacine (**9**), was isolated, in addition to maytanprine and maytanbutine (Kupchan *et al.*, 1977). The uv, ir, and mass spectra of (**9**) indicated that there were two sidechain esters, one of which was an acetate. The nmr spectrum suggested that the second ester was an isobutyrate and not the ester sidechain of maytansine. In addition, signals were observed for an acetate methyl singlet and a one-proton singlet which shifted upfield in the mild hydrolysis product (**10**). This behavior was similar to that observed in colubrinol acetate and indicated that the isobutyrate ester was at C-3, since it was not affected by the hydrolysis. Therefore, the acetate was at C-15. This was confirmed by oxidation of (**10**) to form a conjugated enone system.

2. Maytansides (Maysine **11**, Normaysine **12**, Maysenine **13**, Maytansinol **14**)

When maytansine, maytanprine, maytanbutine, and maytanvaline were isolated from *M. buchananii*, the fractions obtained by the final column chromatography showed that several spots other than maytanside esters were also present on TLC (Kupchan *et al.*, 1974). Repeated preparative TLC of these fractions finally gave, in extremely low yield, maysine (**11**), normaysine (**12**), and maysenine (**13**), all of which showed greatly diminished activity against both P-388 and KB relative to maytanside esters. The mass spectral characteristics of maytansides (**11**)–(**13**) indicated that these compounds have a macrocyclic ring similar to that of maytanside esters but lack the ester sidechain. The nmr spectrum of (**11**) showed the presence of a *trans* α,β-unsaturated amide with no proton in the γ position. Treatment of maytansine with sodium carbonate in aqueous methanol gave one major product which was identical with maysine in all respects. This information, along with the disappearance of the carbonyl ir absorptions of the C-3 ester, established structure (**11**) for maysine.

The mass spectral fragmentation pattern of (12) indicated that it was the N-demethyl homologue of maysine. Its nmr spectrum showed a signal corresponding to the proton on the C-1 nitrogen and lacked the NCH_3 signal of maysine. The mass spectrum of (13) showed that it was a deoxy derivative of (12). The nmr spectrum of (13) showed signals for a vinyl methyl group instead of those for the C-4 methyl and C-5 protons of the 4,5-epoxide system of (12), and showed a downfield shift of the C-2 and C-3 protons relative to (12). The structure of (13) was supported also by the bathochromic shift of its uv and ir carbonyl absorption bands in comparison with those of (12). Chemical interrelation was proven by reductive elimination of the epoxide of (12) with chromous chloride in acetic acid to give maysenine (13).

The presence of such a variety of C-3 esters of maytanside prompted an effort to isolate a possible common precursor, the C-3 alcohol (Kupchan et al., 1975). Synthetic maytansinol (14) for use as a reference in the isolation was prepared by reductive cleavage of maytanbutine, which was treated with lithium aluminum hydride in dry tetrahydrofuran. Repeated column chromatography, preparative TLC, and finally preparative HPLC of the fraction obtained in the course of isolation of maytanacine from P. verrucosa gave pure maytansinol, identical in every respect with synthetic maytansinol.

Maytansinol, which lacks antileukemic activity, was used to prepare several semisynthetic maytanside esters, maytansinol 3-bromoacetate, 3-trifluoro- acetate, 3-crotonate, and 3-propionate, to evaluate the effects of variations in the structure of the ester moiety on the biological activity.

C. Ansamitocins: Maytansinoids of Microorganism Origin

In the course of antitumor antibiotic screening, a group of researchers of Takeda Chemical Industries, Japan, found that the fermented broth of Nocardia has strong antitumor activity. Five active components were isolated and shown to be maytansinoids. Three of them were found to be novel maytansinoids of microorganism origin and named ansamitocins P-3 (16), P-3' (17), and P-4 (18). The mitotic arrest of various ascites tumor cells and the inhibiton of tubulin polymerization were observed with ansamitocins P-3 and P-4. The name was derived from the ansa structure and mitotic inhibition activity. This was the first discovery of naturally occurring maytansinoids of microorganism origin (Higashide et al., 1977).

1. Fermentation, Isolation, and Characterization

The ansamitocine-producing microorganism, strain number C-15003 (N-1), forms many coremia-like bodies and motile heteromorphic cells. This strain, found in Japan, was concluded to be a new species of the genus Nocardia. It was fermented at 28° C for 4 days under aeration and with agitation, in a 2000-liter fermentor containing 1000 liters of a culture medium consisting of

5% dextrin, 3% corn steep liquor, 0.1% peptone, and 0.5% $CaCO_3$. Under the culture conditions, most of the active substances were produced in the extra-cellular medium (Tanida, 1980).

The active components, which have a lipophilic neutral character, were ex-tracted with ethyl acetate from the culture filtrate. Five components were isolated as crystals by silica gel chromatography using a solvent system of chloroform-methanol or ethyl acetate–water. They were tentatively designated as P-1, P-2, P-3, P-3′, and P-4, according to their increasing R_f values on TLC. Their physicochemical properties were as follows:

P-1: m.p. 235–236° C (decomp.), $[\alpha]D$ −121° (CHCl$_3$), $C_{30}H_{39}CIN_2O_9$; MS m/e 545.
P-2: m.p. 188–190° C (decomp.), $[\alpha]D$ −127°, $C_{31}H_{41}CIN_2O_9$; MS m/e 559.
P-3: m.p. 190–192° C (decomp.), $[\alpha]D$ −136°, $C_{32}H_{43}CIN_2O_9$; MS m/e 573.
P-3′: m.p. 182–185° C (decomp.), $[\alpha]D$ −134°, $C_{32}H_{43}CIN_2O_9$; MS m/e 573.
P-4: m.p. 177–180° C (decomp.), $[\alpha]D$ −142°, $C_{33}H_{45}CIN_2O_9$; MS m/e 587.

The fragment peaks of m/e 485, 470, and 450 were observed in all of the components.

They showed the same uv absorption maxima in methanol at 233, 240 (sh.), 252, 280, and 288 nm. The ir absorptions at 1740, 1730, 1670 (C=0) and 1580 cm^{-1} were observed in common. They were positive in Beilstein and Dragen-droff reactions and easily soluble in alcohol, chloroform, acetone, and ethyl acetate.

These five antibiotics had not been reported previously in literature as micro-bial metabolites. However, their characteristic uv spectra were very similar to those of maytansinoids of plant origin, and the physicochemical properties of P-1 were identical to those of maytanacine, isolated from *Putterlickia verrucosa,* and those of P-2 to those of maytansinol propionate (**15**) synthesized from maytan-sinol (**14**).

It was revealed that the remaining three components, P-3, P-3′, and P-4, were novel. They were named ansamitocins P-3, P-3′, and P-4, respectively (Higa-shine *et al.,* 1977; Asai *et al.,* 1979).

2. Structures of Ansamitocins

The structures of ansamitocins P-3 (**16**), P-3′ (**17**), and P-4 (**18**) were elucidated by their physicochemical properties such as uv, ir, mass and nmr spectra and chemical degradation reaction into known compound. Their uv ab-sorption maxima showed that they have the same diene and aromatic chromophore of maytanacine (**6**). Their ir spectra were very similar to that of maytanacine (**6**), with three ester carbonyl absorptions observed in all. They gave the same mass fragments (m/e 485, 470, and 450) as those observed in maytan-sine (**1**) (Kupchan *et al.,* 1972a).

The highest mass numbers of ansamitocin P-3 (**16**) and P-4 (**17**) were m/e 573.2471 (calcd. $C_{31}H_{40}ClNO_7$ = 573.2493) and 587.2626 (calcd. $C_{32}H_{42}ClNO_7$

= 587.2649), which were identical with calculated mass numbers of the molecule formula substracting CH_3NO_2 (H_2O + NHCO). The fragment m/e 485 was assumed to be $M^+ - a - b$; a = (H_2O + NHCO), b = RCOOH. Thus, the structural difference should be present in the fatty-acid moiety of maytansinoid. In the case of maytanacine (P-1) (6) and maytansinol propionate (P-2) (5), subtraction of the fragment ($M^+ - a - b = 485$) from the highest mass number ($M^+ - a = 545.559$) gave 60 and 74 which correspond to acetic and propionic acid, respectively. Ansamitocins P-3 (16), P-3' (17), and P-4 (18) gave 88, 88, and 102 which correspond to butyric, butyric, and valeric acid, respectively. To confirm these fatty acid moieties, ansamitocins P-3 (16), P-3' (17), and P-4 (18) were hydrolyzed with sodium hydroxide; then isobutyric, butyric, and isovaleric acid were identified, respectively, by gas chromatography.

The nmr spectrum of ansamitocin P-3 indicated the presence of an isobutyl-geminal methyl group [-$CH(CH_3)_2$] as doublet. The methine group attached to these geminal methyl groups was assigned by nmr spin-decoupling study. The *n*-butyric methyl group of ansamitocin P-3' (17) was observed as a triplet. Ansamitocin P-4 (18) showed a geminal methyl group as a doublet. From the nmr spin-decoupling study, the chemical shifts, and the coupling constants of ansamitocin P-3 (16), the partial structure from C-1 to C-15 can be well explained by comparison with those of maytansine (1). The mutual stereochemical relations between H_2-H_3, H_5-H_6, H_{10}-H_{11}, and H_{12}-H_{13} in ansamitocin P-3 were the same as those of maytansine according to the coupling constants.

Reductive cleavage of each ansamitocin component with $LiAlH_4$ at low temperature gave the same product, P-0, independently. The physicohemical properties of P-0 were almost identical to those of maytansinol (14). Acetylation of P-0 with acetic anhydride in pyridine yielded a monoacetate which was identical with P-1, maytanacine (6), in m.p., molecular formula, R_f values on silica gel TLC, and ir, mass, and nmr spectra. These findings show that the structures of ansamitocins P-3, P-3', and P-4 correspond, respectively, to those of the novel maytansinoids, maytansinol isobutyrate, butyrate, and isovalerate, shown in Fig. 1 (Asai *et al.*, 1979).

It is interesting from the viewpoints of stereochemistry and biosynthesis that natural maytansinoids were isolated from plants and a microorganism. We believe that these superactive maytansinoids will become key substances in the near future for use in chemical and microbial modifications.

III. CHEMICAL SYNTHESIS OF MAYTANSINOIDS

The chemical synthesis of maytansinoids, particularly maytansine, which is now in phase II of clinical trials, is needed because of its potent antitumor activity and extremely low isolation yield, $10^{-4}\%$ or less, from plant

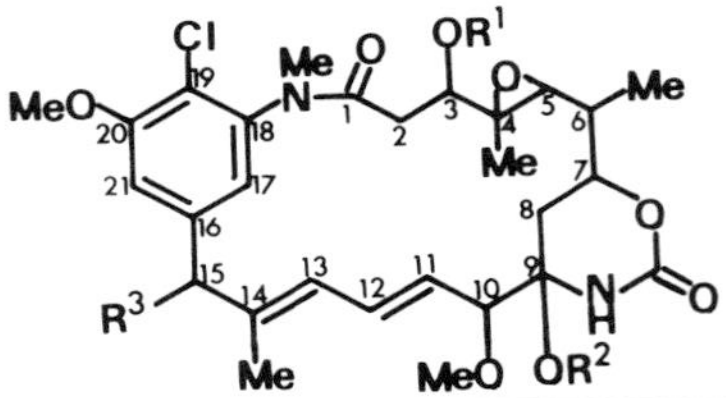

1 Maytansine $R^1 = COCHMeNMeCOMe; R^2 = R^3 = H$
2 Maytansine bromopropyl ether $R^1 = COCHMeNMeCOMe; R^2 = CH_2CH_2CH_2Br; R^3 = H$
3 Maytanprine $R^1 = COCHMeNMeCOCH_2Me; R^2 = R^3 = H$
4 Maytanbutine $R^1 = COCHMeNMeCOCH(Me)_2; R^2 = R^3 = H$
5 Maytanvaline $R^1 = COCHMeNMeCOCH_2CH(Me)_2; R^2 = R^3 = H$
6 Maytanacine $R^1 = COMe; R^2 = R^3 = H$
7 Colubrinol $R^1 = COCHMeNMeCOCH(Me)_2; R^2 = H; R^3 = OH$
8 Colubrinol acetate $R^1 = COCHMeNMeCOCH(Me)_2; R^2 = H; R^3 = OCOMe$
9 Maytanbutacine $R^1 = COCH(Me)_2; R^2 = H; R^3 = OCOMe$
10 Desacetylmaytanbutacine $R^1 = COCH(Me)_2; R^2 = H; R^3 = OH$
14 Maytansinol $R^1 = R^2 = R^3 = H$
15 Maytansinol 3-propionate $R^1 = COCH_2Me; R^2 = R^3 = H$
16 Ansamitocin P-3 $R^1 = COCH(Me)_2; R^2 = R^3 = H$
17 Ansamitocin P-3' $R^1 = COCH_2CH_2Me; R^2 = R^3 = H$
18 Ansamitocin P-4 $R^1 = COCH_2CH(Me)_2; R^2 = R^3 = H$

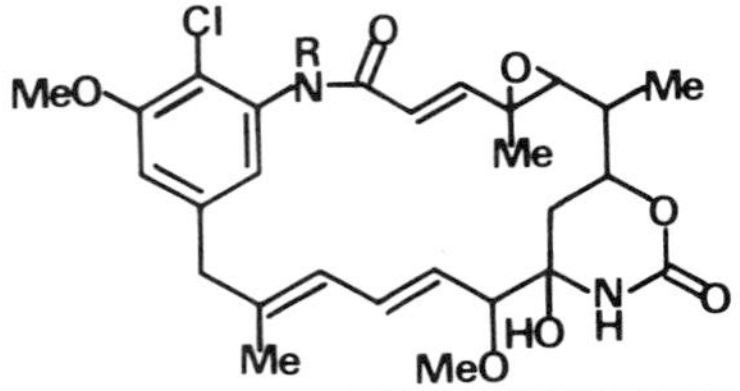

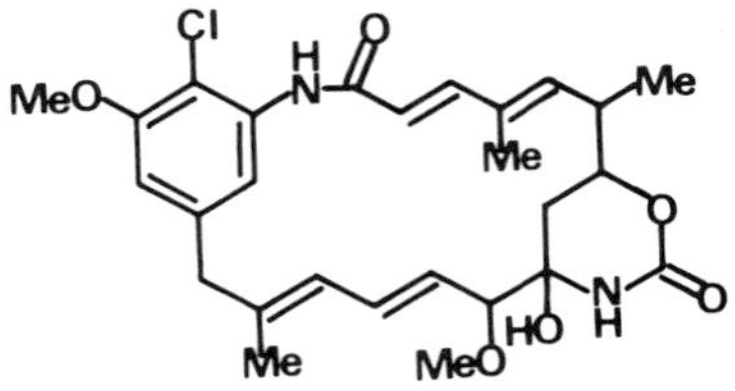

11 Maysine, $R = Me$
12 Normaysine, $R = H$

13 Maysenine

Figure 1.

sources. The recent discovery of a microorganism producing ansamitocins may eliminate the problem of their extreme scarcity. Nevertheless, their complex structures—for example, maytansine has a 19-membered macrocyclic ring, eight asymmetric centers, a conjugated diene (E,E), and a benzene ring substituted with four functional groups—undoubtedly present a formidable synthetic challenge to organic chemists who are trying to obtain not only the natural products themselves but also structurally related substances, which may possess the biological activities of the formers.

The first attempt toward the total synthesis of maytansinoids was reported by Meyers and Shaw (1974), two years after the isolation and structural elucidation of maytansine. Subsequently, a number of groups reported various routes to several key intermediates for the synthesis of maytansinoids. Finally, Corey *et al.* (1978b) succeeded in the total synthesis of ($\pm$)-*N*-methylmaysenine, which was the first formation of a macrocyclic ring related to that of maytansinoids.

In this section, we describe the preparation of the three moieties of maytansinoids and the total synthesis of ($\pm$)-*N*-methylmaysenine. The numbering of the carbon atoms in each compound corresponds to that of the maytansinoids.

A. Approaches to the Synthesis of Maytansinoids

1. Preparation of Parts of the C-5 to C-15 Moiety

The C-5 to C-15 moiety of maytansinoids includes most of the asymmetric centers, a conjugated diene (E,E), and a cyclic carbamate ring.

Meyers and Shaw (1974) treated the tetrahydropyranyl (THP) ether of methallyl alcohol with diborane followed by oxidation to give the primary alcohol, which was further oxidized to the aldehyde (**1**). Condensation of (**1**) with the lithioimine (**2**) prepared from pyruvaldehyde dimethylacetal produced the β-hydroxy ketone (**3**) after hydrolysis. Addition of phosgene to (**3**) gave the chloroformate, which was treated directly with ammonia to afford the desired cyclic carbamate (**4**) (Scheme 1).

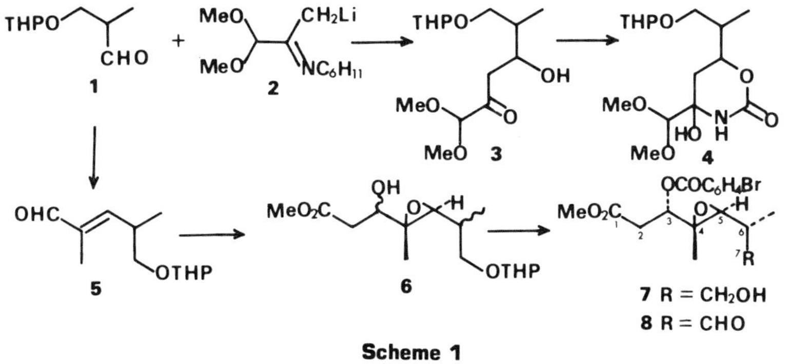

Scheme 1

Treatment (Meyers *et al.*, 1975) of (**1**) with the cyclohexylimine of propional-dehyde followed by dehydration afforded the unsaturated aldehyde (**5**). Further elaboration of (**5**) was accomplished by condensation with lithiomethylacetate to give the β-hydroxy ester as a mixture of diastereomers, which was transformed into the epoxide (**6**) using *t*-butyl hydroperoxide in the presence of vanadium acetylacetone. The mixture of (**6**) was converted to the corresponding *p*-bromo-benzoate, which was deprotected to give the mixture of diastereomeric alcohols. The major product (**7**) was isolated by PTLC from the mixture, and its stereochemistry was determined by x-ray crystallography. Oxidation of (**7**) gave a single aldehyde (**8**), which was reduced back to the carbinol (**7**) without any epimerization of the C-6 methyl group. The aldehyde (**8**) synthesized stereoselec-tively had all the attending stereochemistry corresponding to the contiguous carbon chain, C-1 to C-7, of maytansine (Scheme 1).

Furthermore, Meyers and Brinkmeyer (1975) synthesized an advanced model (**13**) possessing the appropriately functionalized carbon chain from C-7 to C-15, including the fused cyclic carbamate and the benzene ring. The THP ether (**9**) of dithiane alcohol prepared by alkylation of 2-lithio-1,3-dithiane with 1,2-epoxybutane was converted to the methyl ether (**10**) with only a single homogeneous spot on TLC, by addition of the monoacetal of fumaraldehyde to the lithio salt of (**9**) followed by methylation *in situ*. Selective cleavage of the acetal group of (**10**) furnished the aldehyde, which was treated by Wittig's aldol method for a three-carbon homologation, then dehydrated directly from the hy-droxy intermediate to yield the *E,E*-dienal (**11**). Treatment of (**11**) with phenyl-magnesium bromide followed by oxidation afforded the phenyl ketone (**12**). This ketone (**12**), after removal of the dithiane and THP protecting groups in a single step using mercuric chloride, was made to react with phosgene, then methanolic ammonia, to give the cyclic carbamate (**13**). Reduction of (**13**) at C-15 with NaBH₄ gave the benzyl alcohol derivative, which corresponded to the C-5 to C-15 chain of colubrinol and desacetylmaytanbutacine (Scheme 2).

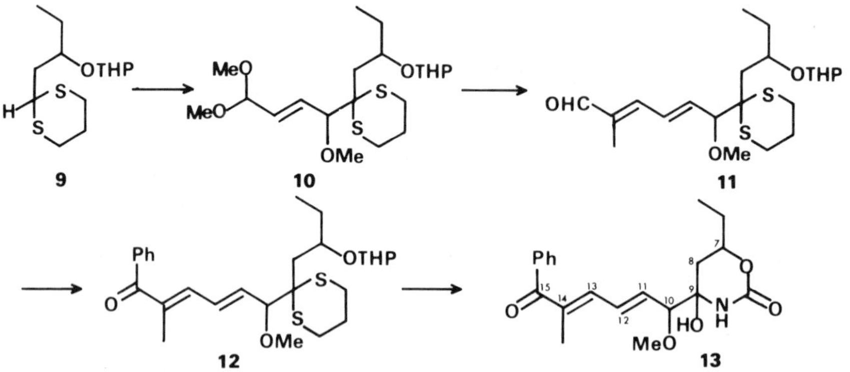

Scheme 2

On the other hand, Corey and Bock (1975) synthesized the acyclic intermediate (**19**) under efficient stereocontrolling. The acetonide of *cis*-2-buten-
1,4-diol was converted to the expoxide, which upon treatment with dimethylcopperlithium stereospecifically gave the hydroxy ketal (**14**). Brief exposure of (**14**)
to boron trifluoride etherate led quantitatively to the isomeric hydroxy ketal (**15**),
the hydroxy function of which was protected by conversion to the methyl-
thiomethyl ether (**16**). After hydrolysis of the acetonide function of (**16**), the diol
obtained was converted via the monotosylate to the epoxide (**17**). This was
treated with 2-lithio-1,3-dithiane to afford the hydroxy dithiane (**18**). Reaction of
(**18**) with *t*-butylidimethylsilyl chloride and imidazole gave the silyl ether (**19**).
The possible elaboration of (**19**) by attachment of carbons 10–15 and the aromatic ring to C-9 was tested using sorbaldehyde as a model, and the dienol (**21**)
obtained is converted to the corresponding methyl ether. The alternative elaboration of (**19**) by the addition of a C-1 to C-4 unit to C-5 required the selective
cleavage of the methylthiomethylether to form the corresponding alcohol (**20**).
This was readily accomplished under neutral conditions in the presence of mercuric ion without affecting both the dithiane unit and the silyl ether (Scheme 3).

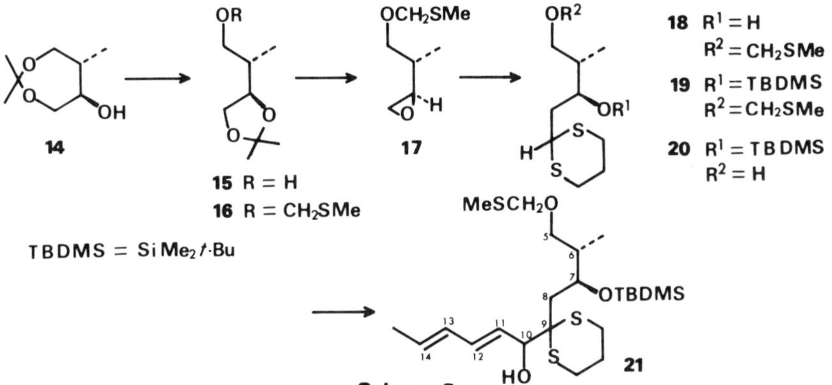

Scheme 3

The facile removal of the *t*-butyldimethylsilyl group from oxygen by fluoride
ion allowed selective deprotection of the hydroxyl group in the presence of the
dithiane unit, as has been demonstrated with the model system (**22**), which was
smoothly converted to the alcohol (**23**). Reaction of (**23**) sequentially with
sodium hydride, phosgene, and ammonia led to the urethane (**24**), which underwent facile loss of the dithiane unit when treated under neutral conditions in the

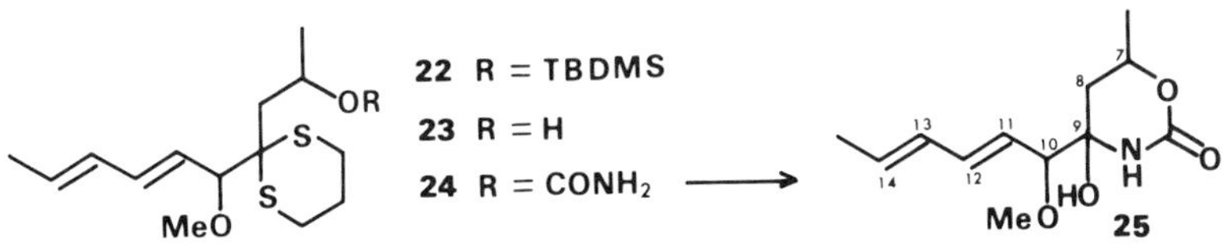

Scheme 4

presence of mercuric ion to give the cyclic carbamate (25), which possesses the characteristic structure of the C-7 to C-14 part of maytansinoids. The urethane function in this reaction may accelerate cleavage of the trimethylene thioketal (Scheme 4).

From the same intermediate (15) in Scheme 3, Elliott and Fried (1976) synthesized compound (29), which represents a fragment corresponding to carbons 5–12 of the maytansinoid ring skeleton. The benzyl ether of (15) was hydrolyzed to the glycol (26), which was converted to the monoacetate (27) by treatment with acetic anhydride and pyridine. When (27) was treated with sodium cyanate and trifluoroacetic acid, the desired urethane (28) was formed. Condensation of (28) with styrylglyoxal gave the adduct (29) (Scheme 5).

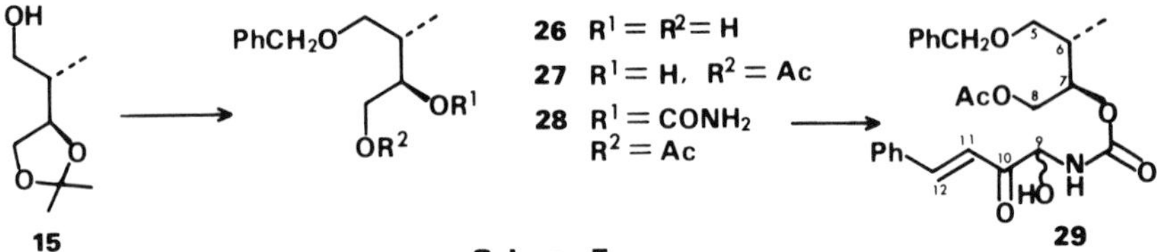

26 $R^1 = R^2 = H$
27 $R^1 = H$, $R^2 = Ac$
28 $R^1 = CONH_2$
    $R^2 = Ac$

Scheme 5

Edwards and Ho (1977) reported a relatively new approach to the cyclic carbamate (39) which involved the successful introduction of the required four asymmetric centers of maytansinoids with complete stereospecificity. The readily available 3,4-epoxycyclohexane was converted to the alcohol (30) by reaction with methyl lithium. Oxidation of (30) by a modified Lemieux periodate–permanganate method followed by heating with acetic acid afforded the γ-lactone (31). This substance was treated with ethyl chloroformate and triethylamine, followed by reduction of the resulting anhydride with $NaBH_4$ to yield the hydroxy lactone (32). The hydroxy group of (32) was protected by treatment with p-anisyldiphenylmethylchloride and pyridine to give the lactone (33). Further elaboration of (33) was accomplished by condensation of acetaldehyde with the lithium enolate derived from (33), producing a mixture of isomeric alcohols (34). This mixture was dehydrated via the mesylates to give the desired α,β-unsaturated lactone (35) after separation from a minor isomer by chromatography. Osmylation in the presence of pyridine, followed by reductive workup using aqueous sodium hydrogen sulfite, converted (35) to the diol (36). This product was converted to the acetonide followed by treatment with hydrazine to yield the unstable hydroxy hydrazide (37). Immediate reaction of (37) with N_2O_4 afforded the sensitive hydroxy azide (38), which was converted to the cyclic carbamate (39) by Curtius rearrangement (Scheme 6).

Bonjouklian and Ganem (1977) also synthesized the carbons 6–14 moiety (45) with a cyclic carbamate by cyclization during the Curtius rearrangement of γ-hydroxyacid azide. The dianion of phenoxyacetic acid was condensed readily with propylene oxide to give a mixture of the stereoisomeric γ-lactone (40)

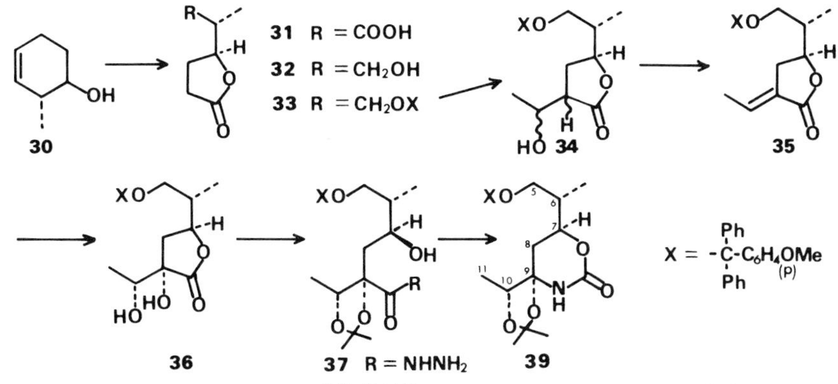

31 R = COOH
32 R = CH₂OH
33 R = CH₂OX

$X = -\overset{\overset{\displaystyle Ph}{|}}{\underset{\underset{\displaystyle Ph}{|}}{C}}-C_6H_4OMe$ (P)

Scheme 6

(*cis:trans*, 1:1) which was alkylated with sorbaldehyde after generation of the lactone enolate, and further with methyl iodide *in situ*, to yield the dimeric lactone (**41**). Exposure of (**41**) to excess 95% hydrazine furnished the hydroxyhydrazide (**42**). Immediate treatment of (**42**) with N₂O₄, then trimethylsilylation of the hydroxy group, gave the silyloxy azide (**43**), which was transformed to the corresponding silyloxyisocyanate by Curtius rearrangement. Desilylation of this substance using tetra-*n*-butylammonium fluoride resulted in spontaneous cyclization of the intermediate hydroxyisocyanate to produce the cyclic carbamate (**44**). The phenoxy substituent of (**44**) was readily exchanged by acid-catalyzed elimination-hydration, yielding (**45**) (Scheme 7).

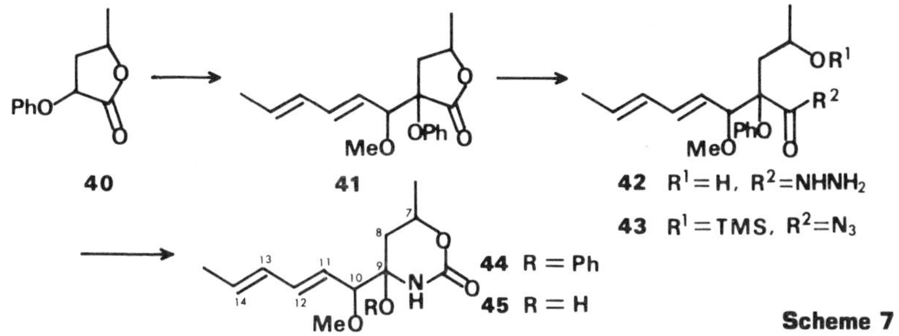

40
41
42 R¹ = H, R² = NHNH₂
43 R¹ = TMS, R² = N₃
44 R = Ph
45 R = H

Scheme 7

Samson *et al.* (1977) described the stereocontrolled synthesis of the carbons 5–9 parts (**49**) and (**53**) of maytansinoids. Treatment of 4-cumyloxy-2-cyclopentenone with dimethylcopperlithium followed by trimethylsilylation gave the silylenolether (**46**), whose ozonide was reduced with NaBH₄ and then esterified with CH₂N₂ to yield the acyclic product (**47**). After protection of the hydroxy group by silylation, (**47**) was reduced with diisobutylaluminumhydride, giving the aldehyde (**48**), which was converted the potential acylanion equivalent (**49**) by treatment with trimethylorthothioborate under neutral conditions. In a similar way, 4-*t*-butyldimethylsilyloxy-2-cyclopentenone was led via the silylenolether

(**50**) to the alcohol (**51**), the hydroxy group of which was protected with methylthiomethoxymethyl ether before reduction to obtain the aldehyde (**52**). In the present case, the formation of another potential acylanion equivalent was investigated by treating (**52**) with trimethylsilylcyanide to obtain a mixture of two diastereoisomers of the protected cyanohydrin (**53**) (Scheme 8).

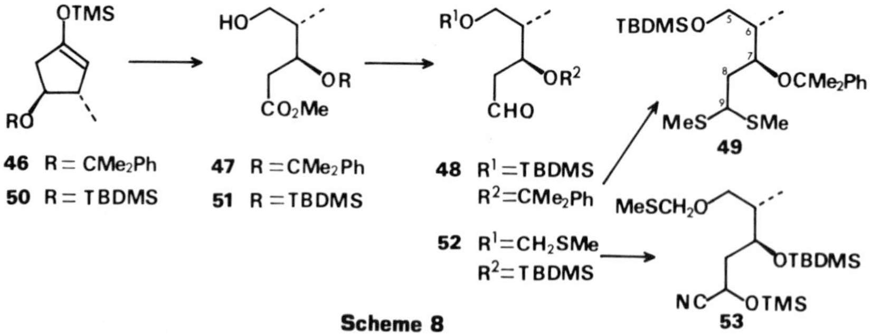
OTMS
RO
HO
OR
CO_2Me
R^1O
OR^2
CHO
TBDMSO
$OCMe_2Ph$
MeS
SMe
49
46 R= CMe_2Ph
50 R= TBDMS
47 R = CMe_2Ph
51 R = TBDMS
48 R^1=TBDMS
R^2=CMe_2Ph
52 R^1=CH_2SMe
R^2=TBDMS
$MeSCH_2O$
OTBDMS
NC
OTMS
53

Scheme 8

2. Preparation of the Aromatic Ring Moiety

The aromatic ring moiety of maytansinoids contains an unusual aromatic substitution array—i.e., three different contiguous hetero substituents—and furthermore, in the total synthesis of maytansinoids, suitable functional groups must be introduced to form the 19-membered macrocyclic ring.

Kane and Meyers (1977) reported the synthesis of the appropriately substituted aromatic rings (**59**) and (**60**). Methyl vanillate was nitrated to give the nitro derivative (**54**), which was treated with thionyl chloride and DMF under reflux to yield the chloride (**55**). The ready formation of (**55**) provided a key aromatic intermediate possessing the requisite substitutions mentioned above. Reduction of (**55**) gave the aniline (**56**), which was monomethylated using the Kadin method or through the trifluoroacetyl derivative to produce the *N*-methyl derivative (**59**). On the other hand, (**55**) could be elaborated further by hydrolysis of the ester group and treatment with HgO–Br_2 to give the bromide (**58**). The aniline derivative obtained by reduction of (**58**) was monomethylated, as above, to (**60**) (Scheme 9).

Foy and Ganem (1977) started with 5-methylcyclohexane-1,3-dione, which was condensed with methylamine to yield (**63**). Chlorination of (**63**) with

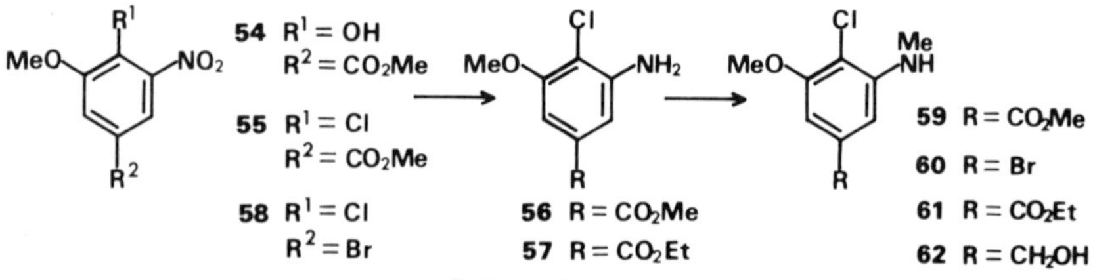
R^1
MeO
NO_2
R^2
54 R^1 = OH
R^2 = CO_2Me
55 R^1 = Cl
R^2 = CO_2Me
58 R^1 = Cl
R^2 = Br
Cl
MeO
NH_2
R
56 R = CO_2Me
57 R = CO_2Et
Cl
Me
NH
MeO
R
59 R = CO_2Me
60 R = Br
61 R = CO_2Et
62 R = CH_2OH

Scheme 9

N-chlorosuccinimide followed by bromination gave a mixture of the monochlorobromide (64). This substance was treated without purification with acetic anhydride and TsOH to produce the aromatized derivative (65). Saponification of (65) and further methylation *in situ* gave the methoxyacetanilide (66), which was brominated with N-bromosuccinimide to yield the bromide (67) (Scheme 10).

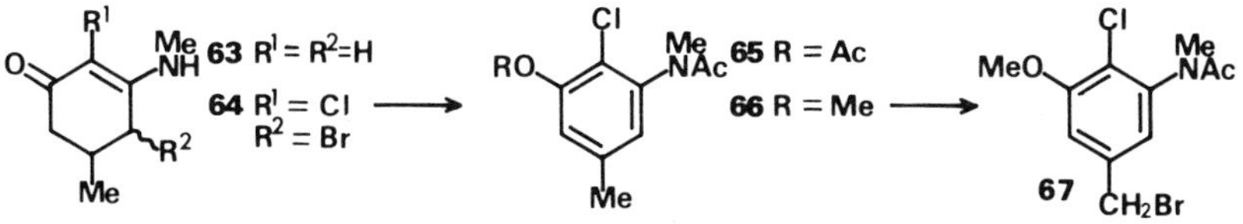

Scheme 10

Another aromatic moiety, (61) in Scheme 9, was prepared by Corey *et al.* (1977). Birch reduction of gallic acid followed by workup with aqueous acid and reaction with acidic ethanol gave the enone ester (68), which was treated with N-methylbenzylamine to give the enamino ketone (69), Reaction of (69) with *t*-butylhypochlorite furnished the chloro derivative (70). This substance was aromatized by reaction successively with lithium diethylamide and benzene selenylbromide to afford the benzoic ester (71). Methylation of (71) gave the phenolic methyl ether (72), which underwent hydrogenolysis to give the desired amino ester (61) (Scheme 11).

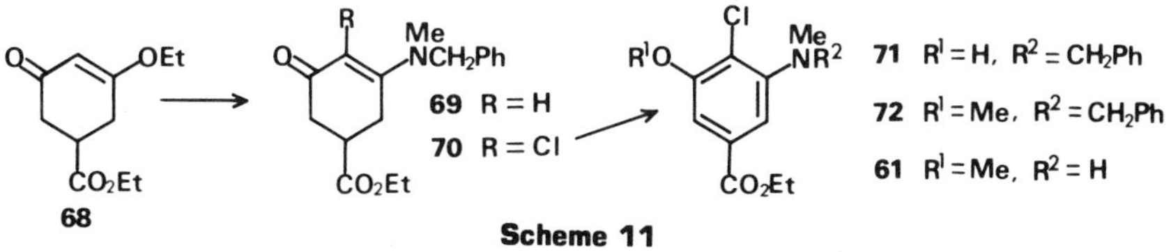

Scheme 11

Götschi *et al.* (1977) synthesized the same intermediate (61) from ethyl vanillate by a route similar to that shown in Scheme 9. The starting substance was nitrated, then converted to the chloride by treatment with phosphoryl chloride and DMF in the presence of equimolar amounts of lithium chloride and *s*-collidine. This product was reduced to the aniline derivative (57), then converted to the monomethyl derivatives (61) or (62) by different routes. Formylation of (57) with ethyl or phenyl formate, followed by reduction with LiAlH₄, gave (62). Alternatively, reductive methylation of (57) with formaldehyde in methanol in the presence of hydrogen and Raney nickel afforded the ester (61), and subsequent reduction also yielded (62) (Scheme 9).

3. Preparation of the C-10 to Aromatic Ring Moiety

Three groups have reported the synthesis of the C-10 to aromatic ring moieties, dienals (77), (89), (92), and (95), which are the suitably functionalized intermediates for the total synthesis of maytansinoids.

Götschi *et al.* (1977) oxidized the amino alcohol (**62**), obtained as shown in Scheme 9 with the Cornforth reagent to yield the benzaldehyde (**73**). This product was condensed with propionaldehyde to afford (**74**). The sidechain was elongated by reaction of (**74**) with the Grignard reagent prepared from 2-(2'-bromoethyl)dioxolane to yield the alcohol (**75**). Dehydration of (**75**), then acylation with 2,2,2-trichloroethoxycarbonyl chloride, gave the carbamate (**76**). Hydrolysis of (**76**) furnished a 2:1 mixture of the two stereoisomeric dienals (**77**) and (**78**). The desired product (**77**), obtained by purification, was equilibrated to give a 2:1 mixture of (**77**) and (**78**) under the conditions used for the hydrolysis of (**76**). Further elaboration of (**77**) was attempted by reaction of the mixture of (**77**) and (**78**) with 2-lithio-1,3-dithiane, followed by *in situ* methylation and, finally, reduction of the crude product to produce a 2:1 mixture of the two isomeric methoxy compounds (**79**) and (**80**). Alternatively, the aldehyde (**73**) was condensed with nitroethane to yield the nitro derivative (**81**), which upon reduction and hydrolysis afforded the ketone (**82**). This substance may be another possible intermediate (Scheme 12).

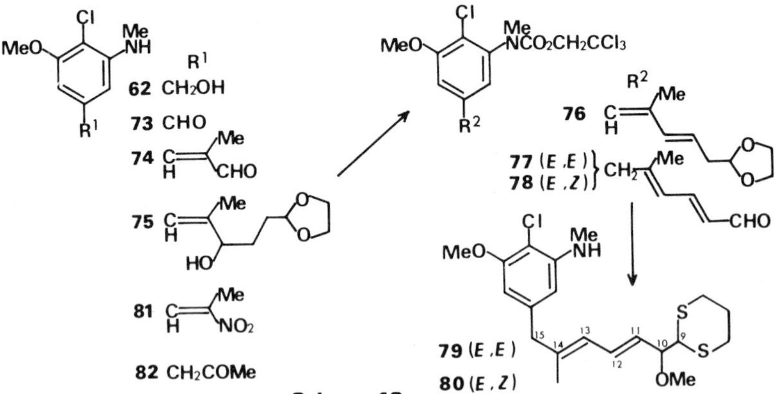

Scheme 12

Corey *et al.* (1978a) converted the amino alcohol (**62**) to the urethane (**83**) by treating the former with methyl chloroformate and subsequently hydrolyzing the crude product. The iodide (**84**) was obtained by mesylation of (**83**), then treatment of the crude product with sodium iodide *in situ*. The iodide (**84**) was transformed efficiently and selectively into the *E*-trisubstituted olefinic derivative (**87**) through a cross-coupling reaction with a specially designed mixed Gilman reagent (**86**), obtained from the bromo alcohol (**85**). Cleavage of THP group of (**87**) and subsequent oxidation with active manganese dioxide produced the corresponding aldehyde (**88**). Elaboration of the enal (**88**) to the dienal (**89**) was carried out using methodology specially developed for maytansine synthesis (Corey *et al.*, 1976). The α-TMS derivative of acetaldehyde *N-t*-butylimine was

converted to the α-lithio derivative by reaction with *sec*-BuLi and then allowed
to react with the aldehyde (**88**), to produce the dienal (**89**) (Scheme 13).

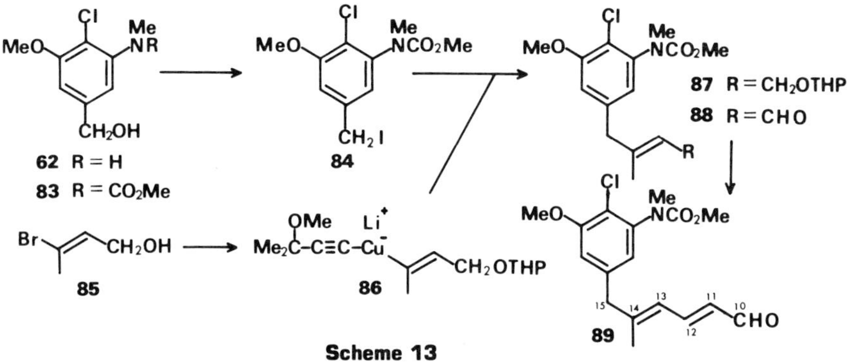

62 R = H
83 R = CO₂Me

84

85

86

87 R = CH₂OTHP
88 R = CHO

89

Scheme 13

On the other hand, Meyers *et al.* (1978) prepared the dienals (**92**) and (**95**)
from the amino ester (**59**). The *N*-benzoyl derivative of (**59**) was methylated,
then reduced to the *N*-benzyl benzyl alcohol (**90**). This product was converted to
the chloride, using mesyl chloride and lithium chloride, then further treated with
lithiated ethyldiisopropyl phosphonate to afford the phosphonate (**91**). The
lithiated derivative of (**91**) was treated with *E*-γ,γ-dimethoxycroton aldehyde,
followed by immediate hydrolysis to give the dienal (**92**). Although (**92**) appeared to be purely the *E,E*-isomer, the presence of a small amount of the
E,Z-isomer could not be precluded. The sequence leading to another dienal also
started with the amino ester (**59**). The benzyl alcohol prepared by reduction of
(**59**) was converted to the carbamate–carbonate with methyl chloroformate.
Selective removal of the carbamate group gave the alcohol (**93**). The benzyl
bromide obtained through mesylation and bromination was alkylated with
lithioethyldiisopropyl phosphonate and then with methyliodide to furnish the
phosphonate (**94**). Lithiation of (**94**), followed by introduction of *E*-
γ,γ-dimethoxycrotonaldehyde and sequentially direct hydrolysis of the product,
gave the dienal (**95**) (Scheme 14).

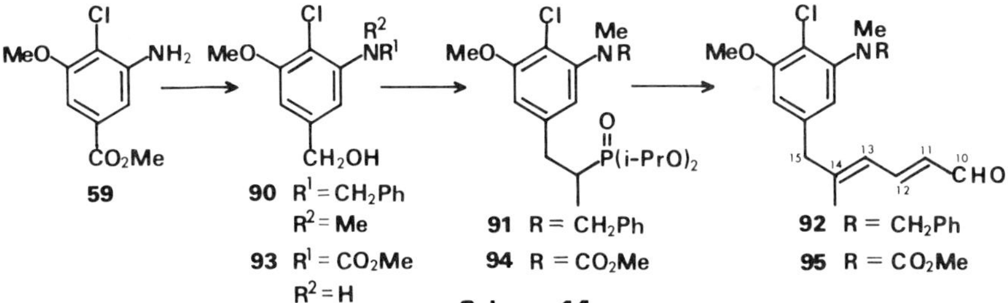

59

90 R¹ = CH₂Ph
    R² = Me
93 R¹ = CO₂Me
    R² = H

91 R = CH₂Ph
94 R = CO₂Me

92 R = CH₂Ph
95 R = CO₂Me

Scheme 14

B. The Total Synthesis of (±)-N-Methylmaysenine

The first total synthesis of maytansinoid—*i.e.*, the first formation of a 19-membered macrocyclic ring—was reported by Corey *et al.* (1978b) (Scheme 15). They connected the two parts, (**20**) in Scheme 3 and (**89**) in Scheme 13, elongated a chain to accommodate all the carbon atoms needed for the 19-membered ring, carried out cyclization of the precursor amino acid by lactamization, and finally prepared a cyclic carbamate ring, to synthesize (±)-*N*-methylmaysenine, which is the *N*-methyl derivative of a natural maytansinoid, maysenine.

The dithiane (**20**) was converted to the 2-methoxy-2-propyl ether (**96**). Reaction of (**96**) with *n*-BuLi and *N,N'*-tetramethylenediamine in THF afforded the 2-lithiodithiane, which after removal of most the THF and tetramethylenediamine, replacement by toluene, and coupling to the dienal (**89**) gave the adduct (**97**) as a mixture (55:45) of diastereoisomers at C-10. Although this product routinely was used directly in the next step of the synthesis, the diastereomeric diols obtained by removal of the 2-methoxypropyl group of (**97**) have been separated chromatographically and individually carried through the synthesis. The less predominant diol corresponds in relative stereochemistry to the maytansine series. Methylation of (**97**) with methyl iodide, followed by acid hydrolysis, afforded the diastereomeric methoxy alcohols (**98**). Oxidation of (**98**)

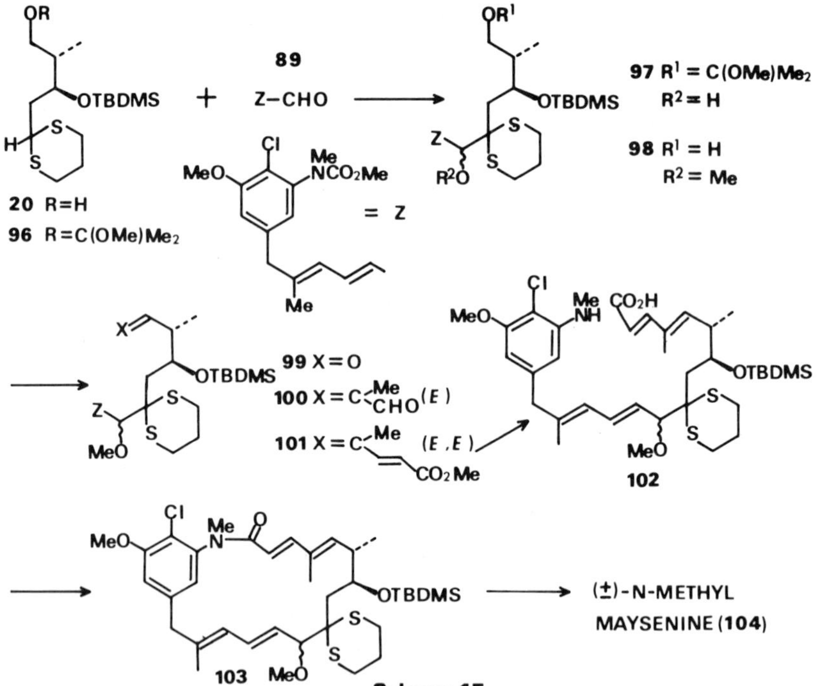

Scheme 15

using dimethyl-sulfoxide–benzene with diethylcarbodiimide, pyridine, and tri-fluoroacetic acid gave the aldehydes (99). The elaboration of (99) to the enal (100) was accomplished by the method previously designed for this application and also used to prepare the dienal (89) from the aldehyde (88). The α-TMS derivative of propionaldehyde N-t-butylimine was α-lithiated with sec-BuLi, and the resulting reagent was treated with the aldehydes (99), yielding (100) with a small amount of the isomeric Z-enal which was isomerizable to the E form. Chain extension of the E-enal (100) to form the conjugated α,β-E, γ,δ-E-unsaturated ester (101) was effected smoothly by reaction of (100) with the lithio derivative of dimethyl methoxycarbonylmethane phosphonate. Hydrolysis of the ester (101), followed by reaction with lithium n-propylmercaptide, gave the amino acid (102). This product was converted to the soluble tetra-n-butylammonium salt by reaction with tetra-n-butylammonium hydroxide, and then azeotropically dried. A benzene solution of the salt was reacted with a benzene solution of excess mesitylenesulfonyl chloride and diisopropyl amine to yield the macrolactam (103) admixed with comparable amounts of the C-10 epimer, from which it was readily separated by PTLC. Macrolactam (103) was transformed into (±)-N-methylmaysenine (104) by the sequence of three steps used to prepare (25) from (22) as shown in Scheme 4. Synthetic (±)-N-methyl-maysenine was shown to be spectroscopically and chromatographically identical with authentic N-methylmaysenine prepared from maytansine. The C-10 epimer of (104) was synthesized from the C-10 epimer of (103) in a similar manner.

IV. BIOLOGICAL ACTIVITIES OF MAYTANSINOIDS

A. Antitumor Activity against Experimental Tumors

1. Antitumor Activity of Maytansine

Maytansine (1) was the first plant product found of the ansa structure. Ansamycin is a class of antibiotic with ansa structure that includes the rifamy-cins, streptovaricins, tolypomycins, and geldanamycin. They are known for their inhibition of bacterial DNA-dependent RNA polymerase and viral RNA-directed DNA polymerase, but do not show significant antitumor activity in $vivo$.

However, maytansine showed an excellent antitumor activity against mouse lymphocytic leukemia P-388 upon intraperitoneal (i.p.) injection as shown in Fig. 2. T/C is the ratio (%) of the median survival time of the treated group of mice divided by that of the control group. Compounds are considered active by the NCI protocol if T/C ≥ 125%. Maytansine was effective over a wide dosage range from 0.4 to 50 μg/kg of animal weight when given daily for 9 days when administered i.p. It had a potentially safety of 50- to 100-fold the dosage range. The highest T/C value observed was 220% on the optimally effective dosage of

1 Maytansine — $R^1 = COCHMeNMeCOMe; R^2 = OH; R^3 = H$

2 Maytanprine — $R^1 = COCHMeNMeCOCH_2Me; R^2 = OH; R^3 = H$

3 Maytanbutine — $R^1 = COCHMeNMeCOCH(Me)_2; R^2 = OH; R^3 = H$

4 Maytanvaline — $R^1 = COCHMeNMeCOCH_2CH(Me)_2; R^2 = OH; R^3 = H$

5 Maytansinol — $R^1 = H; R^2 = OH; R^3 = H$

6 Maytanacine — $R^1 = COMe; R^2 = OH; R^3 = H$

7 Maytansinol 3-propionate — $R^1 = COCH_2ME; R^2 = OH; R^3 = H$

8 Maytansinol 3-bromoacetate — $R^1 = COCH_2Br; R^2 = OH; R^3 = H$

9 Maytansinol 3-trifluoroacetate — $R^1 = COCF_3; R^2 = OH; R^3 = H$

10 Maytansinol 3-crotonate — $R^1 = COCH=CHMe; R^2 = OH; R = H$

11 Maytanbutacine — $R^1 = COCH(Me)_2; R^2 = OH; R^3 = OCOCH_3$

12 Ansamitocin P-3 — $R^1 = COCH(Me)_2; R^2 = OH; R^3 = H$

13 Ansamitocin P-3' — $R^1 = COCH_2CH_2Me; R^2 = OH; R^3 = H$

14 Ansamitocin P-4 — $R^1 = COCH_2CH(Me); R^2 = OH; R^3 = H$

15 Maytansine 9-methyl ether — $R^1 = COCHMeNMeCOMe; R^2 = OMe; R^3 = H$

16 Maytansine 9-ethyl ether R^1 = COCHMeNMeCOMe; R^2 = OCH$_2$Me; R^3 = H
17 Maytansine 9-propyl thioether R^1 = COCHMeNMeCOMe; R^2 = SCH$_2$CH$_2$Me; R^3 = H
18 Maytansine 9-bromopropyl ether R^1 = COCHMeNMeCOMe; R^2 = OCH$_2$CH$_2$CH$_2$Br; R^3 = H
19 Maytanbutine 9-methyl ether R^1 = COCHMeNMeCOCH(Me)$_2$; R^2 = Me; R^3 = H
20 Maytanbutine 9-ethyl ether R^1 = COCHMeNMeCOCH(Me)$_2$; R^2 = OCH$_2$Me; R^3 = H
21 Maytanbutine 9-propyl thioether R^1 = COCHMeNMeCOCH(Me)$_2$; R^2 = SCH$_2$CH$_2$Me; R^3 = H
22 Colubrinol R^1 = COCHMeNMeCOCH(Me)$_2$; R^2 = H; R^3 = OH
23 Colubrinol acetate R^1 = COCHMeNMeCOCH(Me)$_2$; R^2 = H; R^3 = OCOMe

26   Maysenine

24 Maysine   R = Me
25 Normaysine   R = H

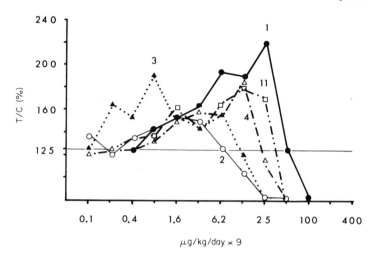

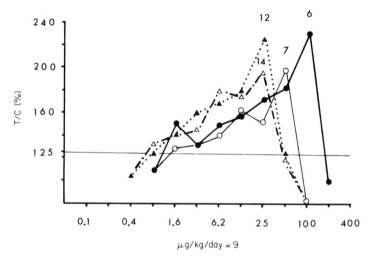

Figure 2.   Antileukemic activity of maytansinoids against P-388 leukemia with i.p. administration.

25 μg/kg (Venditti and Wolpert-DeFilippes, 1976; Kupchan *et al.*, 1978). The influences of variation in treatment schedule and drug route on the therapeutic effect of maytansine were investigated (Table I). Although the differences were not profound, daily treatment once a day for 9 days seemed less effective than intensive intermittent i.p. treatment of eight times at 3-hour intervals on day 1;

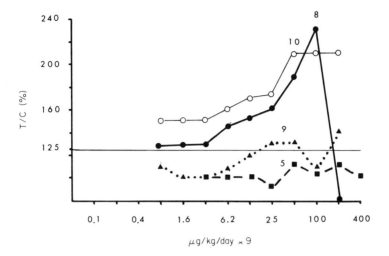

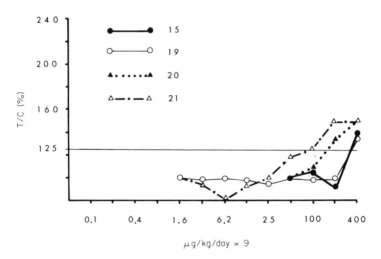

Figure 2 (*Continued*)

eight times on days 1, 5, and 9; eight times on days 1 and 9; or two widely spaced treatments on day 1 and 9 or once on day 1. Subcutaneous (SC) daily injection appeared to be as effective as daily i.p. injection. Given orally, maytansine was ineffective agains P-388. Against Melanoma B-16, it caused a moderate increase in the survival time of mice. The effective dose was from 4 to 16 μg/kg/day (T/C

TABLE I

Influence of Route and Schedule on Maytansine Activity against Mouse Leukemia P-388 and Melanoma B-16

Tumor	Route	Total no. of times	Day of treatment	Dose range (μg/kg)	O.E.D.[e] (μg/kg)	T/C (%)	Dose range (μg/kg)	O.E.D.[e] (μg/kg)	T/C (%)
				Experiment 1			Experiment 2		
P-388	i.p.	Once	Day 1	1.0–256	128	154	128–1034	512	154
		×9	D1–9[a]	0.3–64	16	186	2.0–32	8.0	159
		×3	D1, 5, 9[b]	0.5–128	128	190	128–1024	128	163
		×2	D1, 9[c]	1.0–256	256	181	128–1024	256	181
		×8	D1[d]	0.1–32	32	200	16–128	32	181
		×16	D1, 9[d]	0.1–32	16	190	8.0–128	16	181
		×24	D1, 5, 9[d]				16–128	16	195
	SC	×9	D1–9[a]				64–512	256	172
B-16	i.p.	×9	D1–9[a]	4–32	16	145	4–32	16	157

[a] Daily for 9 days.
[b] Every fourth day.
[c] With an 8-day interval.
[d] Eight times in 1 day at 3-hour intervals.
[e] Optimally effective dose.

value was 157% at 16 μg/kg) (Venditti and Wolpert-DeFilippes, 1976). Maytansine also showed antitumor activity *in vivo* against Walker carcinosarcoma 256, mastocytoma P-815, and plasmocytoma YPC 1, as for P-388 and B-16 at daily doses of 25 μg/kg. It was inactive against P-388 transplanted intracerebrally (Adamson *et al.*, 1976) and showed relatively weak activity against leukemia L-1210 (Adamson *et al.*, 1976; Venditti and Wolpert-DeFilippes, 1976). Maytansine was also effective against vincristine-sensitive P-388 (original line) as described above [i.p., optimally effective dose (O.E.D.) 25 μg/kg/day × 9, T/C 195%], but was ineffective against the vincristine-resistant line of the P-388 (i.p., O.E.D. 25 μg/kg, T/C 101%) (Wolpert-DeFilippes *et al.*, 1975a).

Colubrinol (**22**) and colubrinol acetate (**23**), which have hydroxy and acetoxy groups at C-15, respectively, exhibited confirmed activity against P-388 leukemia at the microgram per kilogram level (Wani *et al.*, 1973).

2. Antitumor Activity of Ansamitocins

Ansamitocins P-3 (**12**) and P-4 (**14**) were effective over a wide dosage range from 0.8 to 50 μg/kg by i.p. administration as shown in Fig. 2. The highest T/C values were 223% and 195%, respectively, with the optimally effective dose of 25 μg/kg. When ansamitocin P-3 or P-4 was injected intravenously (i.v.) for five consecutive days starting one day after P-388 transplantation into BDF$_1$ mice, it had a marked effect on survival time of tumor-bearing mice, as shown in Fig. 3. Also, excellent therapeutic efficacies of ansamitocins P-3 and P-4 against B-16 were found upon i.p. administration of a dose of 1.6 to 50 μg/kg, as shown in Fig. 3. Ansamitocin P-3 was also significantly active against sarcoma 180, Ehrlich carcinoma, and P-815 mastocytoma as shown in Table II, but less active against leukemia L-1210, MOPC 104E myeloma, and C-1498 leukemia. The highest T/C values against these tumors were 130%, 133%, and 138%, respectively, at the optimal effective dose of 50 μg/kg (Ootsu *et al.*, 1980). Two maytansinoids, maytansine (**1**) and ansamitocin P-3 (**12**), are undergoing further study in detail on their antitumor activities.

3. Structure-Activity Relationship

Structural requirements for antileukemic activity among the naturally occurring and semisynthetic maytansinoids were investigated by Kupchan *et al.* (1978). Maytansinoids (**1**), (**2**), (**3**), (**4**), (**6**), and (**11**), which each have an amino acid or fatty acid ester group at C-3, showed excellent activity against P-388 leukemia as shown in Fig. 2. Colubrinol (**22**) and colubrinol acetate (**23**), which each have an amino acid ester group at C-3 and a hydroxy or acetoxy group at C-15, were significantly active against P-388 leukemia (Wani *et al.*, 1973). And ansamitocins P-3 (**12**) and P-4 (**14**), which each have a fatty acid ester group at C-3, showed strong activity against P-388 leukemia (Ootsu *et al.*, 1980). In contrast, maytansinol (**5**), maysine (**24**), normaysine (**25**), and maysenine (**26**),

Yasuo Komoda and Toyokazu Kishi

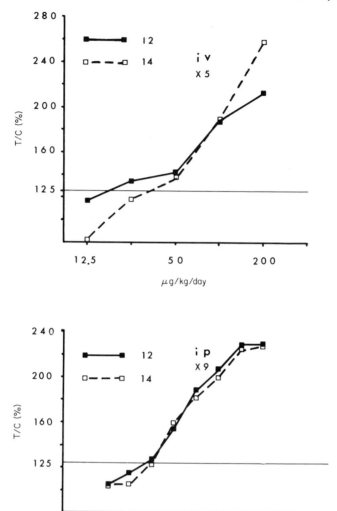

Figure 3. Effect of ansamitocins on survival time. Upper: against P-388 leukemia transplanted mice by i.v. treatment for 5 consecutive days; lower: against B-16 melanoma transplanted mice by i.p. treatment for 9 consecutive days.

TABLE II

Effect of Antitumor Activity of Ansamitocin P-3 against Sarcoma 180, Ehrlich Carcinoma, and P-815 Mastocytoma

Drug	i.p. Dose (μg/kg/day) given for 7 days	Tumor (T/C %) S-180	Ehrlich	P-815
ASM P-3 (12)	50	284	480	100
	25	258	393	241
	12.5	131	480	205
	6	152	202	188

which lack the C-3 ester moiety, showed much less activity against P-388 leukemia. The C-3 ester moiety seems to be very important for significant antitumor activity. Several modified C-3 esters were synthesized from maytansinol (5). Maytansinoids (6), (7), (12), and (14), in which the C-3 amino acid ester is replaced by a simple alkyl ester, and (8) and (10), in which it is replaced by an alkyl ester having potential alkylating sites, showed strong activity against P-388 leukemia. The C-3 ester group appears to be necessary, but the amino nitrogen atom, carbon number, and potential sites showed little effect on the activity. The lack of antileukemic activity exhibited by the C-3 trifluoroacetate ester (9) is presumably due to *in vivo* hydrolysis to inactive maytansinoids. When the C-9 alcohol of the carbinolamide is converted to an ether, as in the case of (15), (16), (17), (18), (19), (20), and (21), a marked decrease in the activity is observed as shown in Fig. 2. Compounds showing less activity (T/C < 125%: 16, 17, 18, 24, 25, and 26) were not included in the figure.

Therefore, the C-3 ester is apparently needed for significant activity, while variation of the ester group does not cause marked change in the activity. The presence of a C-15 ester or hydroxy group does not affect the activity. However, elimination of the ester group at C-3, as in maytansinol (5), maysine (24), normaysine (25), and maysenine (26), significantly decreases the activity against P-388 leukemia. Blockage of the C-9 carbinolamide via etherification markedly reduced antileukemic activity. Thus, a free carbinolamide or free hydroxy group at C-9 is advantageous for optimal activity. The presence of a C-4, 5 epoxide group is assumed to affect the activity (Kupchan *et al.*, 1978). The structure–activity relationship of maytansinoids will be revealed in the course of the total synthesis of maytansinoids and the chemical modification of ansamitocins.

B. Clinical Studies of Maytansine

Preliminary clinical trial of maytansine began in December 1975 in the United States of America by NCI. A phase I trial of maytansine was conducted in

21 adult and six pediatric patients, using a single i.v. dose every 3 weeks in adults and every 1-2 weeks in children with acute lymphocytic leukemia. Dose-limiting toxicity was not reached at up to 0.8 mg/m^2 from 0.03 mg/m^2. No hepatic or renal toxicity was reported. But distal extremity paresthesia was noted in two patients at 0.045 mg/m^2 and in one patient at 0.07 mg/m^2. At higher doses, one patient developed diarrhea for 24 hours on the fourth day after receiving 0.8 mg/m^2. Two antitumor responses were seen: one patient with ovarian cancer had 25–50% shrinkage during seven courses of 0.07 mg/m^2, and a child with acute lymphocytic leukemia had a clearing of blasts from the blood and marrow during 15 weekly courses of 0.07 mg/m^2 (Chabner et al., 1977).

Maytansine was evaluated by the Mayo Clinic group with various solid tumors in 40 adult patients. Severe nausea and vomiting, sometimes associated with watery diarrhea and abdominal cramps, and liver function abnormalities, mainly elevation of SGOT levels, were found. Mild hematologic toxicity, neurotoxicity, and possibly cardiac toxicity were also noted. No tumor responses were seen. Most treated tumors, colorectal and lung types, were relatively drug-resistant to single-agent chemotherapy. The recommended starting dosage for phase II studies was 2.25 mg/m^2 total dose per course with i.v. doses of 0.750 mg/m^2 on days 1, 3, and 5 repeated every 4 weeks (Egan et al., 1978).

At the University of Texas System Cancer Center, maytansine was administered to 60 patients. Doses from 0.01 to 0.9 mg/m^2 were given for 3 days. The toxic effects encountered were principally nausea, vomiting, diarrhea, and occasionally stomatitis and alopecia. Superficial phlebitis was also encountered and occurred when the drug was diluted with < 250 ml of saline. Myelosuppression occurred infrequently; it was almost always associated with abnormal liver function. Antitumor response was detected in one patient each with melanoma, breast carcinoma, and head and neck clear-cell carcinoma of 19 patients given doses of 0.3 mg/m^2. In patients with normal liver function tests, the recommended dose for phase II studies was 0.5 mg/m^2 for 3 days. At this dose, maytansine appeared to be nonmyelosuppressive (Cabanillas et al., 1978a).

In another test, dose-limiting toxicity was observed at 2 mg/m^2, manifested by profound weakness, diarrhea, nausea, and vomiting. Symptoms persisted for 3–14 days after drug administration. No consistent myelosuppression occurred at any dose up to 2 mg/m^2. Antitumor responses occurred in a patient with non-Hodgkin's lymphoma and one with ovarian cancer, who had been treated on the 3-week schedule, as well as in two patients with acute lymphocytic leukemia treated with single weekly doses. Three of the four responding patients had received extensive prior treatment with vincristine, and two were clearly resistant to it; this differed from the result of animal tests (Chabner et al., 1978).

Five daily bolus injections were repeated in 38 adult solid-tumor patients at 21-day intervals over a dose range of 0.1–0.8 mg/m^2/day × 5 days. Gastrointestinal toxicity was dose-related, and dose-limiting at doses of ≥ 0.5 mg/m^2.

Dose-related neurotoxicity was also observed. Neither drug-related myelosuppression nor a change in serum creatinine level was seen. Hepatic toxicity was subclinical and reversible. Of 16 patients evaluated for response, two with breast cancer exhibited demonstrable antitumor effect. For phase II studies of maytansine, a starting dose of 2.0–2.5 mg/m² repeated at 21-day intervals was recommended (Blum and Kahlert, 1978).

Of the 17 patients, 11 with melanoma and 6 with breast carcinoma were treated and evaluated. Two patients with melanoma exhibited a response and two a mixed response. Another patient with melanoma was stable for 3 months and one with breast cancer for 4 months. Toxicity consisted of nausea and/or vomiting in eight patients, diarrhea in eight, paresthesia in three and adynamic ileus in two patients (Cabanillas *et al.*, 1978b).

The maximal tolerated dose of maytansine was 0.5 mg/m²/day × 5 days in another group of 21 patients. A more convenient schedule of 2 mg/m² at 21-day intervals was used with 32 patients. Gastrointestinal and neurologic toxicities were observed, but not myelosuppression. Another 41 patients were evaluated (colon 6, sarcoma 8, breast 7, lung 1, renal 5, bladder 2, other 6) and two exhibited better than 50% response of more than 2 months duration in a bladder case and 2 months in an oad cell case. Two minor responses were seen with breast cancer (Blum *et al.*, 1978).

Maytansine has recently entered phase II clinical trials at several cancer centers in the United States under a variety of dosage schedules, and preliminary data will be available soon (Douros, 1978).

C. Inhibition of Cell Growth and Cytological Effect

Among the natural products which have the ansa structure and were isolated from fermented broth of microorganisms and plants, ansamycins displayed growth inhibition of prokaryotic cells such as bacteria, but none of the common bacteria were susceptible to inhibition by maytansinoids. Maytansinoids did not show any appreciable growth inhibition of prokaryotic cells, but were strongly active against eukaryotic cells such as fungi, protozoa, tumor cells, cultured mammalian cells and egg cells, and higher plant cells.

When maytansine was first discovered, its isolation from plants was guided by assay against KB cells derived from human carcinoma of nasopharynx and P-388 leukemia.

In the case of ansamitocins, inhibition of growth and mitosis of P-388 and activity against *Tetrahymena* were the major indicators of their biological activities during the course of the early stage purification.

1. Mammalian Cells

Maytansine inhibited the growth of leukemia L-1210, L-5178Y, and P-388 cells in suspension culture. Its effective dose (ED$_{50}$) on these murine leukemia cells at 48 hours was $6 \times 10^{-10}M$ against P-338, $2 \times 10^{-9}M$ against L-1210, and $1.5 \times 10^{-9}M$ against L-5178Y. The P-388 line was the most sensitive, and the effect could be reversed by removal of the drug after incubation at a concentration of $10^{-9}M$. When the level was higher than $10^{-8}M$, growth inhibition could not be reversed by washing (Wolpert-DeFilippes et al., 1975a). When incubated with KB cells for 72 hours, maytansine exhibited a growth-inhibitory ED_{50} of 6.1×10^{-6} μg/ml (Kupchan et al., 1978). Other maytansinoids such as maytanside esters also displayed activity against KB, mouse 3T3, CHO, human CEM, and rat AC cells as described below.

2. Other Eukaryotic Cells

Maytansine irreversibly prevented cell division in eggs of sea urchins and clams at $6 \times 10^{-8}M$, which was approximately 100 times more potent than vincristine (Remillard et al., 1975). Maytansine prevented spindle formation or dispersed an already formed spindle in marine eggs (Schnaitman et al., 1975).

Penicillium avellaneum (Talaromyces avellaneus, Hamigera avellanea) was the most susceptible to maytansine among several fungi and protozoa (Hanka and Bernett, 1974a,b). Ansamitocins (**12**) and (**14**) inhibited the growth of some eukaryotic microorganisms such as those of genera *Hamigera, Tricophyton, Cryptococcus,* and *Tetrahymena in vitro* at 1 to 50 μg/ml (Tanida et al., 1980).

Maytansine inhibited the cell division of tobacco pith callus even at 1×10^{-2} μg/ml. It clearly retarded the growth of rice seedlings, even at 1×10^{-1} μg/ml. The effect was marked on the root growth. Not only was the root growth inhibited, but also the root tips were swollen and malformed. The effect of maytansine in the *Avena* straight growth test, which is valid only for compounds causing elongation of cells, was a significant promoting activity on the coleoptile sections at 100 μg/ml, contrary to the anticipation (Komoda and Isogai, 1978).

A reduction in the number of microtubules and a partial blockage of fast axoplasmic transport was observed with 2×10^{-5} and 1×10^{-4} M of maytansine, respectively. It also induced *in vitro* alteration of the neurofibrillar elements concomitant with a partial blockage of the tested axoplasmic transport (Donoso et al., 1978). AC cells, a rat glioma cell line having an epithel-like form in standard culture, were altered morphologically into an astrocyto-like form with many-dendrite structure by exposure to Db-c AMP (Igarashi et al., 1978). This alternation was prevented or apparently reversed with ansamitocin P-3. In addition to dendrite structure, cytoplasmic microtubule networks were also destroyed with ansamitocin P-3 (see Section IV,D,1). These observations suggested that ansamitocin P-3 disintegrated the microtubule system and resulted in inducing morphological changes of AC cells (Tsukamoto et al., 1978).

3. Prokaryotic Cells

Maytansine and its analogues were not effective against common bacteria *in vitro* (Hanka and Bernett, 1974a,b). Ansamitocins were also not active against prokaryotic microorganisms such as various Gram-positive, Gram-negative, and acid-fast bacteria at doses of 100 μg/ml (Tanida *et al.*, 1980).

D. Action Mechanism and Structure-Activity Relationship

Maytansine prevented polymerization of microtubule protein *in vitro*. Its binding site was as same as or very close to that of vincristine. Maytansinoids probably act initially on tumor cells by inhibiting the assembly of microtubule proteins into spindle fibers, thus inducing mataphase arrest and resulting in cytokillings in a manner similar to that of *Vinca* alkaloids.

1. Primary Target of Drug Action: Interaction with Microtubule Systems

Histological examination of L-1210 cells after 24 hours of exposure to 10^{-8} M of maytansine showed 30% with mitotic figures, compared to 3% for the control cells (Wolpert-DeFilippes *et al.*, 1975a). One effect of maytansine on human CEM cells of PHA stimulated lymphocytes *in vitro* was metaphase arrest (Adamson *et al.*, 1976); ansamitocin P-3 or P-4 in an i.p. dose of 1 μg/kg induced minimum mitotic arrest in P-388 leukemia (Ootsu *et al.*, 1980). Histological examination of L-1210 cells revealed that the majority were arrested in the metaphase, suggest that maytansine impairs the function of mitotic spindles (Wolpert-DeFilippes *et al.*, 1975b). Maytansine was extremely effective in producing mitotic arrest of L-1210 and P-388 leukemias *in vivo* over a broad dosage range (i.p., 0.410–0.041 mg/kg). The peak in mitotic arrest occurred between 12 and 20 hours with 60–70% of the cells in mitosis. The extent and duration of stathmokinesis in these cell lines were greater than those of Vinca alkaloids. The ability of maytansine to produce mitotic arrest was not related to its therapeutic activity in these cell lines (Johnson *et al.*, 1978). Maytansine prevented spindle formation or dispersed an already formed spindle in marine eggs at 5 × 10^{-8} M (Schnaitman *et al.*, 1975). Studies of maytansinoids with purified microtubule protein *in vitro* have been done in some laboratories. Maytansine prevented polymerization of tubulins by causing disappearance of 30 S rings and rapidly dispersing already formed microtubules (Schnaitman *et al.*, 1975). Maytansine did not affect the formation of the mitotic organizing center but inhibited *in vitro* polymerization of tubulin which had been purified from rabbit or pig brain (Remillard *et al.*, 1975). It seemed to bind to tubulin at the vincristine binding site (Adamson *et al.*, 1976). Binding of maytansine to rat brain tubulin was temperature- and ionic strength-dependent as was that of vincristine. Maytansine competitively inhibited vincristine binding. The binding of both drugs was at least partly reversible. Both drugs appeared to share a common binding site, although an additional site specific for maytansine seemed to be present, because unlabeled maytansine rapidly displaced the labeled maytansine whereas about

25% of the bound maytansine did not exchange by vincristine (Mandelbaum-Shavit *et al.*, 1976).

Ansamitocins P-3 (**12**) and P-4 (**14**) completely inhibited the polymerization of tubulin isolated from bovine brain at $3.3 \times 10^{-6} M$ and completely depolymerized the polymerized tubulin at $1.6 \times 10^{-5} M$. No nonspecific aggregation of tubulin, like that found with vincristine, was observed with ansamitocins P-3 and P-4. Furthermore, the well-defined cytoplasmic fiber networks of cultured mouse BALB 3T3 cells, visualized with antitubulin antibodies, disappeared shortly after incubation with P-3 at $1.1 \times 10^{-5} M$ (Ootsu *et al.*, 1980). The above observed effects against microtubule systems of the cell of ansamitocins are consistent with the observation that they strongly inhibit the regeneration of cilia in partially deciliated protozoan *Tetrahymena pyriformis* W, for which the assembly of ciliary tubulin is essential (Tanida *et al.*, 1979). Maytansine and the various maytansinoids failed to depress the enzymatic activity of simian sarcoma virus DNA polymerase and of DNA and RNA polymerases prepared from BALB/C mouse embryo cells (O'Conner *et al.*, 1975). The distribution of the DNA content in a population of exponentially growing L-1210 cells within 12 hours of exposure to $10^{-8} M$ maytansine shifted to a single peak which corresponded to cells with a G2 + M DNA content (Wolpert-DeFilippes *et al.*, 1975b). In studies of maytansine action in other site, inhibition of DNA synthesis with L-1210, L-5178Y, and P-388 cells were observed at a drug concentration of $10^{-7} M$. RNA synthesis was inhibited to a lesser degree. In these studies, however, duration of incubation of cells with the drug was 12 hours. *E. coli* RNA polymerase was not inhibited by $10^{-4} M$ (Wolpert-DeFilippes *et al.*, 1975a).

For ansamitocins, similar experiments were carried out except for shorter periods of incubation with drugs. In growing KB cells, no appreciable inhibition on the syntheses of DNA, RNA, and protein was observed until 8 hours incubation with the drug at a concentration of 1.6×10^{-9} mole/liter. Cytotoxicity of maytansine against KB cells (ED_{50}) was 8.8×10^{-12} mole/liter. After 24 hours incubation with ansamitocin P-3, the rates of synthesis of these macromolecule gradually decreased. These results suggest that the effects of ansamitocin P-3 on DNA, RNA, and protein synthesis are presumably secondary or indirect (Tsukamoto *et al.*, 1978). All evidence described above indicates that the primary target of maytansinoid action is the microtubule system of cell, although other possibilities cannot be totally excluded.

2. Structure-Activity Relationship

Maytanside esters (**1**), (**2**), (**3**), (**4**), and (**6**) showed significant cytotoxicity against KB cells at 10^{-6} μg/ml. Esters at C-3 like (**7**), (**8**), (**10**), and (**11**), and ethers like (**15**), (**17**), (**18**), (**19**), and (**21**) decreased the cytotoxicity against KB cells at about 10^{-4} μg/ml. However, some maytansinoids which lack the C-3 ester, like (5), (24), (25), and (26), were less active than (1) (Kupchan *et al.*, 1978). Colubrinol (22) and colubrinol acetate (23) exhibited cytotoxicity (ED_{50}) against KB cells at 10^{-4}–10^{-5} μg/ml (Wani *et al.*, 1973). Maytansine inhibited the growth of murine sarcoma virus transformed mouse 3T3 cells at 7×10^{-4} μg/ml. Other maytansinoids (4) (?), (16), (24), (25), and (26) differed significantly from (1) in focus inhibition and cell toxicity (O'Connor *et al.*, 1975). Maytansine lysed chinese hamster ovary (CHO-Ki) cells in culture at $10^{-8}M$. The cells gradually recovered and normal growth was resumed after 48 hours exposure to (1) at $5 \times 10^{-9}M$ (Schnaitman *et al.*, 1975). Inhibition of murine sarcoma virus focus formation in 3T3 may^r cells, which were resistant to maytansine, occurred only at a tenfold higher concentration of (1) than that in 3T3 cells (Aldrich and O'Conner, 1976).

Maytansine irreversibly prevented cell division in eggs of sea urchins and clams. Several other maytansinoids were assayed using this mitotic inhibition. Of the samples tested, (5), (24), and (25), which lack the ester group at C-3, exhibited reduced potency. The inhibition of sea urchin eggs with these maytansinoids was not comparable to the antitumor activity against P-388 leukemia as shown in the case of (26) (Kupchan *et al.*, 1978).

The activity of ansamitocins against *Tetrahymena* and *Hamigera avellanea* related to the acyl group at C-3 because the order of activities was (14) > (12) > (7) > (6) > (5) (Tanida *et al.*, 1980). These antiprotozoal and antifungal activities of maytansinoids may correlate with their antileukemic activities.

Maytansine (1) and maytanvaline (4) competitively inhibited vincristine binding of rat brain tubulin. Normaysine (25) showed 100 times less inhibition, while maysine (24) displayed no competition. The inhibition of maysenine (26) resembled that of a partial noncompetitive inhibitor (York *et al.*, 1978). Several maytansinoids—(1), (2), (3), (4), (6), (10), and (11)—inhibited the polymerization of brain tubulin. Maytansinol (5), maysine (24), and normaysine (25), which lacked an ester group at C-3, reduced the potency. An important exception was maysenine (26), which significantly inhibited sea urchin egg mitosis and tubulin polymerization (Kupchan *et al.*, 1978).

The study of the structural requirements of maytansinoids for significant cytotoxic, antitubulinic, antimitotic, and antileukemic activity indicated the necessity of the presence of the C-3 ester and the free hydroxyl group at C-9. An exception was activity of the C-9 ethers against sea urchin egg division. Study of the structural requirements of maytansinoids will progress further in the very near future from two aspects of development: the total synthesis of maytansinoids and the chemical modification of ansamitocins. We hope that maytansinoids will be used clinically as antitumor agents, and applied as a specific biochemical reagent.

Acknowledgment

This chapter is dedicated to the memory of the late Professor S. Morris Kupchan. The authors thank Drs. J. M. Venditti and M. K. Wolpert-DeFilippes for the adaptation into Table I. T. Kishi also thanks Dr. E. Ohmura, Director of the Division, Takeda Chemical Ind. Ltd. for his permission, continuing interest, and encouragement.

REFERENCES

Adamson, R. H., Sieber, S. M., Wang-Peng, J., and Wood, H. B. (1976). *Proc. Am. Assoc. Cancer Res.* **17,** 42 (No. 165).

Aldrich, C. D., and O'Connor, T. E. (1976). *Proc. Am. Assoc. Cancer Res.* **17,** 85 (No. 339).

Asai, M., Mizuta, E., Izawa, M., Haibara, K., and Kishi, T. (1979), *Tetrahedron* **35,** 1079–1085.

Blum, R. H., and Kahlert, T. (1978). *Cancer Treat. Rep.* **62,** 435–438.

Blum, R. H. Wittenberg, B. K., Canellos, G. P., Mayer, R. J., Skarin, A. T., Lokich, J. J., Henderson, I. C., and Parker, L. M. (1978). *Proc. Am. Assoc. Cancer Res.* **19,** 399 (No. C-369).

Bonjouklian, R., and Ganem, B. (1977). *Tetrahedron Lett.,* 2835–2838.

Bryan, R. F., Gilmore, C. J., and Haltiwanger, R. C. (1973). *J.C.S. Perkin II,* 897–902.

Cabanillas, F., Rodriguez, V., Hall, S. H., Burgess, M. A., Bodey, G. P., and Freireich, E. J. (1978a). *Cancer Treat. Rep.* **62,** 425–428.

Cabanillas, F., Rodriguez, V., and Bodey, G. P. (1978b). *Proc. Am. Assoc. Cancer Res.* **19,** 102 (No. 406).

Chabner, B. A., Levine, A. S., Adamson, R., Johnson, B. L., Wang-Peng, J., and Young, R. C. (1977). *Proc. Am. Assoc. Cancer Res.* **18,** 129 (No. 515).

Chabner, B. A., Levine, A. S., Johnson, B. L., and Young, R. C. (1978). *Cancer Treat. Res.* **62,** 429–433.

Corey, E. J., and Bock, M. G. (1975). *Tetrahedron Lett.,* 2643–2646.

Corey, E. J., Enders, D., and Bock, M. G. (1976). *Tetrahedron Lett.,* 7–10.

Corey, E. J., Wetter, H. F., Kozikowski, A. P., and Rama Rao, A. V. (1977). *Tetrahedron Lett.,* 777–778.

Corey, E. J., Bock, M. G., Kozikowski, A. P., and Rama Rao, A. V. (1978a). *Tetrahedron Lett.,* 1051–1054.

Corey, E. J., Weigel, L. O., Floyd, D., and Bock, M. G. (1978b). *J. Am. Chem. Soc.* **100,** 2916–2918.

Donoso, J. A., Watson, D. F., Heller-Bettinger, I. E., and Samson, F. E. (1978). *Cancer Res.* **38,** 1633–1637.

Douros, J. D. (1978). Personal communication.

Edwards, O. E., and Ho, P. T. (1977). *Can. J. Chem.* **55,** 371–373.

Egan, R. T., Ingle, J. N., Rubin, J., Frytak, S., and Moertel, C. G., (1978). *J. Natl. Cancer Inst.* **60,** 93–96.

Elliott, W. J., and Fried, J. (1976). *J. Org. Chem.* **41,** 2469–2475.

Foy, J. E., and Ganem, B. (1977). *Tetrahedron Lett.,* 775–776.

Götschi, E., Schneider, F., Wagner, H., and Bernauer, K. (1977). *Helv. Chim. Acta* **60,** 1416–1418.

Hanka, L. J., and Bernett, M. S. (1974a). *Antimicrob. Agents Chemother.* **6,** 651–652.

Hanka, L. J., and Bernett, M. S. (1974b). *Proc. Am. Assoc. Cancer Res.* **15,** 115 (No. 459).

Higashide, E., Asai, M., Ootsu, K., Tanida, S., Kozai, Y., Hasegawa, T., Kishi, T., Sugino, Y., and Yoneda, M. (1977). *Nature (London)* **270**, 721-722.

Igarashi, K., Ikeyama, S., Takeuchi, M., and Sugino, Y. (1978). *Cell Struct. Funct.* **3**, 103-112.

Johnson, R. K., Inouye, T., and Wolpert-DeFilippes, M. K. (1978). *Biochem. Pharmacol.* **27**, 1973-1975.

Kane, J. M., and Meyers, A. I. (1977). *Tetrahedron Lett.*, 771-774.

Komoda, Y. (1974). *Kagaku No Ryoiki* **28**, 887-894.

Komoda, Y., and Isogai, Y. (1978). *Sci. Pap. Coll. Gen. Educ., Univ. Tokyo* **28**, 129-134.

Kupchan, S. M., Komoda, Y., Court, W. A., Thomas, G. J., Smith, R. M., Karim, A., Gilmore, C. J., Haltiwanger, R. C., and Bryan, R. F. (1972a). *J. Am. Chem. Soc.* **95**, 1354-1356.

Kupchan, S. M., Komoda, Y., Thomas, G. J., and Hintz, H. P. J. (1972b). *J.C.S. Chem. Commun.* p. 1065.

Kupchan, S. M., Komoda, Y., Branfman, A. R., Dailey, R. G., Jr., and Zimmerly, V. A. (1974). *J. Am. Chem. Soc.* **96**, 3706-3708.

Kupchan, S. M., Branfman, A. R., Sneden, A. T., Verma, A. K., Dailey, R. G., Jr., Komoda, Y., and Nagao, Y. (1975). *J. Am. Chem. Soc.* **97**, 5294-5295.

Kupchan, S. M., Komoda, Y., Branfman, A. R., Sneden, A. T., Court, W. A., Thomas, G. J., Hints, H. P. J., Smith, R. M., Karim, A., Howie, G. A., Verma, A. K., Nagao, Y., Dailey, R. G., Jr., Zimmerly, V. A., and Sumner, W. C., Jr. (1977). *J. Org. Chem.* **42**, 2349-2357.

Kupchan, S. M., Sneden, A. T., Branfman, A. R., Howie, G. A., Rebhun, L. I., McIvor W. E., Wang, R. W., and Schnaitman, T. C. (1978). *J. Med. Chem.* **21**, 31-37.

Mandebaum-Shavit, F., Wolpert-DeFilippes, M. K., and Johns, D. G. (1976). *Biochem. Biophys. Res. Commun.* **72**, 47-54.

Meyers, A. I., and Brinkmeyer, R. S. (1975). *Tetrahedron Lett.*, 1749-1752.

Meyers, A. I., and Shaw, C. C. (1974). *Tetrahedron Lett.*, 717-720.

Meyers, A. I., Shaw, C. C., Horne, D., Trefonas, L. M., and Majeste, R. J. (1975). *Tetrahedron Lett.*, 1745-1748.

Meyers, A. I., Tomioka, K., Roland, D. M., and Comins, D. (1978). *Tetrahedron Lett.* 1375-1378. 1375-1378.

O'Conner, T. E., Aldrich, C. A., Hadidi, A½, Lomax, N., Okano, P., Sethi, S., and Wood, H. B. (1975). *Proc. Am. Assoc. Cancer Res.* **16**, 29 (No. 114).

Ootsu, K., Kozai, Y., Takeyama, S., Ikeyama, S., Igarashi, K., Tsukamoto, K., Sugino, Y., Tashiro, T., Tsukagoshi, S., and Sakurai, Y. (1980). *Cancer Res.* **40**, 1707-1717.

Remillard, S., Rebhun, L. I., Howie, G. A., and Kupchan, S. M. (1975). *Science* **189**, 1002-1005.

Samson, M., De Clercq, P., De Wilde, H., and Vandewalle, M. (1977). *Tetrahedron Lett.* 3195-3198.

Schnaitman, T., Rebhun, L. I., and Kupchan, S. M. (1975). *J. Cell Biol.* **67**, 388a (No. 775).

Tanida, S., Hasegawa, T., Hatano, K., Higashide, E., and Yoneda, M. (1980). *J. Antibiot.* **33**, 192-198.

Tanida, S., Higashide, E., and Yoneda, M. (1979). *Antimicrob. Ag. Chemother.* **16**, 101-103.

Tsukamoto, K., *et al.* (1978). Unpublished observations.

Venditti, J. M., and Wolpert-DeFilippes, M. K. (1976). *In* "Chemotherapy" (K. Hellmann, and T. A. Connors, eds.), Vol. 7, pp. 129-147. Plenum, New York.

Wani, M. C., Taylor, H. L., and Wall, M. E. (1973). *J.C.S. Chem. Commun.* p. 390.

Wolpert-DeFilippes, M. K., Adamson, R. H., Cysyk, R. L., and Johns, D. G. (1975a). *Biochem. Pharmacol.* **24**, 751-754.

Wolpert-DeFilippes, M. K., Bono, V. H., Jr., Dion, R. L., and Johns, D. G. (1975b). *Biochem. Pharmacol.* **24**, 1735-1738.

York, J., Wolpert-DeFilippes, M. K., and Johns, D. G. (1978). *Proc. Am. Assoc. Cancer Res.* **18**, 110 (No. 438).

CHAPTER 11

Harringtonine and Related Cephalotaxine Esters*

CECIL R. SMITH, JR., KENNETH L. MIKOLAJCZAK, AND RICHARD G. POWELL

*The mention of firm names or trade products does not imply that they are endorsed or recommended by the U.S. Department of Agriculture over other firms or similar products not mentioned.

Anticancer Agents Based on Natural Product Models
Copyright © 1980 by Academic Press, Inc.
All rights of reproduction in any form reserved.
ISBN 0-12-163150-8

I. INTRODUCTION

A. Cephalotaxine and Its Esters

Cephalotaxine (1) and its esters represent a relatively new group of alkaloids with a unique ring system. Ester alkaloids derived from cephalotaxine are of particular interest because of their antitumor activity. The most active compounds in this series are harringtonine (2) and homoharringtonine (3) (Powell *et al.*, 1972). Cephalotaxine and its congeners have been isolated from several species of *Cephalotaxus* (Paudler *et al.*, 1963; Perdue *et al.*, 1970; G. F. Spencer *et al.*, 1976). In addition to (2) and (3), two other active ester alkaloids occur in *Cephalotaxus*—deoxyharringtonine (4) and isoharringtonine (5) (Powell *et al.*, 1972). Cephalotaxine per se is devoid of antitumor activity.

1 R = H

2 R = (structure with OH, HO, CO_2Me)

3 R = (structure with OH, HO, CO_2Me)

4 R = (structure with HO, CO_2Me)

5 R = (structure with HO, HO, CO_2Me)

B. Current Status of Harringtonine and Homoharringtonine

The pharmacological activity of (2) and (3) has been studied extensively in several experimental tumor systems under the auspices of the National Cancer Institute. Homoharringtonine (3), the more thoroughly investigated of the two, shows activity in the following experimental tumor systems: P-388 lymphocytic leukemia, L-1210 lymphoid leukemia, Lewis lung carcinoma, Colon 38 (Corbett *et al.*, 1977), and epidermoid carcinoma of the nasopharynx (KB cell culture). Harringtonine has the same general spectrum of activities, but it has been examined less thoroughly (Powell *et al.*, 1972, and unpublished data). The National Cancer Institute has contemplated the preclinical and clinical testing of (3) for several years. Although laboratory procedures have been adapted to large-scale isolation of (2) and (3) (Powell *et al.*, 1974a), this work has been delayed by a continuing scarcity of plant material needed to provide the requisite amount of pure alkaloid.

Meanwhile, extensive investigations of the *Cephalotaxus* alkaloids are being carried out in the People's Republic of China. The Chinese are advantageously situated to procure and develop *Cephalotaxus,* since the interior of Mainland China is a principal area in which *Cephalotaxus* species occur (Perdue *et al.,* 1970; Dallimore and Jackson, 1967). Using two preparations containing different proportions of (2) and (3), the Chinese group reported favorable results in clinical studies with 41 patients having acute leukemia of various types (Anonymous, 1976a). The Chinese workers also examined the pharmacology of (2) in several experimental tumor systems and found that it is effective against L-615 leukemia, L-7212 leukemia, sarcoma 180, and Walker carcinosarcoma 256. They observed that (2) is not cross-resistant with 6-mercaptopurine (Anonymous, 1977).

The antitumor activity of cephalotaxine esters in combination with the novel structure of (1) has generated considerable research activity in several areas—pharmacology, biochemistry, and various phases of organic chemistry.

II. STRUCTURAL STUDIES ON CEPHALOTAXINE AND ITS ESTERS

A. Characterization of Cephalotaxine

Cephalotaxine (1) was isolated from two *Cephalotaxus* species by Pau-dler and co-workers (1963), who did the first characterization work on this alkaloid. Through a combination of chemical and spectral evidence, they concluded that (1) has the empirical composition $C_{18}H_{21}NO_4$ and that it contains a methylenedioxyphenyl group, a secondary hydroxyl group, a double bond, and a vinyl methoxyl group. In his doctoral thesis, McKay (1966) advanced a specula-tive structure (6) for cephalotaxine which accommodated these features. After it became apparent that the antileukemic constituents of *Cephalotaxus* are deriva-tives of cephalotaxine, Powell and co-workers (1969) reconsidered structure (6). Although most of the spectral characteristics of cephalotaxine appeared consis-tent with (6), Powell *et al.* noted that this structure required a doublet proton

6   7

magnetic resonance (PMR) signal for the olefinic proton but the spectrum instead showed a singlet at δ4.85. Powell *et al.* (1969) suggested two possible revisions, (**1**) and (**7**), which satisfied the structural implications of the PMR spectrum, including placement of the olefinic proton adjacent to a quaternary carbon. Formula (**1**) was shown to represent the correct structure and relative stereochemistry of cephalotaxine through the x-ray crystallographic work of Abraham *et al.* (1969) in a study based on cephalotaxine methiodide.

B. Active Cephalotaxine Esters: Characterization and Confirmatory Syntheses of Their Acyl Groups

Upon alkaline hydrolysis, esters (**2**), (**3**), (**4**), and (**5**) yield (**1**) plus a dicarboxylic acid. Initially, the dicarboxylic acid moieties from (**2**), (**3**), (**4**), and (**5**) were tentatively identified by PMR and mass spectral studies (Powell *et al.*, 1970; Mikolajczak *et al.*, 1972). The structure of (**8**), the dimethyl ester of the acid moiety from deoxyharringtonine (**4**), was verified by synthesis. Methyl isopentyl ketone (**9**) was condensed with diethyl carbonate (**10**) to give ethyl 6-methyl-3-oxo-heptanoate (**11**), which was converted to a cyanohydrin (**12**).

Acid-catalyzed methanolysis of (**12**) provided a racemic dimethyl ester (**8**) identical with the product from (**4**) except for its chiroptical properties (Mikolajczak *et al.*, 1972). This confirmatory synthesis supported structures assigned to the corresponding acid moieties of (**2**), (**3**), and (**5**).

Characterization of these dicarboxylic acids did not establish which of their two available carboxyl groups is attached to the cephalotaxine moiety. This feature was revealed through an attempted synthesis of (**4**): when (**1**) was acy-

13

lated with the requisite half ester having the *primary* carboxyl group in the acid chloride form, the resulting product was an isomer of (**4**): ψ-deoxyharringtonine (**13**). The NMR spectrum of (**13**) was distinctly different from that of deoxyharringtonine, thus providing evidence for structure (**4**), in which the ester sidechain is attached through the *tertiary* carboxyl group. From close parallels of their NMR spectra, Mikolajczak and associates (1972) concluded that the acyl moiety was constituted similarly in (**2**), (**3**), (**4**), and (**5**).

14   15   8

Structures for (**2**), (**3**), (**4**), and (**5**) were reinforced by further syntheses of the dicarboxylic acid moieties. Auerbach and co-workers (1973) carried out an alternative synthesis of (**8**). Methyl itaconate (**14**) was epoxidized with buffered peroxytrifluoroacetic acid, and the resulting epoxide (**15**) was condensed with a reagent prepared from isobutyl lithium and cuprous iodide; the resulting product was (**8**).

Ipaktchi and Weinreb (1973) synthesized two diastereomeric compounds corresponding to the acid moiety of isoharringtonine (**5**). Isoamylacetoacetate (**16**) was converted through a haloform-type reaction to isoamylfumaric acid (**17**) which, in turn, was dehydrated to an anhydride. This anhydride was methanolyzed to a dimethyl maleate derivative that was hydroxylated with osmium tetroxide–hydrogen peroxide in *t*-butanol to give a diol (**18**) identical with the dimethyl ester from (**5**) except for rotation. Thus the relative configuration of

16   17   18

(**18**) was established as erythro, although the absolute configuration had yet to be determined.

The diacid from hydrolysis of harringtonine (**2**) was synthesized by Kelly and co-workers (1973) through an α-keto ester intermediate (**20**). Benzylated hydroxyacetylene (**19**) was treated with butyllithium, and the resulting lithio derivative was condensed with ethyl t-butyl oxalate to provide the α-keto ester (**20**). At this stage, the carbomethoxymethyl sidechain was introduced by condensing (**20**) with lithium methyl acetate (LiCH$_2$CO$_2$Me), a reagent generated from methyl acetate and lithium cyclohexylisopropylamide. Debutylation of the product, (**21**), was effected with trifluoroacetic acid; after methylation (diazomethane treatment) and catalytic reduction with debenzylation, the dimethyl ester (**22**) of the desired diacid was obtained.

The diacid from hydrolysis of homoharringtonine (**3**) was synthesized by Utawanit (1975) through a sequence in which 1-bromo-3-methyl-2-butene (**23**) was condensed with methyl acetoacetate (**24**) to give (**25**). Unsaturated ester (**25**) was converted to the corresponding cyanohydrin (**26**), which was converted by acid hydrolysis to the racemic form of the desired diacid (**27**); hydration of the double bond accompanied hydrolysis of the nitrile group.

C. Absolute Configuration of Cephalotaxine and Its Active Esters

The synthetic and x-ray crystallographic work discussed in preceding sections established the relative stereochemistry of (**1**) and the relative configuration of (**18**), but did not establish their absolute configurations. The crystallo-graphic analysis of Abraham *et al.* (1969) had been carried out with cephalotaxine methiodide (**28a**), and these investigators reported that this deriva-tive was racemic, even though it was prepared from optically active (**1**). Initially, this observation seemed inexplicable, since conversion of (−)-cephalotaxine to its (+)-enantiomer requires inversion of four chiral centers; the racemization was rationalized later by a mechanism involving a series of equilibrating macrocyclic ions such as (**28a**), (**28b**), (**28c**), and (**28d**). Formation of related macrocycles

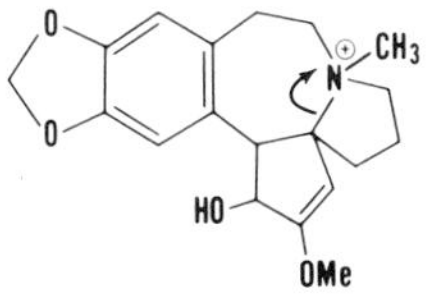

28 a   **28 b**

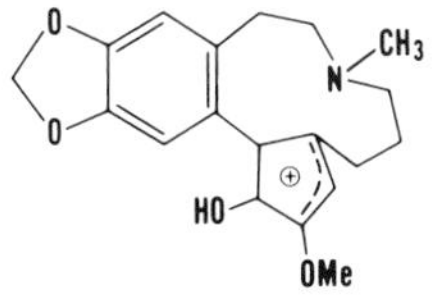

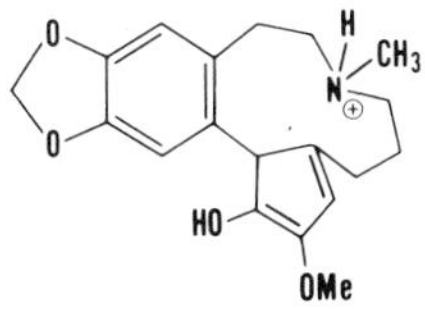

28 d   **28 c**

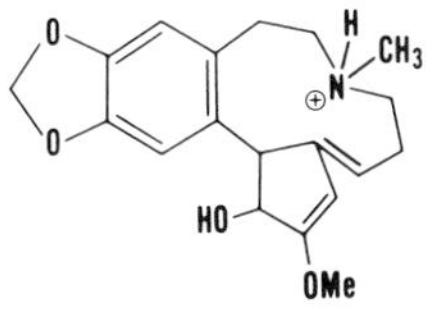

derived from (**1**) via C–N bond cleavage has been documented in certain syn-thetic sequences (Dolby *et al.*, 1972; Schwab *et al.*, 1975).

The absolute stereochemistry of cephalotaxine was determined by Arora and associates (1974) through x-ray crystallography. Working with the *p*-bromo-benzoate of (**1**), they ascertained that natural (−)-cephalotaxine has the 3*S*, 4*S*, 5*R* configuration* (Fig. 1). Arora *et al.* (1976) confirmed their assignment in a subsequent crystallographic investigation of underivatized (**1**).

Absolute configurations of diacids from isoharringtonine (**29**), homohar-ringtonine (**30**), and deoxyharringtonine (**31**) were established through circular

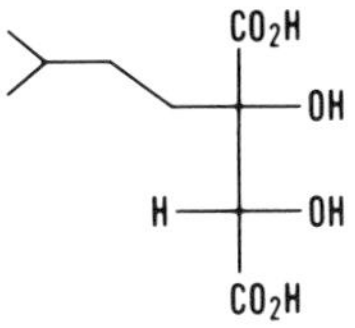

29 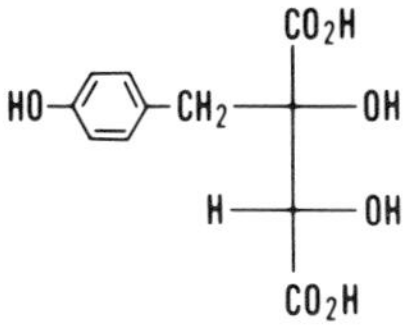   **32** 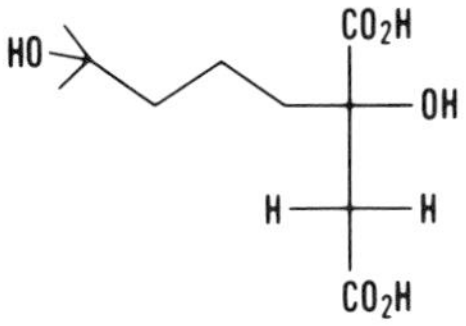   **30** 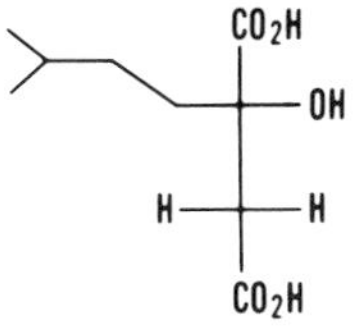   **31**

dichroism (CD) studies. The CD spectrum of the molybdenum complex prepared from (**29**) was similar to those from piscidic acid (**32**) and hexahydropiscidic acid (Fig. 2). On this basis, Brandänge and co-workers (1974b) concluded that (**29**)

*Arora and co-workers (1974, 1976) followed the numbering system for (**1**) used by Powell *et al.* (1969) in their initial paper rather than the revised scheme indicated in this chapter.

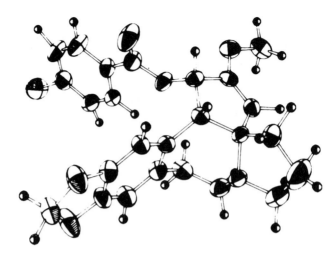

Figure 1. Computer-generated view of cephalotaxine *p*-bromobenzoate. (From Arora *et al.*, 1974, reproduced with permission of the American Chemical Society.)

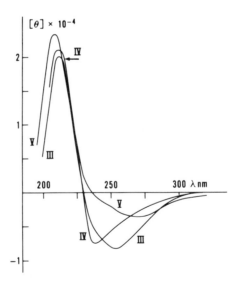

Figure 2. Circular dichroism spectra of molybdate complexes of the acid from isoharringtonine (**29**, curve III), piscidic acid (**32**, curve IV), and the hexahydro derivative of piscidic acid (curve V). The solutions had pH 2.9, 2.9, and 3.1, respectively. (From Brandänge *et al.*, 1974b, reproduced with permission of *Acta Chemica Scandinavica*.)

has the 2*R*, 3*S* configuration in common with (**32**) and its hexahydro derivative. In a similar way, Brandänge *et al.* (1974a) demonstrated that (**30**) and (**31**) have the 2*R* configuration, since their Cotton effect extrema are opposite in sign from those of synthetic (*S*)-citramalic acid and related (*S*)-2-alkylmalic acids. A related chiroptical investigation was carried out subsequently by K. H. Spencer and co-workers (1976); their statement that (**31**) has the (*S*) configuration seems to have been the result of a misunderstanding (S. M. Weinreb, personal communication).

III. SYNTHESIS OF CEPHALOTAXINE AND ITS NATURALLY OCCURRING ESTERS

The novel character of cephalotaxine's ring system and the meager natural occurrence of its active esters have engendered considerable synthetic activity. The relative abundance of (**1**), together with its lack of antitumor activity, makes it particularly desirable to develop procedures for attaching an ester sidechain that would bestow activity upon (**1**).

A. Conversion of Cephalotaxine to Its Naturally Occurring Esters

Attachment of the fully elaborated ester sidechains of (**2**), (**3**), (**4**), and (**5**) by standard acylation procedures might appear straightforward, but attempts in this direction have been uniformly unsuccessful (Mikolajczak *et al.*, 1972; Auerbach *et al.*, 1973). Inspection of a molecular model of (**1**) reveals a cagelike structure with steric features that might severely encumber the approach of a bulky acylating reagent to the hydroxyl group. Among esters (**2**)–(**5**), deoxyharringtonine (**4**) has the simplest acyl group and was the initial target of synthetic efforts.

Mikolajczak and associates (1974a) employed an indirect route to (**4**) in which the lithio derivative of 3-methyl-1-butyne was condensed with ethyl *t*-butyloxalate to give an α-keto ester (**33**). Hydrogenation of (**33**) gave a saturated α-keto ester, which was heated with trifluoroacetic acid to remove the *t*-butyl group and give an α-keto acid (**34**). Acid (**34**) was converted to the corresponding acid chloride with oxalyl chloride, and with this product (**1**) was acylated to give (**35**). This α-keto ester of cephalotaxine was treated with lithium methyl acetate (LiCH₂CO₂Me) to generate the carbomethoxymethylene branch of the ester sidechain and thus afford (**4**) along with an epimeric form. (Here, and subsequently in structural formulas, "ceph" is an abbreviation for the cephalotaxine moiety.) These two diastereomers were separable by chromatography and distinguishable by PMR.

Mikolajczak and Smith (1978) later developed an indirect route to harringtonine (**2**). This sequence proceeded through a series of cyclic ketal or hemiketal intermediates which obviated some previous difficulties with steric interferences as well as in unmasking a hydroxyl group at the final stage. Malonic acid and isobutyraldehyde were condensed to give (**36**) after decarboxylation; (**36**) was esterified and then condensed with diethyl oxalate to give α-keto ester (**37**). During subsequent rigorous hydrolysis with aqueous hydrochloric acid, hemiketal (**38**) was formed as a major product; (**38**) was converted to an ester-ketal (**39**) by treatment with methanolic hydrogen chloride. Saponification of (**39**) under mild conditions produced a sodium salt (**40**) with concurrent elimination of the elements of methanol. Acylation of (**1**) with the acid chloride derived from (**40**) afforded a cephalotaxine ester (**41**), which was hydrated with hydrogen chloride–acetic acid to give (42). The stage was then set for application of the Reformatsky reaction, which provided harringtonine (2).

Investigators in the People's Republic of China have developed procedures for converting (1) to (2) and (4); their synthetic work parallels that of Mikolajczak and co-workers. Cephalotaxine was acylated with an α-keto acid (43) to give an

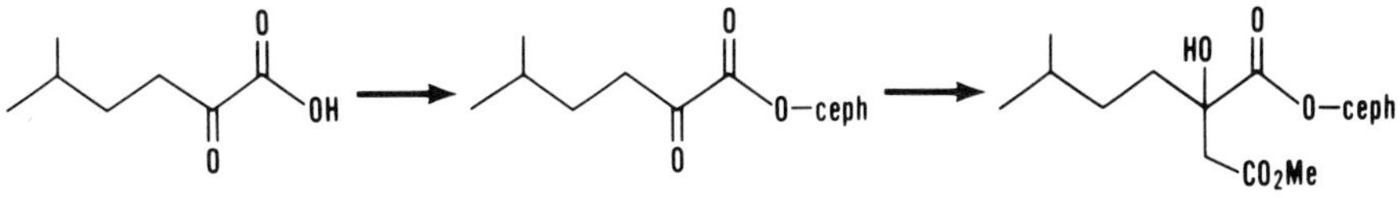

43   44   4

ester (44), which was converted to deoxyharringtonine (4) by a Reformatsky reaction (Li and Dai, 1975; Huang et al., 1976). The ester sidechain of harringtonine (2) was attached in a sequence involving a series of cyclic intermediates (Anonymous, 1975, 1976b). 4-Oxopentanoic acid (45) was converted by a Grignard reaction to a hydroxy acid, which lactonized to give (46).

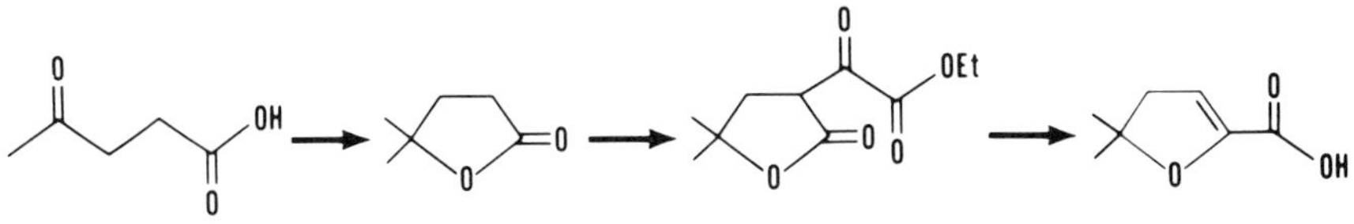

45   46   47   48

γ-Lactone (46), in turn, was condensed with ethyl oxalate to attach an α-keto ester moiety and give (47). Hydrolysis of (47) with dilute alkali and then with hydrochloric acid yielded (48) as the ultimate product; presumably, this transformation proceeds by ring opening, decarboxylation, and then ring closure in a different sense. Unsaturated acid (48) was converted to an acid chloride used in

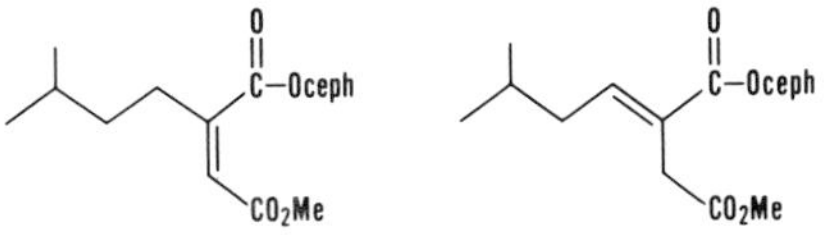

49   50

acylating (1). The remainder of this synthetic sequence involved intermediates (41) and (42); (2) was generated through a Reformatsky reaction in the final step.

Partial synthesis of deoxyharringtonine (4) through such unsaturated ester intermediates as (49) and (50) has been tried without success (Mikolajczak et al., 1974b; Bates et al., 1977).

B. Total Syntheses of Cephalotaxine

Two sophisticated total syntheses of (1) have been carried out. One of these was accomplished by Weinreb and Auerbach (1975), who commenced their syntheses by acylating prolinol (51) with a piperonal-derived acid chloride

(52). The resulting carbinol (53) was oxidized to the corresponding aldehyde, and this product was cyclized by treatment with boron trifluoride to give (54). The cyclic amide, (54), was reduced with lithium aluminum hydride to provide an intermediate (55) for use in an enamine annelation procedure. The mixed anhydride from pyruvic acid and ethyl chloroformate was employed to acylate (55) and give a dicarbonyl intermediate (56). Compound (56) was cyclized by treatment with magnesium methoxide–methanol; the product was desmethyl-cephalotaxinone (57), a cephalotaxine-related alkaloid which occurs naturally (Powell and Mikolajczak, 1973; Asada, 1973). Upon treatment with 2,2-dimethoxypropane and p-toluenesulfonic acid, (57) was converted to cephalotaxinone (58), which was reduced with sodium borohydride to give the target compound, (±)-cephalotaxine (1).

Another total synthesis of (±)-cephalotaxine was carried out by Semmelhack and associates (1975). A key intermediate in their synthesis, spirocycle (61), was constructed from 2-ethoxypyrroline (59). Two three-carbon sidechains were attached to (59) by consecutive reactions with allyl magnesium bromide, and the amine function was blocked with t-butyloxycarbonyl azide to give (60). The allyl groups of (60) were ozonized to give aldehyde functions that were subsequently transformed into carbomethoxy groups to which an acyloin condensation was applied. After successive treatments with sodium–potassium alloy in trichloromethylsilane and with bromine, a 1,2-diketone was obtained which was methylated with diazomethane to provide spirocycle (**61**). The aromatic moiety was constructed from piperonal by transforming it into piperonylformic acid and reducing this product to the corresponding alcohol; after iodination, the alcohol was acylated with *p*-nitrobenzenesulfonyl chloride to give (**62**). The spirocycle (**61**) was attached to (**62**) by a nucleophilic displacement process to give the precursor of anion (**63**). In an earlier version of their synthesis, Semmelhack and co-workers (1972) effected final ring closure through a benzyne intermediate. However, they greatly improved their yield of cephalotaxinone (**58**) by applying a photochemical ring closure after generating the requisite anion (**63**) with potassium *t*-butoxide (Semmelhack *et al.*, 1973, 1975).

Two groups of investigators (Dolby *et al.*, 1972; Weinstein and Craig, 1976) independently undertook total syntheses of (**1**) along lines paralleling the procedure of Weinreb and Auerbach. Both groups designed their synthetic routes to proceed through enamine (**55**), but their work was terminated before completion. By irradiating (**64**), Tse and Snieckus (1976) prepared a precursor (**65**) of this same enamine (**55**). Bryson and co-workers (1977) have investigated routes to various synthons related to (**61**), the spirocyclic intermediate in Semmelhack's synthesis.

Some biomimetic approaches to cephalotaxine synthesis will be mentioned in the following section.

IV. BIOSYNTHESIS OF CEPHALOTAXUS ALKALOIDS

A. Biosynthesis of Cephalotaxine: Relationship to Homoerythrina Series

In addition to cephalotaxine and its antileukemic esters, *Cephalotaxus* species contain certain related alkaloids: 11-hydroxycephalotaxine (**66**), drupacine (**67**), desmethylcephalotaxinone (**57**), and others based on the

66   67   68

cephalotaxine skeleton (Powell *et al.*, 1974b; G. F. Spencer *et al.*, 1976; Asada, 1973; Powell and Mikolajczak, 1973). Furthermore, homoerythrina alkaloids occur in certain *Cephalotaxus* species, especially in *C. wilsoniana* (Powell, 1972; Furukawa *et al.*, 1976; G. F. Spencer *et al.*, 1976); wilsonine (**68**) is a typical example. The discovery that the two groups of alkaloids occur together in *Cephalotaxus* led to speculation that they are derived from a common biogenetic precursor. Previous investigation has established 1-benzyl-1,2,3,4-tetrahydroisoquinoline, presumably formed from tyrosine, as the precursor of erythrina alkaloids through a scheme involving oxidative phenol coupling (Barton and Widdowson, 1969). Fitzgerald and co-workers (1969) extended this concept to the homoerythrina series by postulating a dibenzacine precursor (**69**) derived from 1-phenylethyl-1,2,3,4-tetrahydroisoquinoline; an intermediate such as (**70**) would be formed via path *a*. Powell (1972) pointed out that by an appropriate change in mode of ring closure (path *b*), Fitzgerald's intermediate (**69**) could serve as precursor for cephalotaxine (**1**) via such intermediates as (**72**), (**73**), and (**74**). By a third alternative ring closure, homoerysodienone (**71**) could be generated from (**69**).

These intriguing speculations about the biosynthesis of cephalotaxine have been tested experimentally by Parry and associates (Schwab *et al.*, 1977), and their results do not support a phenol-coupling mechanism of the sort outlined above. They interpreted their results as indicating that one molecule each of

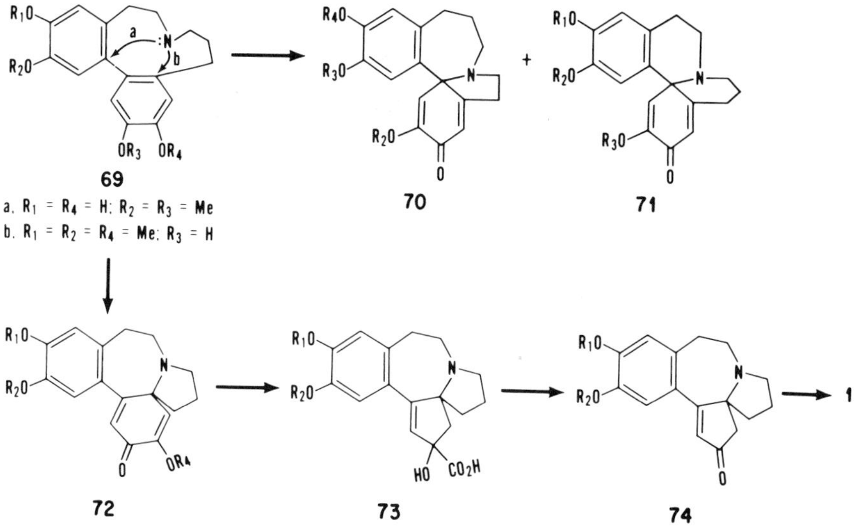

69

a. $R_1 = R_4 = H$; $R_2 = R_3 = Me$
b. $R_1 = R_2 = R_4 = Me$; $R_3 = H$

70   71

72   73   74

tyrosine and phenylalanine are incorporated into (1) in the manner expected for a modified phenylethylisoquinoline alkaloid. However, they also found evidence for an aberrant pathway in which a tyrosine unit is cleaved between the aromatic ring and sidechain; the resulting C_3 unit, but not the C_6 unit, subsequently is incorporated into (1) and provides those carbons in the periphery of the two five-membered rings (C-1 to C-3 and C-6 to C-8).

B. Biosynthesis of Ester Sidechains

Parry and co-workers (1976) have extended their biosynthetic studies to the acyl sidechain of *Cephalotaxus* alkaloids. By radiolabeling experiments, they established [^{14}C]leucine (75) as a precursor of a dicarboxylic acid (77) which is a homolog of the acid moiety (31) of deoxyharringtonine. Most of the label was incorporated in the tertiary carboxyl carbon of (77), an acid which had not been detected previously in *Cephalotaxus* plant material. Parry and associates (1976) regarded this result as evidence for a biosynthetic pathway through keto acid (76), with which acetyl CoA would condense to give (77). By feeding radiolabeled (79) to *Cephalotaxus* plants, the Brandeis group secured evidence for biosynthesis of (31) via intermediates (80) and (34).

C. Approaches to Biomimetic Synthesis of Cephalotaxine

Three research groups have explored biomimetic synthetic routes to cephalotaxine that involved oxidative coupling of phenols or related reactions. Each of these approaches featured phenolic dibenzacine intermediates related to dienone (**69**). Marino and Samanen (1976) oxidized a free amine (**69a**) with potassium hexacyanoferrate in methylene chloride–sodium bicarbonate and obtained two products: methyl ethers (**70**) and (**71**). However, McDonald and Suksamrarn (1975) obtained only a 5:7 fused dienone, (**70**), under similar reaction conditions. In contrast, Kupchan and co-workers (1978) applied similar reaction conditions to a monophenolic benzazocine (**69b**) and generated a product with the ring system of (**72**), one of the postulated precursors of cephalotaxine.

D. Production of Cephalotaxus Alkaloids by Tissue Culture

Tissue from a growing *Cephalotaxus* plant has been cultured successfully (Delfel and Rothfus, 1977; Delfel, 1977, 1978), and it has been demonstrated that cephalotaxine and its active esters are formed both in the callus tissue and the surrounding medium. This development opens up intriguing possibilities for studying the biosynthesis of *Cephalotaxus* alkaloids and for controlling the course of their biosynthesis. Eventual scaled-up production of harringtonine and its congeners by tissue culture is a future possibility.

Delfel and Rothfus (1977) made the interesting observation that the relative amounts of alkaloids in a six-month culture of *Cephalotaxus* resemble those in mature plants with cephalotaxine predominating. In contrast, a culture aged only three months contains a much higher proportion of esters, primarily due to higher levels of deoxyharringtonine (**4**). In addition, a previously unreported alkaloid was detected and tentatively identified as deoxyhomoharringtonine.

V. STRUCTURE–ACTIVITY RELATIONSHIPS AMONG CEPHALOTAXUS ALKALOIDS AND THEIR DERIVATIVES

A. Differences in Activity among Naturally Occurring Cephalotaxus Alkaloids

In view of their wide range of antitumor activities, the structure–activity relationships among the antileukemic esters of cephalotaxine are a matter of prime importance. Harringtonine (**2**) and homoharringtonine (**3**) show about the same level of activity in the P-388 and L-1210 experimental leukemia systems (Tables I and II) (Powell *et al.*, 1972). Thus, insertion of an additional

TABLE I

Activity of Some Cephalotaxus Alkaloids against P-388 Lymphocytic Leukemia[a]

Alkaloid	Dose (mg/kg)	Survivors	Animal weight difference (T − C)	Survival time, T/C (days)	T/C (%)
(**2**)	4.00	2/6	−5.5	5.0/9.0	—
	2.00	6/6	−3.3	18.5/9.0	205
	1.00	6/6	−2.3	36.5/9.0	405
	0.50	6/6	−1.0	26.5/9.0	294
(**3**)	2.00	6/6	−3.8	7.5/9.0	—
	1.00	6/6	−2.8	30.5/9.0	338
	0.50	6/6	−1.8	24.5/9.0	272
	0.25	6/6	−2.2	22.0/9.0	244
(**4**)	4.00	6/6	−3.4	14.0/10.0	140
	2.00	6/6	−3.3	18.0/10.0	180
	1.00	6/6	−2.4	15.5/10.0	155
	0.50	6/6	−1.2	14.5/10.0	145
(**5**)	15.0	6/6	−4.3	9.5/9.0	105
	7.50	6/6	−3.0	24.5/9.0	272
	3.75	6/6	−2.8	15.5/9.0	172
	1.87	5/6	−1.3	13.5/9.0	150

[a] Data presented are representative of results from several assays with different samples of each alkaloid. Materials are considered active if the survival time of animals treated (T) with them is ≥ 125% of that of the controls (C) (i.e., T/C ≥ 125%). Values are quoted from Powell *et al.* (1972).

TABLE II

Activity of Some Cephalotaxus Alkaloids against P-1210 Lymphoid Leukemia[a]

Alkaloid	Dose (mg/kg)	Survivors	Animal weight difference (T − C)	Survival time, T/C (days)	T/C (%)
(1)	220	6/6	−1.2	9.8/9.6	102
	110	6/6	0.2	10.3/9.6	107
	55	6/6	0.5	9.5/9.6	98
(2)	4.00	1/6	−4.1	0.0/9.1	—
	2.00	6/6	−2.4	12.5/9.1	137
	1.00	6/6	−1.2	12.3/9.1	135
	0.50	6/6	−1.0	12.0/9.1	131
(3)	2.00	6/6	−3.0	9.2/9.1	101
	1.00	6/6	−1.4	13.0/9.1	142
	0.50	6/6	−0.8	11.0/9.1	120
	0.25	6/6	−0.5	11.2/9.1	123
(5)	15.00	6/6	−3.5	10.0/9.1	109
	7.50	6/6	−1.3	11.5/9.1	126
	3.75	6/6	−0.5	11.3/9.1	124
	1.87	6/6	−1.1	10.0/9.1	109
(1)	100	6/6	−0.6	9.8/9.6	102
acetate	50	6/6	0.0	10.2/9.6	106
	25	6/6	−0.3	10.2/9.6	106

[a] Data presented are representative of results from several assays with different samples of each alkaloid. Materials are considered active if the survival time of animals treated (T) with them is ≥ 125% of that of the controls (C) (i.e., T/C ≥ 125%). Values are quoted from Powell *et al.* (1972).

methylene group in the terminal portion of the acyl sidechain of (2) has little effect on activity. On the other hand, removal of the hydroxyl group from the penultimate carbon of the acyl moiety (to give deoxyharringtonine, 4) reduces activity by about half. Shifting this same hydroxyl to give isoharringtonine (5) lowers P-388 activity by nearly an order of magnitude.

As indicated previously, natural (−)-cephalotaxine (1) per se is devoid of antileukemic activity (Powell *et al.*, 1972); (31), the acid provided by hydrolysis of (4), and the corresponding dimethyl ester (8), likewise are inactive (K. L. Mikolajczak, unpublished observations). Other inactive companion alkaloids of cephalotaxine include 11-hydroxycephalotaxine (66), drupacine (67), and all members of the homoerythrina series (68 and others) that have been tested (R. G. Powell, unpublished observations). While an ester linkage may be a structural requirement for activity, it is insufficient in the absence of other features since cephalotaxine acetate is inactive (Powell *et al.*, 1972).

Although (22), the dimethyl ester of the dicarboxylic acid derived from har- ringtonine, has not been assayed in the usual tumor systems (P-388, KB), there is evidence based on work with a HeLa cell system that this acid has no inhibitory activity when tested as a separate entity (Huang, 1975).

81

B. Cephalotaxine Esters with "Unnatural" Structural Variations

A considerable number of cephalotaxine derivatives have been prepared which incorporate "unnatural" acyl groups or other structural variations. Some seemingly subtle structural alterations abolish activity; on the other hand, certain active esters of (**1a**) have been prepared whose acyl groups bear no resemblance to those of the harringtonines: (**2**), (**3**), (**4**), and (**5**).

The partial synthesis of deoxyharringtonine (**1k**) provided a diastereomer (**1l**) differing from it in configuration only at C-2′ of the acyl sidechain (Mikolajczak *et al.*, 1974a); (**1l**) appears to be less active than (**1k**) (Table III), although further testing is needed to establish this point definitely (Mikolajczak *et al.*, 1977). Attaching the "wrong" carboxyl group of the acyl moiety to (**1a**) gives an inactive isomer (**1i**) of deoxyharringtonine. A "rearranged" isomer of deoxyhar-ringtonine (**81**), in which the acyloxy group is shifted from C-1 to C-3, likewise has little or no antileukemic activity (Mikolajczak *et al.*, 1975). Removal of the hydroxyl function from C-2′ of the acyl sidechain of deoxyharringtonine (**1k**) produces (**1dd**) and thereby abolishes activity. Cephalotaxine was acylated with

TABLE III

Activity of Various Cephalotaxine Esters against P-388 Lymphocytic Leukemia[a]

Compound	Vehicle[b]	Dose (mg/kg/inj[c])	Animal weight difference, T − C	T/C[d] (%)
(1b)	D	20	0.1	135
	D	20	−0.2	211
	D	13	−0.8	154
	T	20	−0.1	129
(1d)	B	80	−0.7	145
	B	40	−0.9	134
	B	20	−1.0	125
	B	10	−1.2	136
	D	4.4	−1.5	147
	D	1.9	0.5	134
(1g)	D	365	−3.0	198[e]
	D	240	−1.4	169
	D	160	−0.1	183
	D	160	−1.1	167
	D	80	−0.1	173
	D	40	−1.3	135
(1j)	C	20	0.9	131
(1k)	D	5.9	−0.9	184
	A	4	0.0	174
	A	2	0.8	126
(1l)	D	9	−2.2	131
	A	4	0.5	150
(1s)	D	80	0.9	150
	D	40	−1.8	125
	D	20	0.9	130
(1w)	A	320	−2.3	136
	A	160	1.2	154
	A	80	−1.0	138
(1cc)	D	320	−1.0	172
	D	160	−0.9	162
	D	160	−0.3	155
	D	80	−0.9	183
	D	80	−1.3	160
	D	40	−0.4	140
	D	40	0.9	183
	D	20	−2.7	128
	D	20	−1.5	160
	D	20	−1.0	195
	D	13	−0.5	138
	D	8.8	−0.3	170

(Continued)

TABLE III (*Continued*)

Compound	Vehicle[b]	Dose (mg/kg/inj[c])	Animal weight difference, T − C	T/C[d] (%)
Rearranged ester				
(81)	—	29	−0.3	105
	—	20	−1.5	127
	—	10	−0.2	100
	—	5	−0.3	100

[a] Assays were performed under the auspices of Drug Research and Development, National Cancer Institute; protocols are described in Geran *et al.* (1972).

[b] A, Saline; B, water + alcohol + acetone; C, water + acetone; D, water + alcohol; T, saline + Tween 80.

[c] One intraperitoneal injection for 9 days; DBA/2 mice.

[d] T/C, Mean survival time of test animals/mean survival time of control animals; 125% or above considered active. Unaccountable variations in T/C values among duplicate tests were sometimes observed; these may possibly be due to solubility properties of the esters in the vehicles used.

[e] One 30-day cure was reported.

an acid having a hydroxyl group α to a tertiary carboxyl; although the resulting ester (**1ee**) incorporated some steric features of the active esters, it showed no more than marginal activity (Mikolajczak *et al.*, 1974b).

A structurally diverse series of cephalotaxine esters has been prepared with acyl groups that contain conjugated double bonds, aromatic rings (both benzenoid and heterocyclic), chloro, nitro, hydroxy, and sulfonate groups (**1b**)–(**1ee**). Esters showing at least marginal activity are listed in Table III; others are inactive (Mikolajczak *et al.*, 1977). We have noted that some seemingly minor alterations in the acyl moieties of the more potent antileukemic esters of (**1a**) markedly diminish their activity. Accordingly, the structures of some active esters in Table III may occasion some surprise.

No apparent ''rhyme or reason'' emerges to provide a basis for a rational statement of structure–activity relationships for the series of compounds in Table III. One particularly intriguing observation is that methyl cephalotaxyl itaconate (**1g**) prepared from the natural (−)-isomer of (**1**) is active (and exceptionally nontoxic), while its optical antipode (**1h**) is inactive. However, until additional representatives of the (+)-cephalotaxine ester series have been prepared and tested, generalizations about their activity will be premature.

In terms of dose requirements for effective tumor inhibition, harringtonine and homoharringtonine are at least an order of magnitude more active than any of the esters in Table III.

C. Pharmacology and Biochemistry of Harringtonine and Its Congeners

Several groups of researchers have investigated the mechanism by which harringtonine and related alkaloids exert their inhibitory effects. All have concluded that harringtonine and its congeners inhibit protein synthesis in the cell, but there have been some differences in their conclusions about the details of this inhibitory process.

Huang (1975; also see Grollman and Huang, 1973, 1976) ascertained that harringtonine inhibits protein biosynthesis in HeLa cells, in intact rabbit reticulocytes, and in reticulocyte lysates. He concluded that in the reticulocyte lysate, the principal effect of harringtonine is to inhibit the *initiation* of protein synthesis, but contended that it did not inhibit protein chain elongation. Harringtonine does not prevent binding of mRNA or tRNA to reticulocyte ribosomes, nor does it cause accumulation of ribosomal subunits. In addition, Huang found that harringtonine inhibits synthesis of DNA in HeLa cells, while RNA synthesis is largely unaffected. Harringtonine does penetrate the cell membrane, according to Huang's findings.

Tscherne and Pestka (1975) carried out an investigation with HeLa cells and, in accord with Huang, concluded that harringtonine, homoharringtonine, and isoharringtonine inhibit the *initiation* of protein synthesis. They observed that "direct treatment of the cells with harringtonine led to essentially complete disappearance of the polyribosomes and a concomitant appearance of monosomes and subunits."

Conclusions of subsequent investigators are at variance with those of Huang and Tscherne and Pestka. A group of Spanish workers (Fresno *et al.,* 1977) observed that harringtonine, homoharringtonine, and isoharringtonine inhibit protein synthesis in eukaryotic cells. However, they maintained that these alkaloids do not inhibit the initiation of protein synthesis, but instead inhibit initial cycles of the elongation process. Baaske and Heinstein (1977) studied the effect of homoharringtonine on HeLa, KB, and L cells in monolayer cultures, and concluded that protein synthesis was inhibited in the G_1 and G_2 phases.

D. Prospects for Design of Harringtonine Analogues with Improved Properties

As detailed in the preceding section, a number of structural analogues of harringtonine have been prepared. This work has delineated a number of structural variations in the acyl moiety which greatly reduce or abolish activity. Indeed, with these examples alone, one might infer that there are rather rigid structural requirements for activity in the harringtonine series. Nevertheless, activity appears in several cephalotaxine esters having acyl groups with no apparent relationship to that of harringtonine (e.g., **1cc**, the trichloroethoxycarbonyl derivative). Obviously, possible variations in the acyl group of harringtonine are endless, and this is a potential area for future exploration. It should be noted, however, that most of the cephalotaxine esters described by Mikolajczak *et al.* (1977), (**1b**)–(**1cc**), can be prepared without the severe steric constraints that hamper conversion of cephalotaxine to harringtonine. Although the recent synthetic procedure of Mikolajczak and Smith (1978) opens the way towards resolution of this obstacle, considerable work remains to be done before harringtonine-like acyl sidechains can be attached to (**1**) at the pleasure of the chemist.

Except in the case of the "rearranged" ester, (**81**), structure–activity relationships in the harringtonine series have not been extended to the cephalotaxine moiety. There are many structural variations in the cephalotaxine ring system whose physiological effects await investigation, some of the more obvious being: (a) transformation of the methylenedioxy groups at C-18 into dimethoxy, dihydroxy, or other functions; (b) aromatic substitutions at C-14 and/or C-17; (c) inversion of the oxygen function at C-3 to give an epimeric series of esters; (d) quaternarization of the nitrogen (e.g., to form a methiodide); (e) substitution or oxidation at one or more of the methylene groups—C-6, C-7, C-8, C-10, or C-11; (f) skeletal alterations via variations in total synthesis procedures by photorearrangements, or by other approaches.

Mathematical techniques for predicting structure–activity relationships of drugs (Hansch, 1969; Free and Wilson, 1964) may be useful in the harringtonine series, but their application must await a more ready availability of both cephalotaxine and its active esters. Zee-Cheng and Cheng (1970) drew attention to a triangular relationship between two oxygens and one nitrogen, characterized by certain average interatomic distances, that occurs in a variety of antileukemic compounds, including the harringtonines. However, this structural feature does not bestow activity upon cephalotaxine.

REFERENCES

Abraham, D. J., Rosenstein, R. D., and McGandy, E. L. (1969). *Tetrahedron Lett.*, 4085–4086.
Anonymous (1975). *Hua Hsueh T'ung Pao* **20,** 437.
Anonymous (1976a). *Chin. Med. J.* **2,** 263–272.
Anonymous (1976b). *K'o Hsueh T'ung Pao* **21,** 512, 509.
Anonymous (1977). *Chin. Med. J.* **3,** 131–136.
Arora, S. K., Bates, R. B., Grady, R. A., and Powell, R. G. (1974). *J. Org. Chem.* **39,** 1269–1271.
Arora, S. K., Bates, R. B., Grady, R. A., Germain, G., Declercq, J. P., and Powell, R. G. (1976). *J. Org. Chem.* **41,** 551–554.
Asada, S. (1973). *Yakugaku Zasshi* **93,** 916–924.
Auerbach, J., Ipaktchi, T., and Weinreb, S. M. (1973). *Tetrahedron Lett.*, 4561–4564.
Baaske, D. M., and Heinstein, P. (1977). *Antimicrob. Agents Chemother.* **8,** 479–487.

Barton, D. H. R., and Widdowson, D. A. (1969). *In* "Symposiumsberichtes. IV: Internationales Symposium, Biochemie und Physiologie der Alkaloide" (K. Mothes, K. Schreiber, and H. R. Schütter, eds.), pp. 7–19. Akademie-Verlag, Berlin.

Bates, R. B., Cutter, R. S., and Freeman, R. M. (1977). *J. Org. Chem.* **42**, 4162–4164.

Brandänge, S., Josephson, S., and Vallén, S. (1974a). *Acta Chem. Scand., Sect. B* **28**, 1237–1238.

Brandänge, S., Josephson, S., Vallén, S., and Powell, R. G. (1974b). *Acta Chem. Scand., Sect. B* **28**, 1237–1238.

Bryson, T. A., Smith, D. C., and Krueger, S. A. (1977). *Tetrahedron Lett.*, 525–528.

Corbett, T. H., Griswold, D. P., Roberts, B. J., Peckham, J. C., and Schabel, F. M. (1977). *Cancer* **40**, 2660–2680.

Dallimore, W., and Jackson, A. B. (1967). "A Handbook of Coniferae and Ginkgoaceae" (revised by S. G. Harrison), p. 146. St. Martin's, New York.

Delfel, N. E. (1977). *Plant Physiol.* **59**, Suppl. 62.

Delfel, N. E. (1978). *Abstr. Int. Plant Tissue Cell Cult. Conf., 4th, Calgary, Alberta*, p. 124.

Delfel, N. E., and Rothfus, J. A. (1977). *Phytochemistry* **16**, 1595–1598.

Dolby, L. J., Nelson, S. J., and Senkovich, D. (1972). *J. Org. Chem.* **37**, 3691–3695.

Fitzgerald, J. S., Johns, S. R., Lamberton, J. A., and Sioumis, A. A. (1969). *Aust. J. Chem.* **22**, 2187–2201.

Free, S. F., and Wilson, J. (1964). *J. Med. Chem.* **7**, 395–399.

Fresno, M., Jiménez, A., and Vázquez, D. (1977). *Eur. J. Biochem.* **72**, 323–330.

Furukawa, H., Itoigawa, M., Haruna, M., Jinno, Y., Ito, K., and Lu, S.-T. (1976). *Yakugaku Zasshi* **96**, 1373–1377.

Geran, R. T., Greenburg, N. H., MacDonald, M. M., Schumacher, A. M., and Abbott, B. J. (1972). *Cancer Chemother. Rep., Part 3* **3**, 1.

Grollman, A. P., and Huang, M.-T. (1973). *Fed. Proc., Fed. Am. Soc. Exp. Biol.* **32**, 1673–1678.

Grollman, A. P., and Huang, M.-T. (1976). *In* "Protein Synthesis" (E. H. McConkey, ed.), Vol. 2, pp. 125–147. Dekker, New York.

Hansch, C. (1969). *Acc. Chem. Res.* **2**, 232–239.

Huang, M.-T. (1975). *Mol. Pharmacol.* **11**, 511–519.

Huang, W.-K., Li, Y.-L., and Pan, S.-F. (1976). *K'o Hsueh T'ung Pao* **21**, 178.

Ipaktchi, T., and Weinreb, S. M. (1973). *Tetrahedron Lett.*, 3895–3898.

Kelly, T. R., McKenna, J. C., and Christenson, P. A. (1973). *Tetrahedron Lett.*, 3501–3504.

Kupchan, S. M., Dhingra, O. P., and Kim, C.-K. (1978). *J. Org. Chem.* **43**, 4464–4468.

Li, S.-W., and Dai, J.-Y. (1975). *Hua Hsueh Hsueh Pao* **33**, 75–78.

McDonald, E., and Suksamrarn, A. (1975). *Tetrahedron Lett.*, 4425–4428.

McKay, J. (1966). Ph.D. Thesis, Ohio Univ., Athens.

Marino, J. P., and Samanen, J. M. (1976). *J. Org. Chem.* **41**, 179–180.

Mikolajczak, K. L., and Smith, C. R. (1978). *J. Org. Chem.* **43**, 4762–4765.

Mikolajczak, K. L., Powell, R. G., and Smith, C. R. (1972). *Tetrahedron* **28**, 1995–2001.

Mikolajczak, K. L., Smith, C. R., Weisleder, D., Kelly, T. R., McKenna, J. C., and Christenson, P. A. (1974a). *Tetrahedron Lett.*, 283–286.

Mikolajczak, K. L., Smith, C. R., and Powell, R. G. (1974b). *J. Pharm. Sci.* **63**, 1280–1283.

Mikolajczak, K. L., Powell, R. G., and Smith, C. R. (1975). *J. Med. Chem.* **18**, 63–66.

Mikolajczak, K. L., Smith, C. R., and Weisleder, D. (1977). *J. Med. Chem.* **20**, 328–332.

Parry, R. J., Sternbach, D. D., and Cabelli, M. D. (1976). *J. Am. Chem. Soc.* **98**, 6380–6382.

Paudler, W. W., Kerley, G. I., and McKay, J. (1963). *J. Org. Chem.* **28**, 2194–2197.

Perdue, R. E., Spetzman, L. A., and Powell, R. G. (1970). *Am. Hortic. Mag.* **49**, 129–132.

Powell, R. G. (1972). *Phytochemistry* **11**, 1467–1472.

Powell, R. G., and Mikolajczak, K. L. (1973). *Phytochemistry* **12**, 2987–2991.

Powell, R. G., Weisleder, D., Smith, C. R., and Wolff, I. A. (1969). *Tetrahedron Lett.*, 4081–4084.

Powell, R. G., Weisleder, D., Smith, C. R., and Rohwedder, W. K. (1970). *Tetrahedron Lett.*, 815–818.

Powell, R. G., Weisleder, D., and Smith, C. R. (1972). *J. Pharm. Sci.* **61**, 1227–1230.

Powell, R. G., Rogovin, S. P., and Smith, C. R. (1974a). *Ind. Eng. Chem. Prod. Res. Dev.* **13**, 129–132.

Powell, R. G., Madrigal, R. V., Smith, C. R., and Mikolajczak, K. L. (1974b). *J. Org. Chem.* **39**, 676–680.

Schwab, J. M., Parry, R. J., and Foxman, B. M. (1975). *J. C. S. Chem. Commun.* 906–907.

Schwab, J. M., Chang, M. N. T., and Parry, R. J. (1977). *J. Am. Chem. Soc.* **99**, 2368–2370.

Semmelhack, M. F., Chong, B. P., and Jones, L. D. (1972). *J. Am. Chem. Soc.* **94**, 8629–8630.

Semmelhack, M. F., Stauffer, R. D., and Rogerson, T. D. (1973). *Tetrahedron Lett.*, 4519–4522.

Semmelhack, M. F., Chong, B. P., Stauffer, R. D., Rogerson, T. D., Chong, A., and Jones, L. D. (1975). *J. Am. Chem. Soc.* **97**, 2507–2516.

Spencer, G. F., Plattner, R. D., and Powell, R. G. (1976). *J. Chromatogr.* **120**, 335–341.

Spencer, K. H., Khatri, H. N., and Hill, R. K. (1976). *Bioorg. Chem.* **5**, 177–186.

Tscherne, J. S., and Pestka, S. (1975). *Antimicrob. Agents Chemother.* **8**, 479–487.

Tse, I., and Snieckus, V. (1976). *J. C. S. Chem. Commun.* 505–506.

Utawanit, T. (1975). Ph.D. Thesis, Univ. of Illinois at Urbana-Champaign, Urbana, Illinois.

Weinreb, S. M., and Auerbach, J. (1975). *J. Am. Chem. Soc.* **97**, 2503–2506.

Weinstein, B., and Craig, A. R. (1976). *J. Org. Chem.* **41**, 875–878.

Zee-Cheng, K.-Y., and Cheng, C. C. (1970). *J. Pharm. Sci.* **59**, 1630–1634.

CHAPTER 12

Camptothecin

MONROE E. WALL AND MANSUKH C. WANI

I. INTRODUCTION

Camptothecin (**1**) is a novel plant antitumor agent which was isolated by Wall and co-workers (1966) from *Camptotheca acuminata* Decaisne (Nyssaceae). The history of the manner in which this plant became available for chemical and biological screening is interesting and has been described in detail by Perdue *et al.* (1970). Seeds of *C. acuminata,* which is a small tree native to China, were received by the United States Department of Agriculture in 1934 from A. N. Steward, then teaching botany in a Chinese college. After germination, various seedlings were distributed to a number of USDA experimental gardens including the Chico, California, station.

In 1950, a group of scientists at the Eastern Regional Research Laboratory,

417

Anticancer Agents Based on Natural Product Models

USDA, under the direction of M. E. Wall, began a search for sources of steroidal sapogenins that could be converted into cortisone. Thousands of plants were screened in this program—the plants being obtained by the USDA plant collectors. These collections consisted of samples from many of the USDA plant introduction stations including the one in Chico, California, which supplied leaves of *C. acuminata*. Qualitative analysis of this plant gave positive tests for flavonoids, tannins, and sterols, but negative tests for sapogenins and alkaloids (Wall *et al.*, 1954). Although the alcoholic extract of *C. acuminata* was negative for the desired constituent, it was the fortunate custom of the group to save a representative alcoholic extract of every plant. Some years later in 1957, Dr. Jonathan Hartwell, who had initiated the plant antitumor program for the National Cancer Institute, contacted Dr. Wall and requested that he provide several thousand extracts for antitumor testing. Of the several thousand plant extracts which were provided, only one of these, the extract of *C. acuminata*, gave any notable antitumor response. Wall left the Eastern Regional Research Laboratory in 1960 and initiated his studies on plant antitumor agents with *C. acuminata* at the Research Triangle Institute. Interestingly enough, on a visit to China years later, Wall was informed that *C. acuminata* was never used in Chinese folk medicine. Since the isolation of camptothecin (Wall *et al.*, 1966), the Chinese have conducted many studies on the chemistry and pharmacology of this compound (Wall, 1977).

II. CAMPTOTHECIN AND ITS NATURAL AND SEMISYNTHETIC ANALOGUES

A. Camptothecin

The isolation of (**1**) was unique in that *C. acuminata* was possibly the only plant antitumor agent with activity so potent that it could be isolated with the assistance of *in vivo* screening in the L-1210 mouse leukemia system. Even crude extracts showed very high T/C values in the neighborhood of 200% or more. This permitted the isolation of camptothecin using Craig countercurrent distribution techniques. The details of this isolation have been published (Wall *et al.*, 1976). The structure of the alkaloid (**1**) was established by a combination of chemical and spectroscopic methods, including x-ray crystallography. It has been shown that (**1**) is related to the indole alkaloids (Heckendorf *et al.*, 1976). The structure of camptothecin can be depicted either by the linear formula (**1**) or the angular formula (**1A**). Although the latter representation may be desirable in order to emphasize the biogenetic relationship of camptothecin to indole alkaloids, we have chosen to represent camptothecin by the linear formula (**1**) in the interest of conserving space.

Camptothecin is a high-melting compound which is insoluble in water but also has only limited solubility in most organic solvents. Some of the more important physical properties of camptothecin are shown in Table I. The compound has an unusually broad spectrum of activity in a variety of leukemia and solid tumor systems; camptothecin has also been the subject of intensive investigation in regard to its mode of action on mammalian cells and their viruses (Horwitz, 1975). In particular, camptothecin is a potent inhibitor of nucleic acid synthesis (Horwitz, 1975). The biological activities of camptothecin and its natural and synthetic analogues will be discussed in greater detail in Sections IV and V.

B. Naturally Occurring Analogues

Naturally occurring analogues of camptothecin (Fig. 1) have been relatively rare. 10-Hydroxycamptothecin (2) and 10-methoxycamptothecin (3) were isolated from *C. acuminata* by Wani and Wall (1969). 9-Methoxycamptothecin (4) was isolated from *Mappia foetida* by Govindachari and Viswanathan (1972). All of these ring A hydroxylated substances are found only as trace constituents along with the major product, camptothecin. The hydroxylated compounds may well be produced as a result of further plant metabolism. The 9- and 10-oxygenated compounds show the same degree of broad-spectrum activity as the parent compound, with 10-hydroxycamptothecin undoubtedly being the most potent (cf. Section IV).

C. Semisynthetic Analogues

1. Modifications of Rings E and B

This section deals with a number of derivatives (Fig. 2) which were prepared from (1) in the course of its structural determination (Wall *et al.*, 1966; Wall, 1969). Prolonged treatment of (1) with acetic anhydride and pyridine at 25° gave the 20-acetate (5). Compound (5) exhibited markedly reduced activity in L-1210. Treatment of (1) with thionyl chloride gave 20-chlorocamptothecin (6); the latter could be readily reduced to give 20-deoxycamptothecin (7). The 20-chloro analogue (6) was inactive, and the 20-deoxy analogue (7) was only slightly active in the L-1210 *in vivo* leukemia tests. Treatment of (1) with *m*-chloroperbenzoic acid gave the 1-*N*-oxide (8), a compound with considerably less activity in the L-1210 leukemia system than (1). Reduction of (1) under mild conditions with sodium borohydride resulted in the lactol (9), a compound which had greatly reduced activity compared to the parent compound. This information has caused us to formulate the hypothesis (Wall, 1969) that the α-hydroxy lactone system in camptothecin is absolutely required for antitumor activity. It should, however, be noted that analogues lacking this functionality (e.g., 20-

TABLE I

Properties of Camptothecin

Light yellow needles, m.p. 264–267°(dec.)
Intense blue fluorescence under uv
$[\alpha]_D^{25°}$, + 31.3°
M⁺ at m/e, 348.1117
Calcd. for $C_{20}H_{16}N_2O_4$: 348.1111

Qualitative reactions

1. Negative phenol ($FeCl_3$).
2. Negative Indole tests.
3. Negative Dragendorff and Mayer tests.
4. Gave no crystalline salts with a variety of acids.
5. Could not be methylated with diazomethane or dimethyl sulfate under a variety of conditions.
6. No reactions with bicarbonate or carbonate, but is quantitatively converted to the sodium salt with NaOH at room temperature. On acidification, Na salt regenerates camptothecin.

deoxycamptothecin, (7) are quite effective in inhibiting RNA synthesis or in causing the depolymerization of DNA (Horwitz, 1975). Further discussion on this subject will be found in Section V.

The lactone moiety of camptothecin is very reactive. Thus, reaction of (1) with methyl amine gave the N-methyl amide (10), a compound which has about 3/5 the *in vivo* activity of (1) (Wall, 1969). Recently, Adamovich and Hutchinson (1979) have prepared some N-isopropylamide derivatives of (1), such as (11) and the corresponding 17-acetate (12). Compound (12) has reduced activity in the *in vivo* leukemia system. Under acid conditions, these amides could be readily recyclized to the parent compound (1). It is of interest that Adamovich and

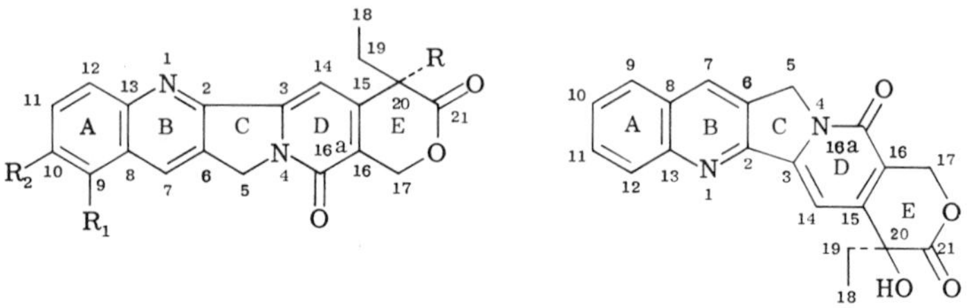

I   R = OH, R_1 = R_2 = H   IA
2   R = R_2 = OH, R_1 = H
3   R = OH, R_2 = OMe, R_1 = H
4   R = OH, R_1 = OMe, R_2 = H

Figure 1.   Camptothecin and its natural analogues.

Hutchinson (1979) prepared a compound (**13**) which has a carbon–nitrogen bond at C-17 which cannot be hydrolyzed and recyclized to camptothecin in the presence of dilute acid. This compound is completely inactive in the P-388 *in vivo* leukemia system. When camptothecin was treated with sodium hydroxide under very mild conditions, it was readily converted to the sodium salt (**14**) (Wall, 1969). Compound (**14**) is soluble in water and recyclizes to (**1**) in dilute acid. Another compound (**15**) which lacks the lactone ring was synthesized by Sugasawa *et al.* (1976). It was said to have considerable *in vivo* activity (Table II). We would rationalize that the activity is probably due to recyclization after *in vivo* hydrolysis, thus regenerating the α-hydroxy lactone moiety.

2. Water-Soluble Ring A Analogues

Oncologists prefer the use of water-soluble antitumor agents so that they may be administered under controlled conditions by the intravenous route. Since camptothecin has been shown to be clinically inactive when administered as the sodium salt (**14**) (cf. Section IV), it was hoped that the preparation of a water-soluble salt from a ring A derivative would give the desirable combination of water solubility and the intact α-hydroxy lactone ring. Accordingly, the two water-soluble ethers, (**16**) and (**17**) (Fig. 2), were prepared from 10-hydroxy-camptothecin (**2**) (Wani *et al.*, 1980).

Attempts at preparing water-soluble esters were vitiated by the instability of these compounds. The sodium salt of the 10-carboxymethyloxy derivative was prepared by treating 10-hydroxycamptothecin (**2**) with ethyl bromoacetate in the presence of potassium carbonate. The 10-carboethoxymethyloxy ether thus obtained was hydrolyzed to the acid, which was then quantitatively converted to the water-soluble sodium salt (**16**). This compound was completely inactive in the P-388 *in vivo* system. Treatment of (**2**) with 2-diethylaminoethyl hydrochloride in the presence of potassium carbonate yielded the 2'-diethylaminoethoxy derivative, which on treatment with hydrogen chloride was converted to (**17**). The salt was completely water soluble, and as shown in Table II had considerable activity in the P-388 leukemia system, although it was considerably less potent than the parent compound in terms of the dose required to produce a maximal antitumor effect. Compound (**17**) is rather unstable, yielding the parent 10-hydroxy-camptothecin (**2**) on standing in aqueous solution.

3. 12-Substituted Analogues

The Chinese workers (Pan *et al.*, 1975) have reported the preparation of a number of 12-substituted derivatives of camptothecin. Nitration of camptothecin in sulfuric acid followed by catalytic hydrogenation gave the 12-amino derivative (**18**). Diazotization of (**18**) followed by suitable transformations gave the desired analogues.

TABLE II

Antileukemic Activity and Cytotoxicity of Camptothecin and Analogues

Compound no.	NSC[a] no.	Tumor system	Dose range (mg/kg)	Optimal T/C	Optimal dose (mg/kg)	Lowest toxic dose (mg/kg)	Therapeutic index[b]	9KB ED$_{50}$ (μg/ml)
(1)	94600	L-1210	3.2–0.2	196	1.6	3.2	4	2×10^{-2}[c]
		P-388	8–0.5	197	4.0	8.0	8	
(2)	107124	L-1210	4–0.25	229	2.0	4.0	8	2×10^{-2}[c]
		P-388	8–0.5	314	4.0	8.0	8	
(3)	111533	L-1210	20–1.25	250	10.0	20.0	1	2×10^{-2}[c]
		P-388	4–0.5	145	0.5	1.0	2	
(4)	176323	L-1210	2.25–1.0	158	2.25	—	2	N.T.[d]
		P-388	4–0.5	195	1.0	4	4	
(5)	95382	L-1210	2–0.13	Inactive				N.T.
(6)	101833	L-1210	6–1.0	Inactive				N.T.
(7)	105132	L-1210	15–3.0	136	15.0	—	1	N.T.
(8)	106748	L-1210	16–1.0	144	2.0	8.0	2	N.T.
(9)	106621	L-1210	24–8.0	139	24.0	—	1	N.T.
(10)	106609	L-1210	5.3–1.0	155	2.3	5.3	2	N.T.
(12)		L-1210	11.1–0.4	147	1.2	3.7	3	4×10^{-1}
(13)		L-1210	62.5–5.1	Inactive				1×10^{-2}
(14)	100880	L-1210	160–5	222	80	160	8	2×10^{-1}[c]
		P-388	80–2.5	212	40	80	4	
		L-1210	25–2.0	>270	25.0	—	12.5	N.T.
(15)		P-388	32–4.0	Inactive				$>1 \times 10^{0}$[c]
(16)	302992	P-388	32–2.0	234	32	—	16	$>1 \times 10^{0}$[c]
(17)	302993	P-388	16–1.0	222	8.0	16.0	8	1×10^{-1}[c]
(27)	302991	P-388	32–2.0	175	32	—	2	$>1 \times 10^{0}$[c]
(30)	302995	P-388	32–1.0	198	16	—	4	2×10^{-1}[c]
(32)	305983	P-388	50–2.0	345	25	50	12	N.T.
(39)	302994	P-388	8–0.5	160	4.0	8.0	2	2×10^{-1}[c]

[a] NSC denotes the identification number employed by the National Cancer Institute.
[b] Therapeutic index was obtained by dividing the highest active dose by the lowest active dose.
[c] These data were obtained under standardized conditions using the same control.
[d] N.T. = Not tested.

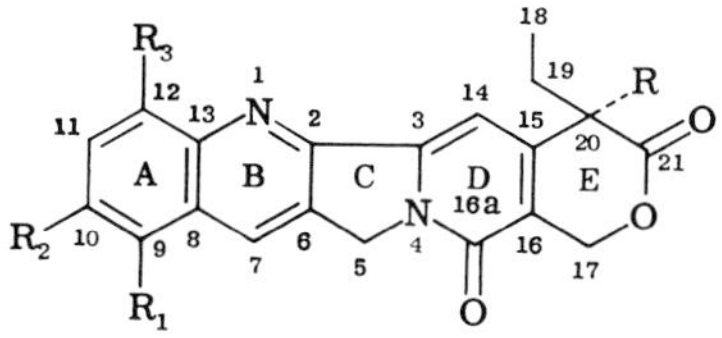

5 R = OAc, $R_1 = R_2 = R_3 = H$
6 R = Cl, $R_1 = R_2 = R_3 = H$
7 $R = R_1 = R_2 = R_3 = H$
8 Same as (1) except N → O at position 1
16 R = OH, $R_1 = R_3 = H$, $R_2 = OCH_2CO_2Na$
17 R = OH, $R_1 = R_3 = H$, $R_2 = OCH_2CH_2NEt_2 \cdot HCl$
18 R = OH, $R_1 = R_2 = H$, $R_3 = NH_2$
19 R = OH, $R_1 = R_2 = H$, $R_3 = Cl$
20 $R = R_3 = OH$, $R_1 = R_2 = H$
21 R = OH, $R_1 = R_2 = H$, $R_3 = OMe$

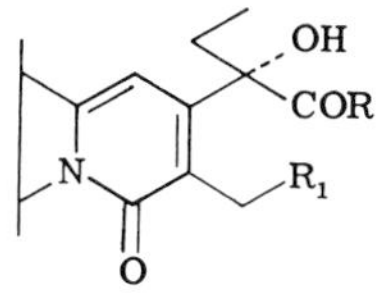

10 R = HNMe, $R_1 = OH$
11 R = HNCHMe$_2$, $R_1 = OH$
12 R = HNCHMe$_2$, $R_1 = OAc$
13 R = HNCHMe$_2$, $R_1 = c\text{-}C_4H_8N\text{-}$
14 R = ONa, $R_1 = OH$

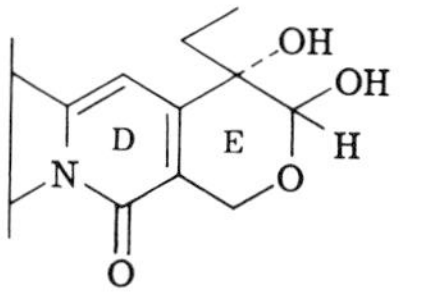

9

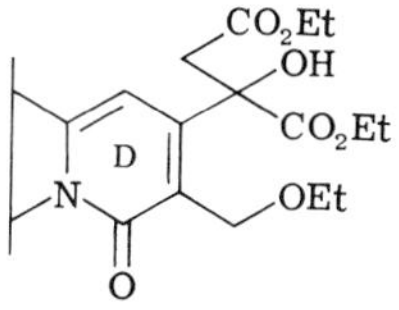

15

Figure 2. Semisynthetic analogues.

Among these analogues, the chloro compound (19) was reported to be very active in L-615 leukemia screening. The 12-hydroxy analogue (20) and the 12-methoxy analogue (21) were more active against Ehrlich ascites carcinoma than camptothecin.

III. TOTALLY SYNTHETIC ANALOGUES

The fact that the α-hydroxy lactone moiety is an absolute requirement for the antitumor activity of camptothecin has resulted in the synthesis of a number of bicyclic, tricyclic, pentacyclic, and hexacyclic analogues containing this moiety. Although detailed descriptions of these syntheses are beyond the scope of this chapter, it is interesting to note that widely different strategies were employed in accomplishing these objectives.

A. Ring DE Analogues

The synthesis of the bicyclic analogue (**22**) was first reported from our laboratory (Wall *et al.*, 1972). Since then, the syntheses of a number of other bicyclic analogues (Fig. 3) with substituents at the 1 and 6 positions of the pyridone moiety have been published (Danishefsky *et al.*, 1973b; Lyle and Kane, 1973; Plattner *et al.*, 1974; Sugasawa *et al.*, 1974). Most of these syntheses are the results of model studies directed towards the total synthesis of camptothecin.

The biological activity of only a few of these analogues has been reported. The bicyclic analogue (**22**) was only weakly cytotoxic, its activity being 1/100 that of camptothecin (Wall *et al.*, 1972). It was found to be inactive against L-1210 leukemia system (unpublished observations). The 1-*n*-butyl-6-methyl analogue (**23**) was ineffective in the leukemia system as well as in the inhibition of nucleic acid synthesis *in vitro* (Danishefsky *et al.*, 1973b; Horwitz, 1975) (Table III). The *N*-β-naphthylmethyl analogue (**24**) (Bristol *et al.*, 1975) was also inactive against the L-1210 leukemia system and did not cause the degradation of DNA in the Hela cells. The analogue (**24**) differs from camptothecin in two respects. The nitrogen at position 1 is replaced by carbon, and the bond between positions 2 and 3 is missing. Therefore, the naphthalene ring is capable of free rotation. Bristol *et al.* (1975) have suggested the necessity of the 2–3 bond for biological activity, for this would allow better orbital overlap of the two π systems, thus increasing the barrier to rotation and creating an essentially planar aromatic system. It would be interesting to synthesize such an analogue to test this hypothesis.

B. Ring CDE Analogues

The synthesis of the tricyclic analogue (**25**) (Fig. 4) was first reported by Plattner *et al.* (1972, 1974). This synthesis, which was based on the total synthe-

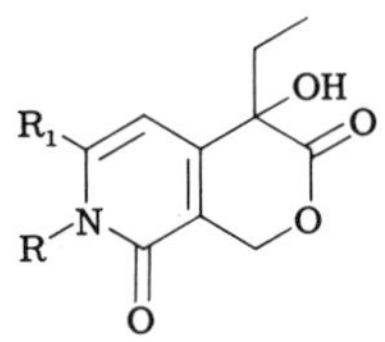

22 R = H, R_1 = Me
23 R = *n*-C_4H_9, R_1 = Me
24 R = β-Naphthylmethyl,
     R_1 = H

Figure 3. Ring DE analogues.

sis of camptothecin by the same group (Tang and Rapoport, 1972; Tang *et al.*, 1975), was rather long and involved a series of rearrangements, hydrogenolyses, and dehydrogenations. Danishefsky and Etheredge (1974) have reported an expeditious synthesis of the same analogue. These authors have also reported on the biological activity of (**25**). This analogue was inactive against L-1210 leukemia system and caused no inhibition of macromolecular synthesis at concentrations where camptothecin showed 50% inhibition. It was concluded on the basis of these results that rings A and B may be essential for activity.

Recently, we have completed a high-yield synthesis of the tricyclic analogue (**26**) (Fig. 4) required in the synthesis of various pentacyclic and hexacyclic analogues (Wani *et al.*, 1980).

C. Pentacyclic and Hexacyclic Analogues

1. Ring A Modified Analogues

These analogues, which were obtained by the Friedlander condensation of (**26**) with appropriate *o*-amino aldehydes followed by hydroxylation, include D,L-camptothecin (**27**), D,L-10-methoxycamptothecin (**28**), D,L-10-hydroxycamptothecin (**29**), 12-aza-analogue (**30**), thiophene analogue (**31**), and the hexacyclic analogue (**32**) (Fig. 5) (Wani *et al.*, 1980). In analogues (**30**) and (**31**), ring A has been replaced by a heterocyclic ring, and in (**32**) an additional benzene ring has been fused onto ring A.

The screening data on these analogues are shown in Table II. D,L-Camptothecin was as active as (**1**) but less potent when both were tested in the same control series, thus indicating that the *S*-configuration at C-20 is required for maximal antitumor activity. D,L-10-Methoxy (**28**) and D,L-10-hydroxy (**29**) analogues were not evaluated, but it is reasonable to suggest that these would be one-half as potent as the corresponding natural alkaloids. The D,L-12-aza analogue (**30**) was synthesized to determine the effect of replacing a benzene ring with a pyridine ring. This analogue was less active and much less potent than D,L-camptothecin (**27**). It is likely that DNA binding may be inhibited. It is

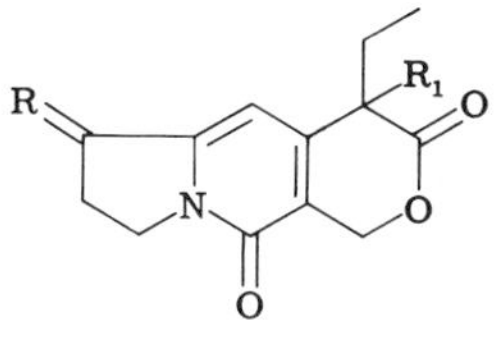

25 R = H, R$_1$ = OH
26 R = O, R$_1$ = H

Figure 4.   Ring CDE analogues.

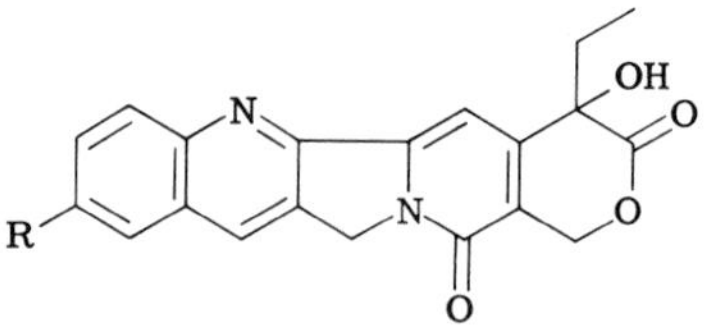

27 R = H
28 R = OMe
29 R = OH

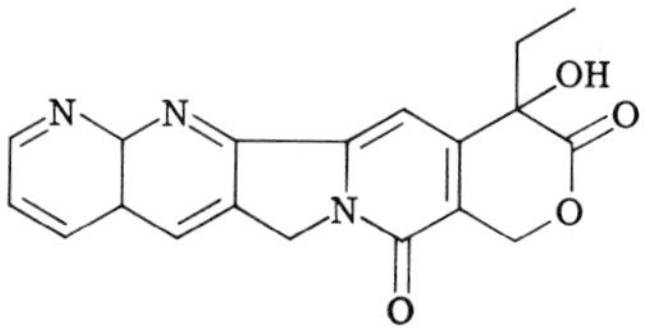

30

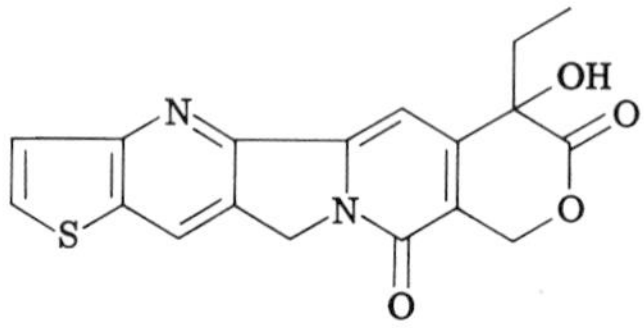

31

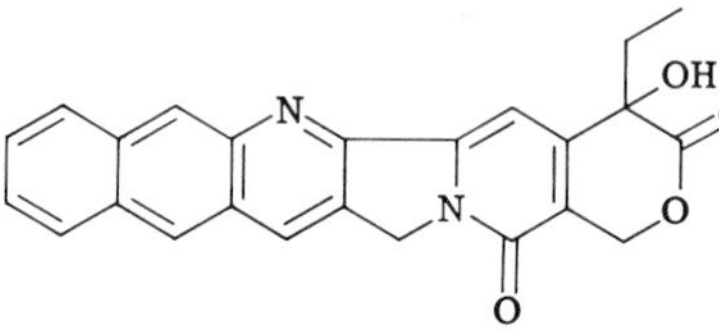

32

Figure 5. Ring A modified analogues.

known that 1,8-naphthyridines form complexes with metal ions (Hendricker and Bodner, 1970). Therefore, one possible explanation for the reduced activity of the 12-aza analogue (**30**) may lie in the formation of a complex which prevents intercalation between the two base pairs of DNA for steric reasons. The biological activity of the thiophene analogue (**31**) is not available. When compared with (**1**) in the same control series, the D,L-hexacyclic analogue (**32**) was found to be of the same order of activity but about one-half the potency in P-388 leukemia.

2. Ring B Modified Analogues

The 7-chloro (**33**), 7-acetoxy (**34**) (Ohlendorf et al., 1976), and 7-methoxy (**35**) (Bauxmann and Winterfeldt, 1978) analogues (Fig. 6) were obtained from the tetracyclic intermediate (**36**) (Krohn and Winterfeldt, 1975),

The object of the synthesis of (**35**) was to study the effect of the 7-methoxy substituent on the inhibition of RNA synthesis. However, the biological properties of (**33**), (**34**), and (**35**) were not reported. In view of the reported (Wall, 1969; Pan et al., 1975) improvement in the antileukemic activity of camptothecin upon introduction of the chloro or hydroxy substituent in ring A, it would be interesting to test the corresponding 7-substituted analogues.

3. Ring E Modified Analogues

Sugasawa and co-workers (1976) have reported the synthesis of four analogs of D,L-camptothecin in which the ethyl sidechain at C-20 was replaced by other groups such as allyl, propargyl, benzyl, and phenacyl. These were obtained from the key pentacyclic intermediate (37) (Fig. 7) via alkylation followed by deformylation, decarboxylation, oxygenation, and aromatization. All these analogues showed activity against the L-1210 leukemia system, the allyl analogue (38) being more active than D,L-camptothecin and other analogues.

Bauxmann and Winterfeldt (1978) have also reported the synthesis of similar analogues of 7-methoxycamptothecin (35) from the intermediate (36) mentioned in Section III,C,2. However, the biological activity of these analogues was not reported.

Recently, we have synthesized the 18-methoxy analogue (39) by the Friedlander condensation of a derivative of o-aminobenzaldehyde with the intermediate (40) (Wani et al., 1980). In L-1210 leukemia screening, this analogue exhibited activity comparable to D,L-camptothecin.

The analogues (41) (Boch et al., 1972) and (42) (Kende et al., 1973) (Fig. 7) were obtained during the total syntheses of D,L-camptothecin. Horwitz (1975) has studied the effects of these analogues on the inhibition of RNA synthesis and degradation of DNA (Table III). This will be discussed in Section V.

4. Iso- and Homo-D,L-camptothecin Analogues

Isocamptothecin (43), homocamptothecin (44), and isohomo-camptothecin (45) (Fig. 8) were obtained as by-products during the total synthesis of D,L-camptothecin (Danishefsky et al., 1973a), which involved lactomethylation of the tetracyclic ester (46) (Quick, 1977) with paraformaldehyde in the presence of sulfuric acid.

The activity of these analogues against the L-1210 leukemia system was not reported; however, the effects of analogues (43), (44), and (46) on the synthesis

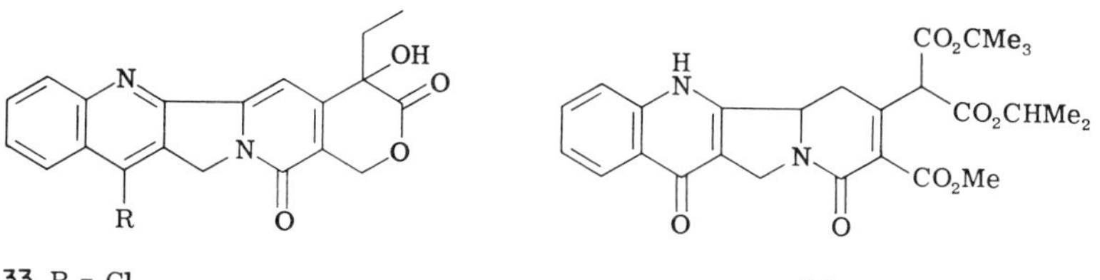

33 R = Cl
34 R = OAc
35 R = OMe

36

Figure 6. 7-Substituted analogues.

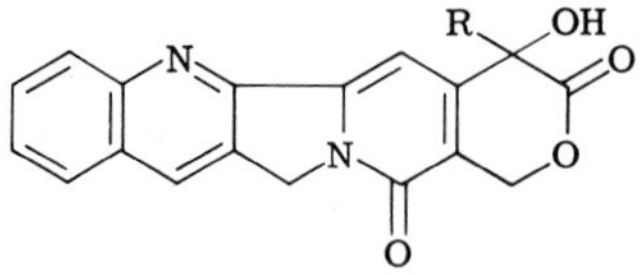

38 R = CH₂CH = CH₂
39 R = CH₂CH₂OMe

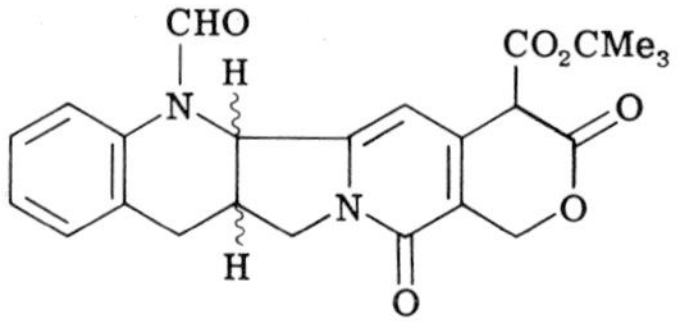

37

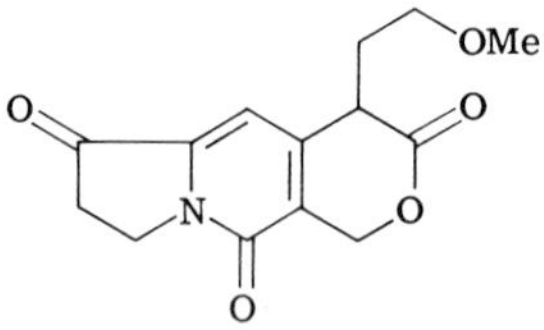

40 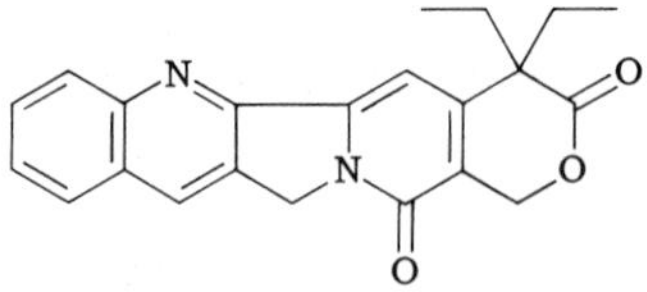   41

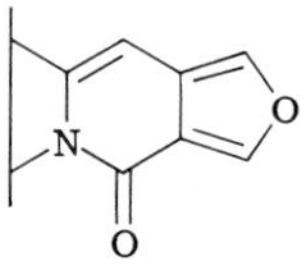

42

Figure 7. Ring E modified analogues.

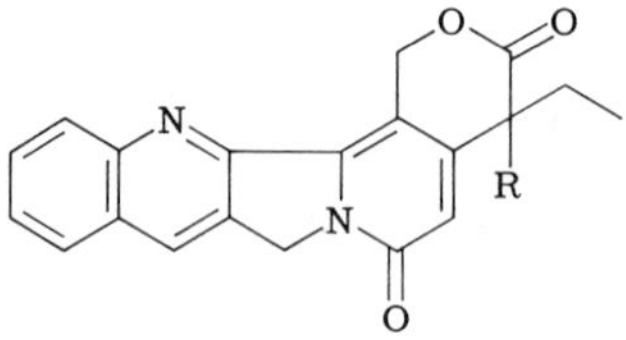

43 R = OH
45 R = CH₂OH

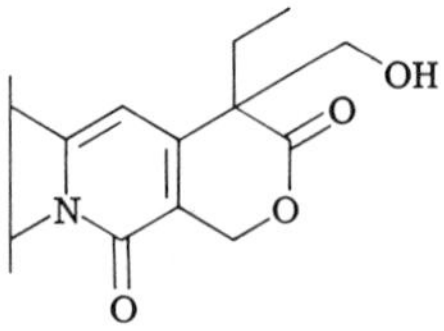

44

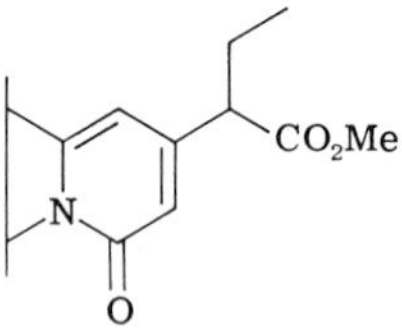

46

Figure 8. Iso- and homo-D,L-camptothecin analogues.

and degradation of nucleic acids (Table III) were investigated by Horwitz (1975). This will be discussed in Section V.

IV. ANTITUMOR ACTIVITY AND STRUCTURE-ACTIVITY RELATIONSHIPS OF CAMPTOTHECIN AND ITS ANALOGUES

A. Antitumor Spectrum

Both camptothecin (1) and 10-hydroxycamptothecin (2) have high activity against L-1210 and P-388 mouse leukemia, with T/C values in the range of 200 to 300% (Table II). Both compounds are active in certain solid tumor systems. Thus (1) and (2) have good activity against B-16 melanoma. The former has also been tested against Walker 256 carcinosarcoma showing very high activity over a wide dose range, with a therapeutic index of 13. Both (1) and (2) are inactive against the Lewis lung tumor.

The water-soluble sodium salt (14) has received more extensive animal leukemia and solid tumor tests than camptothecin or any of the other analogues of (1). The spectrum of antitumor activity found when (14) is administered by the intraperitoneal (i.p.) route is similar to (1), but in general (14) is somewhat less active and always less potent than the parent compound. However, (14) was inactive against L-1210 leukemia in mice by the intravenous (i.v.) route.

B. Structure-Activity Relationships (SAR)

Before discussing SAR in the camptothecin series, it must be emphasized that (1) and its analogues have been studied over a 12-year period in several leukemia systems. In addition to the studies conducted at RTI, researches at several other laboratories will be discussed. This means the results obtained by a number of biological screening groups must be compared. Hence one must be cautious in attempting to determine quantitative relationships. Nevertheless, we believe that valid information can be derived from these studies. We were particularly fortunate in that many of our analogues were compared to (1) against P-388 leukemia under the same control number, thus permitting excellent SAR studies to be made. The *in vivo* data are summarized in Table II; the *in vitro* data in Table III are adapted from the review by Horwitz (1975).

1. Number of Rings

The bicyclic (ring DE) and tricyclic (ring CDE) analogues, such as (22) and (25), respectively, both of which contain the pyridone and α-hydroxy lactone moieties found in camptothecin, are inactive in the *in vivo* leukemia systems, in 9KB cytotoxicity, in inhibition of RNA synthesis, and in depolymerization of DNA. On the basis of molecular orbital analysis, Flurry and Howland (1971) predicted that a tetracyclic system containing rings BCDE would be the minimum-size structure compatible with biological activity in the camptothecin series. Although we have attempted the synthesis of such a ring system, our efforts to date have not been successful.

2. Effects of Modifications in Ring E

As mentioned in Section II,C,1, the acetate (**5**), the chloro analogue (**6**), and the deoxy analogue (**7**) are essentially inactive in L-1210 leukemia. Conversion of the lactone to the lactol (**9**) also greatly reduces *in vivo* antitumor activity.

As noted previously, the lactone moiety is very reactive toward nucleophilic attack, resulting in cleavage of the lactone ring to yield compounds such as the sodium salt (**14**) which has had clinical trial, and amides such as (**10**) and (**12**). All of these compounds have *in vivo* antitumor activity but have considerably reduced intrinsic activity and/or potency as compared to (**1**). Recently, Adamovich and Hutchinson (1979) have raised the question of whether the various lactone cleavage compounds described above may have intrinsic activity, and that the α-hydroxy lactone moiety may not be an absolute requirement for activity. This is a rather "gray" issue since all of these compounds are readily relactonized under acid conditions to yield (**1**). Indeed, compound (**13**), which cannot be cleaved under acid conditions, is inactive in several *in vivo* leukemia systems. To us, this finding emphasizes the importance of the intact α-hydroxy lactone ring E or a structure such as (**15**) containing the required constituents so that the α-hydroxy lactone ring can be regenerated under *in vivo* conditions.

Sugasawa *et al.* (1976) have shown that replacement of the C-20 ethyl moiety by alkyl, propargyl, and phenacyl groups is compatible with retention of activity in the *in vivo* L-1210 mouse leukemia (Section III,C,3). Sugasawa *et al.* (1976) also prepared the sodium salt of (**38**) and noted that this compound was less active in L-1210 than the parent compound when administered by the i.v. route and was inactive when administered subcutaneously. Sugasawa *et al.* stated that their data confirmed the earlier studies of Wall (1969) which stressed the importance of the α-hydroxy lactone ring for antitumor activity.

The inactivity of the water-soluble sodium salt of camptothecin or its allyl analogue against L-1210 mouse leukemia when administered parenterally (i.v. or subcutaneous) casts *doubt* on the rationale of administering the sodium salt (**14**) by the i.v. route in clinical trials. The pH of blood is 7.2, and at this pH the sodium salt of camptothecin cannot regenerate the α-hydroxy lactone ring required for antitumor activity.

3. Modifications in Rings A and B

Few analogues of camptothecin with structural modifications in ring B have been reported. Biological data on such compounds are meager. Conversion of the pyridine nitrogen in (1) to the N-oxide (8) resulted in reduced activity and potency in L-1210 leukemia. Unfortunately, there is no antitumor information on the various C-7 analogues (cf. Section III,C,2) reported by Bauxmann and Winterfeldt (1978).

Introduction of certain substituents in ring A at the 9, 10, and 12 positions is not only compatible with activity but may even result in increased activity. The naturally occurring 10-hydroxy analogue (2) is considerably more active and equipotent to (1). A number of 12-substituted derivatives (cf. Section II,C,3) have been reported by Pan et al. (1975). Unfortunately, no data for these compounds in L-1210 or P-388 leukemia are available so that a realistic assessment of the activity of 12-substituted ring A analogues cannot be made at this time.

Two products prepared by total synthesis (Wani et al., 1980) have permitted additional evaluation of the effects of structural changes in ring A. We had anticipated that the D,L-12-aza analogue (30) would have activity and potency similar to (1) or (27) because the shape and structural parameters are similar for all three compounds. To our initial surprise, (30) had considerably lower activity and potency in in vivo leukemia tests and was inactive in the 9KB cytotoxicity test. As discussed in Section III,C,1, (30) may complex with metal ions to form a bulky product which could inhibit DNA binding.

As mentioned in Section III,C,1, the D,L-hexacyclic analogue (32) has activity comparable to (1) but has possibly one-half the potency of (1). Since (32) is racemic, it is conceivable that the intrinsic potency and activity of the compound would be of the same order as (1).

In Section II,C,2, we have mentioned the rationale behind the preparation of the water-soluble ethers (16) and (17). Although both (16) and (17) had excellent water solubility, their antitumor activity and/or potency was disappointing. Compound (16) was nontoxic and inactive in P-388. The compound was also inactive in 9KB. Electrostatic interactions between the negatively charged carboxylate anion of (16) and negatively charged phosphate groups may inhibit DNA binding.

The water-soluble diethylaminoethoxy hydrochloride analogue (17) was found to have the same order of activity as (1) at high dose levels, but had much lower potency in P-388 leukemia. When (17) was compared to the sodium salt of camptothecin (14) in the same control series, both compounds had the same order of activity and potency by the i.p. route. Compound (17) is inactive in 9KB and is probably intrinsically inactive as an antitumor agent, due conceivably to steric or electronic interference of the diethylaminoethoxy moiety with DNA binding.

The observed activity of (17) is probably due to slow hydrolysis of the ether to (2), a process which occurs *in vitro*. Hence (17) may be regarded as a water-soluble pro-drug form of (2) and may have useful clinical applications.

4. Structure–Activity Relationships Derived from in Vitro 9KB Studies

As shown in Table II, the inhibition of the growth of 9KB cells has good correlation with *in vivo* activity and/or potency observed for (1) and its analogues. It must be emphasized that valid comparisons can only be made when (1) and its analogues are compared in the same control series. This has been carried out for all samples prepared at RTI, and only these will be considered in this section. Thus (1), (2), and (3), which have maximal antitumor activity, show maximal 9KB inhibition, on the order of $ED_{50} = 1 \times 10^{-2}$ μg/ml. The sodium salt (14), which has the same order of activity as (1) but is less potent, gives a value of 2×10^{-1}. D,L-Camptothecin (27), which has activity equal to (1) but is less potent, gives similar values. On the other hand, inactive compounds or those with low activity and/or low potency gave values greater than 1×10^{0}. Such results were found for the inactive sodium salt (16), the active but very-low-potency diethylaminoethoxy hydrochloride salt (17), and the relatively inactive, low-potency 12-aza analogue (30). The inactive bicyclic analogue (22) in earlier testing at our laboratory also was found to be inactive in 9KB. It is apparent that in the camptothecin series, the 9KB cytotoxicity test has *excellent predictive* value for *in vivo* antitumor activity. This is in contrast to *in vitro* inhibiton of RNA or DNA by (1) and its analogues, which will be discussed in Section V.

V. STRUCTURE–ACTIVITY RELATIONSHIPS OF CAMPTOTHECIN AND ITS ANALOGUES DERIVED FROM DNA AND RNA STUDIES

A. Effects on DNA and RNA

Camptothecin (1) in the form of its sodium salt (14) has become an important tool for molecular biologists studying macromolecular processes because of its rapid onset of action, rapid reversibility, selective inhibition of high-molecular-weight RNA synthesis, and its sparing of mitochondrial macromolecular processes (Horwitz *et al.,* 1971; Kessel, 1971a,b). In addition, (1), (14), and certain other analogues have an extraordinary effect on DNA, producing smaller pieces in alkaline sucrose, a process which is also reversible (Horwitz and Horwitz, 1971; Spotorv and Kessel, 1972). The situation in regard to the DNA and RNA studies has an element of confusion because although the sodium salt of camptothecin (14) was the *actual* compound used in the great majority of the foregoing cases, it is frequently referred to as "camptothecin." It is evident that many of the workers were unaware that under neutral or alkaline conditions (1) and (14) are different compounds with different biological properties. In particular, (14) will not have the ability of generating a positive charge at C-21 in the manner of (1). As a result, (14) would not be expected to react with nucleophiles at pH 7.0 or above; this is the pH at which most of the RNA and DNA studies have been conducted.

B. Structure-Activity Relationships

The most extensive SAR studies of (1), (14), and other analogues have been made by Horwitz et al. (1971) and Horwitz (1975) (Table III). Two types of structure–activity relationships can be noted. Inhibition of RNA synthesis seems to be a general property of camptothecin and all analogues which have the aromatic ABCD rings; the bicyclic compound (23) and the lactol (9) were the only compounds showing low activity in this test. Conversion of cellular DNA to lower-molecular-weight species seems to be more sensitive to alterations in ring E. Only compounds with structures close to (1) are active. This test may also have good predictive value for in vivo antitumor activity, since deoxycamptothecin was the only in vivo inactive compound to give good activity in this test.

VI. CONCLUSIONS

An extensive review of the in vivo antitumor activity and various in vitro activities of (1) and its natural and synthetic analogues has been presented. The structural requirements for activity vary depending on the test system employed. For the in vitro inhibition of RNA synthesis, the requirements primarily are the presence of the planar ABCD ring system, and appear to be flexible as far as ring E structure is concerned. The requirements for DNA fragmentation activity are more rigid; essentially, (1) or closely related compounds with the α-hydroxy lactone ring E moiety are required. The inhibition of the growth of 9KB cells is probably the best in vitro test with excellent predictive value for in vivo antitumor activity.

The requirements for antitumor activity are the presence of the planar ABCD ring system together with the α-hydroxy lactone moiety of ring E or compounds which can regenerate the ring E system. Replacement of the alkyl substituent in ring E by certain other sidechains, particularly the allyl group, is not only compatible with activity but leads to increased activity. The optically active form of (1) has maximal activity and potency, with racemic (1) being less potent. Substitution in ring A by small groups, such as 10-hydroxy (2) and 10-methoxy (3) is permissible, with (2) being considerably more active than (1). However,

TABLE III

Effect of Camptothecin and Analogues on Inhibition of RNA Synthesis and DNA Molecular Weight[a]

Inhibitor added	50% Inhibition of RNA synthesis (μM)	High-molecular-weight DNA (%)
None	—	93
Camptothecin (**1**)	1	14
Deoxycamptothecin (**7**)	2	20
10-Methoxycamptothecin (**3**)	5	5
Camptothecin lactol (**9**)	30	77
Homocamptothecin (**44**)	1	62
Diethylcamptothecin (**41**)	6	88
Isocamptothecin (**43**)	8	84
Furan derivative (**42**)	4	81
Camptothecin monoester (**46**)	5	89
Camptothecin analogue-1 (**23**)	[b]	93

[a] Taken from Horwitz (1975), with permission of the author and Springer-Verlag.
[b] 40% Inhibition at 70 μM.

bulky groups substituted in ring A such as the 10-carboxymethyloxy analogue (**16**) which was inactive, or the 10-diethylaminoethoxy analogue (**17**) which has low potency, lead to inactivation, probably because of steric or electronic inhibition of binding to DNA or RNA.

Since the antitumor activity of (**1**) is due to both the planar aromatic structure of rings ABCD and the α-hydroxy moiety in ring E, an attractive hypothesis is that (**1**) binds to nucleic acids in a manner similar if not identical to intercalation. The lactone moiety in ring E may then be located in a favorable orientation to form a covalent bond with a nucleophilic group appropriately located on the nucleic acid.

Acknowledgments

We wish to thank Dr. Peter Ronman for synthesizing many of the camptothecin analogues reported. The expert technical assistance of H. L. Taylor for 9KB data and of J. B. Thompson for semisynthetic analogues is also gratefully acknowledged. We are happy to acknowledge the National Cancer Institute for continual support of these studies on Contract No. NO1-CM-67089 and Grant No. 5 RO1 CA20050-03 MCHA. We wish to thank John Douros and Matthew Suffness, National Cancer Institute, for encouraging the analogue studies and facilitating the *in vivo* testing of these compounds. We are indebted to Drs. R. K. Johnson and J. Wodinsky of Arthur D. Little, Inc. for stimulating discussion in connection with certain of the analogue studies.

REFERENCES

Adamovich, J. A., and Hutchinson, C. R. (1979). *J. Med. Chem.* **22**, 310.

Bauxmann, E., and Winterfeldt, E. (1978). *Chem. Ber.* **111**, 3403.

Boch, M., Korth, T., Nelke, J. M., Pike, D., Radunz, H., and Winterfeldt, E. (1972). *Chem. Ber.* **105**, 2126.

Bristol, J. A., Comins, D. L., Davenport, R. W., Kane, M. J., Lyle, R. E., Maloney, J. R., Portlock, D. E., and Horwitz, S. B. (1975). *J. Med. Chem.* **18**, 535.

Danishefsky, S., and Etheredge, S. J. (1974). *J. Org. Chem.* **39**, 3430.

Danishefsky, S., Volkman, R., and Horwitz, S. B. (1973a). *Tetrahedron Lett.*, 2521.

Danishefsky, S., Quick, J., and Horwitz, S. B. (1973b). *Tetrahedron Lett.*, 2525.

Flurry, R. L., and Howland, J. C. (1971). *Am. Chem. Soc., Natl. Meet., 162nd, Washington, D.C.* Abstr. MEDI 30.

Govindachari, T. R., and Viswanathan, N. (1972). *Indian J. Chem.* **10**, 453.

Heckendorf, A. H., Mattes, K. C., Hutchinson, C. R., Hagaman, E. W., and Wenkert, E. (1976). *J. Org. Chem.* **41**, 2045.

Hendricker, D. G., and Bodner, R. L. (1970). *Inorg. Chem.* **9**, 273.

Horwitz, M. S., and Horwitz, S. B. (1971). *Biochem. Biophys. Res. Commun.* **45**, 723.

Horwitz, S. B. (1975). *In* "Antibiotics. III. Mechanism of Action of Antimicrobial and Antitumor Agents" (J. W. Corcoran and F. E. Hahn, eds.), pp. 48–57, Springer-Verlag, Berlin and New York.

Horwitz, S. B., Chang, C., and Grollman, A. P. (1971). *Mol. Pharmacol.* **7**, 632.

Kende, A. S., Bentley, T. J., Draper, R. W., Jenkins, J. K., Joyeux, M., and Kubo, I. (1973). *Tetrahedron Lett.*, 1307.

Kessel, D. (1971a). *Cancer Res.* **31**, 1883.

Kessel, D. (1971b). *Biochim. Biophys. Acta* **246**, 255.

Krohn, K., and Winterfeldt, E. (1975). *Chem. Ber.* **108**, 3030.

Lyle, R. E., and Kane, M. J. (1973). *J. Org. Chem.* **38**, 3740.

Ohlendorf, H. W., Stranghöner, R., and Winterfeldt, E. (1976). *Synthesis* p. 741.

Pan, P. C., Pan, S. Y., Tu, Y. H., Wang, S. Y., and Owen, T. Y. (1975). *Hua Hsueh Hsueh Pao* **33**, 71.

Perdue, R. E., Smith, R. L., Wall, M. E., Hartwell, J. L., and Abbott, B. J. (1970). *U.S. Dept. Agric., Agric. Res. Serv., Tech. Bull.* No. 1415.

Plattner, J. J., Gless, R. D., and Rapoport, H. (1972). *J. Am. Chem. Soc.* **94**, 8613.

Plattner, J. J., Gless, R. D., Cooper, G. K., and Rapoport, H. (1974). *J. Org. Chem.* **39**, 303.

Quick, J. (1977). *Tetrahedron Lett.*, 327.

Spotorv, A., and Kessel, D. (1972). *Biochem. Biophys. Res. Commun.* **48**, 643.

Sugasawa, T., Sasakura, K., and Toyoda, T. (1974). *Chem. Pharm. Bull.* **22**, 763.

Sugasawa, T., Toyoda, T., Uchida, N., and Yamaguchi, K. (1976). *J. Med. Chem.* **19**, 575.

Tang, C., and Rapoport, H. (1972). *J. Am. Chem. Soc.* **94**, 8616.

Tang, C. S. F., Morrow, C. J., and Rapoport, H. (1975). *J. Am. Chem. Soc.* **97**, 159.

Wall, M. E. (1969). *In* "Symposiumsberichtes. IV: Internationales Symposium, Biochemie and Physiologie der Alkaloide" (K. Mothes, K. Schreiber, and H. R. Schütte, eds.), pp. 77–87. Akademie-Verlag, Berlin.

Wall, M. E. (1977). *In* "Oral Contraceptives and Steroid Chemistry in the Peoples Republic of China (SCPRC Report)" (J. Fried, K. J. Ryan, and P. J. Tsuchitani, eds.), pp. 62–65. National Adademy of Sciences, Washington, D. C.

Wall, M. E., Krider, M. M., Krewson, C. F., Eddy, C. R., Willaman, J. J., Correll, D. S., and Gentry, H. S. (1954). *U.S. Dep. Agric., Circ.* AIC-363.

Wall, M. E., Wani, M. C., Cook, C. E., Palmer, K. H., McPhail, A. T., and Sim, G. A. (1966). *J. Am. Chem. Soc.* **88**, 3808.
Wall, M. E., Campbell, H. F., Wani, M. C., and Levine, S. G. (1972). *J. Am. Chem. Soc.* **94**, 3632.
Wall, M. E., Wani, M. C., and Taylor, H. L. (1976). *Cancer Treat. Rep.* **60**, 1011.
Wani, M. C., and Wall, M. E. (1969). *J. Org. Chem.* **34**, 1364.
Wani, M. C., Ronman, P. E., Lindley, J. T., and Wall, M. E. (1980). *J. Med. Chem.* **23**, 554.

CHAPTER 13

Microbial Transformations as an Approach to Analogue Development

JOHN P. ROSAZZA

I. MICROBIAL TRANSFORMATIONS

Introduction and Background

Microorganisms are ubiquitous in nature, and they thrive by utilizing diverse substrates which serve to support their growth and metabolic needs. All nutrients like the lignins, cellulose, and even simpler organic compounds undergo metabolism through series of individual enzymatic transformation steps which may ultimately lead to total mineralization of the nutrient. This may occur by the action of complementary enzyme systems of single and/or mixed microbial populations. The technology of microbiological transformations is concerned with attempting to harness this tremendous enzymatic potential through the proper selection of conditions conducive to the accumulation of intermediates or metabolites. This technology has been successfully applied in the steroid field and it has shown considerable promise with other classes of physiologically active compounds.

This chapter will review the features of microbial transformations which render them attractive as tools in the development and design of new, highly active

437

Anticancer Agents Based on Natural Product Models

antitumor compounds. Emphasis will be placed on the use of microorganisms for preparing quantities of potentially active metabolites; providing difficult-to-synthesize drug metabolites which can serve to simplify drug metabolism studies in mammalian systems; elucidating novel pathways of metabolism of structurally complicated antitumor agents which may provide clues regarding toxicity and/or mechanisms of action; and establishing a firmer working basis for further structure–activity relationship studies.

1. Microbial Transformations of Organic Compounds

Peterson and Murray described one of the first successful applications of microbial transformations in the steroid field (Peterson and Murray, 1952). The fungus *Rhizopus arrhizus* was reported to affect the highly selective 11-α-hydroxylation of progesterone in 70–90% yield. Introduction of a hydroxyl group into the nonactivated 11-position was extremely difficult to achieve by routine chemical procedures, and the microbial conversion signaled the wide application of microbial transformations in the synthesis of steroid hormone products. Reactions of importance included the introduction of 11-β- and 16-α-hydroxyl functions; A-ring aromatization reactions through 1-dehydrogenation, and later 19-hydroxylation; and, more recently, the conversion of cheap and abundant sterols to C-19 steroids, including 4-androstene-3,17-dione and 1,4-andro-stadiene-3,17-dione (Marsheck, 1971; Marsheck *et al.*, 1972). An extensive body of literature describing microbial transformations in the steroid field now exists (Charney and Herzog, 1967; Wallen *et al.*, 1959; Iizuka and Naito, 1967; Laskin and Lechevalier, 1974; Sebek, 1977; Jones, 1973). It may be conservatively said that without the new technology of microbial conversions, convenient syntheses of steroid hormones would not have been realized for many years.

Successes in the steroid field prompted the use of microbial transformations with other classes of compounds. These include the alkaloids (Iizuka and Naito, 1967; Rosazza, 1978; Fonken and Johnson, 1972; Kieslich, 1976; Tamm, 1962), the antibiotics (Sebek, 1974, 1975, 1977), and a variety of other natural products (Wallen *et al.*, 1959; Tamm, 1962, 1974; Kieslich, 1969, 1976). A valuable result of extensive efforts with structurally complicated organic compounds has been the use of microbial transformations as a tool in synthetic organic chemistry (Sih and Rosazza, 1976; Sih *et al.*, 1977; Kieslich, 1969; Fonken and Johnson, 1972). Numerous fermentation-type reactions have now been catalogued, and many of these have considerable potential in the preparation of analogues of antitumor compounds. Those with the greatest potential utility include (Kieslich, 1976):

Oxidations: Aromatic and aliphatic hydroxylations; epoxidation; dehydrogenation; Baeyer–Villiger oxidation; decarboxylation; ring fission, N-oxidation; sulfoxidation; O-, N-, and S-dealkylations; alcohol dehydrogenase.

Reductions: Ketone reduction; hydrogenation of double bonds; NO$_2$, NO, NHOH, N-oxide reductions.

Hydrolytic reactions: Hydration of olefins; ester hydrolysis; amide hydrolysis, glycoside cleavage; epoxide hydrolysis.

Dehydrations: Alcohols to olefins.

Acylations: Esterification; amide formation.

Glycosidation: Glucosidation; ribosidation.

Decarboxylation.

Isomerizations and rearrangements: Dismutations; ring contraction; ring formation; *cis–trans* isomerization.

Condensations: Acyloin condensation.

Many of these reactions occur quite predictably with a given substrate (Sih and Rosazza, 1976), and progress in the microbial transformation field has moved sufficiently far that it is now possible to select microorganisms for the purpose of achieving specific types of reactions. This is especially true in the case of aromatic hydroxylation, O- and N-dealkylations, reductions of ketones, steroid hydroxylations, and ester or amide hydrolyses.

2. Characteristics of Microbial Transformation Reactions

Several features of microbial transformation reactions make them particularly useful for preparing analogues of antitumor compounds. They are all mediated by enzymes which are chiral catalysts with rigid-substrate binding characteristics. Thus, high degrees of regio- and stereoselectivities occur under extremely mild reaction conditions. Highly selective enzymatic transformations may occur with polyfunctional substrates without the need for protecting groups normally used in synthetic organic chemistry.

All useful microbial transformation reactions are first elaborated on a laboratory scale, either in shaken-flask culture or with stirred bench-top fermentors. If desirable, these kinds of reactions may be conducted on very large industrial scale as well. The technology for conducting large-scale fermentations is well in hand, as evidenced by the growing application of microbial cells or enzymes in several industrially important process (Abbott, 1976, 1977; Bernath *et al.,* 1977). This is perhaps best evidenced by the tremendous success obtained with glucose isomerase (more than a billion pounds of high-fructose corn syrup produced in the United States alone in 1976; see Bernath *et al.,* 1977), and by other enzymatic or microbial processes which are being used or are being considered for industrial use, including: L-amino acid production; various steroid or sterol conversions; production of 6-APA for semisynthetic penicillin production; and oxidation of sorbitol to sorbose for the production of vitamin C.

3. Methods of Microbial Transformations

The methodology of microbial transformations has been addressed in several recent reviews (Smith *et al.,* 1977; Smith and Rosazza, 1975b; Marsheck, 1971; Perlman, 1976; Kieslich, 1976). Some aspects of the methodology which are pertinent to the production and accumulation of microbial metabolites will be discussed here.

The fermentation protocol followed in our laboratories is described as follows. Surface growth from fresh (7-day-old) agar-slants of microorganisms are suspended in sterile medium, and the resulting vegetative and/or spore suspension is used to inoculate culture medium held in steel-capped DeLong culture flasks. Stage I cultures are incubated with shaking at 250 rpm at 27° on gyrotory shakers for 72 hours, during which time nearly all cultures approach maximum growth. Actively growing Stage I cultures are used to inoculate Stage II cultures containing the same, or slightly altered, medium. Normally the inoculum volume represents 5–10% of the total volume of medium held in Stage II culture flasks. Stage II cultures are incubated for 24–36 hours before receiving the substrate normally dissolved in a vehicle such as dimethylformamide, dimethylsulfoxide, acetone, methylene chloride, 0.001% Tween 80, or others. Samples are taken from substrate containing incubations at time intervals for analysis. All of our screening experiments are performed in this general way in 25 ml of medium held in 125-ml Erlenmeyer flasks. Preparative-scale incubations are achieved in numerous larger-sized Erlenmeyer flasks holding several liters of medium, or in stirred bench-top fermentors each holding 10 liters of medium.

Microbial transformation experiments begin with a screening process usually involving many microorganisms. The initial goal is to identify microorganisms capable of providing metabolites of antitumor drug substrates. Samples of incubations are routinely assayed for metabolites by thin-layer, gas-liquid, or high-performance liquid chromatographic methods. Although in principle, the screening process is simple to describe and perform, great care must be put into devising sensitive analytical procedures which will provide a true picture of the microbial transformation process. Important steps include determinations of the stability of antitumor drug substrates under varying conditions of pH; the extractability of the substrate and possible metabolites from aqueous media with organic solvents under various conditions of pH; sensitive and well-resolved chromatographic systems capable of detecting as little as 1% bioconversions; and spray reagents for thin-layer chromatography which distinguish *bona fide* microbial transformation products from secondary microbial metabolites.

Cultures used in microbial transformations may be obtained from standard culture collections (Martin and Skerman, 1972), or from rich natural environments like the soil. Standard catalogued cultures are easier to use because literature precedence quite often exists for specified microorganisms to perform useful biotransformation reactions. Experiments with standard catalogued cultures are more likely to be reproducible because such cultures are routinely available, and it is possible to work with "pure" cultures under relatively defined conditions. The isolation of new cultures from soil or other natural habitats is a more complicated and time-consuming process. Nevertheless, cultures may be obtained rather selectively by the so-called enrichment culture technique. By this process, the antitumor compound, or perhaps a less toxic analogue, may be added to a basal mineral-salts medium as the sole source of carbon and/or nitrogen. When this medium is inoculated with a mixture of cultures from soil, for example, only those organisms capable of utilizing the antitumor compound (or its analogue) will grow. The population is selectively enriched in these cultures by repeated serial transfer from spent to fresh medium. Pure cultures are finally obtained by repeated plating techniques. One drawback to the use of the enrichment culture technique is that isolated cultures may represent those which are merely resistant to the action of toxic antitumor agents.

4. Improvement of Metabolite Yields

Although microorganisms produce the enzymes which accomplish desired biotransformation reactions, they may also accomplish numerous other undesirable reactions at the same time. When this results in the production of unwanted products, it may be necessary to adjust conditions of the microbial transformation experiment to favor the production of a single compound. Parameters which are easily varied are the obvious adjustments in pH, temperature, medium composition, aeration of cultures, substrate concentrations, and the proper time for addition of substrates. The results obtained by variations in experimental conditions have been well discussed and documented elsewhere (Smith and Rosaza, 1975b; Perlman, 1976).

The use of actively growing cultures in microbial transformation experiments may result in unanticipated experimental difficulties. For example, the purification of microbial transformation products from complex fermentation media may be extremely tedious since growing cultures produce quantities of lipids, proteins, polysaccharides, phenolics, and other secondary metabolites which may co-extract with microbial transformation products and complicate reaction work-up; or they may aid in the formation of intractable emulsions. Furthermore, growing cultures may be "poisoned" by highly toxic antitumor compounds. One method of bypassing these difficulties is the use of resting-cell suspensions of microbial cells to achieve transformation reactions. Resting cells are those which are not in an active state of growth. Suspensions of these cells may be obtained by harvesting fermentations by filtration or centrifugation and by resuspending the resulting cell concentrates in buffers, or in buffers containing small amounts of selected nutrients.

When it is possible to use resting cells (some cultures are not amenable to this technique due to the instability of enzyme systems), several advantages become apparent. Microbial cells may be produced in large quantity and stored in the cold until required for use. In this way, they resemble "reagents" which may be used simply by suspending them in solvent (buffers). Other advantages obtained with resting cell suspension incubations include ease of isolating metabolites;

stability of transformation systems; little possibility of contamination with other microorganisms; enhanced rates of transformation due to higher cell (enzyme) concentration; ease of maintaining conditions affecting the production of microbial metabolites such as aeration, pH, and various co-metabolism substrates; and greater reproducibility. Much recent work has been directed toward entrapping resting microbial cells and/or their enzymes into insoluble polymeric matrices. The resulting "bio-reactors" often metabolize substrates to products in a highly efficient manner. Many such applications have been reviewed (Smith *et al.*, 1977; Abbott, 1977; Bernath *et al.*, 1977).

5. Microbial Models of Mammalian Metabolism

New drug development studies all require an understanding of their metabolism. Attempts are made to determine how they are absorbed, distributed, and excreted in the mammalian systems, and how they might be changed by the action of drug metabolism enzymes. Valuable information concerning the mechanism of action, and possibly the toxicity of antitumor compounds, may be ascertained by establishing metabolic pathways and by elucidating the structures of drug metabolites.

Many recently discovered antitumor compounds are obtained from natural sources, and they fall into classes of compounds like the terpenes, alkaloids, and antibiotics. The fact that these compounds are structurally complex presents difficult practical problems for the pharmacologist interested in performing drug metabolism studies. Some of these difficulties relate to the inability of most mammalian systems to produce sufficient quantities of drug metabolites to enable complete structure elucidation, and to the need for developing means of obtaining quantities of metabolites for pharmacological/toxicological testing. A prime example in the antitumor drug area which has been frustrated by these kinds of concerns is that of the *Catharanthus* alkaloids.

It was recently suggested (Smith and Rosazza, 1975a,b; Smith *et al.*, 1977) that a microbial transformation system could mimic many of the biotransformations normally observed in mammalian systems. The system would typically consist of several selected microorganisms which together would produce a profile of drug metabolites similar to that obtained with mammalian drug-metabolizing systems. We called these systems "microbial models of mammalian metabolism" (Smith and Rosazza, 1975a,b). Although we suggested that it might be unlikely for a single microorganism to mimic all of the biotransformations achieved by mammalian systems, single cultures have now been described which perform an amazing array of metabolic transformations like N- and O-demethylation, aromatic hydroxylation, and other mammalian-type reactions (Ferris *et al.*, 1976; Duppel *et al.*, 1973).

Advantages to the employment of microbial models of antitumor drug metabolism are numerous. Ideally, drug metabolism studies would be performed simultaneously using both mammalian and microbial metabolic systems. Where metabolites common to both kinds of systems are observed, the microbial system would be used to produce gram quantities of metabolites through the use of routine fermentation optimization and scale-up techniques. In this way, difficult-to-synthesize metabolites of antitumor drugs would be readily obtained for structure elucidation and biological testing. Nascent or short-lived metabolites produced in mammalian systems usually remain undetected. Due to the nature of the microbial transformation system (cells suspended in liquid fermentation medium), the likelihood of isolation and characterizing such compounds is increased. Minor but significant mammalian metabolites are often produced in relatively large quantities by microorganisms. An additional useful feature of the typical microbial transformation system is the better reproducibility of fermentations when compared to their mammalian counterpart.

Support for the "microbial models of mammalian metabolism" concept derives from several areas of research. The biochemical basis for invoking similarities between microbial and mammalian systems is very strong. Mammals and microbes both contain cytochrome P-450 monooxygenase enzymes with essentially the same components (i.e., cytochrome P-450 monooxygenase; a flavoprotein; nonhemeiron protein; NADPH or NADH cofactor requirement) (Cerniglia and Gibson, 1978; Berg et al., 1976; Ferris et al., 1976). Microbes and mammals both achieve hydroxylation of aromatic compounds through the now well-known NIH Shift mechanism (Auret et al., 1971; Ferris et al., 1976; Boyd et al., 1976). Profiles of metabolites of several drugs which have been metabolized in both microbial and mammalian metabolic systems are remarkably similar. These include the antitumor drugs acronycine (Betts et al., 1974; Brannon et al., 1974), and ellipticine (Chien et al., 1978); and other drugs including Danazol (Rosi et al., 1977), spironolactone analogues (Marsheck and Karim, 1973), papaverine (Rosazza et al., 1977), miscellaneous antiinflamatory agents (Kishimoto et al., 1976), and prostaglandin analogues (Sebek et al., 1976; Lanzilotta et al., 1976). Many of the mammalian metabolites have been produced in relatively large amounts by microbial transformations. In the case of Danazol, for example, four of the human metabolites of the drug were produced in 20–70 gm amounts by the actions of several selected microorganisms. The application of this concept has been verified in several other cases as well (Beukers, 1972; Sehgal and Vezina, 1967). A related area of importance which has received only little attention to date is the influence of the microbial flora of the intestinal tract on the metabolism and disposition of antitumor drugs.

II. REVIEW OF MICROBIAL TRANSFORMATIONS OF ANTITUMOR AGENTS

Numerous new structural classes of antitumor compounds from higher plants and from microorganisms have been characterized. The extensive efforts

supported through the National Cancer Institute have resulted in the identification
of alkaloids, steroids, terpenes, enzymes, antibiotics, and other classes of compounds which possess high levels of antitumor activity. Because many of these
compounds are also toxic, their use as therapeutic agents is restricted, and the
complexities of their structures render the usual simple chemical modifications
relatively difficult to achieve.

Microbial transformations will best contribute to the development of useful
analogues of antitumor agents through the production of potentially useful
metabolites with less toxicity than the parent compounds, and/or through the
determination of their metabolic fates.

The utility of microbial transformations may be illustrated by comparisons
within two structural grups of antitumor compounds: the *Catharanthus* alkaloids, and the anthracycline antibotics. The *Catharanthus* alkaloids vinblastine
(37) and vincristine (38) differ from one another in the state of oxidation of
the carbon attached to the indole nitrogen substituent of the vindoline part of
the molecule. Likewise, the anthracycline antibiotics daunomycin (6) and adriamycin (7) differ only in the state of oxidation of the 14-C atom of their respective
ethyl sidechains. Although these structural differences are quite subtle, they are
sufficient to cause profound differences in the toxicity, potency, and spectrum
of activity. Microbial transformations may be expected to cause similar subtle
changes in the structures of other antitumor compounds which will result in
favorable changes in antitumor activity as well.

Many antitumor compounds have now been studied in microbial transformation experiments, and the results of this work are highlighted here. For conve-
nience, the classes of antitumor compounds have been categorized as to their
origin as either fermentation products, or plant antitumor agents.

A. Fermentation Products

1. Anguidine

Anguidine (1) is a scirpenol derivative with demonstrated activity
against a number of neoplasms. This compound was transformed by resting cells
of *Streptomyces griseus* and *Mucor mucedo;* and by growing cultures of
Acinetobacter calcoaceticus (Claridge and Schmitz, 1978). Several derivatives
of the antibiotic were obtained by a combination of chemical and microbiological
hydrolysis or esterification reactions. *M. mucedo* achieves 3-acetylation reactions with anguidine to provide triacetoxyscirpene (2), and it also converts 15-
acetoxyscirpen-3,4-diol (3) and 4-acetoxyscirpen-3,15-diol (4) to their respective
3-acetylated derivatives. The reaction occurs best with resting cells at pH 5.0,
and it is markedly stimulated in the presence of 0.1% maltose. Scirpenentriol did
not react under these conditions. *Streptomyces griseus* selectively removed the

	R_3	R_4	R_{15}
1	H	Ac	Ac
2	Ac	Ac	Ac
3	H	H	Ac
4	H	Ac	H
5	Ac	Ac	H

C-15 acetoxyl group of (2), and it converts (1) to 4-acetoxyscirpen-3,15-diol (4). This reaction occurs at pH 7.5, and it is not increased in the presence of co-metabolism substrates. *A. calcoaceticus* causes the removal of the 3-acetoxy function, and the reaction occurs best under growing culture conditions. Cytotoxicity testing with Hela and L-929 cells indicated that 15-acetoxyscirpen-3,4-diol was the only anguidine derivative possessing activity similar to anguidine itself.

This work is an excellent example of the selective hydrolysis and acetylation obtainable by microorganisms, and it made available nearly the entire series of scirpenol, or ester derivatives for biological evaluation.

2. Anthracycline Antibiotics

Microbial transformation experiments have been conducted with several of the anthracycline antibiotics. The major objectives or goals in this work have been to provide difficult-to-synthesize derivatives of (6) and (7) such as carcinomycin (8). Florent and Lunel (1975) first described the conversion of daunomycin (6) to daunomycinol (9). This 13-dihydroderivative of daunomycin is also a major mammalian metabolite (Aszalos *et al.*, 1977; Bachur *et al.*, 1976). Unfortunately, this ubiquitous microbial anthracycline metabolite is less active than daunomycin itself as an antitumor compound. Karnetova *et al.* (1976) found the 13-ketone reduction reaction to be prevalent with the aglycone daunomycinone (10) with *Streptomyces aureofaciens*.

Streptomyces nogalater and *S. peucetius* var. *caesius* both cause the reduction of the 1-ketone of a related anthracycline, steffimycinone (Wiley *et al.*, 1977), and cell-free preparations demonstrated the reduction to be NADPH-linked. Wiley and Marshall (1975) also studied anthracycline conversions using microaerophilically grown cultures and cell-free extracts of *Aeromonas hydrophila*.

	R_4	R_7	R_{13}	R_{14}
6	CH_3	Daunosamine	=O	H
7	CH_3	Daunosamine	=O	OH
8	H	Daunosamine	=O	H
9	CH_3	Daunosamine	H, OH	H
10	CH_3	H	=O	H
11	CH_3	H	(ethylene ketal)	H
12	H	H	=O	H

A reductive glycoside cleavage reaction occurs, resulting in the production of 7-deoxyaglycones. The reaction is catalyzed by NADH, and the mammalian parallel reaction has been reported with solubilized preparations from rat-liver microsomes (Oki *et al.*, 1977). *Mucor spinosis* causes reduction of the 13-ketofunction of daunomycin, and the reaction is catalyzed by NADPH-dependent enzymes (Marshall *et al.*, 1978). These same workers also described the NADH-dependent cleavage of daunosamine to yield 7-deoxyaglycone derivatives of daunomycin and adriamycin.

We have attempted to bypass what appears to be the major difficulty with microbial conversions of anthracyclines by chemically protecting the 13-ketone position of compounds like daunomycinone (**10**). This approach has met with success, in that several microorganisms, including *Streptomyces griseus* and *Beauveria sulfurescens,* achieve the selective 4-*O*-demethylation of (**11**) to produce either carcinomycinone (**12**) or its 13-ethyleneketal derivative (Rosazza *et al.*, 1978a). Several microorganisms have demonstrated the ability to cleave the ethylene–ketal protecting group as well.

In general, microbial transformations have provided excellent mimics of mammalian metabolism with the anthracyclines; and it has been possible to direct fermentation reactions to sites other than the 13-ketone by preparing suitable derivatives of the substrate.

3. Bleomycin

The bleomycins are structurally related antitumor antibiotics produced by strains of *Streptomyces verticillus,* differing in the nature of the terminal amine moiety attached to bleomycinic acid (**13**) (Umezawa *et al.*, 1973). Microbial transformations were investigated as a possible means of converting bleomy-cin B$_2$ (14) into (13) for use in semisynthesis of new bleomycin derivatives. This thinking was analogous to that employed in the development of semisynthetic penicillin derivatives. Bleomycin B$_2$ (14) was selectively hydrolyzed to (13) by whole cultures, or by cell-free preparations of *Fusarium anguioides*. The en-zyme performing the transformation was highly specific for the terminal amine portion of bleomycin B$_2$, but not for bleomycinic acid. Thus, other bleomycins did not serve as substrates for the enzyme. This process has not been used because the more abundant bleomycin A$_2$ undergoes facile chemical cleavage to (13) (Takita *et al.*, 1973).

4. Aureolic Acid Derivatives

Antibiotics of the aureolic acid class, including olivomycin-A (15) and chromomycin A-2 (16), were transformed by a culture of *Whetzelinia sclerotiorum* ATCC 10939 (Schmitz and Claridge, 1978). Deisobutyryl olivo-mycin-A (17) was the major transformation product of olivomycin, while minor amounts of desisobutyryldesacetyl olivomycin-A were also obtained. Chromomycin A$_2$ and A$_3$ gave identical mixtures of microbial transformation products with *W. sclerotiorum*. The major metabolite with chromomycin A$_2$ was identified as the deisobutyryl derivative (18). Antimicrobial activities of the two new microbial metabolites were determined: (17) was less active than olivo-mycin-A; and (18) was nearly as active as chromomycin A$_2$. Antitumor activities

	R_1	R_2	R_3
15	H	CH_3CO	$COCH(CH_3)_2$
16	CH_3	CH_3CO	$COCH(CH_3)_2$
17	H	CH_3CO	H
18	CH_3	CH_3CO	H

were not given for the transformation products. This was the first report of microbial transformations within the aureolic acid group of antibiotics.

We are conducting studies with mithramycin (also known as aureolic acid). Screening of more than 200 cultures using various media and culture conditions resulted in the identification of several basidiomycetes with the capacity for transforming mithramycin into several new derivatives (Rosazza et al., 1978c). The microbial metabolites are of unknown structure at this time, but several of them display antibacterial activity against Bacillus subtilis and Sarcina lutea as test organisms. The enzyme which performs the bioconversion is extracellular, and it possesses activities indicative of a laccase.

5. Streptonigrin

The antitumor antibiotic streptonigrin (**19**) has been known for many years. The biosynthesis of this compound has been the subject of recent study (Gould and Chang, 1978) but much work remains to be done concerning the structure–activity relationships of this antibiotic. We have examined the microbial metabolism of streptonigrin and have obtained several microorganisms capable of performing transformations with the antibiotic.

19 $R_1 = R_2 = H$
20 $R_1 = COCH_3$, $R_2 = H$
21 $R_1 = H$, $R_2 = COCH_3$

Penicillium brevi-compactum cultures produce a streptonigrin metabolite which has been isolated and tentatively identified as the N-acetyl derivative (**20**) (Rosazza et al., 1978d). The metabolite is obtained in high yield (1.1 gm from 1.5 gm of streptonigrin), and it possesses approximately one-half of the antibacterial activity of streptonigrin itself against *Sarcina lutea* on plate-disc assays. Attempts to chemically acetylate streptonigrin (acetic anhydride–pyridine) lead to exclusive production of the other N-acetyl streptonigrin (**21**). *P. brevicompactum* thus produces a streptonigrin derivative which is not easily obtained by simple chemical procedures. This metabolite has been submitted for antitumor testing.

The possibility exists of using the acylating potential of the microorganism to prepare several other derivatives of the antibiotic. Such acylated derivatives of streptonigrin may be useful as pro-drugs, and they may increase adsorption of the antibiotic.

At least two other microorganisms produce significant and reproducible quantitites of streptonigrin metabolites. Efforts are underway to elucidate the structures of these compounds.

6. Actinomycins

Microbial degradations of actinomycin D (**22**) were first reported to occur with *Achromobacter* species (Katz and Pienta, 1957). This microorganism converted the antitumor antibiotic into antibacterially inactive red pigments. Later work with resting cells and a cell-free preparation of *Actinoplanes* (IMRU No. F3-15) succeeded in identifying actinomycin monolactone (**23**) and actinomycinic acid (**24**) as major microbial transformation products (Perlman *et al.*, 1966). The lactonase enzyme of *A. missouriensis* was inducible with actinomycin and other compounds, and its properties were determined (Hou and Perlman, 1970). None of the reported conversion products are active.

Numerous derivatives of the actinomycins have been prepared by chemical derivatization or syntheses, or by guided biosynthetic processes. Structure–activity relationships among the actinomycins have been extensively reviewed (Perlman, 1977). Potentially useful microbial transformations of the actinomycins would only be those which accomplish modifications of selected sites of the peptide sidechains of the actinomycin D structure, since chemical and biochemical variations have been extensively examined at other sites of the molecule. Microbial transformations of the actinomycins may provide useful information regarding pathways of metabolism other than those documented by Perlman and co-workers.

7. Mycophenolic Acid

In one of the most extensive microbial transformation studies of any antitumor antibiotic, mycophenolic acid (**25**) was converted into 19 novel an-

tibiotic derivatives (Jones *et al.*, 1970). Structures of each metabolite were determined by spectroscopic methods, and they were all chemically related to mycophenolic acid. All the observed transformations, including reduction or oxidation of the olefinic part, demethylation of the methoxyl group, oxidation of the methyl group attached to the aromatic ring, and lengthening or shortening of the terpene sidechain, resulted in disappearance of reduction of antitumor activity. Sixty-five derivatives and analogues of the antitumor antibiotic, including several of the microbial transformation products (Jones and Mills, 1971), and an additional 108 derivatives at the phenolic hydroxyl and/or carboxyl sites of mycophenolic acid (Suzuki and Mori, 1976) have all been examined for antitumor activity. Nearly all chemical or biochemical modifications in the structure of mycophenolic acid result in loss of antibiotic activity. The one active derivative which resulted from all of this work is the carbamoyl-mycophenolic acid ethyl ester which is orally active (Suzuki and Mori, 1976). Microbial transformations contributed to the structure–activity relationships of mycophenolic acid by providing new leads for several novel compounds. Although a few of the originally discovered microbial metabolites have not been examined for antitumor activity, it is probably unlikely that any of these will prove superior to mycophenolic acid itself.

8. Other Antibiotics

Microbial transformation studies have been performed with a selection of other antibiotics with known antitumor activity. Mitomycin C apparently undergoes degradation or at least inactivation by an enzyme found in the mycelium of *Streptomyces caespitosus*, the antibiotic-producing microorganism (Gourevitch *et al.*, 1961). Thus it became necessary to separate the mycelium from fermentation beers prior to isolation of the antibiotic. The structures of presumed mitomycin metabolites are unknown. It is conceivable that a multistep transformation is involved, and that active microbial metabolites might be obtained by adjusting conditions of incubation.

Showdomycin undergoes an isomerization with growing cultures and washed cells of a streptomycete (Ozaki *et al.*, 1972). The metabolite is an isomer of the parent antibiotic. Formycin is converted into formycin-B, an antibiotic derivative with signifcantly less activity against the Yoshida rat sarcoma (Umezawa *et al.*, 1965). Rifamycin derivatives also undergo a series of biotransformations with the producing microorganism (Lancini *et al.*, 1967, 1969; Lancini and Hengeller, 1969).

B. Plant Products

1. Acronycine

Acronycine (**26**) is one of many brilliant yellow acridone alkaloids isolated from the stem bark of *Acronychia baueri* Schott. The alkaloid displayed significant activity in several tumor test systems, and it was reported to be orally active. Mammalian metabolic studies revealed that several hydroxylated ac-

ronycine derivatives were produced, including 9-, 11-, 9,11-di-, and 3,11-dihydroxyacronycines, with several test animals (Sullivan *et al.*, 1970). Microbial transformation experiments with acronycine were reported from two laboratories. In our work, screening revealed 10 cultures capable of producing metabolites of the alkaloid. Of these, seven were *Cunninghamella* species. *Cunninghamella echinulata* (NRRL 3655) provided 30% yields of 9-hydroxyacronycine as the major metabolite when incubated with 4 gm of the alkaloid (Betts *et al.*, 1974), and the structure of the highly insoluble phenolic metabolite was determined by proton magnetic resonance (PMR) spectral examination of the acetate derivative. 9-Hydroxyacronycine is much more insoluble than acronycine itself. Both 9-hydroxy- and 9-acetoxyacronycine were inactive when tested against the L-1210 test system, undoubtedly due to the high insolubility of the compounds. Workers at Eli Lilly and Company reported that other cultures could achieve 11- and 3-hydroxylations of acronycine (Brannon *et al.*, 1974). Microorganisms produced several of the known mammalian metabolites in sufficient quantity for biological testing, and the microbial systems served as excellent models for mammalian metabolism.

2. Ellipticine

Alkaloids isolated from several *Ochrosia* species demonstrate potent antitumor activity. Among these, the alkaloid ellipticine (**27**) has been well studied (Lecointe *et al.*, 1978; Li and Cowie, 1974; Bhuyan *et al.*, 1972). The yellow and fluorescent alkaloid is metabolized by rat-liver preparations mainly to 9-hydroxyellipticine (**28**) (Lesca *et al.*, 1976, 1977), but a recent report indicates that 7-hydroxylation of ellipticine to (**29**) represents a minor pathway (Lalleman *et al.*, 1978).

$$27 \quad R_7 = R_8 = R_9 = H$$
$$28 \quad R_7, R_8 = H, R_9 = OH$$
$$29 \quad R_8, R_9 = H, R_7 = OH$$
$$30 \quad R_7, R_9 = H, R_8 = OH$$

Microbial transformations of ellipticine have been conducted in our laboratory (Chien *et al.*, 1978) with the objective of producing several novel derivatives of the alkaloid. Through screening experiments, many cultures were found to produce several ellipticine metabolites, including *Aspergillus alliaceus* (UI 315), *Aspergillus fumigatus* (UI 51), and *Penicillium purpurogenum* (UI 193). *Aspergillus alliaceus* gave two phenolic metabolites which were isolated and characterized. 9-Hydroxyellipticine (**28**) was obtained in 10% yield, and its structure determination was based on PMR and mass spectral analyses, and on comparison with 9-hydroxyellipticine prepared by chemical demethylation of 9-methoxyellipticine. 8-Hydroxyellipticine (**30**) was obtained in 2.5% yield, and its structure was determined primarily on the basis of PMR spectral analysis. In the PMR spectrum of this metabolite, protons of the D ring were displayed as an ABX system with signals at 8.15 ppm (d, J = 8Hz), 6.94 (d, J = 2Hz) and 6.71 ppm (dd, J = 2HZ and J = 8Hz). The ABX pattern was different from that obtained for (**28**), and all other proton signals were comparable to those of the parent compound. The 8-hydroxyellipticine derivative is a new metabolite of the alkaloid, and it has not yet been tested for antitumor activity. So far, the

9-hydroxyellipticine derivative appears to be the most active form of the alkaloid.

It is interesting to think that the 9- and 8-hydroxyellipticines could arise though a common arene oxide intermediate with an epoxide at positions 8 and 9 of the ring. As with naphthalene, an ellipticine epoxide might give rise to disproportionately greater amount of 9- than 8-hydroxyellipticine. Although it has not yet been reported, it is likely that the 8-hydroxy derivative will be observed in mammalian systems. Numerous other ellipticine derivatives have been detected, and we are in the process of producing more of them now. Again, the microbial systems have served as an excellent model for mammalian metabolism.

3. D-Tetrandrine

D-Tetrandrine (31) is a bisbenzyltetrahydroisoquinoline alkaloid isolated from several plant species, and the alkaloid has demonstrated antitumor activity against the Walker 256 test system. Microbial transformation experiments with the alkaloid were successful in producing two metabolites in good yield during preparative-scale incubations: *Streptomyces griseus* (UI 1158) gave 50% yields of N'-nor-D-tetrandrine (32) (also known as cycleanorine) (Davis and Rosazza, 1976). This derivative of D-tetrandrine shows no activity against the

31 R = R' = CH$_3$

32 R=CH$_3$, R' = H

33 R=H, R' = CH$_3$

Walker test system, thus suggesting the importance of an N-alkyl functional group. *Cunninghamella blakesleeana* (gives 20% yields of the N-nor-D-tetrandrine derivative (33) (Davis *et al.*, 1977). Both reactions occur in noncomplicated fashion, and the microbial demethylations are superior to chemical N-demethylation procedures (Davis *et al.*, 1977). The mammalian metabolism of D-tetrandrine has not yet been studied, but it is possible that the N-demethylation reactions will be observed there as well. This work demonstrates the high regioselectivity obtainable in N-demethylation of a polyfunctional substrate such as D-tetrandrine. Other reactions which are being examined now include N-oxidation and possible O-demethylation of the alkaloid.

4. Thalicarpine

The plant *Thalictrum dasycarpum* has yielded a number of alkaloids, one of which is the antitumor compound thalicarpine (**34**). Thalicarpine is active in the Walker 256 test system, and it has been examined for antitumor activity in the clinic. Several microorganisms provided relatively low yields of thalicarpine

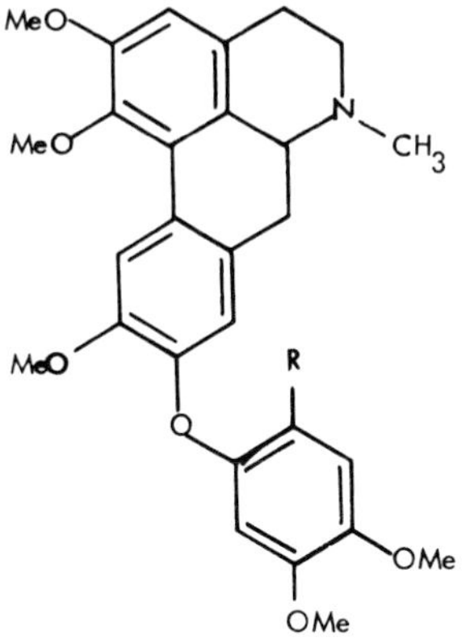

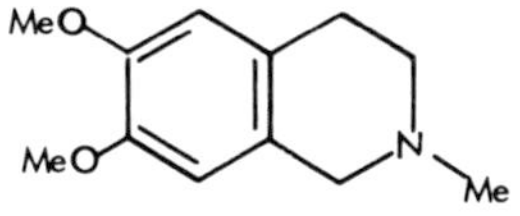

34 R =

35 R = CHO

36 R = CH$_2$OH

metabolites in microbial transformation experiments (Nabih *et al.*, 1977). *Streptomyces punipalus* (NRRL 3529) provided 10% yields of a metabolite which was identified as hernandalinol (**36**) on the basis of spectral and chemical synthetic methods. It was established that thalicarpine most likely undergoes oxidative cleavage of the isoquinoline ring to provide hernandaline (**35**), an aldehyde, and that the aldehyde is reduced by a reductase of the microorganism. Urine samples from patients receiving thalicarpine were examined for the possible presence of hernandaline and hernandalinol by high-performance liquid chromatographic procedures (Smellie *et al.*, 1978); however, none of these compounds could be detected. Efforts are now in progress to prepare hernandalinol and other related compounds in sufficient quantity for biological evaluation.

5. Vinblastine

Vinblastine (**37**) and vincristine (**38**) are two potent antitumor alkaloids from *Catharanthus roseus* species. Vinblastine has been studied for two reasons: the semisynthesis of the less abundant vincristine, and the preparation of novel derivatives. *Streptomyces albogriseolus* was capable of performing the N-demethylation of (**37**) at the indole nitrogen of the vindoline part of the molecule. The resulting product (**39**) may be formylated to give vincristine (Brannon and Neuss, 1975). Vincristine may be prepared more advantageously by chemical oxidation of (**37**) with chromium trioxide in acetone at low temperature. Other streptomycetes have provided two major metabolites of (**37**) (Neuss *et al.*, 1974a). These include VLB-Ether, and Hydroxy-VLB, both differing

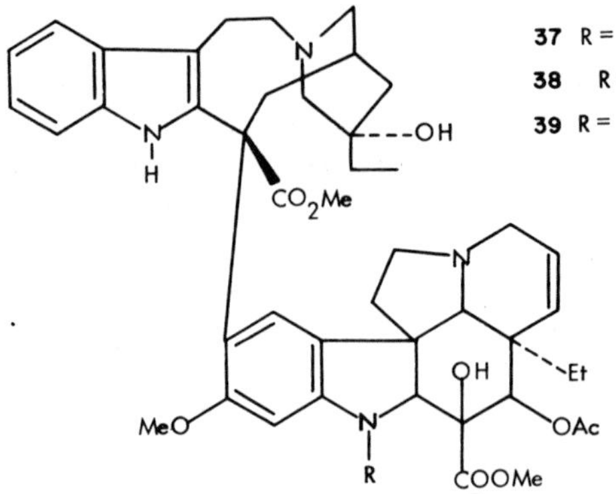

37 R = CH₃ → 37 $R = CH_3$
38 $R = CHO$
39 $R = H$

from the parent alkaloid in the indolenine portion of the molecule. No activities of the derivatives were reported.

6. Vindoline

Several interesting microbial transformation studies have been reported for the alkaloid vindoline (**40**), a monomeric *Catharanthus* alkaloid, and one of the most abundant compounds present in the plant. This alkaloid is not an active antitumor compound in itself, but it is present as half of the structures of the active dimeric alkaloids (**37**) and (**38**). Vindoline undergoes highly selective O-demethylation to provide the phenol (**41**) in good yield. The reaction is uncomplicated by the formation of side-products, and it makes available a "synthon" which could be used to prepare unusual derivatives of the active *Catharanthus* alkaloid dimers (Wu *et al.*, 1979).

Several microbial metabolites of vindoline were reported by workers at Eli Lilly. Metabolic products obtained included *N*-demethyl vindoline (**42**) from *S. albogriseolus* (Neuss *et al.*, 1974b); 17-des-acetylvindoline (**43**) (Mallett *et al.*,

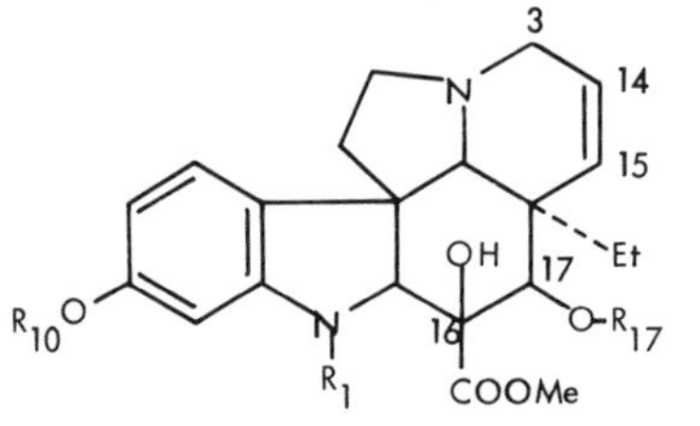

40 $R_1 = R_{10} = CH_3$, $R_{17} = COCH_3$

41 $R_1 = CH_3$, $R_{10} = H$, $R_{17} = COCH_3$

42 $R_1 = H$, $R_{10} = CH_3$, $R_{17} = COCH_3$

43 $R_1 = R_{10} = CH_3$, $R_{17} = H$

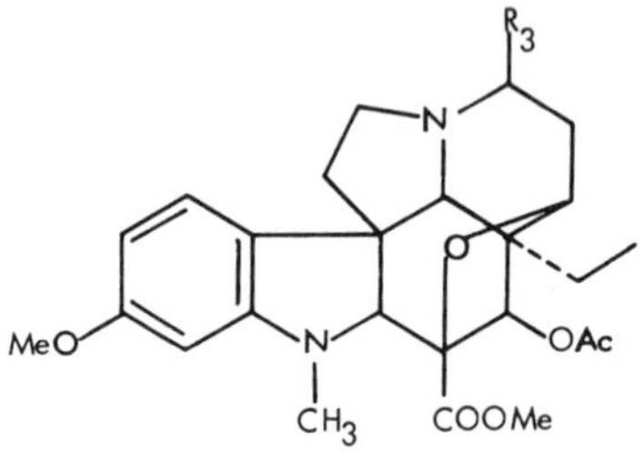

44 $R_3 = H$

45 $R_3 = CH_2-CO-CH_3$

1964); and the production of several unusual ether derivatives such as dihydrovindoline ether (DHVE) (**44**) and 3-acetonyldihydrovindoline ether (**45**) (Neuss *et al.*, 1973).

46

Studies in our laboratory focused on the mechanism of formation of compounds like (**44**), and culminated in the identification of a dihydrovindoline ether "dimer" (**46**), a metabolic product of vindoline formed by *Streptomyces griseus* (Nabih *et al.*, 1978). The structure of the unusual metabolite was determined by a combination of ^{13}C-NMR, mass, and PMR spectroscopic methods. The structural nature of the dimer provided a clue as to how it and dihydrovindoline ether might be formed. It was suggested (Nabih *et al.*, 1978) that a logical sequence would involve the N-oxidation of vindoline, which after metabolic conjugation and elimination would produce the immonium species (**47**). The driving force for cyclization of the ether between positions 16 and 15 was thus provided to give the enamine (**48**). Since this enamine could isomerize to (**49**), both precursors for an enamine dimerization reaction were provided; and it would be possible for (**49**) to undergo reduction by the microorganism with NADH (or NADPH) to DHVE (**44**) or to react with a suitable precursor to give 3-acetonyldihydrovindoline ether (**45**). Very recent work from our laboratory (Gustafson and Rosazza, 1979) has resulted in the isolation and characterization of (**48**) from cultures of *S. griseus* being grown in the presence of vindoline. Further studies concerning the nature of early metabolites in the conversion of vindoline to the dimer, and other microbial products are in progress.

The significance of all of the work with vindoline lies in the fact that this alkaloid is found structurally intact in the active *Catharanthus* alkaloids vincristine and vinblastine. We now know that at least one position on the vindoline portion of these dimeric alkaloids may be metabolically converted into a chemically reactive position. It is very tempting to think that similar bioconversion reactions will be uncovered in mammalian systems as well.

Further interesting studies may be expected with compounds like catharanthine, and some of the dimeric *Catharanthus* alkaloids.

7. Lapachol

This naphthoquinone occurs widely in nature and it has demonstrated antitumor activity. Lapachol (**50**) undergoes transformation by several microorganisms, and the most common microbial metabolite was isolated and characterized as (**51**) (Otten and Rosazza, 1978). This compound, which was produced by *Penicillium notatum* (UI 1602), is probably formed by initial epoxidation followed by hydrolytic ring opening. Interestingly, a chemical analogy exists for this reaction in that (**51**) had been previously characterized as an intermediate in the Hooker oxidation. The involvement of a peroxidase was ruled out in this transformation.

Lapachol is also transformed into dehydro-α-lapachone (**52**) by *Curvularia lunata* (NRRL 2178). The compound was isolated and identified by comparison with synthetic (**52**). This metabolite may form either by radical or ionic mechanisms, and interestingly, (**52**) has a broader spectrum of activity than lapachol itself (Otten and Rosazza, 1979).

8. Withaferin-A

The withanolides display antitumor activity in several test systems, and withaferin-A (**53**) occurs abundantly in the plant *Withania somnifera*. Several microorganisms accumulate a variety of metabolites of withaferin-A, and *Cunninghamella elegans* (NRRL 1393) produced two major metabolites. A polar metabolite was produced in 30% yields in preparative-scale incubations, and its structure was determined as 14-α-hydroxywithaferin-A (**54**) by PMR and mass spectral analysis (Rosazza *et al.*, 1978b). This microbial metabolite was equally as effective against the Sarcoma 180 test system as withaferin-A itself, but a minor metabolite displayed about twice the activity of (**54**). The identity of the minor metabolite is under investigation.

53 R = H
54 R = OH

C. Other Compounds

1. Hycanthone

Miracil D (**55**) was a well-established drug in the treatment of schistosomiasis, and it was determined that the compound exhibited interesting biological activity in a series of test animals. It was assumed that the compound underwent metabolic conversion to a more active form, and it was ultimately discovered that (**55**) was oxidized to Hycanthone (**56**). The addition of Miracil-D to fermentation media of *Aspergillus sclerotiorum* led to its rapid oxidation to Hycanthone and the corresponding aldehyde and carboxylic acid derivatives (Rosi *et al.*, 1967). Hycanthone was three times more effective against *Schistosoma mansoni* infections than Miracil-D, and it is about equally as effective in the treatment of experimental tumors. Hycanthone is being examined further for its potential as an antitumor agent.

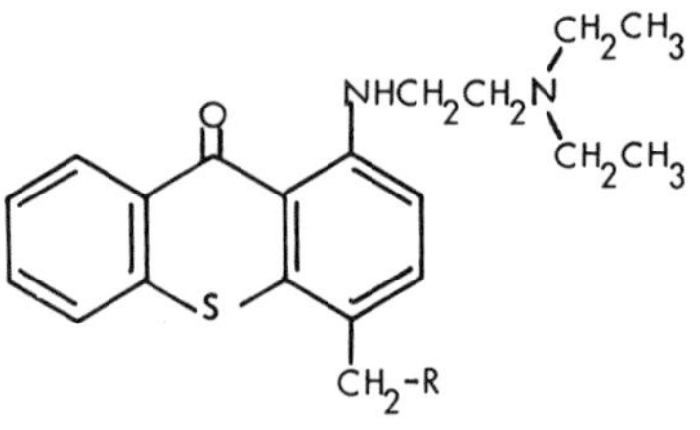

55 R = H

56 R = OH

Studies have also been conducted with the alkaloids related to colchicine (Zeitler and Niemer, 1969), where transformations including deacetylation and oxidative deamination occur. Such compounds present interesting substrates because microorganisms have been reported to selectively *O*-demethylate the alkaloid. Indolizidine alkaloids like monocrotaline and europine N-oxide have been studied with little success (Thede and Rosazza, 1979). The major reaction with these compounds is either their N-oxidation, or the reduction of the N-oxides. These reactions may have their parallel in mammalian systems. 9-Methoxyellipticine has also shown numerous metabolites with selected microorganisms (Chien *et al.*, 1979).

III. SUMMARY AND CONCLUSIONS

It is safe to say that when many new antitumor compounds of natural and synthetic origin become available, routine microbial metabolism studies will demonstrate their utility in the drug development process. It is the author's opinion that considerable effort in the area of microbial transformations remains to be completed, even among some of the compounds already studied. The studies cited in this chapter illustrate the potential for microorganisms to provide difficult-to-synthesize antitumor derivatives, and underlines their capabilities in the determination of novel metabolic pathways. The field of microbial transformations of novel and structurally complex antitumor alkaloids, antibiotics, terpenes, and other classes is relatively new and developing. Technological advances in the isolation and stabilization of microbial enzymes capable of performing the same or similar transformations as those found with whole microbial cells are sure to enhance the significance of this field into the development of new and useful analogues of antitumor drugs.

Acknowledgments

I am grateful for the financial support provided through NIH grant CA-13786, and National Cancer Institute contract NCI-CM-77176. I want also to thank the many colleagues and students who contributed in many ways to the completion of much of the work described in this chapter, including Millie Chien, Wanda Peczynska-Czoch, Patrick J. Davis, Thomas Nabih, Allan W. Nicholas, Geng-Shuen Wu, Ronald Betts, Fran Eckenrode, Mark E. Gustafson, Sharee Otten, Jack Schaumberg, Mark Smellie, Bruce Thede, and Leisa Youel.

REFERENCES

Abbott, B. J. (1976). *Adv. Appl. Microbiol.* **20**, 203–257.

Abbott, B. J. (1977). *Annu. Rep. Ferment. Processes,* **1**, 205–233.

Aszalos, A., Bachur, N. R., Hamilton, B. K., Langlykke, A. F., Roller, P. P., Sheikh, M. Y., Sutphin, M. S., Thomas, M. C., Wareheim, D. A., and Wright, L. H. (1977). *J. Antibiot.* **30**, 50–58.

Auret, B. J., Boyd, D. R., Robinson, P. M.. Watson, C. G., Daly, J. W., and Jerina, D. M. (1971). *J. Chem. Soc. D* 1585–1587.

Bachur, N. R., Steele, M., Meriweather, W. D., and Hildebrand, R. (1976). *J. Med. Chem.* **19**, 651–654.

Berg, A., Gustafsson, J.-A., and Ingelman-Sundberg, M. (1976). *J. Biol. Chem.* **251**, 2831–2838.

Bernath, F. R., Venkatasubramanian, K., and Vieth, W. R. (1977). *Annu. Rep. Ferment. Processes* **1**, 235–266.

Betts, R. E., Walters, D. E., and Rosazza, J. P. (1974). *J. Med. Chem.* **17**, 599–602.

Beukers, R. (1972). *In* "Drug Design" (E. J. Ariens, ed.), Vol. 3, pp. 1–131. Academic Press, New York.

Bhuyan, B. K., Fraser, T. J., and Li, L. H. (1972). *Cancer Res.* **32**, 2538–2544.

Boyd, D. R., Campbell, R. M., Craig, H. C., Watson, C. G., Daly, J. W., and Jerina, D. M. (1976). *J.C.S. Perkin I,* 2438–2443.

Brannon, D. R., and Neuss, N. (1975). Ger. Patent 2,440,931 [*C.A.* **83**, 7184k].

Brannon, D. R., Horton, D. R., and Svoboda, G. H. (1974). *J. Med. Chem.* **17**, 653–654.

Cerniglia, C. E., and Gibson, D. T. (1978). *Arch. Biochem. Biophys.* **156**, 121–127.

Charney, W., and Herzog, H. L. (1967). "Microbial Transformations of Steroids." Academic Press, New York.

Chien, M., Markovetz, A., and Rosazza, J. P. (1978). *Abstr., ASP/PSNA Meet., Stillwater, Okla.* AC-13, p. 50.

Chien, M., Markovetz, A. J., and Rosazza, J. P. (1979). Unpublished observations.

Claridge, C. A., and Schmitz, H. (1978). *Appl. Environ. Microbiol.* **36**, 63–67.

Davis, P. J., and Rosazza, J. P. (1976). *J. Org. Chem.* **41**, 2548–2551.

Davis, P. J., Wiese, D. R., and Rosazza, J. P. (1977). *Lloydia* **40**, 239–246.

Duppel, W., Lebeault, J. M., and Coon, M. J. (1973). *Eur. J. Biochem.* **36**, 583–592.

Ferris, J. P., MacDonald, L. J., Patrie, M. A., and Martin, M. A. (1976). *Arch. Biochem. Biophys.* **175**, 443–452.

Florent, J., and Lunel, J. (1975). Ger. Patent 2,456,139 [*C.A.* **83**, 112355q (1975)].

Fonken, G., and Johnson, R. S. (1972). "Chemical Oxidations with Microorganisms." Dekker, New York.

Gould, S. J., and Chang, C. C. (1978). *J. Am. Chem. Soc.* **100**, 1624–1626.

Gourevitch, A., Pursiano, T. A., and Lein, J. (1961). *Arch. Biochem. Biophys.* **93**, 283–285.

Gustafson, M. E., and Rosazza, J. P. (1979). *J. Chem. Res.* **5**, 166–167.

Hou, C. T., and Perlman, D. (1970). *J. Biol. Chem.* **245**, 1289–1295.

Iizuka, H., and Naito, A. (1967). "Microbial Transformations of Steroids and Alkaloids." Pennsylvania State Univ. Press, University Park.

Jones, D. F., and Mills, S. D. (1971). *J. Med. Chem.* **14**, 305–311.

Jones, D. F., Moore, R. H., and Crowley, G. C. (1970). *J. Chem. Soc.* C 1725–1737.

Jones, E. R. H. (1973). *Pure Appl. Chem.* **33**, 39–52.

Karnetova, J., Mateju, J., Sedanora, P., Vokourn, J., and Vanek, Z. (1976). *J. Antibiot.* **29**, 1199–1202.

Katz, E., and Pienta, P. (1957). *Science* **126**, 402–403.

Kieslich, K. (1969). *Synthesis* 120–134, 147–157.

Kieslich, K. (1976). "Microbial Transformations of Non-Steroid Cyclic Compounds." Wiley (Interscience), New York.

Kishimoto, S., Sugikno, H., Tanaka, K., Kakinuma, A., and Noguchi, S. (1976). *Chem. Pharm. Bull.* **24**, 584–590.

Lalleman, J. Y., Lemaitre, P., Beeley, L., Lesca, P., and Mansuy, D. (1978). *Tetrahedron Lett.*, 1261.

Lancini, G. C., and Hengeller, C. (1969). *J. Antibiot.* **22**, 637–638.

Lancini, G. C., Thiemann, J. E., Sartori, G., and Sensi, P. (1967). *Experientia* **23**, 899–900.

Lancini, G. C., Gallo, G. G., Sartori, G., and Sensi, P. (1969). *J. Antibiot.* **22**, 369–377.

Lanzilotta, R. P., Bradley, D. G., McDonald, K. M., and Tokes, L. (1976). *Appl. Environ. Microbiol.* **32**, 726–728.

Laskin, A. I., and Lechevalier, H., eds. (1974). "CRC Handbook of Microbiology," Vol. 4. CRC Press, Cleveland, Ohio.

Lecointe, P., Lesca, P., Cros, S., and Paoletti, C. (1978). *Chem. Biol. Interact.* **20**, 113.

Lesca, P., Lecointe, P., Paoletti, C., and Mansuy, D. (1976). *C. R. Acad. Sci., Ser. D* **282**, 1457.

Lesca, P., Lecointe, P., Paoletti, C., and Mansuy, D. (1977). *Biochem. Pharmacol.* **26**, 2169.

Li, L. H., and Cowie, C. H. (1974). *Biochim. Biophys. Acta* **353**, 375–384.

Mallett, G. E., Fukuda, D. S., and Gorman, M. (1964). *Lloydia* **27**, 334–339.

Marshall, V. P., Reisender, E. A., Reineke, L. M., Johnson, J. H., and Wiley, P. F. (1976). *Biochemistry* **15**, 4139–4145.

Marshall, V. P., McGovren, J. P., Richard, F. A., Richard, R. E., and Wiley, P. F. (1978). *J. Antibiot.* **31**, 336–342.

Marsheck, W. J. (1971). *Prog. Ind. Microbiol.* **10**, 49–103.

Marsheck, W. J., and Karim, A. (1973). *Appl. Microbial.* **25**, 647–649.

Marsheck, W. J., and Miyano, M. (1973). *Biochim. Biophys. Acta* **316**, 363–365.

Marsheck, W. J., Kraychy, S., and Muir, R. D. (1972). *Appl. Microbiol.* **23**, 72–77.

Martin, S. M., and Skerman, V. B. D. (1972). "World Directory of Collections of Cultures of Microorganisms." Wiley, New York.

Nabih, T., Davis, P. J., Caputo, J. F., and Rosazza, J. P. (1977). *J. Med. Chem.* **20**, 914–917.

Nabih, T., Youel, L., and Rosazza, J. P. (1978). *J.C.S. Perkin I*, 757–762.

Neuss, N., Fukuda, D. S., Mallett, G. E., Brannon, D. R., and Huckstep, L. L. (1973). *Helv. Chim. Acta* **56**, 2418–2426.

Neuss, N., Mallett, G. E., Brannon, D. R., Mabe, J. A., Horton, H. R., and Huckstep, L. L. (1974a). *Helv. Chim. Acta* **57**, 1887–1891.

Neuss, N., Fukuda, D. S., Brannon, D. R., and Huckstep, L. L. (1974b). *Helv. Chim. Acta* **57**, 1891–1893.

Oki, T., Komiyama, T., Tone, H., Inui, T., Takeuchi, T., and Umezawa, H. (1977). *J. Antibiot.* **30**, 613–615.

Otten, S., and Rosazza, J. P. (1978). *Appl. Environ. Microbiol.* **35**, 554–557.

Otten, S., and Rosazza, J. P. (1979). *Appl. Environ. Microbiol.* **38**, 311–313.

Ozaki, M., Kariya, T., Kato, H., and Kimura, T. (1972). *Agric. Biol. Chem.* **36**, 451–456.

Perlman, D. (1976). *In* "Application of Biochemical Systems to Organic Synthesis" (B. Jones, D. Perlman, and C. J. Sih, eds.), pp. 47–68. Wiley, New York.

Perlman, D., ed. (1977). "Structure-Activity Relationships among the Semisynthetic Antibiotics." pp. 427–529. Academic Press, New York.

Perlman, D., Mauger, A. B., and Weissbach, H. (1966). *Biochem. Biophys. Res. Commun.* **24,** 513–518.

Peterson, D. H., and Murray, H. C. (1952). *J. Am. Chem. Soc.* **74,** 1871–1873.

Rosazza, J. P. (1978). *Lloydia* **41,** 397–411.

Rosazza, J. P., Kammer, M., Youel, L., Smith, R. V., Erhardt, P. W., Truong, D. H., and Leslie, S. W. (1977). *Xenobiotica* **7,** 133–143.

Rosazza, J. P., Wu, G. S., Gard, T., and Markovetz, A. J. (1978a). Unpublished observations.

Rosazza, J. P., Nicholas, A. W., and Gustafson, M. E. (1978b). *Steroids* **31,** 671–679.

Rosazza, J. P., Peczynska-Czoch, W., and Sutton, M. (1978c). Unpublished observations.

Rosazza, J. P., Ervin, J. E., Wu, G. S., and Chien, M. M. (1978d). Unpublished observations.

Rosi, D., Peruzzotti, G., Demms, E. W., Berberian, D. A., Freele, H., Tullar, B. F., and Archer, S. (1967). *J. Med. Pharm. Chem.* **10,** 867.

Rosi, D., Neuman, H. C., Christiansen, R. G., Schane, H. P., and Potts, G. O. (1977). *J. Med. Chem.* **20,** 349–352.

Schmitz, H., and Claridge, C. A. (1978). *J. Antibiot.* **30,** 635–38.

Sebek, O. K. (1974). *Lloydia* **37,** 115–133.

Sebek, O. K. (1975). *Acta Microbiol. Acad. Sci. Hung.* **22,** 381–388.

Sebek, O. K. (1977). *In* "Biotechnological Application of Protein and Enzymes" (Z. Bolak and N. Sharon, eds.), pp. 203–219. Academic Press, New York.

Sebek, O. K., Lincoln, F. H., and Schneider, W. P. (1976). *Proc. Int. Ferment. Congr. Symp., 5th, Berlin* Abstr. 17.05.

Sehgal, S. N., and Vezina, C. (1967). *Can. Microbiol. Annu. Meet., 17th, Hamilton, Ont.* Pap. No. 37.

Sih, C. J., and Rosazza, J. P. (1976). *In* "Applications of Biochemical Systems to Organic Synthesis" (J. B. Jones, D. Perlman, and C. J. Sih, eds.), pp. 69–106. Wiley, New York.

Sih, C. J., Abushanab, E., and Jones, J. B. (1977). *Annu. Rep. Med. Chem.* **12,** 298–308.

Smellie, M., Corder, M., and Rosazza, J. P. (1978). *J. Chromatogr.* **155,** 439–442.

Smith, R. V., and Rosazza, J. P. (1975a). *Biotechnol. Bioeng.* **17,** 785–814.

Smith, R. V., and Rosazza, J. P. (1975b). *J. Pharm. Sci.* **64,** 1737–1759.

Smith, R. V., Acosta, D., and Rosazza, J. P. (1977). *Adv. Biochem. Eng.* **5,** 70–96.

Sullivan, H. R., Billings, R. E., Occolowitz, J. L., Boaz, H. E., Marshall, F. J., and McMahon, R. E. (1970). *J. Med. Chem.* **13,** 904–909.

Suzuki, S., and Mori, T. (1976). *J. Antibiot.* **29,** 286–291.

Takita, T., Fujii, A., Fukuoka, T., and Umezawa, H. (1973). *J. Antibiot.* **26,** 252–256.

Tamm, C. (1962). *Angew. Chem., Int. Ed. Engl.* **1,** 178–195.

Tamm, C. (1974). *FEBS Lett.* **48,** 7–21.

Thede, B., and Rosazza, J. P. (1979). Unpublished observations.

Umezawa, H., Sawa, T., Fukagawa, Y., Koyama, G., Murase, M., Hamada, M., and Takeuchi, T. (1965). *J. Antibiot.* **18,** 178–181.

Umezawa, H., Takahashi, Y., Fujii, A., Saino, T., Shirai, T., and Takita, T. (1973). *J. Antibiot.* **26,** 117–119.

Wallen, L. L., Stodola, F. H., and Jackson, R. W. (1959). "Type Reactions in Fermentation Chemistry," ARS Bull. ARS-71-13. Peoria, Illinois.

Wiley, P. F., and Marshall, V. P. (1975). *J. Antibiot.* **28,** 838–840.

Wiley, P. F., Koert, J. M., Elrod, D. W., Reisender, E. A., and Marshall, V. P. (1977). *J. Antibiot.* **30,** 649–651.

Wu, G.-S., Nabih, T., Peczynska-Czoch, W., Youel, L., and Rosazza, J. P. (1978). *Antimicrob. Agents Chemother.* **14,** 601–604.

Zeitler, H. J., and Neimer, H. (1969). *Hoppe-Seyler's Z. Physiol. Chem.* **350,** 366–372.

CHAPTER 14

Miscellaneous Natural Products with Antitumor Activity

MATTHEW SUFFNESS AND JOHN DOUROS

I. INTRODUCTION

The purpose of this chapter is to draw the attention of the reader to a variety of compounds which have shown antitumor activity but which are not covered in other chapters of this book, either due to limitations of space or because they are structurally unique and do not fall into larger classes of compounds which are the subjects of the other chapters.

Anticancer Agents Based on Natural Product Models
Copyright © 1980 by Academic Press, Inc.
All rights of reproduction in any form reserved.
ISBN 0-12-163150-8

Some of the compounds discussed are quite recently discovered, while others have been known for many years, but it is desirable to present all of them and give a brief evaluation of both positive and negative aspects as a guide to future research.

II. ANTITUMOR AGENTS FROM HIGHER PLANTS

A. Ellipticine

Ellipticine (Fig. 1) is a compound which possesses marked antitumor activity in a variety of mouse tumor systems and has excellent potential for analog development.

Ellipticine was originally isolated from *Ochrosia elliptica* (Goodwin *et al.*, 1959) and has subsequently been isolated from several other species of *Ochrosia* and *Bleekeria vitiensis* (synonym *Excavatia vitiensis*). There have been many syntheses of ellipticine and derivatives performed, as a result of interest stimulated by the antitumor activity reported. A variety of approaches have been used, as can be seen in the papers by Woodward *et al.* (1959), Cranwell and Saxton (1962), Govindachari *et al.* (1963), Loder (1966), Dalton *et al.* (1967), Le Goffic *et al.* (1972, 1973), Guthrie *et al.* (1975), Sainsbury and Webb (1974), Sainsbury *et al.* (1975), Sainsbury and Schinazi (1976), Besselievre *et al.* (1975), Langlois *et al.* (1975), Kilminster and Sainsbury (1972), Rousselle *et al.* (1977), Kozikowski and Hasan (1977), Jackson *et al.* (1977), and Bergman and Carlsson (1977).

Ellipticine derivatives which retain antitumor activity include 9-methoxy-ellipticine (Fig. 1) (Le Men *et al.*, 1970), 9-hydroxyellipticine (Le Pecq *et al.*, 1973), 9-aminoellipticine (Sainsbury *et al.*, 1975), 6-methyl-ellipticine (Dalton *et al.*, 1967) and, interestingly, several members of a series of 9-substituted 2-alkylellipticinium derivatives (Le Pecq *et al.*, 1975; Paoletti *et al.*, 1978). Ellipticine and derivatives exert their antitumor effect at a molecular level through intercalation of DNA, as has been demonstrated in studies by Garcia-Giralt and Macieira-Coelho (1970), Festy *et al.* (1971), Saucier *et al.* (1971), Snyder *et al.* (1971), Kann and Kohn (1972), Bhuyan *et al.* (1972), Li (1973), Kohn *et al.* (1975), and Pelaprat *et al.* (1976).

The antitumor activity of ellipticine demonstrated in murine tumor systems at the Natural Cancer Institute is quite remarkable (Table I). The broad spectrum of activity shown in both leukemias and solid tumors is unusual, and the results in the two colon tumors studied are of high interest since human colon tumors present a difficult clinical problem. The correlation of the Colon 26 and Colon 38 murine tumors with clinical effectiveness is not yet established, and will have to be proven in clinical trials.

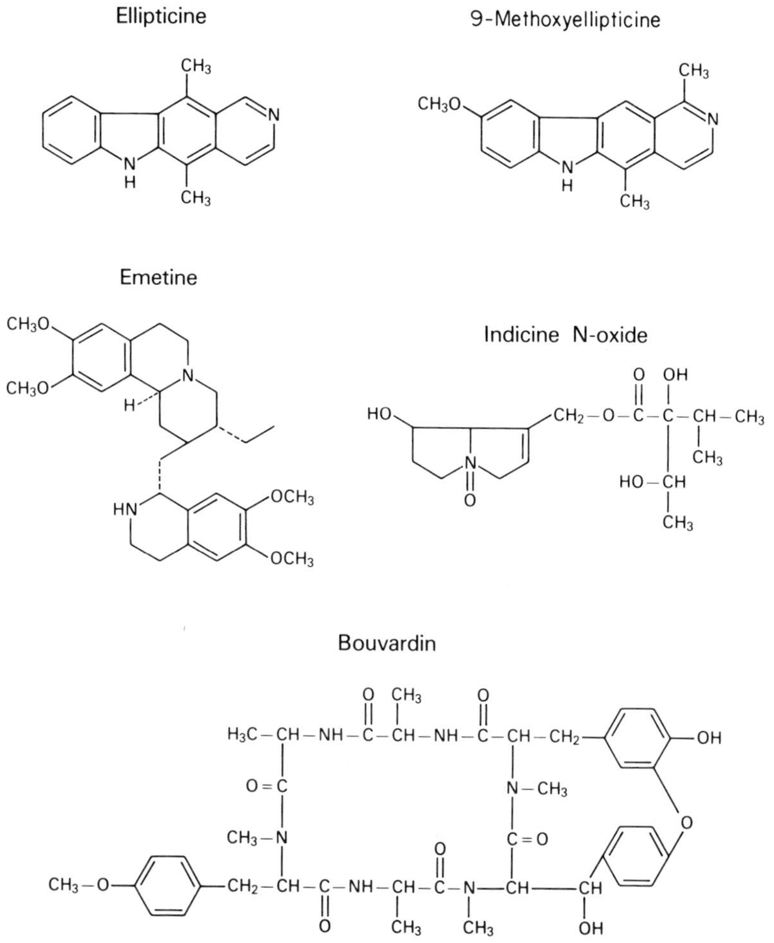

Ellipticine

9-Methoxyellipticine

Emetine

Indicine N-oxide

Bouvardin

Figure 1. Antitumor agents from higher plants.

Neither ellipticine nor any of its derivatives have been in clinical trial in the United States, primarily due to toxicity problems. Herman and co-workers (1971) found alterations of the cardiovascular system in dogs, and subsequently (Herman *et al.*, 1974a) found that ellipticine produced hemolysis, hypotension, and bradycardia when injected into monkeys. Further studies by Herman *et al.* (1974b) showed that other ellipticine derivatives, including 9-methoxyellipticine, have similar cardiovascular effects. The cardiovascular effects of the ellipticines appear to be mainly elicited after parenteral administration, and since ellipticine is active orally in murine tumor systems, investigation of the oral route as a means of administration of ellipticine has been undertaken by NCI. Cur-

TABLE I

Anticumor Activity of Ellipticine in Mice

Tumor system	Parameter	Activity criteria	Treatment schedule[a]	Best T/C[b]	Optimal dose (mg/kg/inj.)	Reproduced T/C
B-16 melanocarcinoma	Survival time	T/C >125	Days 1–9	147	40	136
Colon 26	Survival time	T/C >140	Days 1,5,9	263	25	209
Colon 38	Tumor inhibition	T/C <42	Days 2,9,16	0	200	0
CD8F₁ mammary	Tumor inhibition	T/C < 42	Days 1,8,15,22,29	8	100	9
L-1210 leukemia	Survival time	T/C >125	Day 1	357	128	331
P-388 leukemia	Survival time	T/C >120	Days 1–9	204	25	195

[a] The day of tumor implantation is taken as day zero.
[b] T/C is the ratio of the parameter studied in test to control animals, expressed as a percentage.

rently, radiolabeled ellipticine has been prepared and experiments are underway to determine the parameters of absorption, tissue distribution, and pharmacokinetics of oral ellipticine. If these studies demonstrate that the absorption and distribution of orally administered compound are consistently predictable, it is likely that a clinical trial will be undertaken.

Clinical trials of ellipticine derivatives have been undertaken in Europe starting in 1970. Mathé and co-workers (1970) and Ansari and Thompson (1975) reported that 9-methoxyellipticine lactate was useful in treatment of acute myeloid leukemia (AML), and, more recently, 9-hydroxy-2-methyl ellipticinium has been studied by Juret *et al.* (1978), who reported beneficial effects in 8 of 18 patients treated.

In summary, the high degree and broad spectrum of activity of ellipticine in murine tumor systems, combined with the encouraging results of the limited clinical trials undertaken with ellipticine derivatives, make this compound an extremely attractive target for further analogue studies.

B. Emetine

Emetine (Fig. 1) is a well-known alkaloid isolated from a number of plants in the family Rubiaceae and which has been known for many years as one of the active ingredients in Syrup of Ipecac, an excellent emetic. Emetine salts are also used clinically in treatment of amebiasis.

Emetine (NSC-33669) showed good activity against the L-1210 and P-388 leukemia systems in mice but is inactive against solid tumors in murine systems, including a variety of melanoma, carcinoma, and sarcoma systems tested at the National Cancer Institute.

Based on its activity in murine tumor systems and two early reports of clinical use of emetine in malignancy (Lewisohn, 1918; Van Hoosen, 1919), phase I trials were done by Mastrangelo *et al.* (1973) and Panettiere and Coltman (1971). Panettiere and Coltman observed two partial remissions in patients with lung cancer in a phase I study, and this led to phase II clinical trials by Moertel *et al.* (1974) in advanced gastrointestinal cancer, Siddiqui *et al.* (1973) in a variety of solid tumors, and Kane *et al.* (1975) in bronchiogenic carcinoma. None of these trials showed useful results.

Surprisingly, no clinical trials have ever been undertaken with emetine in leukemia and lymphoma. It would seem that, based on the good effectiveness against murine leukemias, these tumors would have been a logical place to start clinical trials. NCI has recently indicated a willingness to try emetine in leukemia and lymphoma, but no investigators have come forward.

There have also been reports of the use of dihydroemetine in treatment of cancer (Abd-Rabbo, 1960, 1966; Wyburn-Mason, 1966), which have not been followed up with full studies. Dihydroemetine is claimed to be less toxic than emetine and may present clear advantages. Other related compounds such as tubulosine and cephaline have similar activity to emetine in murine tumor systems used at the NCI.

There is not a great deal of current interest in developing either emetine or derivatives, since there are currently a variety of extremely effective drugs for treatment of leukemia and lymphoma. An emetine analogue would have to show significant activity against solid tumors and hence a broader spectrum of activity to be of interest for further development.

Since the mechanism of action of emetine is well established as being inhibition of protein synthesis, it follows that this effect will be greater on very rapidly growing tumor cells such as are found in leukemia and relatively smaller against the slower growing solid tumor cells. This is a typical problem in analogue development, where, if one takes the mechanism of action into account, it becomes clearer how to proceed. The approach to emetine analogues at this point should be to assume that there is little possibility of finding analogues with solid tumor activity unless the present molecule is incorporated into a molecule which has a carrier portion which will direct distribution to specific tissues.

C. Indicine N-Oxide

Indicine N-oxide (Fig. 1) is a pyrrolizidine alkaloid, isolated from *Heliotropium indicum,* which is interesting in terms of analogue development already done. The pyrrolizidine alkaloids are well known as causative agents of cirrhosis of the liver and hepatic tumors and cause problems worldwide as livestock poisons and human toxins when ingested (McLean, 1970).

A considerable number of pyrrolizidine alkaloids have shown antitumor activity in NCI mouse screens, as can be seen in the papers of Culvenor (1968), Kupchan and Suffness (1967) and Kupchan *et al.* (1964). None of these pyrrolizidines had good therapeutic ratios, probably as a result in part of hepatotoxicity, and no further progress was made for several years until the isolation of indicine N-oxide from *H. indicum* and the demonstration of its efficacy and low toxicity (Kugelman *et al.,* 1976).

Indicine N-oxide was studied in three phase I studies at the Mayo Clinic, University of Kansas, and Mount Sinai School of Medicine, and was found to be tolerated at the rather high dose of 4.0 gm/m^2 on a weekly administration. As of February, 1979, the compound has been cleared for phase II clinical trials.

Development of further analogues should center about the reasons for the lack of hepatotoxicity seen with indicine N-oxide and also clinical pharmacology data concerning the mechanism of action of this compound. At the moment it is not clear that indicine N-oxide itself is the active material *in vivo,* since the doses are extremely high for an antitumor agent (optimal dose is 600–800 mg/kg/day in tumored mice), and the antitumor activity could be due to a metabolite. Clinical pharmacology studies now ongoing should resolve some of these questions and give a better information base from which to approach analogue synthesis.

D. Bouvardin

Bouvardin (Fig. 1) is an unusual cyclic peptide which has been isolated from the Mexican plant *Bouvardia ternifolia* (family Rubiaceae) by Cole and collaborators at the University of Arizona (Jolad *et al.*, 1977).

Bouvardin was selected for further development at the National Cancer Institute in 1977, based on its unusual structure and its activity against the B-16 melanosarcoma and the P-388 lymphocytic leukemia in mice. Further testing in tumored mice showed negative results in multiple tests in other mouse tumor systems, and in 1979 it was decided to discontinue further development of bouvardin because of its narrow spectrum of activity.

Bouvardin acts through a mechanism of inhibition of protein synthesis (Johnson and Chitnis, 1978), and its detailed effects on the cell cycle were examined by Tobey *et al.* (1978).

Although bouvardin does not have a broad spectrum of activity, it is a unique entity chemically which merits attention in terms of analogue synthesis, especially since no work on analogues has been reported to date.

E. Thalicarpine and Tetrandrine

Thalicarpine and tetrandrine (Fig. 2) are both dimeric benzylisoquinoline-derived alkaloids isolated from plants. Tetrandrine was isolated from *Cyclea peltata* by Kupchan and co-workers (1961), while thalicarpine was isolated and characterized by Kupchan *et al.* (1963), Kupchan and Yokoyama (1963), and Tomita *et al.* (1965). A total synthesis of thalicarpine has been described by Kupchan and Liepa (1971).

Both tetrandrine and thalicarpine were selected for development at the National Cancer Institute based on their reproducible activity against the Walker 256 intramuscular carcinosarcoma. Neither of the compounds was active in any of a wide variety of other *in vivo* murine systems in NCI screening. Kupchan and Altland (1973) have examined some parameters of structure–activity relationships in these types of compounds.

Investigational New Drug applications were filed for thalicarpine in 1972 and tetrandrine in 1973, and phase I trials were completed on both compounds. Data from phase I trials with thalicarpine have been published by De Conti *et al.* (1975), who observed no antitumor effects in 32 patients.

The Walker 256 tumor system is now considered to be too sensitive to a variety of compounds which are not active in other tumor systems to make activity in this system a meaningful criterion for drug development. The Walker

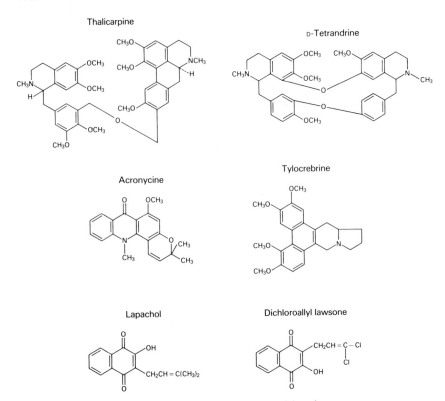

Figure 2. Antitumor agents from higher plants.

256 carcinosarcoma system does pick up activity of compounds known to be clinically useful, but also gives a high proportion of false leads and therefore has been dropped from the NCI panel of routine tumor screens. Development of tetrandrine or thalicarpine analogues therefore appears to be a dead end. Both of these compounds were dropped in 1977 due to lack of further clinical interest.

F. Acronycine

Acronycine (Fig. 2) is an acridone alkaloid isolated from the Australian plant *Acronychia baueri,* family Rutaceae (Lahey and Thomas, 1949). Early structural studies by Drummond and Lahey (1949) were subsequently amplified by MacDonald and Robertson (1966), Govindachari *et al.* (1966), and Diment *et al.* (1969), who established the current structure.

Svoboda *et al.* (1966) reported on the antitumor activity of acronycine and found significant tumor inhibitory activity in 12 of the 17 tumor systems tested, including leukemias, sarcomas, carcinomas, and melanoma lines in mice. Syntheses of acronycine were reported by Beck *et al.* (1968) and Hlubucek *et al.* (1970). Sullivan and co-workers (1970) studied the mammalian metabolism of acronycine and found that the predominant metabolites were produced by hydroxylation of the ring and/or sidechain. Studies of microbial transformation of acronycine by Betts *et al.* (1974) and Brannon *et al.* (1974) showed that the same major metabolite, 9-hydroxyacrononycine, that was produced in mammals was also made by microorganisms. Schneider *et al.* (1972) reported on the synthesis of sixteen analogues of acronycine and found that none of these was superior to the parent compound. Liska (1972) synthesized analogues in four series related to partial structures of acronycine, and Svoboda (1966) established some preliminary structure–activity relationships.

Eli Lilly and Co. undertook development of acronycine towards clinical trials, and an Investigational New Drug (IND) application was approved by the Food and Drug Administration. Working in cooperation with Lilly, the NCI received IND approval in 1970. Phase I clinical trials were performed by Lilly and NCI. No significant antitumor activity has been reported from clinical studies.

Acronycine showed no activity in most of the murine tumor systems studied at NCI, including the B-16 melanoma, L-1210 leukemia, P-388 leukemia, Lewis lung carcinoma, and Walker 256 carcinosarcoma, but was active in the LP-1 plasma-cell tumor and the Ridgway osteogenic sarcoma. This is in obvious incongruity with the high degree of activity reported by Svoboda *et al.* (1966) and points out one of the major difficulties in development of antitumor agents, the question of which tumors to use for screening. The tumor systems used currently at NCI for primary and early secondary screening have been verified in the sense that these tumors are known to be sensitive to nearly all of the known clinically effective anticancer drugs and, based on the inactivity of acronycine in these tumor systems, it is likely that if acronycine were a new compound being screened for the first time now, it would not be selected for further development.

G. Tylocrebrine

This phenanthroindolizidine alkaloid (Fig. 2) was isolated by Gellert and co-workers (1962) from the plant *Tylophora crebrifolia*, family Asclepiadaceae. Gellert and Rudzats (1964) reported on the antileukemic action of tylocrebrine. The compound was sufficiently active for further development, and animal toxicology was completed in 1965 and an IND application was filled with FDA the same year. Phase I clinical trials were initiated in 1965 but were halted in 1966, due to central nervous system (CNS) toxicity of the compound, manifested as ataxia and disorientation.

The activity of tylocrebrine in murine tumor systems appears to be confined to the leukemias. Increases in life span of about 50% were observed in the P-388 and L-1210 leukemic systems. On a comparative basis with other agents being developed, the activity is unremarkable and no further studies are planned.

Modifications of tylocrebrine which would increase the polarity of the molecule and hence decrease CNS penetration would be of interest if antitumor activity were found to be retained.

H. Lapachol

Lapachol (Fig. 2) is a simple quinone which has a long history, in that plants known to contain this compound have been used in folk medicine for treatment of malaria and cancer, particularly in South America. Lapachol came into the NCI program as a result of its isolation from the Walker 256 active extract of the Indian plant *Stereospermum suaveolens* (family Bignoniaceae) by Rao *et al.* (1968).

Lapachol was entered in phase I clinical trials on the basis of its activity in the Walker 256 tumor system in 1967, and doses up to 4000 mg/day were administered orally. Some toxicity was observed, primarily nausea, vomiting, and antivitamin K activity, but no therapeutic responses were seen. It was determined that satisfactory blood levels could not be obtained by oral administration, and, due to the insolubility of the compound, parenteral administration of suitable doses was impossible. The IND for the drug was closed in 1970.

A closely related compound, dichlorolapachol (dichloro-allyllawsone), which had better activity in the Walker 256 system and which could be formulated as a sodium salt, was selected as a replacement for lapachol and an IND was approved in 1975. Like lapachol, this compound acts as an inhibitor of oxidative phosphorylation. In toxicology studies prior to clinical trials, some cardiac toxicity was observed, and consequently a very cautious dose escalation was proposed for the phase I trials which are currently ongoing.

The very narrow antitumor spectrum seen in mouse and rat screening systems is not encouraging, and unless the current phase I trial and possibly a subsequent phase II trial show antitumor activity in man, there is little point to analogue development.

I. Nitidine Chloride

Nitidine chloride (Fig. 3) is a benzophenanthridine alkaloid isolated from *Fagara macrophylla* and other species of *Fagara* (family Rutaceae). The antitumor activity of this compound in murine tumor systems studied at the NCI has been limited to the P-388 leukemia, where increases in lifespan (ILS) of

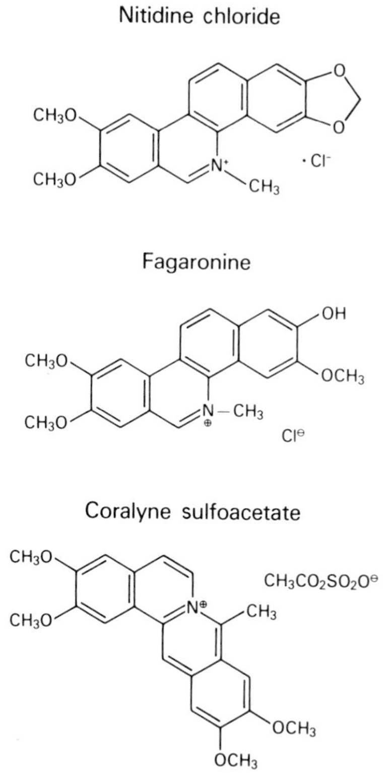

Nitidine chloride

Fagaronine

Coralyne sulfoacetate

Figure 3. Benzophenanthridine alkaloids.

100% were observed, and the L-1210 leukemia, which showed an ILS of 40–50%. Tests in solid tumor systems, including the B-16 melanoma, Colon 38, Lewis lung carcinoma, and Walker 256 carcinosarcoma, were all negative for antitumor activity.

Nitidine chloride was selected for development at the NCI based on its antileukemic activity in 1971, but the compound was subsequently dropped in 1976 after an extensive study of dose response relationships. The compound, being a quaternary amine, is quite soluble in water, but precipitation in biological fluids resulted in erratic dose responses.

Several related alkaloids and derivatives, including fagaronine (Fig. 3), coralyne chloride, coralyne sulfoacetate (Fig. 3), and dihydromethoxynitidine, have shown similar biological activity. Due to the narrow spectrum of activity shown in animal tumor systems, this group of alkaloids does not appear to be very promising for further development.

III. ANTITUMOR AGENTS FROM MICROBIAL SOURCES

A. AT 125 (Acivicin) (α-amino-3-chloro-4,5-dihydro-5-isoxazoloacetic acid)

This isoxazole (Fig. 4) antineoplastic agent was isolated from a *Streptomyces svicus* fermentation by the Upjohn Company (Hanka and Dietz, 1973). The compound was discovered through an antimetabolite screen, as a result of its activity as a glutamine and histidine antagonist (Neil *et al.,* 1975). Cooney and co-workers (1974) showed that AT 125 also inhibits L-asparagine synthetase.

AT 125 showed useful activity against the P-388 and L-1210 murine leukemia systems and against human mammary and colon tumors transplanted as xenografts in athymic mice. Schedule dependency formulation and toxicology studies have been completed at the NCI; a phase I clinical trial is expected to begin in 1979.

To date, only one close analogue of this compound, hydroxy AT 125 (Fig. 4), has been studied, and it shows similar activity. The large-scale fermentation of AT 125 has been very difficult, and a synthetic route to this compound has been developed which should make more analogues available for study in the near future.

B. L-Alanosine, Sodium Salt

L-Alanosine (Fig. 4) is an *N*-nitroso-*N*-hydroxy propionic acid derivative which was discovered at the Lepetit Company (Coronelli *et al.,* 1966). Thus

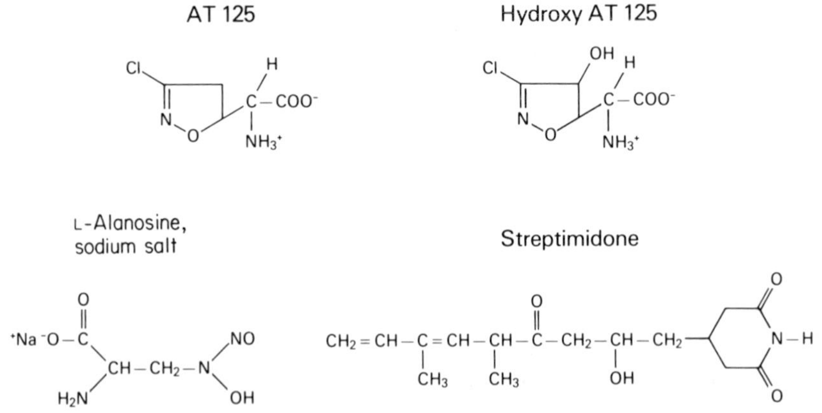

Figure 4.   Antitumor agents from microbial sources.

far, the activity in murine systems appears to be confined to the leukemias, although marginal activity in mammary tumor systems has also been observed.

L-Alanosine is reported to act as an antagonist to L-aspartic and L-glutamic acids (Anandaraj *et al.*, 1977), and other aspects of the mode of action have been studied by Gale and Schmidt (1968).

Toxicology studies in dogs and monkeys showed the dose-limiting toxicities to be hemorrhagic diarrhea and depression of reticulocyte levels, but toxicities were dose-related and manageable. Clinical studies began in 1978 when the IND application was approved, and phase I trials are ongoing at the Mayo Clinic, University of Kansas, M.D. Anderson Tumor Institute, and Mount Sinai Hospital.

The synthesis of alanosine has been reported by Lancini *et al.* (1966), and this group has also reported the synthesis of a series of homologues (Lancini *et al.*, 1969).

C. Neocarzinostatin (Zinostatin)

This protein, produced by *Streptomyces carzinostaticus,* was first described by Ishida *et al.* (1965). Zinostatin (Fig. 5) is a protein of molecular weight 10,717 and consists of 109 amino acids, of which the N-terminal amino acid is alanine and the C-terminal amino acid is asparagine (Maeda *et al.*, 1974). Neocarzinostatin has been shown to act through inhibition of DNA synthesis (Ono *et al.*, 1966) and through DNA degradation (Ono *et al.*, 1968; Ohtsuki and Ishida, 1975). Tatsumi and Nishioka (1977) observed that the inhibitory effect of neocarzinostatin was greater in a DNA-repair-deficient strain of *E. coli* compared to a repair-proficient strain, further indicating that the mode of action of neocarzinostatin is through interaction with DNA.

Clinical trials with neocarzinostatin were begun in Japan in 1965, and a useful review of the data from these trials has been compiled by Legha *et al.* (1976). Phase I and phase II trials have been carried out in the U.S. recently, and data have been reported by Griffin *et al.* (1978), who observed preliminary activity in hepatoma; Rivera *et al.* (1978), who noted activity in pediatric leukemias; and Van Echo and Wiernik (1978), who completed a phase I trial in acute leukemia. Sakamoto *et al.* (1978) observed preliminary activity in bladder cancer in combination with surgery. Neocarzinostatin has been inactive in tumors of the lung, colon tumors and gastric tumors. Further clinical studies are in progress.

D. Macromomycin

Macromomycin is a protein of unknown structure, with a molecular weight of approximately 15,000, which is produced by *Streptomyces macromomyceticus* (Chimura *et al.*, 1968). The original samples of this compound contained a chromophore which was later removed and which increased the

Ala – Ala – Pro – Thr – Ala – Thr – Val – Thr – Pro – Ser – Ser – Gly – Leu – Ser – Asp – Gly – Thr – Val – Val – Lys – Val – Ala – Gly – Ala – Gly –
 5   10   15   20   25

Leu – Gln – Ala – Gly – Thr – Ala – Tyr – Asp – Val – Gly – Gln – Cys – Ala – Ser – Val – Asn – Thr – Gly – Val – Leu – Trp – Asn – Ser – Val – Thr –
 30   35   40   45   50

Ala – Ala – Gly – Ser – Ala – Cys – Asx – Pro – Ala – Asn – Phe – Ser – Leu – Thr – Val – Arg – Arg – Ser – Phe – Glu – Gly – Phe – Leu – Phe – Asp –
 55   60   65   70   75

Gly – Thr – Arg – Trp – Gly – Thr – Val – Asx – Cys – Thr – Thr – Ala – Ala – Cys – Gln – Val – Gly – Leu – Ser – Asp – Ala – Ala – Gly – Asp – Gly –
 80   85   90   95   100

Glu – Pro – Gly – Val – Ala – Ile – Ser – Phe – Asn
 105

Figure 5. Structure of neocarzinostatin.

toxicity of the compound (Yamashita *et al.*, 1976). The compound containing the chromophore is now called auromomycin. Macromomycin is a fairly stable protein, but it is formulated with maltose as a diluent which aids in extending shelf life. At present, amino acid sequencing of this protein is being conducted at the Sidney Farber Cancer Institute.

It is believed that macromomycin exerts its antitumor effects through interaction with the cell membrane of neoplastic cells (Kunimoto *et al.*, 1971, 1972; Roth, 1974) and through DNA strand scission (Beerman, 1978). In murine tumor systems, macromomycin has shown activity against the P-388 leukemia and the B-16 melanoma.

E. Largomycin F-II

The largomycin complex was originally isolated by Yamaguchi and co-workers (1970a) from a strain of *Streptomyces pluricolorescens,* and the active component, largomycin F-II, was characterized as an acidic chromoprotein of molecular weight 25,000 (Yamaguchi *et al.*, 1970b). The antitumor activity of largomycin F-II was also reported by Yamaguchi *et al.* (1970c), who found significant increases in survival in the Ehrlich ascites, Sarcoma-180, and SN-36 leukemia tumor system in mice.

In subsequent testing at the National Cancer Institute, largomycin F-II showed significant reproducible antitumor activity in the B-16 melanoma, Colon 26, Colon 38, P-388 leukemia, and CDF_1 mammary tumor systems carried in mice. Scale-up processing is underway, and development of this agent towards clinical trials is planned.

F. Streptimidone

Streptimidone (Fig. 4) is a glutarimide antibiotic isolated from a strain of *Streptomyces rimosus* by Frohardt *et al.* (1959). It was found to be active against a broad spectrum of fungi, as well as having activity against *Entamoeba histolytica* and *Trichomonas vaginalis* (Kohberger *et al.*, 1960). The correct structure was established through the work of van Tamelen and Haarstad (1960) and Woo *et al.* (1961). Antitumor testing conducted at the NCI in 1975 showed streptimidone to have moderate activity in the P-388 leukemia system. However, further testing has shown negative results in six other murine tumor systems.

Saito *et al.* (1974) have shown that 9-methylstreptimidone has antiviral properties. Analogues of streptimidone with a broader spectrum of activity would be of interest.

G. Echinomycin

Echinomycin (Quinomycin A) is an unusual cyclic peptide (Fig. 6) which is believed to act as an antitumor agent through a bifunctional intercalation mechanism (Lee and Waring, 1978). The revised structure of echinomycin was published by Martin *et al.* (1975). The compound is very toxic, and optimal antitumor doses are in the range of 30–120 μg/kg in murine systems on a daily injection schedule. Thus far, reproducible activity at moderate levels has been seen in the P-388 leukemia and B-16 melanoma systems in mice, while mam-

Echinomycin

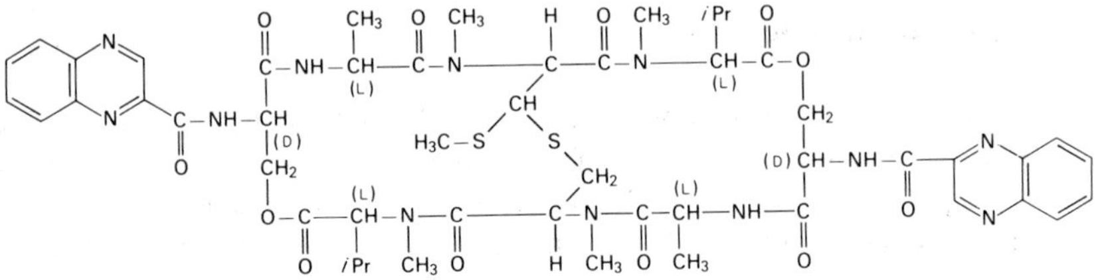

Valinomycin

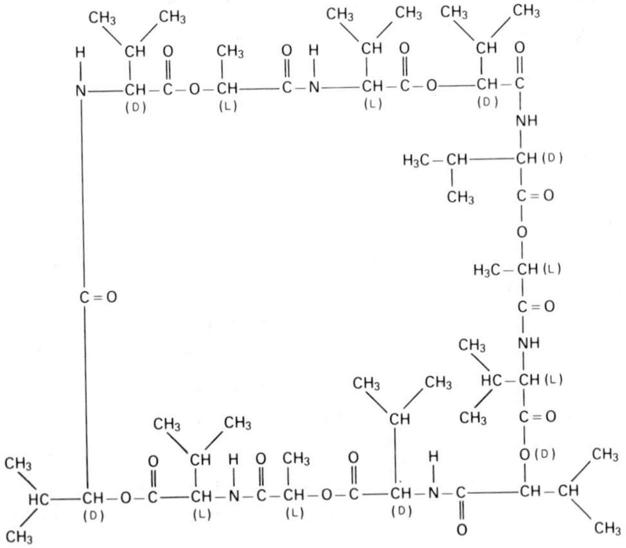

Figure 6. Peptides with antitumor activity.

mary, colon, lung, and other leukemia systems are negative. Procurement of additional material has been completed and further tumor testing is planned.

There has been little structure–activity work done in this series of compounds, and such work could clearly open up a new area of bifunctional intercalating agents for cancer chemotherapy.

H. Valinomycin

Valinomycin (Fig. 6) is too well known to belabor its chemistry and biological properties. The well-established ability of valinomycin and related ionophore antibiotics to bind potassium and cause "leakage" from the cell obviously affects cellular energy metabolism by sodium/potassium-dependent ATPase enzymes. If indeed there is some selective distribution of valinomycin to tumor cells, it follows that energy balance in these cells would be affected, leading to disruption of growth and possibly viability. This could provide a novel means to attack the tumor cell and lead to a new variety of chemotherapeutic agent.

Valinomycin has shown moderate activity at the NCI against the B-16 melanoma, Colon 26, Colon 38, L-1210 leukemia, and P-388 leukemia tumor systems in mice, and further development of this agent is in process.

The synthesis of related cyclic peptides for antitumor testing is an effort which should be undertaken.

I. Mithramycin

Mithramycin (Fig. 7) is an antibiotic, originally isolated from a culture of *Streptomyces tanashiensis* by Rao *et al.* (1962), which is a member of the aureolic acid class. Closely related compounds include the chromomycins and olivomycins, which differ with respect to the carbohydrate moiety.

Mithramycin is a commercially available drug which is considered a treatment of choice for certain types of testicular tumors. Dose-limiting toxicities associated with use of mithramycin include a severe hemorrhagic syndrome and thrombocytopenia. The mechanism of action involves binding of mithramycin to DNA in the presence of a divalent cation, which results in inhibition of DNA-directed RNA synthesis.

Derivatives of mithramycin with a broader spectrum of antitumor activity and/or less toxicity would be of interest to clinicians.

J. Piperazinedione 593A

This compound (Fig. 7) is an extremely powerful alkylating agent which has shown excellent activity against a wide variety of murine tumors at the

Mithramycin

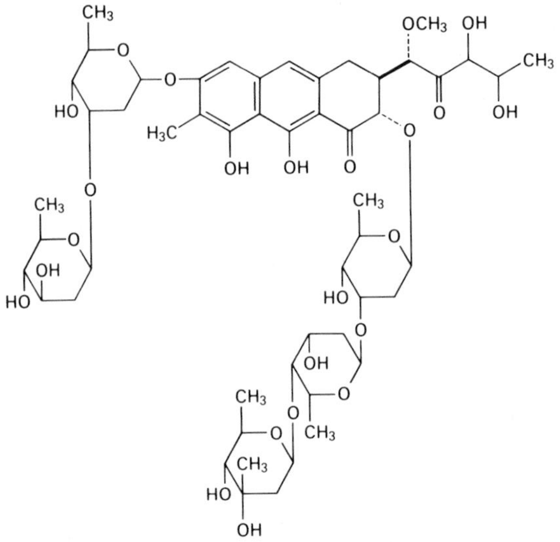

Piperazinedione 593A

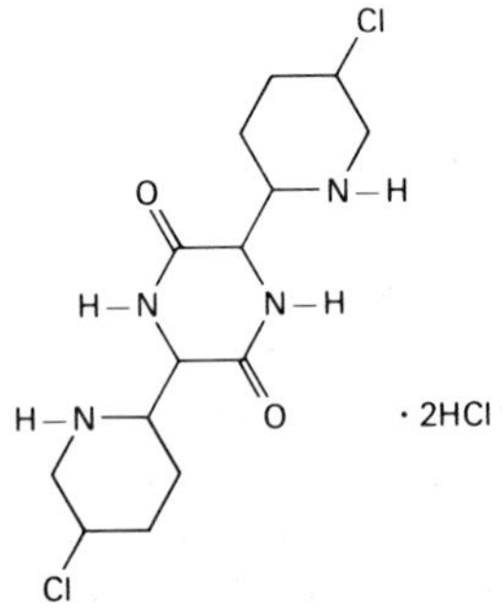

Figure 7. Structures of mithramycin and piperazinedione 593A.

NCI. It probably has the broadest antitumor spectrum with good activity of any agent studied in recent years. The screening data are presented in Table II. Piperazinedione 593A was discovered at Merck and Co. by Gitterman *et al.* (1970) and the structure was established by Arison and Beck (1973).

Animal toxicology showed that myelosupression was dose-limiting, and this was confirmed in phase I and phase II clinical trials conducted under the auspices of the NCI. These trials were very disappointing in that the myelosupression observed was so severe that the agent could not be used effectively due to the very low tolerated dose (Currie *et al.*, 1979; Palmer *et al.*, 1977; Pasmantier *et*

TABLE II

Antitumor Activity of Piperazinedione 593A in Mice

Tumor system	Parameter	Activity criteria	Treatment schedule[a]	Best T/C[b]	Optimal dose (mg/kg/inj.)	Reproduced T/C
B-16 melanocarcinoma	Survival time	T/C >125	Day 1	154	16	142
Colon 26	Survival time	T/C >140	Days 1,5,9	171	6	145
Colon 38	Tumor inhibition	T/C <42	Days 2,9,16	6	5	34
CD8F$_1$ mammary	Tumor inhibition	T/C <42	Days 1,8,15,22,29	0	6	
L-1210 leukemia	Survival time	T/C >125	Day 1	655	10	501
P-388 leukemia	Survival time	T/C >120	Day 1	532	10	505
Ependymoblastoma	Survival time	T/C >125	Days 1–9	272	1	262
MX-1 breast xenograft	Tumor inhibition	T/C <42	Q4D × 3[c]	1	10	5

[a] The day of tumor implantation is taken as day zero.

[b] T/C is the ratio of the parameter studied in test to control animals, expressed as a percentage.

[c] Treatment is started when tumor weight reaches 100–300 mg.

al., 1977; Al-Sarraf *et al.*, 1978; Jones *et al.*, 1977). More recently, the compound has been used by Dicke *et al.* (1979) in studies of autologous bone-marrow transplantation. In this technique, marrow of leukemia patients in remission is stored, and if relapse occurs the patient is treated with piperazinedione 593A and radiation to completely eliminate new cell formation in the marrow, and the patient's own cells stored previously are reintroduced. The initial report indicates some success with this technique.

From a chemical point of view, synthesis of analogues should be achievable, and a related compound which retains the spectrum and level of antitumor activity with less myelosuppression would be exciting.

REFERENCES

Abd-Rabbo, H. (1960). *J. Trop. Med. Hyg.* **72**, 287-290.
Abd-Rabbo, H. (1966). *Lancet* **1**, 1161-1162.
Al-Sarraf, M., Thigpen, T., Groppe, C. W., Haut, A., and Padilla, F. (1978). *Cancer Treat. Rep.* **62**, 1101-1103.
Anandaraj, M. P., Jayaram, H. N., Olney, J. W., and Cooney, D. A. (1977). *Proc. Am. Assoc. Cancer Res.* **18**, 219.
Ansari, B. M., and Thompson, E. N. (1975). *Postgrad. Med. J.* **51**, 103-105.
Arison, B. H., and Beck, J. L. (1973). *Tetrahedron* **29**, 2743-2746.
Beck, J. R., Kwok, R., Booker, R. N., Brown, A. C., Patterson, L. E., Bronc, P., Rockey, B., and Pohland, A. J. (1968). *J. Am. Chem. Soc.* **90**, 4706-4710.
Beerman, T. A. (1978). *Biochem. Biophys. Res. Commun.* **83**, 908-914.
Bergman, J., and Carlsson, R. (1977). *Tetrahedron Lett.*, 4663-4666.
Besselievre, R., Thal, C., Husson, H.-P., and Potier, P. (1975). *J.C.S. Chem. Commun.* 90-91.
Betts, R. E., Walters, D. E., and Rosazza, J. P. (1974). *J. Med. Chem.* **17**, 599-602.
Bhuyan, B. K., Fraser, T. J., and Li, L. H. (1972). *Cancer Res.* **32**, 2538-2544.
Brannon, D. R., Horton, D. R., and Svoboda, G. H. (1974). *J. Med. Chem.* **17**, 653-654.
Chimura, H., Ishizuka, M., Hamada, M., Hori, S., Kimura, K., Iwanaga, T., Takeuchi, T., and Umezawa, H. (1968). *J. Antibiot.* **21**, 44-49.
Cooney, D. A., Jayaram, H. N., Ryan, T. A., and Bono, V. H. (1974). *Cancer Chemother. Rep.* **58**, 793-802.
Coronelli, C., Pasqualucci, C. R., Tamoni, G., and Gallo, G. G. (1966). *Farmaco, Ed. Sci.* **21**, 269-277.
Cranwell, P. A., and Saxton, J. E. (1962). *J. Chem. Soc.* 3842-3847.
Culvenor, C. C. J. (1968). *J. Pharm. Sci.* **57**, 1112-1117.
Currie, V., Woodcock, T., Tan, C., Krakoff, I., and Young, C. (1979). *Cancer Treat. Rep.* **63**, 73-76.
Dalton, L. K., Demerac, S., Elmes, B. C., Loder, J. W., Swan, J. M., and Teitei, T. (1967). *Aust. J. Chem.* **20**, 2715-2727.
De Conti, R. C., Muggia, F., Cummings, F. J., Calabresi, P., and Creasy, W. A. (1975). *Proc. Am. Assoc. Cancer Res.* **16**, 96.
Dicke, K. A., Zander, A., Spitzer, G., Verma, D. S., Peters, L., Vellekoop, L., and McCredie, K. B. (1979). *Lancet* **1**, 514-517.

Diment, J. A., Ritchie, E., and Taylor, W. C. (1969). *Aust. J. Chem.* **22,** 1721-1730.

Drummond, L. J., and Lahey, F. N. (1949). *Aust. J. Sci. Res.* **A2,** 630-637.

Festy, B., Poisson, J., and Paoletti, C. (1971). *FEBS Lett.* **17,** 321-323.

Frohardt, R. P., Dion, H. W., Jakubowski, Z. L., Ryder, A., French, J. C., and Bartz, Q. R. (1959). *J. Am. Chem. Soc.* **81,** 5500-5506.

Gale, G. R., and Schmidt, G. B. (1968). *Biochem. Pharmacol.* **17,** 363-368.

Garcia-Giralt, E., and Macieira-Coelho, A. (1970). *Rev. Eur. Etud. Clin. Biol.* **15,** 539-541.

Gellert, E., and Rudzats, R. (1964). *J. Med. Chem.* **7,** 361-362.

Gellert, E., Govindachari, T. R., Lakshmikantham, M. V., Ragade, I. S., Rudzats, R., and Viswanathan, N. (1962). *J. Chem. Soc.* 1008-1014.

Gitterman, C. O., Rickes, E. L., Wolf, D. E., Madas, J., Zimmerman, S. B., Stoudt, T. H., and Demny, T. C. (1970). *J. Antibiot.* **23,** 305-310.

Goodwin, S., Smith, A. F., and Horning, E. C. (1959). *J. Am. Chem. Soc.* **81,** 1903-1908.

Govindachari, T. R., Rajappa, S., and Sundarsanam, V. (1963). *Indian J. Chem.* **1,** 247-251.

Govindachari, T. R., Pai, B. R., and Subramaniam, P. S. (1966). *Tetrahedron* **22,** 3245-3252.

Griffin, T. W., Comis, R. L., Lokich, J. J., Blum, R. H., and Canellos, G. P. (1978). *Cancer Treat. Rep.* **62,** 2019-2025.

Guthrie, R. W., Brossi, A., Mennona, F. A., Mullin, J. G., Kierstead, R. W., and Grunberg, E. (1975). *J. Med. Chem.* **18,** 755-760.

Hanka, L. J., and Dietz, A. (1973). *Antimicrob. Agents Chemother.* **3,** 425-431.

Herman, E., Vick, J., and Burka, B. (1971). *Toxicol. Appl. Pharmacol.* **18,** 743-751.

Herman, E. H., Lee, I. P., Mhatre, R. M., and Chadwick, D. P. (1974a). *Cancer Chemother. Rep., Part 1* **58,** 171-179.

Herman, E. H., Chadwick, D. P., and Mhatre, R. M. (1974b). *Cancer Chemother. Rep., Part 1* **58,** 637-643.

Hlubucek, J., Ritchie, E., and Taylor, W. C. (1970). *Aust. J. Chem.* **23,** 1881-1889.

Ishida, N. K., Miyazaki, K., Kumagai, K., and Rikimara, M. (1965). *J. Antibiot.* **18,** 68-76.

Jackson, A. H., Jenkins, P. R., and Shannon, P. V. R. (1977). *J.C.S. Perkin I,* 1698-1704.

Johnson, R. K., and Chitnis, M. P. (1978). *Proc. Am. Assoc. Cancer Res.* **19,** 218.

Jolad, S. D., Hoffmann, J. J., Torrance, S. J., Wiedhopf, R. M., Cole, J. R., Asora, S. K., Bates, R. B., Garguilo, R. L., and Kriek, G. R. (1977). *J. Am. Chem. Soc.* **99,** 8040-8044.

Jones, S. E., Tucker, W. G., Haut, A., Tranum, B. L., Vaughn, C., Chase, E. M., and Durie, B. G. M. (1977). *Cancer Treat. Rep.* **61,** 1617-1622.

Juret, P., Tanguy, A., Girard, A., Le Talaer, J. Y., Abbatucci, J. S., Dat-Xuong, N., Le Pecq, J. B., and Paoletti, C. (1978). *Eur. J. Cancer* **14,** 205-206.

Kane, R. C., Cohen, M. H., Broder, L. E., Bull, M. I., Creaven, P. J., and Fossieck, B. E., Jr. (1975). *Cancer Chemother. Rep., Part 2* **59,** 1171-1172.

Kann, H. E., Jr., and Kohn, K. W. (1972). *Mol. Pharmacol.* **8,** 551-560.

Kilminster, K. N., and Sainsbury, M. (1972). *J.C.S. Perkin I,* 2264-2267.

Kohberger, D. L., Fisher, M. W., Galbraith, M. M., Hillegas, A. B., Thompson, P. E., and Ehrlich, J. (1960). *Antibiot. Chemother.* **10,** 9-16.

Kohn, K. W., Waring, M. J., Glaubiger, D., and Friedman, C. A. (1975). *Cancer Res.* **35,** 71-76.

Kozikowski, A. P., and Hasan, N. M. (1977). *J. Org. Chem.* **42,** 2039-2040.

Kugelman, M., Liu, W.-C., Axelrod, M., McBride, T. J., and Rao, K. V. (1976). *Lloydia* **37,** 125-128.

Kunimoto, T., Hori, M., and Umezawa, H. (1971). *J. Antibiot.* **24,** 203-205.

Kunimoto, T., Hori, M., and Umezawa, H. (1972). *Cancer Res.* **32,** 1251-1256.

Kupchan, S. M., and Altland, H. W. (1973). *J. Med. Chem.* **16,** 913-917.

Kupchan, S. M., and Liepa, A. J. (1971). *J. Chem. Soc. D* 599-600.

Kupchan, S. M., and Suffness, M. (1967). *J. Pharm. Sci.* **56,** 541-543.

Kupchan, S. M., and Yokoyama, N. (1963). *J. Am. Chem. Soc.* **85**, 1361.

Kupchan, S. M., Yokoyama, N., and Thyagarajan, B. S. (1961). *J. Pharm. Sci.* **50**, 164-167.

Kupchan, S. M., Chakravarti, K. K., and Yokoyama, N. (1963). *J. Pharm. Sci.* **52**, 985-988.

Kupchan, S. M., Doskotch, R. W., and Vanevenhoven, P. W. (1964). *J. Pharm. Sci.* **53**, 343-345.

Lahey, F. N., and Thomas, W. C. (1949). *Aust. J. Sci. Res. Ser. A* **2**, 423-426.

Lancini, G. C., Diena, A., and Lazzari, E. (1966). *Tetrahedron Lett.*,1769-1772.

Lancini, G. C., Lazzari, E., and Diena, A. (1969). *Farmaco, Ed. Sci.* **24**, 169-178.

Langlois, Y., Langlois, N., and Potier, P. (1975). *Tetrahedron Lett.* 955-958.

Lee, J. S., and Waring, M. J. (1978). *Biochem. J.* **173**, 115-128.

Legha, S. S., Von Hoff, D. D., Rozencwieg, M., Abraham, D., Slavik, M., and Muggia, F. (1976). *Oncology* **33**, 265-270.

Le Goffic, L., Gouyette, A., and Ahond, A. (1972). *C. R. Acad. Sci., Ser. C* **274**, 2008-2009.

Le Goffic, F., Gouyette, A., and Ahond, A. (1973). *Tetrahedron* **29**, 3357-3362.

Le Men, J., Hayat, M., Mathé, G., Guillon, J. C., Chenu, E., Humblot, M., and Masson, Y. (1970). *Rev. Eur. Etud. Clin. Biol.* **15**, 534-538.

Le Pecq, J. B., Gosse, C., Dat-Xuong, N., and Paoletti, C. (1973). *C. R. Acad. Sci., Ser. D* **277**, 2289-2291.

Le Pecq, J. B., Gosse, C., Dat-Xuong, N., and Paoletti, C. (1975). *C. R. Acad. Sci., Ser. D* **281**, 1365-1367.

Lewisohn, R. (1918). *J. Am. Med. Assoc.* **70**, 9-10.

Li, L. H. (1973). *Proc. Am. Assoc. Cancer Res.* **14**, 110.

Liska, K. R. (1972). *J. Med. Chem.* **15**, 1177-1179.

Loder, J. W. (1966). *Aust. J. Chem.* **19**, 1947-1950.

MacDonald, P. L., and Robertson, A. V. (1966). *Aust. J. Chem.* **19**, 275-281.

McLean, E. K. (1970). *Pharmacol. Rev.* **22**, 429-483.

Maeda, H., Glaser, C., Kuomizu, K., and Meinhofer, J. (1974). *Arch. Biochem. Biophys.* **164**, 379-385.

Martin, D. G., Mizsak, S. A., Biles, C., Stewart, J. C., Baczynskyj, L., and Muelman, P. A. (1975). *J. Antibiot.* **28**, 332-336.

Mastrangelo, M. J., Grage, T. B., Bellet, R. E., and Weiss, A. J. (1973). *Cancer (Philadelphia)* **31**, 1170-1175.

Mathé, G., Hayat, M., De Vassal, F., Schwarzenberg, L., Schneider, M., Schlumberger, J. R., Jasmin, C., and Rosenfeld, C. (1970). *Rev. Eur. Etud. Clin. Biol.* **15**, 541-545.

Moertel, C. G., Schutt, A. J., Hahn, R. G., and Reitemeier, R. J. (1974). *Cancer Chemother. Rep., Part 1* **58**, 229-232.

Neil, G. L., Berger, A. E., Blowers, C. L., and Kuentzel, S. L. (1975). *Proc. Am. Assoc. Cancer Res.* **16**, 114.

Ohtsuki, E., and Ishida, N. (1975). *J. Antibiot.* **28**, 143-148.

Ono, Y., Watanabe, Y., and Ishida, N. (1966). *Biochim. Biophys. Acta* **119**, 46-58.

Ono, Y., Ito, Y., Maeda, H., and Ishida, N. (1968). *Biochim. Biophys. Acta* **115**, 616-618.

Palmer, R. L., Samal, B. A., Vaughn, C. B., and Tranum, B. L. (1977). *Cancer Treat. Rep.* **61**, 1711-1712.

Panettiere, F., and Coltman, C. A. (1971). *Cancer (Philadelphia)* **27**, 835-841.

Paoletti, C., Le Pecq, J.-B., Dat-Xuong, N., Lesca, P., and Lecointe, P. (1978). *Curr. Chemother., Proc. Int. Congr. Chemother., 10th* **2**, 1195-1197.

Pasmantier, M. W., Coleman, M., Kennedy, B. J., Eagan, R., Carolla, R., Weiss, R., Leone, L., and Silver, R. T. (1977). *Cancer Treat. Rep.* **61**, 1731-1732.

Pelaprat, D., Oberlin, R., Roques, B.-P., and Le Pecq, J.-B. (1976). *C. R. Acad. Sci., Ser. D* **283**, 1109-1112.

Rao, K. V., Cullen, W. P., and Sobin, B. A. (1962). *Antibiot. Chemother.* **12**, 182-186.

Rao, K. V., McBride, T. J., and Oleson, J. J. (1968). *Cancer Res.* **28**, 1952–1954.

Rivera, G., Howarth, C., Aur, R. J. A., and Pratt, C. B. (1978). *Cancer Treat. Rep.* **62**, 2105–2107.

Roth, J. (1974). *Hippokrates* **45**, 262–263.

Rousselle, D., Gilbert, J., and Viel, C. (1977). *C. R. Acad. Sci., Ser. C* **284**, 377–380.

Sainsbury, M., and Schinazi, R. F. (1976). *J.C.S. Perkin I*, 1155–1160.

Sainsbury, M., and Webb, B. (1974). *J.C.S. Perkin I*, 1580–1584.

Sainsbury, M., Webb, B., and Schinazi, R. F. (1975). *J.C.S. Perkin I*, 289–298.

Saito, N., Kitame, F., Kikuchi, M., and Ishida, N. (1974). *J. Antibiot.* **27**, 206–214.

Sakamoto, S., Ogata, J., Ikegami, K., and Maeda, H. (1978). *Cancer Treat. Rep.* **62**, 453–454.

Saucier, J. M., Festy, B., and Le Pecq, J.-B. (1971). *Biochimie* **53**, 973–980.

Schneider, J., Evans, E. L., Grunberg, E., and Fryer, R. I. (1972). *J. Med. Chem.* **15**, 266–270.

Siddiqui, S., Firat, D., and Olshin, S. (1973). *Cancer Chemother. Rep., Part 1* **57**, 423–428.

Snyder, A. L., Kann, H. E., and Kohn, K. W. (1971). *J. Mol. Biol.* **58**, 555–565.

Sullivan, H. R., Billings, R. E., Occolowitz, J. L., Boaz, H. E., Marshall, F. J., and McMahon, R. F. (1970). *J. Med. Chem.* **13**, 904–909.

Svoboda, G. H. (1966). *Lloydia* **29**, 206–224.

Svoboda, G. H., Poore, G. A., Simpson, P. J., and Boder, G. B. (1966). *J. Pharm. Sci.* **55**, 758–768.

Tatsumi, K., and Nishioka, H. (1977). *Mutat. Res.* **56**, 91–94.

Tobey, R. A., Orlicky, D. J., Deaven, L. L., Rall, L. B., and Kissane, R. J. (1978). *Cancer Res.* **38**, 4415–4421.

Tomita, M., Furukawa, H., Lu, S.-T., and Kupchan, S. M. (1965). *Tetrahedron Lett.*, 4309–4316.

Van Echo, D. A., and Wiernik, P. H. (1978). *Cancer Treat. Rep.* **62**, 1363–1365.

Van Hoosen, B. (1919). *Women's Med. J.* **29**, 101–116.

van Tamelen, E. E., and Haarstad, V. (1960). *J. Am. Chem. Soc.* **82**, 2974.

Woo, P. W., Dion, H. W., and Bartz, Q. R. (1961). *J. Am. Chem. Soc.* **81**, 425–431.

Woodward, R. B., Iacobucci, G. A., and Hochstein, F. A. (1959). *J. Am. Chem. Soc.* **81**, 4434–4435.

Wyburn-Mason, R. (1966). *Lancet* **1**, 1266–1267.

Yamaguchi, T., Furumai, T., Sato, M., Okuda, T., and Ishida, N. (1970a). *J. Antibiot.* **28**, 369–371.

Yamaguchi, T., Kashida, T., Nawa, K., Yajima, T., Miyagishima, T., Ito, Y., Okuda, T., Ishida, N., and Kumagi, K. (1970b). *J. Antibiot.* **28**, 373–381.

Yamaguchi, T., Sato, M., Omura, Y., Arai, Y., Enomoto, K., Ishida, M., and Kumogai, K. (1970c). *J. Antibiot.* **28**, 383–387.

Yamashita, T., Naoi, N., Watanabe, K., Takeuchi, T., and Umezawa, H. (1976). *J. Antibiot.* **29**, 415–423.

Index

MEDICINAL CHEMISTRY
A Series of Monographs

EDITED BY

GEORGE DESTEVENS

Department of Chemistry
Drew University
Madison, NJ 07940

Volume 1. GEORGE DESTEVENS. Diuretics: Chemistry and Pharmacology. 1963

Volume 2. RODOLFO PAOLETTI (ED.). Lipid Pharmacology. Volume I. 1964. RODOLFO PAOLETTI AND CHARLES J. GLUECK (EDS.). Volume II. 1976

Volume 3. E. J. ARIENS (ED.). Molecular Pharmacology: The Mode of Action of Biologically Active Compounds. (In two volumes.) 1964

Volume 4. MAXWELL GORDON (ED.). Psychopharmacological Agents. Volume I. 1964. Volume II. 1967. Volume III. 1974. Volume IV. 1976

Volume 5. GEORGE DESTEVENS (ED.). Analgetics. 1965

Volume 6. ROLAND H. THORP AND LEONARD B. COBBIN. Cardiac Stimulant Substances. 1967

Volume 7. EMIL SCHLITTLER (ED.). Antihypertensive Agents. 1967

Volume 8. U. S. VON EULER AND RUNE ELIASSON. Prostaglandins. 1967

Volume 9. G. D. CAMPBELL (ED.). Oral Hypoglycaemic Agents: Pharmacology and Therapeutics. 1969

Volume 10. LEMONT B. KIER. Molecular Orbital Theory in Drug Research. 1971

Volume 11. E. J. ARIENS (ED.). Drug Design. Volumes I and II. 1971. Volume III. 1972. Volume IV. 1973. Volumes V and VI. 1975. Volume VII. 1976. Volume VIII. 1978. Volume IX. 1979. Volume X. 1980.

Volume 12. PAUL E. THOMPSON AND LESLIE M. WERBEL. Antimalarial Agents: Chemistry and Pharmacology. 1972

Volume 13. ROBERT A. SCHERRER AND MICHAEL W. WHITEHOUSE (Eds.). Antiinflammatory Agents: Chemistry and Pharmacology. (In two volumes.) 1974

Volume 14. LEMONT B. KIER AND LOWELL H. HALL. Molecular Connectivity in Chemistry and Drug Research. 1976